Earthquake Engineering Technology and Its Application

Earthquake Engineering Technology and Its Application

Editors

Su Chen

Chong Xu

Shuang Li

MDPI

Basel • Beijing • Wuhan • Barcelona • Belgrade • Novi Sad • Cluj • Manchester

Editors

Su Chen
College of Architecture and Civil Engineering
Beijing University of Technology
Beijing
China

Chong Xu
Geological Hazards Research Center
National Institute of Natural Hazards
Beijing
China

Shuang Li
School of Civil Engineering
Harbin Institute of Technology
Harbin
China

Editorial Office
MDPI AG
Grosspeteranlage 5
4052 Basel, Switzerland

This is a reprint of articles from the Special Issue published online in the open access journal *Sustainability* (ISSN 2071-1050) (available at: www.mdpi.com/journal/sustainability/special_issues/Earthquake_Engineering_Technology).

For citation purposes, cite each article independently as indicated on the article page online and as indicated below:

Lastname, A.A.; Lastname, B.B. Article Title. *Journal Name* **Year**, *Volume Number*, Page Range.

ISBN 978-3-7258-2116-7 (Hbk)
ISBN 978-3-7258-2115-0 (PDF)
doi.org/10.3390/books978-3-7258-2115-0

Contents

sustainability

MDPI

Editorial

Earthquake Engineering Technology and Its Application

Su Chen [1,*], Chong Xu [2] and Shuang Li [3]

1 Key Laboratory of Urban Security and Disaster Engineering of China Ministry of Education, Beijing University of Technology, Beijing 100124, China
2 National Institute of Natural Hazards, Ministry of Emergency Management of China, Beijing 100085, China; chongxu@ninhm.ac.cn
3 School of Civil Engineering, Harbin Institute of Technology, Harbin 150006, China; shuangli@hit.edu.cn
* Correspondence: chensuchina@bjut.edu.cn

1. Introduction

Recent earthquakes, such as the Osaka–Kobe earthquake (1995) and the Wenchuan earthquake (2008), have challenged earthquake engineering and its application. This Special Issue focuses on current developments in earthquake engineering, including ground motion characteristics, soil and foundation dynamics, wave propagation, behavior of structures, and methods for earthquake resilience and the retrofit of structures that are germane to practicing engineers. In this Special Issue, we have 19 works that improve the understanding of earthquake engineering from theory to application. This combination of both modeling and validation will strengthen our understanding of the impact that earthquakes have on foundations and structures and will help to improve the safety of inhabitants.

2. Seismogeology and Seismic Activity

Detailed geological and geomorphological evidence has suggested that the Yilan-Yitong fault (YYF), one of the key branches of the Tancheng-Lujiang fault zone in northeastern China, has been an active fault since the Holocene that has extended from Liaoning Province to far-eastern Asia. However, there are no clear fault traces or late Quaternary active features northeast of Tangyuan County. They carried out shallow seismic reflection exploration, field geological investigation, and trench excavation across the YYF north of Tangyuan. The results revealed that the YYF is composed of two main branches: the west YYF branch is a late Pleistocene active fault, and the east one is a middle-to-early Pleistocene fault.

The rapid assessment of post-earthquake building damage for rescue and reconstruction is a crucial strategy to reduce the enormous number of human casualties and economic losses caused by earthquakes. Conventional machine learning (ML) approaches for this problem usually employ one-hot encoding to cope with categorical features, and their overall procedure is neither sufficient nor comprehensive. They proposed a three-stage approach, which can directly handle categorical features and enhance the entire methodology of ML applications.

A comprehensive earthquake hazard database is crucial for comprehending the characteristics of earthquake-related losses and establishing accurate loss prediction models. Tongyan Zheng compiled the earthquake events that have caused losses since 1949, and established and shared a database of earthquake hazard information for the Chinese mainland from 1949 to 2021. On this basis, they preliminarily analyzed the spatiotemporal distribution characteristics of 608 earthquake events and the associated losses. Southwest China is in the Circum-Pacific seismic zone where earthquake hazards are highly frequent. The results can provide fundamental data for developing earthquake-related loss prediction models.

Cunpeng Du analyzed the effects of the 2011 Mw 9.0 Tohoku-Oki Earthquake on the locking characteristics and seismic risk of the Yishu fault zone in China. Combined with the b value and strain field characteristics, the properties of deformation of the Yishu fault

Citation: Chen, S.; Xu, C.; Li, S. Earthquake Engineering Technology and Its Application. *Sustainability* **2024**, *16*, 6703. https://doi.org/10.3390/su16156703

Received: 25 July 2024
Accepted: 30 July 2024
Published: 5 August 2024

zone before and after the earthquake are comprehensively analyzed. The results show that before the 2011 Tohoku-Oki Mw 9.0 earthquake, the locking degree and depth of the northern segment of the Yishu fault zone were higher, while the locking depth of the southern segment of the fault was shallower. The 2011 Tohoku-Oki earthquake produced different coseismic effects on the southern and northern sections of the Yishu fault zone. The extension on the southern section and compression on the northern section caused the strain release in the southern sections of the Yishu fault zone following the earthquake. After it, the regional locking degree of the southern section of the Yishu fault zone was relieved. However, the locking degree of the northern segment of the fault zone was still high and the depth was deep, at about 26 km.

3. Experimental Investigation of the Seismic Response of Infrastructures

The shaking table tests of a Seismic–Soil–Pile–Superstructure Interaction (SSPSI) in medium-soft and hard base soil were carried out. Silted clay with a unit weight of 1.70 g/cm^3 and a shear wave velocity of 175 m/s was adopted to simulate the medium-soft soil, while the composite soil obtained by adding 20% quicklime to silted clay with a unit weight of 1.75 g/cm^3 and a shear wave velocity of 300 m/s was adopted to simulate the hard soil in the tests. By inputting the artificial seismic motion time history with different amplitudes synthesized by the RG1.60 response spectrum commonly used in nuclear power engineering to the models, the dynamic interaction characteristics and seismic response laws of the soil–pile–nuclear island structure in the medium-soft and hard base soil were compared, the internal force and deformation distribution characteristics of the pile foundation under different ground conditions were analyzed, and the site conditions and mechanisms of seismic failure of the pile group foundation were described. The research results can provide a reference for site selection and seismic design of a nuclear power plant.

In order to compare and analyze the seismic response characteristics of a safety-related nuclear structure on a non-rock site in the condition of raft and pile group foundations under unidirectional and multidirectional seismic motion input, a large-scale shaking table test of the soil–nuclear structure system was carried out in this paper. In the test, the soil was uniform silted clay, and the shear wave velocity was 213 m/s. The results of the test show that the acceleration response of the safety-related nuclear plant is affected by the directions of the input seismic motion and the forms of the foundation. When the seismic motion is input simultaneously in three directions, the acceleration responses of the horizontal motion and vertical rocking of the safety-related plant are larger than those of the single-direction input. The acceleration response of the horizontal motion and vertical rocking of the safety-related structure with the pile group foundation is smaller than that with the raft foundation. The values of most frequency bands in the horizontal acceleration Fourier amplitude spectrum at the top of the pile–foundation structure are smaller than those at the top of the raft–foundation structure, while the displacement is basically the same as those of the raft–foundation structure.

In the catastrophe insurance industry, it is impractical for a catastrophe model to simulate millions of sites' environments in a short time. Hence, the attenuation relation is often adopted to simulate the ground motion on account of calculation speed, and both ground motion expectations and uncertainties must be calculated. Due to the vulnerability curves of our model being based on simulations with a large number of deterministic ground motions, it is necessary but not efficient for loss assessment to analyze all possible ground motion amplitudes and their corresponding loss rates. They developed a simplified method to rapidly simulate loss expectations and uncertainties. In this research, Chinese masonry buildings are the focus. The result shows that the modified method gives accurate loss results quickly.

4. Engineering Site and Geological Hazard

*Vs*30 (the equivalent shear-wave velocity of soil layers within a depth of 30 m underground) is widely used in the field of seismic engineering; however, due to the limitation of funds, time, measuring devices, and other factors, the depth for testing shear-wave velocity in an engineering site rarely reaches 30 m underground. Therefore, it is necessary to predict $Vs30$ effectively. They analyzed the existing models using 343 boreholes with depths greater than 30 m in Tangshan, China. This analysis shows that the topographic slope method is not suitable for predicting $Vs30$ in Tangshan. The Boore (2011) model overestimates, while Boore (2004) underestimates $Vs30$ in Tangshan, while Junju Xie's (2016) model has ideal prediction results. We propose three new models in this paper, including the bottom constant velocity (BCV) model, linear model, and conditional independent model. We find that the BCV model has limited prediction ability, and the linear model is more suitable when $z \leq 18$ m, while the conditional independent model shows good performance under conditions where $z > 18$ m. We propose that the model can be accurately and effectively applied in Tangshan and other regions with low shear-wave velocity.

At 17:00 (UTC+8) on 1 June 2022, an Ms 6.1 reverse earthquake struck Lushan County, Ya'an City, Sichuan Province. This earthquake event had a focal depth of 10 km and the epicenter was located at 30.37° N and 102.94° E. Shao X documented a comprehensive coseismic landslide inventory for this event and analyzed the distribution pattern and factors controlling the landslides. The landslide area density (LAD) increased with an increase in the above factors and is explained by an exponential relationship, indicating that the occurrence of coseismic landslides in this area was more easily affected by topographic factors than seismic factors. Most small-scale landslides were clustered in the ridge area, which shows the seismic amplification effects of mountain slopes. Due to the impact of the direction of seismic wave propagation, hillslopes facing northeast-east (NE-E) were more prone to collapse than southwest-facing ones. Based on the distribution pattern of the landslides, we suggest that the seismogenic fault of this event was NW dipping. These findings indicate that it is effective to identify the dipping of seismogenic faults using the spatial distribution pattern of coseismic landslides.

Velocity changes (dv/v) during and after earthquakes are important indicators for understanding the earthquake-induced mechanical damage evolution of rock slopes. However, studying slope responses associated with various seismic loading still remains challenging due to limited in situ observations. Huang H. conduct a 20 min temporal resolution monitoring of dv/v at the frequency band between 2 and 20 Hz by applying ambient noise interferometry on the Pubugou rock slope in Southwest China. They observe an instantaneous ~0.41% dv/v drop in the slope caused by the 2021 M 7.4 Maduo earthquake at a distance of 780 kM, following a characterized logarithmic recovery process of ~31.39 h towards its pre-earthquake state. Moreover, the dv/v in five narrow frequency bands show a similar drop and subsequently increased recovery times associated with the deceased frequencies due to the long-distance earthquake shaking. We discuss two possible mechanisms related to the heterogeneous rock slope excited by the long-distance earthquake at a low frequency. This study motivates the damage assessment of the rock slope using in situ dv/v and furthers the understanding of subsurface geological risks under diverse seismic loadings.

Post-earthquake road traversability is a critical factor that affects traffic conditions. Therefore, a post-earthquake traffic simulation method considering road traversability was proposed. First, the impact ranges of the earthquake-induced building collapse and the post-earthquake fire spread of buildings were analyzed, and road traversability was determined accordingly. Subsequently, the post-earthquake traffic flow was predicted based on building characteristics, and micro-level vehicle behaviors were simulated considering post-earthquake road traversability to determine the traffic conditions. In addition, the simulation model was validated using actual data. Finally, a segment of the Tongzhou road network in Beijing was selected as a case study to analyze post-earthquake road traversability and simulate traffic conditions on critical road sections. The proposed method can provide post-earthquake traffic conditions, which benefits decision-making during post-earthquake evacuation and rescue.

5. Dynamic Constitutive Behavior of Materials (Soil, Concrete, etc.)

In the seismic response analysis of liquefiable sites, it is challenging to simulate saturated sand's post-liquefaction deformation with the existing soil dynamic constitutive model, and the current pore-water pressure buildup model cannot reflect the decrease in the actual pore-water pressure under unloading stress. They aim at these problems to propose a feasible and straightforward time-domain post-liquefaction deformation constitutive model through experimental analysis and theoretical research, consisting of reversible pore-water pressure. According to the dynamic triaxial test data, the regularities of large deformation stress and strain behavior of the saturated sand after liquefaction are obtained, and the corresponding loading and unloading criteria are summarized. Combined with the effective stress constitutive model proposed by the author, a soil dynamic constitutive that can describe saturated sand's post-liquefaction deformation path is obtained. According to the test results, the model can simulate the deformation of saturated sand during the whole liquefaction process. The self-developed program Soilresp1D realized the dynamic response analysis of the liquefiable site, and the results were compared with the experimental results. It shows that the model based on the effective stress-modified logarithmic dynamic skeleton and post-liquefaction deformation constitutive can be directly applied to the dynamic response analysis of the liquefiable site.

Twelve site models were established based on the analysis of the influence of site conditions on earthquake damage and the influence of the soft soil layer on-site seismic response. The equivalent linearization site seismic response analysis is carried out at different input ground motion levels to discuss the influence of soft soil layer thickness and buried depth. The results show that the characteristic period of the response spectrum exhibits a gradual increase as the buried depth or thickness of the soft soil layer increases. Furthermore, the characteristic period of the response spectrum also increases with the rise in the input ground motion peak. Moreover, according to the influence that the characteristics of soft soil thickness, buried depth, and input ground motion intensity have on the characteristic period of the site acceleration response spectrum, a method for adjusting the characteristic period of the site acceleration response spectrum with a soft soil layer is put forward.

As an indispensable part of the lifeline for the offshore gas and oil industry, submarine pipelines under long-term marine environmental loadings have historically been susceptible to earthquakes. This study investigates the impact of trench backfilling on the residual liquefaction around a pipeline and the induced uplift of a pipeline under the combined action of an earthquake, ocean wave, and current loading. A fully coupled nonlinear effective stress analysis method, which can consider the nonlinear hysteresis and the large deformation after liquefaction of the seabed soil, is adopted to describe the interaction between the seabed soil and the submarine pipeline. Taking a typical borehole in the Bohai strait as the site condition, the nonlinear seismic response analysis of the submarine pipeline under the combined action of seismic loading and ocean wave and current is carried out. The numerical results show that trench backfilling has a significant impact on the seismic response of the pipeline. The existence of trench backfilling reduces the accumulation of the residual excess pore water pressure, so that the seabed liquefaction around the pipeline is mitigated and the uplift of the pipeline is also decreased.

Overloaded trucks and earthquakes have become two main factors responsible for bridge damage; consequently, the combination of heavy trucks and seismic loads as a typical occurrence of extreme events is likely to lead to bridge collapse or destructive damage, in which the crucial issues of coupling load model, dynamic equations, and bridge responses have not been adequately addressed. A simplified vehicle–bridge model consisting of many containers is established to simulate vehicle passage, and the dynamic equations are derived for a 5-axle truck on a simply supported beam as an illustration. Then, five ground motions selected from PEER with appropriate peak ground accelerations and durations and the three truck models specified in American Association of State Highway and Transportation Officials, Caltrans and Chinese codes are applied to the finite element model of a typical reinforced concrete continuous girder bridge, in which the vehicle speed, number of trucks, ground motion and vehicle type are assumed to be random variables and their influences on dynamic responses of the bridge are analyzed. The results show that the seismic load is the governing factor in dynamic responses, but truck load may change displacement shapes.

6. Performance-Based Seismic Design of Structures and Earthquake-Resilient Cities

Globally, tanks play a major part in the provision of access to clean drinking water to the human population. Beyond aiding in the supply of fresh water, tanks are also essential for ensuring good sanitary conditions for people and for livestock. Many countries have realized that a robust water supply and a robust sanitation infrastructure are necessary for sustainable growth. Therefore, there is large demand for the construction of storage tanks. Furthermore, liquid storage tanks are crucial structures which must continue to be operational even after a catastrophic natural event, such as an earthquake, to support rehabilitation efforts. From an engineering point of view, the various forces acting on the tanks and the behavior of the tanks under various loads are important issues which need to be addressed for a safe design. Analyses of these tanks are challenging due to the interaction between the fluid and tank wall. Thus, researchers have conducted several investigations to understand the performance of storage tanks subjected to earthquakes by considering this interaction. They discuss the historical development of various modeling techniques of storage tanks. The interaction with the soil also influences the behavior of the tanks, and hence, in this paper, various modeling approaches for soil structure interaction are also reviewed. Further, a brief history of various systems of base isolation and modeling approaches of base-isolated structures are also discussed.

Incremental dynamic analysis (IDA) is applied to calculate the exceeding probability of rail displacement under different earthquake excitations. A finite element model (FEM) of a high-speed railway track-bridge system is established, which consists of a CRTS II ballastless track of finite length laid on a five-span simply supported girder bridge. Records from five stations in the PEER NGA–West2 strong ground motion dataset are selected as seismic excitation. Based on the simulation, the characteristics of the vertical displacement of the rail under different seismic excitations are investigated, and the probability of the vertical displacement of the rail exceeding the allowable standard is calculated using IDA. The exceeding probability of the rail above the mid-span is larger than that above other parts of the bridge. Within the mid-span, the exceeding probability of the rail is the largest above the center of the bridge.

Borehole shrinkage and collapse are likely to occur when downhole testing is conducted in soft or loose sandy soils, resulting in testing interruption. To prevent this situation from occurring, installing casing in the borehole is a common approach. However, in actual testing, the quality of the signal obtained from measuring points within the depth of the casing is often not ideal, and there is still no clear and unified justification for the causes of interference generated by the casing. Therefore, the team attempts to investigate and elucidate the impact of casing through on-site experiments and numerical simulations. By numerically simulating different casing materials, the contact state between the casing and the hole wall, and the presence of low wave velocity filling soil around the casing, the variation patterns of the affected measurement point signals in the time and frequency domains are investigated. Furthermore, combined with the measured data, the impact characteristics of the casing on the results of the wave velocity testing using the downhole method are systematically explored. This research can provide some insights for the application and data interpretation of signals in the downhole methods of cased wells.

7. Conclusions and Prospects

We are delighted to see that the articles featured in this Special Issue have contributed new knowledge to the field of earthquake engineering technology and its application, provided innovative ideas for future development, and opened new opportunities for further collaboration and innovation. We extend our gratitude to the staff of Sustainability for their tremendous support to our Guest Editor Team in organizing this Special Issue, as well as their assistance with the review, revision, and verification process. We also thank all authors for their insightful contributions to this Special Issue and all reviewers for their valuable comments and suggestions on the submitted manuscripts.

Funding: This work was supported by the Major Program of National Natural Science Foundation of China (52192675).

Conflicts of Interest: The authors declare no conflicts of interest.

List of Contributions

1. Wei, Q.; Ha, G.; Min, W.; Zhu, M. Geometry and Kinematics of Northmost Yilan-Yitong Fault Zone, China: Insights from Shallow Seismic Data and Field Investigation. *Sustainability* **2024**, *16*, 1943.
2. Li, Y.; Jia, C.; Chen, H.; Su, H.; Chen, J.; Wang, D. Machine Learning Assessment of Damage Grade for Post-Earthquake Buildings: A Three-Stage Approach Directly Handling Categorical Features. *Sustainability* **2023**, *15*, 13847.
3. Li, G.; Zang, M.; Qi, S.; Bo, J.; Yang, G.; Liu, T. An infinite slope model considering unloading joints for spatial evaluation of coseismic landslide hazards triggered by a reverse seismogenic fault: A case study of the 2013 Lushan earthquake. *Sustainability* **2023**, *16*, 138.
4. Du, C.; Yin, H.; Yu, S.; Yang, L.; Jia, Y. Effects of the 2011 Mw 9.0 Tohoku-Oki Earthquake on the Locking Characteristics and Seismic Risk of the Yishu Fault Zone in China. *Sustainability* **2023**, *15*, 4321.
5. Sun, Y.; Wang, G.; Wang, Y.; Tu, J.; Jing, L.; Qi, W. Comparative Study on Shaking Table Tests for a Pile–Nuclear Island Structure under Different Soil Conditions. *Sustainability* **2023**, *15*, 11988.
6. Lu, X.; Jing, L.; Ma, Y.; Yang, J.; Qi, W. Shaking Table Test for Seismic Response of Nuclear Power Plant on Non-Rock Site. *Sustainability* **2023**, *15*, 10366.
7. Zhao, J.; Li, X.; Liu, C. A Structure Economic Loss Optimization Method with the Uncertainty of Ground Motion Amplitude for Chinese Masonry Building. *Sustainability* **2022**, *14*, 13860.
8. Fang, Y.; Li, H.; Li, Y.; Chen, G.; Lv, Y.; Peng, Y. Vs30 Prediction Models Based on Measured Shear-Wave Velocities in Tangshan, China. *Sustainability* **2023**, *15*, 3282.
9. Shao, X.; Xu, C.; Ma, S. Preliminary Analysis of Coseismic Landslides Induced by the 1 June 2022 Ms 6.1 Lushan Earthquake, China. *Sustainability* **2022**, *14*, 16554.
10. Huang, H.; Dai, S.; Yu, Y.; Xie, F. Observing Earthquake-Induced Velocity Change on the Rock Slope Following the 2021 M 7.4 Maduo Earthquake 780 km Away. *Sustainability* **2022**, *14*, 9345.
11. Wu, Y.; Xu, Z.; Liang, C.; Song, R. Post-Earthquake Traffic Simulation Considering Road Traversability. *Sustainability* **2022**, *14*, 11145.
12. Dong, Q.; Zhou, Z.; Li, X.; Hao, B.; Jin, L. Soil Dynamic Constitutive Considering Post-Liquefaction Deformation and Reversible Pore-Water Pressure. *Sustainability* **2022**, *14*, 16512.
13. Li, Y.; Hao, B.; Chen, Z.; Zhou, Z.; Bian, Z.; Han, Y.; Peng, C. Discussion on Adjustment Method of the Characteristic Period of Site Response Spectrum with Soft Soil Layer. *Sustainability* **2023**, *15*, 8837.
14. Wang, Q.; Bian, J.; Huang, W.; Lu, Q.; Zhao, K.; Li, Z. Seabed Liquefaction around Pipeline with Backfilling Trench Subjected to Strong Earthquake Motions. *Sustainability* **2022**, *14*, 12825.
15. Liu, L.; Yang, X.; Yan, B.; Miao, S. Dynamic Responses of RC Girder Bridge under Heavy Truck and Seismic Loads Combined. *Sustainability* **2022**, *14*, 9263.
16. Chaithra, M.; Krishnamoorthy, A.; Avinash, A.R. A Review on the Modelling Techniques of Liquid Storage Tanks Considering Fluid–Structure–Soil Interaction Effects with a Focus on the Mitigation of Seismic Effects through Base Isolation Techniques. *Sustainability* **2023**, *15*, 11040.
17. Xing, H.; Sun, X.; Liu, Y.; Lin, J.; Song, H. Exceeding Probability of Earthquake-Induced Dynamic Displacement of Rail Based on Incremental Dynamic Analysis. *Sustainability* **2022**, *14*, 11871.
18. Yan, S.; Yu, Y.; Zheng, W.; Su, J.; Zhou, Z. Influence of Borehole Casing on Received Signals in Downhole Method. *Sustainability* **2023**, *15*, 9805.

sustainability

Article

Geometry and Kinematics of Northmost Yilan-Yitong Fault Zone, China: Insights from Shallow Seismic Data and Field Investigation

Qinghai Wei [1], Guanghao Ha [2,*], Wei Min [2] and Menghao Zhu [2,3]

1 Heilongjiang Earthquake Agency, Harbin 150090, China
2 Key Laboratory of Seismic and Volcanic Hazards, Institute of Geology, China Earthquake Administration, Beijing 100029, China
3 China Geology Survey, Beijing 100037, China
* Correspondence: haguanghao@163.com or haguanghao@ies.ac.cn

Abstract: Detailed geological and geomorphological evidence has suggested that the Yilan-Yitong fault (YYF), one of the key branches of the Tancheng-Lujiang fault zone in northeastern China, has been an active fault since the Holocene that has extended from Liaoning Province to far-eastern Asia. However, there are no clear fault traces or late Quaternary active features northeast of Tangyuan County. In this study, we carried out shallow seismic reflection exploration, field geological investigation, and trench excavation across the YYF north of Tangyuan. The results revealed that the YYF is composed of two main branches: the west YYF branch is a late Pleistocene active fault, and the east one is a middle-to-early Pleistocene fault. In Heli Town, the west branch of YYF presents fault scarps with heights of ~0.6 m. Across the scarps, we excavated a trench, and we propose that the YYF displaced the late Pleistocene to Holocene deposits, as this was indicated by the geochronological data. The seismic reflection data and sedimentary sequence revealed that the YYF north of Tangyuan is composed of three tectonic belts: the western depression, the central bulge, and the eastern depression. Each tectonic belt is composed of several small folds formed from the end of the Paleogene to the beginning of the Neogene. After the Neogene, different subsidence and uplift events occurred in various parts of the YYF, and after the early Pleistocene, the fault showed a consistent subsidence.

Keywords: Yilan-Yitong fault; geometry and kinematics; seismic reflection profiles; late Quaternary active characteristics

Citation: Wei, Q.; Ha, G.; Min, W.; Zhu, M. Geometry and Kinematics of Northmost Yilan-Yitong Fault Zone, China: Insights from Shallow Seismic Data and Field Investigation. *Sustainability* **2024**, *16*, 1943. https://doi.org/10.3390/su16051943

Academic Editors: Chong Xu, Su Chen and Shuang Li

Received: 11 September 2023
Revised: 23 December 2023
Accepted: 23 February 2024
Published: 27 February 2024

1. Introduction

The giant Tancheng-Lujiang fault zone (TLFZ), striking northeast, is located in the eastern margin of the Asian continent, which extends from southern China through the eastern part of the North China craton into the far east of Russia [1–5]. Due to its key location in a transition zone with a N–S compression tectonic regime between the Indian and Eurasian plates to the southwest and a nearly E–W extensional regime between the Pacific and Eurasian plates to the east, numerous geological, geophysical, and geochemical studies have been carried out focusing on the beginning, development, dynamic mechanism, metallogenic deposits, and earthquake disasters of this fault [6–18].

In northeast China, the Yilan-Yitong fault (YYF) and the Dunhua-Mishan fault (DMF) are considered to be the main branches of the TLFZ (Figure 1). More than 10 strong historical earthquakes have occurred along the TLFZ south of the Bohai Bay, whereas no $M \geq 6$ earthquakes have been recorded along the YYF or DMF. Moreover, the YYF was considered to be inactive since the late Quaternary due to the geographical conditions and artificial reconstruction until two paleoearthquake events were reported along the Fangzheng and Shulan segments of the YYF [10]. Then, research concerning the segmentation, paleoearthquakes, and late Quaternary slip rates of the YYF arose, which suggested that the YYF was an active dextral Holocene fault and could generate strong earthquakes ($M \geq 7$) [14]. South of Tangyuan County in Heilongjiang Province, the YYF is characterized by an assemblage of landforms including linear scarps and troughs, offset or deflected streams, small horsts and grabens, and especially linear sag ponds [14]. North of the Tangyuan, however, the Luobei segment of the YYF has a large, unclear surface compared with the segment south of Tangyuan, which makes it difficult to define the location and late Quaternary active features of the Luobei segment.

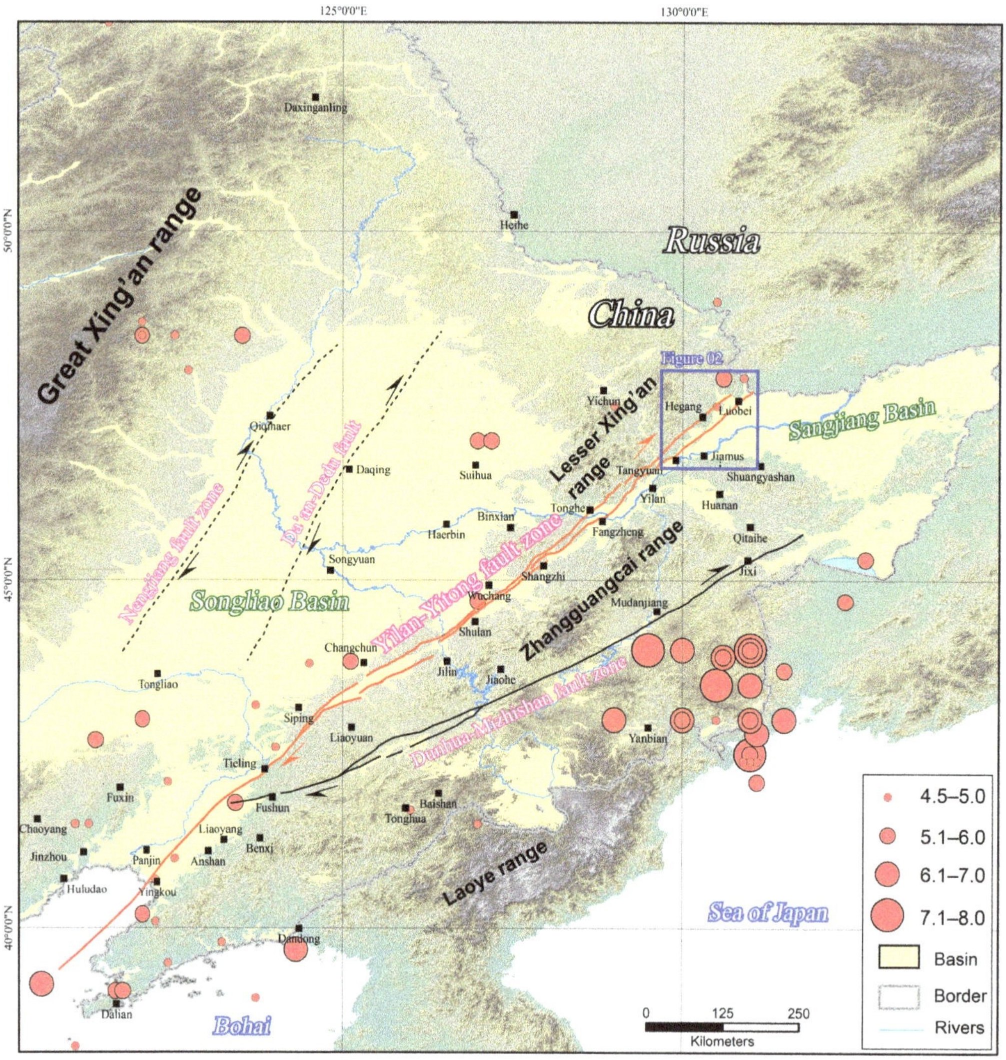

Figure 1. The extension and background of YYF and NE China. The blue rectangle shows the studied segment of YYF.

In this study, we conducted shallow seismic exploration and trench excavation across the YYF north of Tangyuan to map the geometry and kinematics of the fault.

2. Geological Background

As the largest fault in eastern China, the TLFZ controls the development and evolution of the crust and faults within the region, as well as the seismic activity [18,19]. According to the structures and active seismic features, the TLFZ can be divided into three segments: northern, central, and southern [19]. North of the Bohai Bay, the northern segment of TLFZ splits into the Yilan-Yitong fault to the north-northeast (NNE) and into the Dunhua-Mishan fault to the east-northeast (ENE). The southern-to-central TLFZ originated and experienced sinistral slip movement during the Triassic due to the collision between the North China craton and the South China block. Then, the TLFZ developed into NE China, forming the YYF in the early Cretaceous [19,20].

The NE-trending YYF runs from Shenyang, Changchun, and Harbin to far-eastern Asia and Russia. It is composed of two parallel branches with spacing of 10–30 km, resulting in a series of narrow grabens and uplifts (Figure 1) [20–22]. Four main grabens constitute the YYF; these are the Tangyuan, Fangzheng, Shulan, and Yilan grabens, bound by Songhuajiang, Shangzhi, and Yilan uplift from north to south, respectively [22]. The grabens are filled up with Paleogene to Neogene sandstone and mudstone overlaying lower-Cretaceous volcanics with angular unconformity [23–25]. During the Cenozoic, the YYF experienced a four-stage evolution progress: Paleogene rifting, compression-induced inversion at the end of the Paleogene, Neogene subsidence, and Quaternary compression. This was revealed by structural data and oil exploration results [22]. Although the proposed principal direction of compressive stress is still the subject of debate, the YYF was an oblique extensional graben in the Quaternary [7,22,26].

Geological and geomorphological evidence has revealed that the YYF has been active since the late Quaternary, presenting as a reverse and dextral strike–slip fault [9,14,15]. Trench excavation and geochronology identified that the latest paleoearthquake could have occurred in about AD 1810, with a magnitude of Ms 6.8–7.5 and seismic recurrence behavior, followed by a quasi-periodic pattern with a very long interval of 10–20 ka [14]. Geological and geomorphologic studies have suggested that the dextral slip rate of the YYF would have been ~0.2–0.3 mm/y, which is consistent with present geodetic observations [11].

At the northmost point, the YYF develops within the Tangyuan fault depression (TFD), which is a small petroleum basin west of the Sangjiang basin and northeast of China (Figure 2). The Cenozoic strata of TFD are occupied by Paleocene to Pliocene mudstone and sandstone and Quaternary sand and gravel [27]. A deep seismic reflection profile at the southern TFD demonstrated that the YYF was a lithospheric scale fault, with the Moho offset by the fault at a depth of 23–39 km [14]. Around TFD, the northern YYF is composed of two branches: the western YYF and eastern YYF, which separate the Zhangguangcai range massif to the northwest from the Jiamusi massif to the southeast (Figure 2).

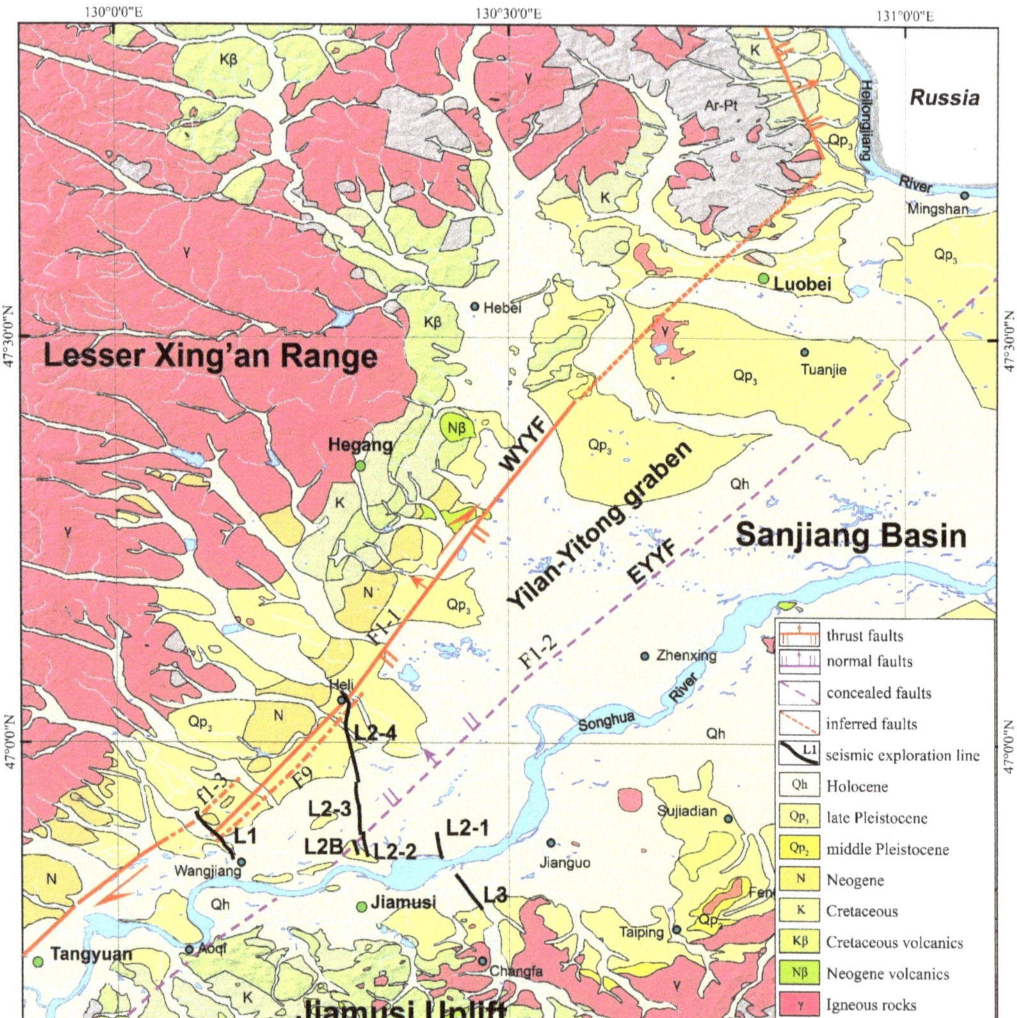

Figure 2. Geological map of YYF along the Tangyuan depression. WYYF: west Yilan-Yitong fault branch; EYYF: east Yilan-Yitong fault branch. The solid black lines show the seismic reflection profiles in this study.

3. Methods and Data

3.1. Seismic Reflection Data

To locate the northern YYF, we carried out 3 seismic reflection profile explorations across the fault zone using the longitudinal wave reflection method (Figure 2). The seismic reflection data were collected by a G3i digital seismometer made in Canada and two parallel 60 Hz longitudinal wave detector strings. The seismic source utilized a KZ-28 vibrator vehicle. The seismic dataset was acquired using a KZ-28 vibroseis vibrator at the center of each receiving station, with bilateral asymmetric reception. A vibrator spacing of 10 m and move-up distance of 2 m were utilized. The sweep frequency was 10–112 Hz, with a length of 15 s. The record length was 1500 ms, with a 0.5 ms sample rate, and the receiving channels were 240.

The seismic reflection lines were conventionally processed by utilizing spectrum analysis, frequency filtering, speed sweep, and NMO correction (normal move-out) stacking to obtain a preliminary reflection seismic time image for quality monitoring during the collection of field data. Then, the seismic data were processed using predictive deconvolution, multiple velocity analysis, refractive static correction, residual static correction, migration, and noise suppression to highlight effective waves and suppress interference waves, thereafter forming the final reflection seismology time profile.

3.2. Trench Excavation

Aerial photos taken by a DJI Phantom 4 quadcopter were used to map lineaments that would potentially reveal the locations of active fault traces. We conducted unmanned aerial vehicle (UAV) surveys at a typical flying altitude of 100 m in Heli Town, a possible site of the YYF identified by seismic reflection data (Figure 2). We constructed dense point clouds, orthophotos, and high-resolution digital elevation models (DEM) from the photographs using the structure-from-motion method with Agisoft software (version 1.4.5). Based on our interpretation of the aerial photos and DEM, the YYF presented as a linear fault scarp across which we excavated a trench to reveal the fault traces of YYF. The depth of the trench was limited to 1.5 m. The walls of the trench were cleaned, photographed, and logged, and samples for radiocarbon and OSL dating were also collected from the trench to constrain the ages of faulted and unfaulted trench layers.

4. Results

4.1. Seismic Interpretation

4.1.1. Seismic Velocity Model

As a potential petroliferous basin, much of the seismic data were collected in the Sanjiang Basin [8,25–28], and we referred to these data to estimate the seismic velocity of different sedimentary sequences in this study (Table 1).

Table 1. Seismic velocity of different units in Tangyuan depression.

Strata		Lithology	P-Wave Velocity (m/s)
Covering layers	Quaternary	Sub-clay	1400
		Silty fine sand	1550
		Gravel	1800
	Neogene	Clay, mudstone, glutenite	1850
Bedrock	Paleogene	Mudstone, sandstone	2200
	Cretaceous	Andesite, rhyolite, tuff	>3000

To obtain the seismic velocity parameter used for the time depth conversion, the following methods were adopted: (1) A velocity spectrum was picked up every 50 CDP while analyzing the reflection data. (2) According to the strength distribution of the energy groups, the best superposition speed curve was selected, and then the superposition speed was calculated and converted to the average speed. (3) The time depth summary table of the whole area was obtained by comprehensively averaging the average velocities obtained from the shallow seismic data of each survey line (Table 2).

Table 2. The time depth summary table of this study.

Time (ms)	Depth (m)	Time (ms)	Depth (m)	Time (ms)	Depth (m)	Time (ms)	Depth (m)
10	7	270	249	530	558	790	930
30	23	290	271	550	584	810	962
50	38	310	293	570	611	830	994
70	54	330	316	590	638	850	1026
90	70	350	339	610	665	870	1058
110	87	370	362	630	693	890	1090
130	105	390	385	650	721	910	1122
150	124	410	409	670	750	930	1153
170	144	430	433	690	779	950	1183
190	164	450	457	710	809	970	1210
210	185	470	482	730	839	990	1238
230	206	490	507	750			
250	227	510	532	770			

4.1.2. Seismic Stratigraphy

The number of effective reflection wave groups that could be identified in the P-wave reflection profiles was different for each profile due to the influence of the geological structure and thickness of the overburden layers.

The regional geological map shows that the strata exposed in the study area were mainly composed of Neogene and Quaternary, of which the Quaternary were widely distributed and constituted the main covering layers in the area (Figure 2). Drilling revealed that the Quaternary strata consisted of Upper, Middle, and Lower Pleistocene, as well as Holocene strata. The thickness and composition of each stratum were relatively stable, and the interface was clear (Figure 3). Together with our seismic reflection data, we suggest 1–4 groups of effective reflection waves (T1, T2, T3, and Tg, Figure 4) for the seismic time profiles.

T1 developed between 50 and 70 ms, showing flattening in the whole area with relatively slight unevenness in some parts, and the reflection events could be traced continuously. T2 developed between 160 and 200 ms, which was relatively gentle across the whole area with slight fluctuations in some parts. The energy of T2 was strong, with obvious dual phases, and the reflection events had good continuity that could be tracked continuously. T3 developed between 260 and 300 ms, had strong energy and was biphasic locally. The reflection event had good continuity and could be tracked continuously. Tg developed between 410–450 ms and gradually deepened from the south to the north in the whole area, with strong amplitude. The event was biphasic and could be tracked continuously.

The collected borehole ZK1 was located 270 m south end of the profile L1, and the ZK2 was 100 m away from the southwest side of 1820CDP of L2-3 (Figure 2). The borehole data are displayed in Figure 3, in which the strata and geological attributes between each reflection group are given. According to the above seismic stratigraphic sequence calibration, we determined the corresponding relationship between the seismic sequence (T1, T2, T3, and Tg) and stratigraphic structure in the area, as shown in Table 3.

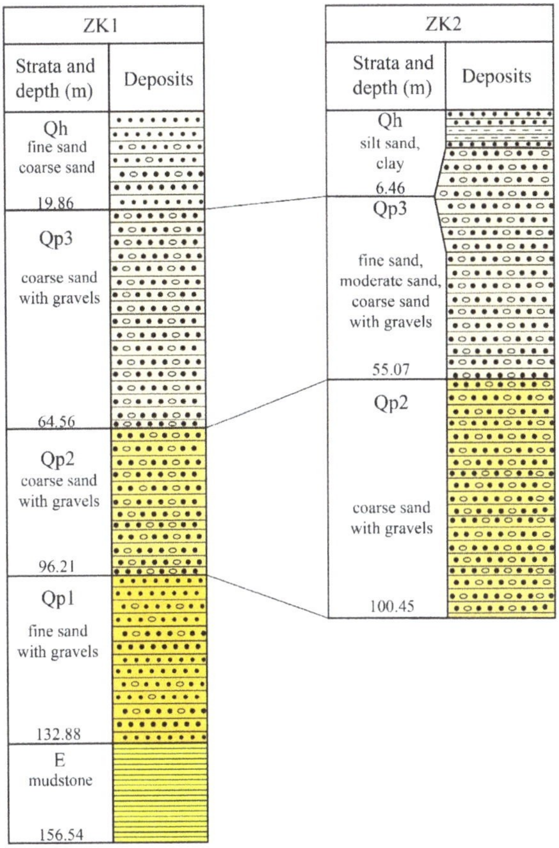

Figure 3. Drill column diagram of ZK1 and ZK2 with location indicated in Figure 2.

Table 3. The corresponding relationship between seismic sequences and stratigraphic structures.

	Strata	Seismic Reflection Groups	Lithology
Covering layers	Late Pleistocene to Holocene (Qp3-Qh)		Middle-to-fine sand, middle-to-coarse sand, coarse sand with gravels, sub-clay
		T1	
	Early-to-middle Pleistocene (Qp1-2)		Middle-to-coarse sand, coarse sand with gravels, sand–gravel
		T2	
	Pliocene (N2)		Clay, silt
		T3	
	Miocene (N1)		Sandstone, mudstone, conglomerate
		Tg	
Bedrock	Paleogene (E)		Sandstone, mudstone
	Cretaceous (K)		Andesite, tuff, rhyolite

4.1.3. Faults Revealed by Seismic Data

Four seismic profiles were created across the TFD (L1–L4, Figure 2), among which 10 fault planes were observed on L1, L2-2, L2-3, and L2-4, while no distinct abnormal reflection wave groups were observed on L2-1 or L3 (Section 4.2 and Figure 6).

(1) Faults on the L1

Three faulting points were observed on the L1: F1-1, F1-2, and F1-3. The F1-1 was located near 2960CDP of L1, where the seismic time profile showed that the reflection wave groups of bedrock were offset (Figure **??**). At F1-1, the fault dipped in the NW direction, with a gentle angle of the Paleogene sandstone and an offset of about 10 ms (~9 m). The F1-1 was covered by early-to-middle Pleistocene (Qp1-2) sediments at the top, with a buried depth of about 125 ms (~110 m), which suggests that the F1-1 has not been active since the early-to-middle Pleistocene.

The F1-2 was located near 4120CDP of L1, causing intense deformation. It was offset at the top of Paleogene and early-to-middle Pleistocene reflectors, as well as at the bottom of the late Pleistocene to Holocene reflectors (Figure **??**). Similarly to the adjacent F1-1, the F1-2 dipped in the NW direction, with a displacement of about 4 ms (~3 m) of the late Pleistocene sediments.

The F1-3 was located near 6310CDP of L1, resulting in the displacement of reflections in the Paleogene sediments (Figure **??**). In contrast to the F1-1 and F1-2, the F1-3 presented as a normal fault with a steep angle dipping in the NW direction, which displaced the top of the Paleogene reflectors and those on the interior of the bedrock with offsets of 6 ms (~5 m) and 54 ms (~47 m), respectively. The top of F1-3, however, terminated in the early-to-middle Pleistocene, indicating that it is an inactive fault.

(2) Fault on L2-2

The F2-2 was located near 2420CDP of L2-2, as indicated by the break of reflections in the bedrock (Figure 8). The fault finished in the early-to-middle Pleistocene sediment, with a depth of ~120 ms (~104 m) into the ground, appearing as a steep and north-west-dipping normal fault with an offset of about 5 ms (~3 m) of bedrock (Figure 8). Different reflectors developed around 1500CDP south of the F2-2, the north of which is characterized by clear reflections, while to the south, no distinct reflections developed. Surface exposures suggest an angular, unconformable contact between the Paleogene to Neogene strata and Cretaceous volcanic rocks, and this bedrock, without obvious reflection wave groups, may be these volcanic rocks. We inferred that the different reflectors around the 1500CDP could be the unconformity between the Paleogene sandstones and mudstones and the Cretaceous volcanic sedimentary rocks which separated the Jiamusi Uplift from the Yilan-Yitong graben.

(3) Fault on L2B

The F2B was located near 1360CDP of L2B, reflected by the offset of Tg that may have been caused by the fault. Within L2B, the F2B appeared to be a normal fault, dipping NW with an offset of approximately 4 ms (3 m), as indicated by the height difference in the bedrock's surface (Figure 9). The top of F2B was covered by Quaternary deposits with a depth of about 120 ms (105 m), which suggests that the F2B has been inactive since the Quaternary.

At CDP900, south of the F2B, a group of gently dipping reflectors divided the Paleogene mudstone and sandstone from the Cretaceous volcanics, which was inferred to be the unconformity (Figure 9). North of the unconformity, the reflectors were well developed in the Paleogene, while no distinct reflectors were developed south of the unconformity in the Cretaceous.

(4) Fault on L2-3

The F2-3 was indicated by a break and drop of reflectors within the bedrock at around 4680CDP on the L2-3 profile, appearing as a steep, SE-dipping normal fault (Figure 10). Up to the ground's surface, the F2-3 could be traced down to a depth of about 120 ms (103 m). Above the bedrock's surface (Tg), the reflections were continuous, which suggests that the F2-3 have been inactive since the Quaternary. North of the F2-3, the reflectors were in disarray and discontinuous; these could be the Cretaceous volcanic sedimentary rocks comprising the central uplift of the YYG. Those south of these were continuously well developed, and may be the Paleogene sandstone and mudstone constituting the eastern depression of the YYG.

(5) Faults on L2-4

Four faults were revealed by line L2-4 (F2-4-1, F2-4-2, F2-4-3, F2-4-4; Figure **??**). The F2-4-1 was characterized by the displacement of reflectors in the bedrock at a depth of about 260 ms (245 m), presenting as a steep, NW-dipping normal fault (Figure **??**). Upward, the reflections were continuous, without displacement in the Neogene or Quaternary, and the bedrock surface (Tg) was slightly twisted, suggesting that the F2-4-1 was active before the Neogene. North of the F2-4-1, the reflectors were well developed and clear; these could be Paleogene sandstone and mudstone as part of the western depression of the YYG. On the other hand, the reflectors were chaotic and unclear on the southern side and could be Cretaceous volcanic sediment belonging to the central uplift of the YYG.

The F2-4-2 was interpreted at around 6520CDP of L2-4, and was reflected by the discontinuity and offset of the reflectors in the bedrock (Figure **??**). The fault dipped southeast with a gentle angle, presenting as a thrust fault. The top of F2-4-2 was limited in the bedrock, with a depth of about 170 m (200 ms), suggesting that this fault branch has been inactive since the Quaternary.

At around 10750CDP of L2-4, the underlying bedrock reflectors were deformed and displaced, showing a NW-dipping thrust of F2-4-3 with an offset of 6 ms (5 m) of the bedrock surface (Figure **??**). The upper termination point of F2-4-3 was buried by the continuous reflections of T1 down to a depth of 140 ms (116 m), indicating that the fault was mainly active before the early-to-middle Quaternary. Similarly to the F2-4-3, the F2-4-4 was also characterized by the deformation and breakage of reflectors at 11250CDP of L2-4 (Figure **??**). Contrastingly, the upper faulted point of F2-4-4 extended into the late Pleistocene sediments, demonstrating that the fault has been active since the late Pleistocene.

4.2. Paleoseismic Evidence

The YYF at Heli Town was visible in the late Quaternary sediment, as evidenced by fault scarps (Figure 12a). The vertical displacement of the geomorphic surface was determined by extracting elevation profiles perpendicular to the strike of the fault scarps from high-resolution DEM generated by UAV. The topographic profile showed that the fault scarp of the late Quaternary sediments was about 0.6 m in height. (Figure 12b,c). Trench excavation was conducted across the fault scarp in this study, which revealed seven distinct sedimentary units (Units 1–7) based on their composition, sedimentary facies, and structure (Figure 13). Unit 1 was composed of grayish-black sand and soil, and it constitutes the main plowlands in the northeastern China. Yellow earth and coarse sand with some gravel constituted unit 2. Samples from unit 2 showed ages of 5650 $\pm$ 30 y B.P. to 11.77 $\pm$ 0.51 ka (samples JMS-1 to JMS-3, as listed in Tables 4 and 5). Unit 3 was composed of blackish-gray soils that presented as lens-shaped, undulating layers. Small, light-gray gravels and sandy gravels made up Unit 4. Unit 5 was composed of cross-bedding alluvial gravels. Grayish-yellow, coarse gravels made up unit 6, which was interbedded with 10 cm of light-gray clay at the top between units 5 and 6. Unit 7 was composed of grayish-yellow alluvial gravels.

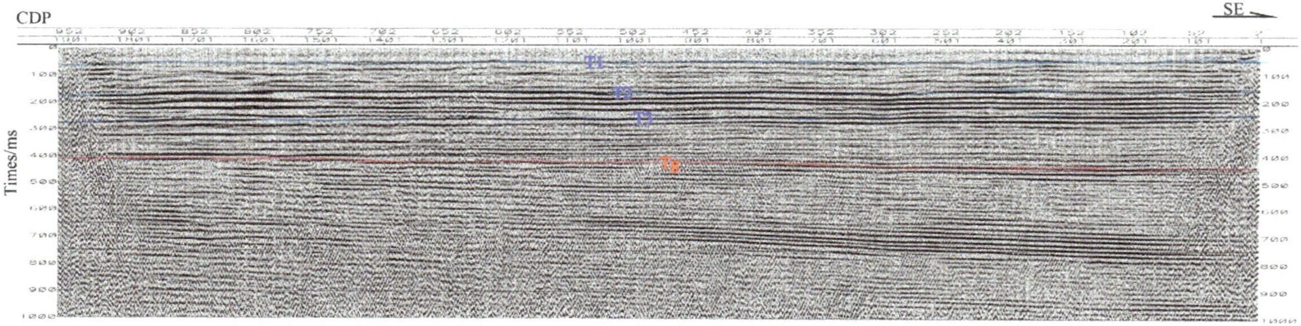

Figure 4. Seismic reflection profiles showing the four groups of reflection waves.

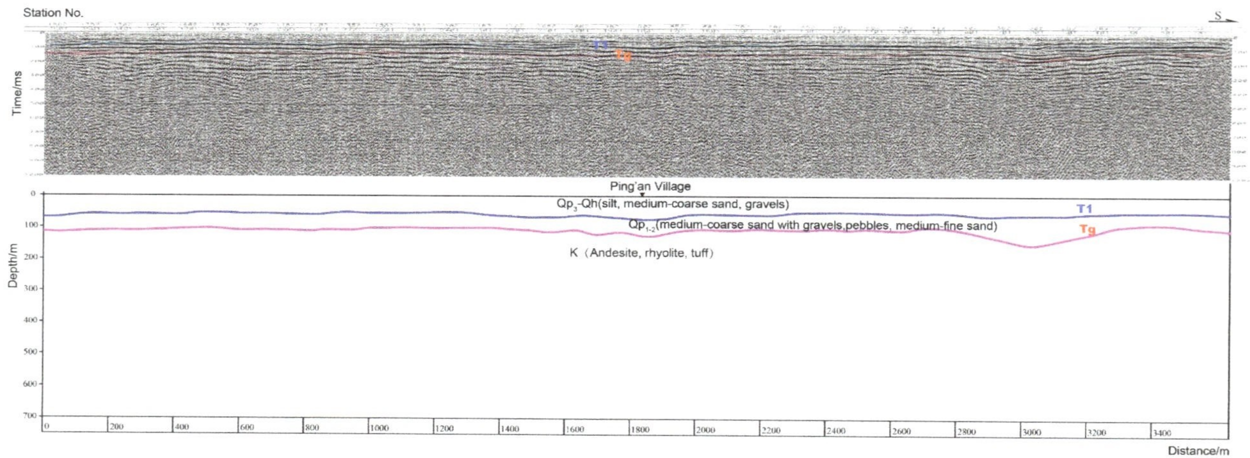

Figure 5. Seismic profiles of Line L2-1 and its interpretation.

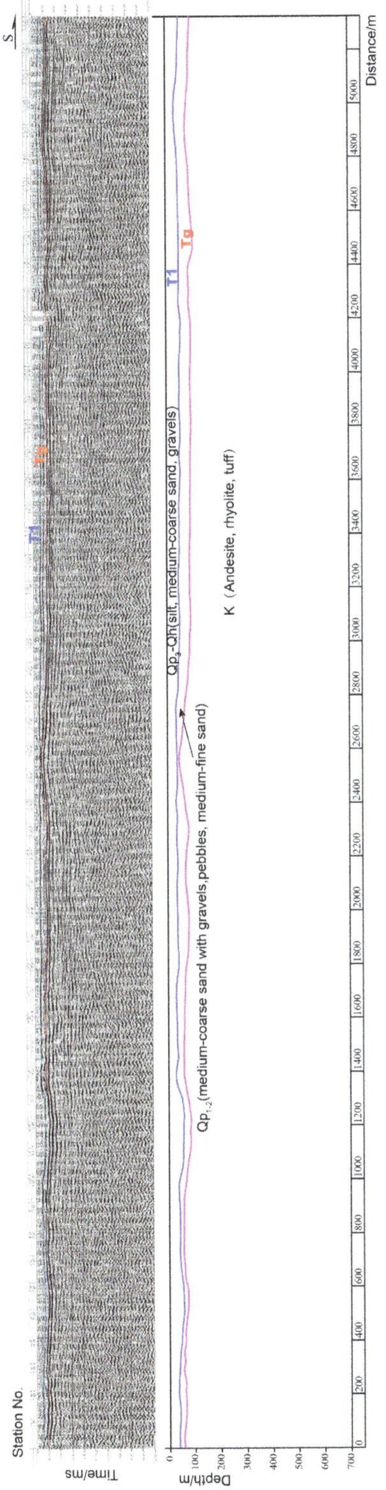

Figure 6. Seismic profiles of Line L3 and its interpretation.

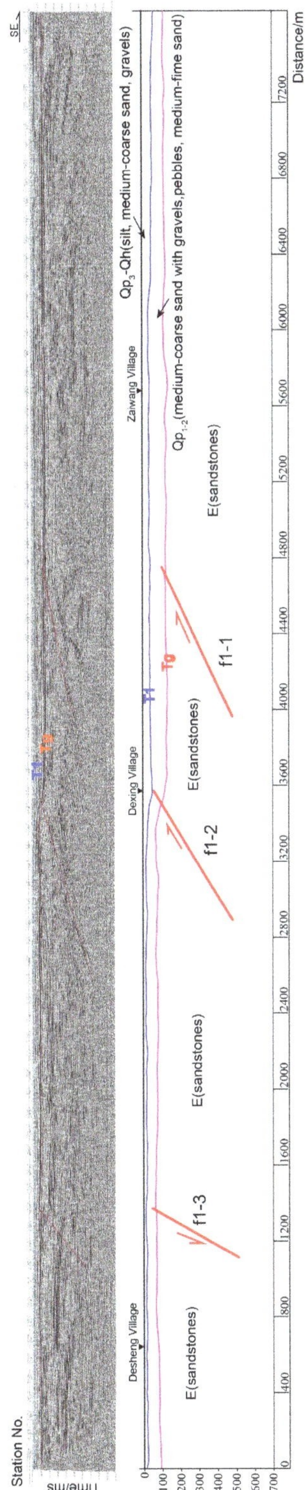

Figure 7. Seismic profiles of Line L1 and its interpretation.

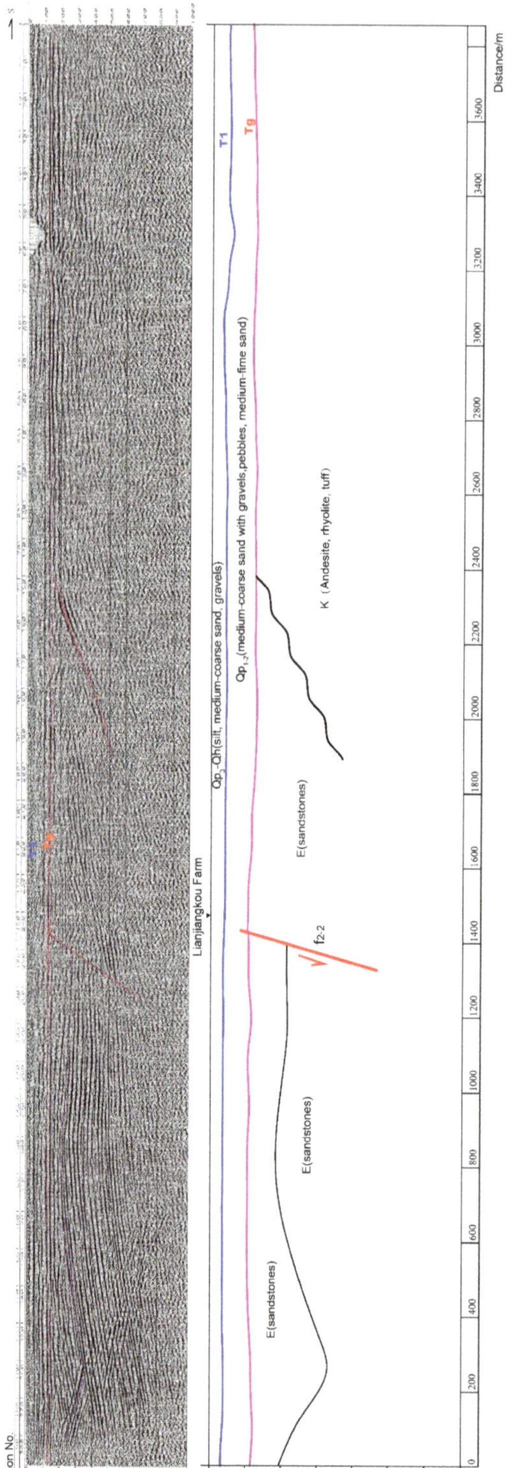

Figure 8. Seismic profiles of Line L2-2 and its interpretation.

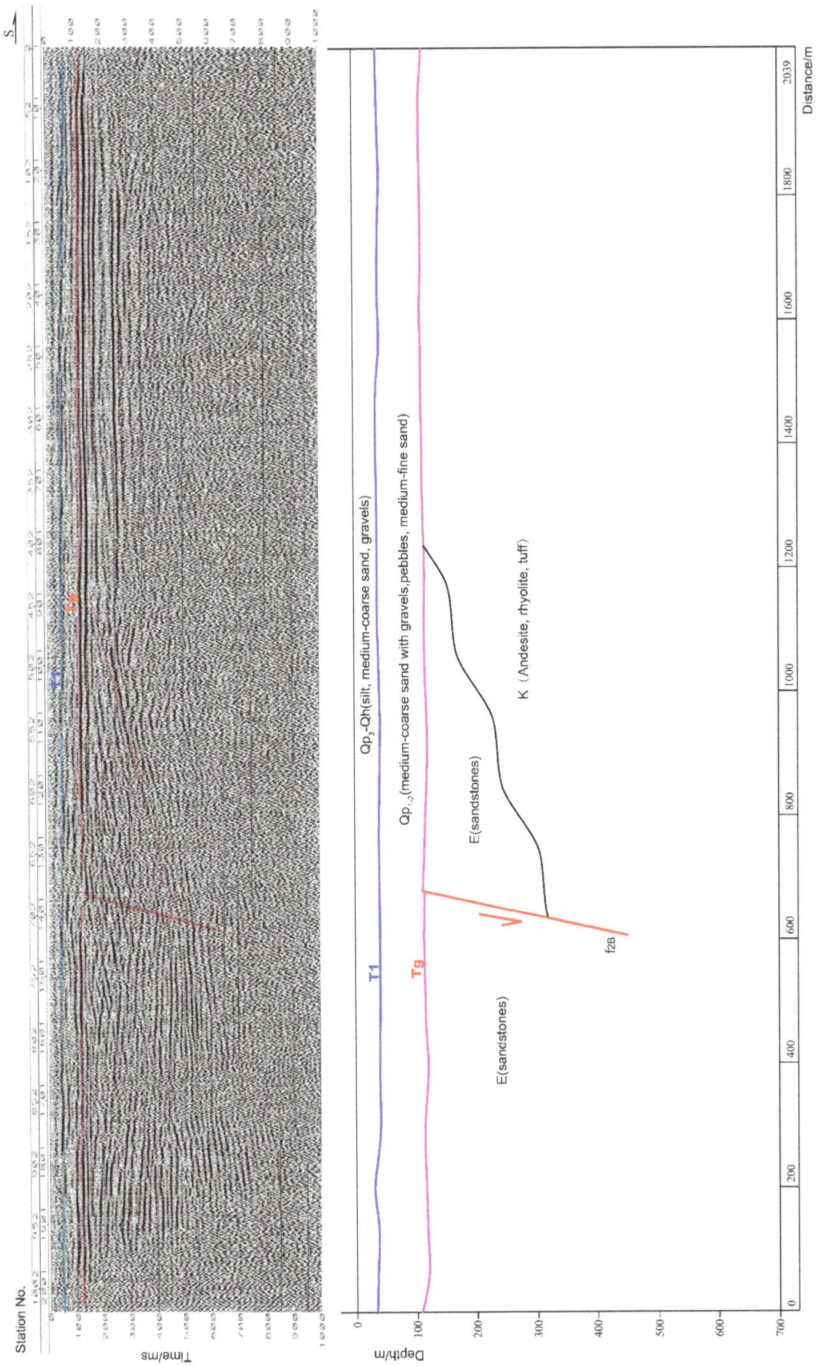

Figure 9. Seismic profiles of Line L2B and its interpretation.

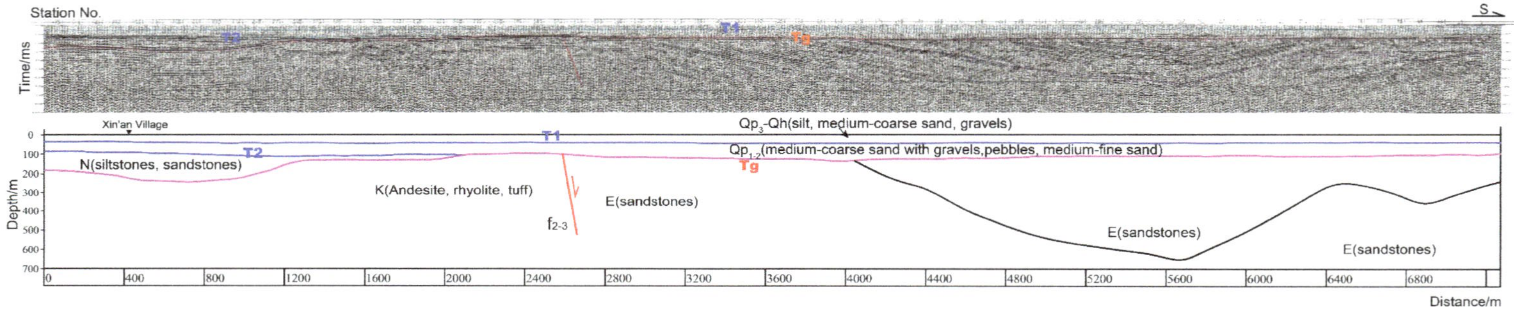

Figure 10. Seismic profiles of Line L2-3 and its interpretation.

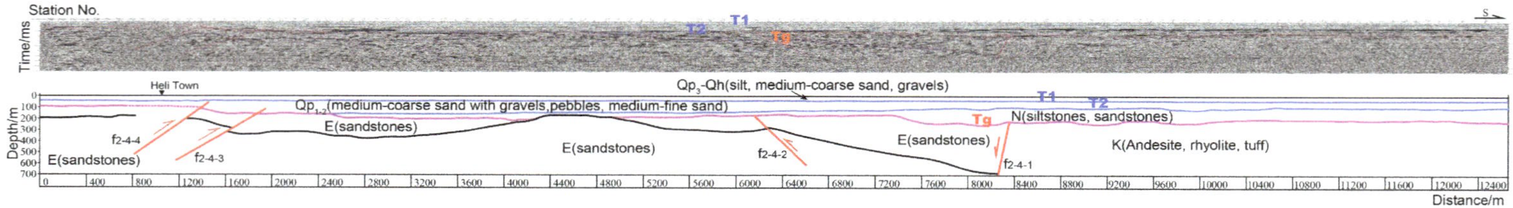

Figure 11. Seismic profiles of Line L2-4 and its interpretation.

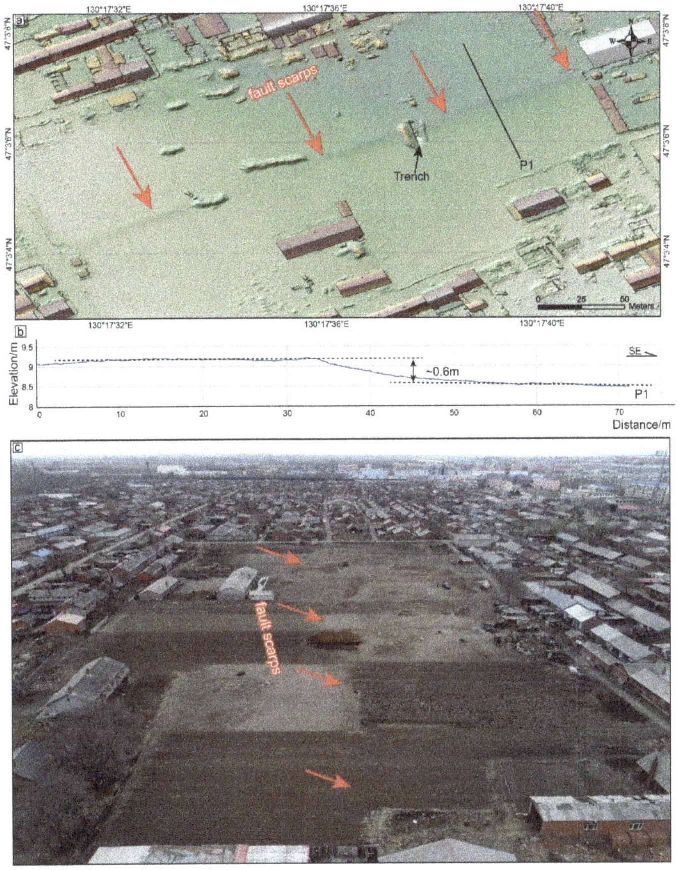

Figure 12. Site of the trench. (**a**) Fault scarps of the WYYF at Heli Town presented on the DEM derived from a UAV. (**b**) Topographic profile across the scarp with a height of about 0.6 m. (**c**) Photo of the fault scarp taken from a UAV.

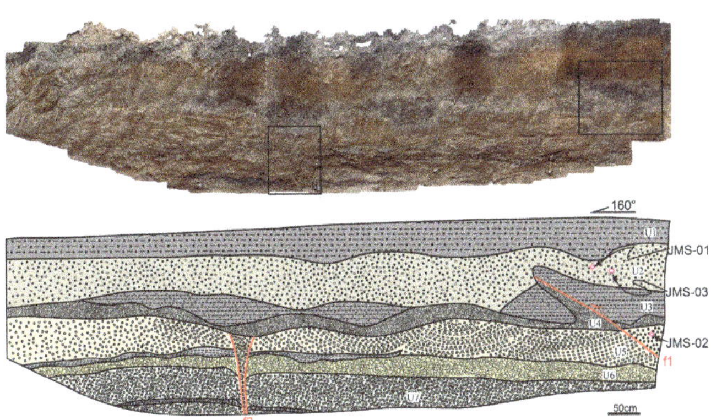

Figure 13. Trench photos and interpretation with the location of samples. The a and b show the detailed the fault features in Figure 14.

Figure 14. (**a**) Photograph of fault F1 revealed by the trench; (**b**) structure features of F2 indicated by oriented gravels.

Table 4. OSL dating results of samples from the trench.

Sample	Depth (m)	U (μg/g)	Th (μg/g)	K(%)	Water (%)	D (Gy/ka)	De (Gy)	Age (ka)
JMS-1	0.5	3.08 ± 0.1	13.8 ± 0.28	2.69 ± 0.02	15.12	3.92 ± 0.15	46.16 ± 1.04	11.77 ± 0.51
JMS-2	2	1.18 ± 0.01	5.37 ± 0.24	4.16 ± 0.03	5.00	4.85 ± 0.22	119.57 ± 7.12	24.64 ± 1.85

Table 5. Radiocarbon results of the sample from the trench.

Sample	Beta ID *	Conventional Age † (yr B.P.)	Calibrated Age ‡ (Cal B.P.)	Calibrated Calendar Age (B.C.)	Description
JMS-3	613471	5650 ± 30	6495-6390	4546-4441	Organic soil

* Samples were analyzed at Beta Analytic Inc., Miami, FL, USA. † Radiocarbon ages were measured using accelerator mass spectrometry (AMS) and are referenced to the year A.D. 1950. The analytical uncertainties are reported at 2σ. ‡ Dendrochronologically calibrated calendar age by OxCal 4.3 [29,30].

The trench wall also revealed the presence of two fault branches (F1–F2) (Figure 13). Fault F1 was characterized by the deformation of U3 and U4 (Figure 14). Fault F2 presented distinct fissures and bending gravels within U5, U6, and U7, causing the displacement of the gray clay interbedded in U6 (Figure 14).

5. Discussion

5.1. Geometry and Kinematics of the Northmost Yilan-Yitong Fault

Based on the shallow seismic profiles, four fault branches (F1-1, F9, F1-3 and F1-2, Figure 2) were identified at the northmost point of the YYF, described as follows:

The F1-2 and F2-4-4 are both northwest-dipping reverse faults with similar strikes in their spatial distributions, which could be connected to F1 fault, as is consistent with the western branch of the YYF. The NE-trending reverse F1 fault was steep at the top and gentle at the bottom, as shown by the shallow seismic reflection data. According to the seismic sequence, the F1 fault has been active since the late Pleistocene.

The northwest-dipping reverse faults F1-1 and F2-4-3 could be linked to fault F9, with a burial depth of about 110-116 m, which suggests that the F9 has been inactive since the early Pleistocene. Based on the geometry revealed by the seismic data, we inferred that the F1 and F9 would be the two main branches of the WYYF, and that they would connect into one fault at a certain depth.

The northwest-dipping normal fault F1-3 faulted in the early-to-middle Pleistocene strata, which indicates that F1-3 has been inactive since the early-to-middle Pleistocene. Fault F1-2, the eastern branch of YYF, comprised the steep, northwest-dipping normal fault F2-2-1 as well as F2B-1, with a burial depth of about 105 m. This suggests that the F1-2 has been an inactive fault since the early Pleistocene.

We identified two fault branches (F1 and F2) on the trench wall based on the displacement of sedimentary units and structural deformation. The F1 cut unit 3, but did not displace the bottom of unit 2, while the F2 was covered by unit 4, which suggests that F1 occurred after F2. We estimated that the most recent paleoearthquake event occurred before the deposition of unit 2 and after that of unit 3. We obtained two possible ages for unit 2 by OSL and 14C dating, with ages of about 11 ka and 4.5 ka, respectively (Tables 4 and 5). Based on the regional Quaternary strata distribution, we suggest that the 4.5 ka estimate may be close to the real age of unit 2. The depth of samples JMS-01 and JMS-03 was about 0.5 m, corresponding to the Holocene. The dating results suggest that the most recent paleoearthquake occurred before 4.5 ka, which is different from other segments of YYF [11]. Previous studies have suggested that the earthquake recurrence interval could be several to tens of thousands of years [15]. Compared with the earthquake recurrence interval, the Tangyuan segment of YYF faces a relatively low seismic risk. However, the seismic hazard assessment requires more evidence, and more research is needed in the future.

5.2. Cenozoic Evolution of Yilan-Yitong Graben

The seismic time section revealed significantly different reflect signals within the bedrock at around 1500CDP of L2-2 and 900CDP of L2B, south of which no obvious reflectors developed, while dense reflectors developed to the north within the bedrock. The different reflectors were bounded by a nonconforming surface, which appears as a group of gently dipping reflections in the seismic time section. The unconformity strikes NEE and divides the Jiamusi uplift from the Yilan-Yitong graben, the west of which is dominated by Paleogene sandstone and mudstone in the Yilan-Yitong graben, and the east of which is occupied by the Cretaceous volcanic sedimentary rocks of the Jiamusi uplift.

Three NE-trending structures exist from the northwest to southeast in the Yilan-Yitong graben: the western depression, central rise, and eastern depression (Figure 15). Based on the seismic profiles, the eastern depression is 6.3 km wide, the central rise is 6.7 km wide, and the width of the western depression exceeds 8.3 km. The western and eastern depressions are composed of Paleogene strata, and the central rise is dominated by the Cretaceous. Additionally, the sand and mudstone sequence in the eastern and western depressions appeared as gentle folds, as revealed by the reflectors in the bedrock. The western depression is composed of two anticlines and two synclines distributed alternately on the seismic data (Figure 15), while the eastern depression is composed of two anticlines sandwiched by a syncline (Figure 15). Only the southern-most anticline shows a local bend located at the 350CDP of L2-3 and the 2650CDP of L2-2, which represents the overlapping parts of the two survey lines, that is, they are conformal bends on the same anticline. The deformed Paleogene and overlying Neogene strata suggest that the folds in the Yilan-Yitong graben were formed after the end of the Paleogene.

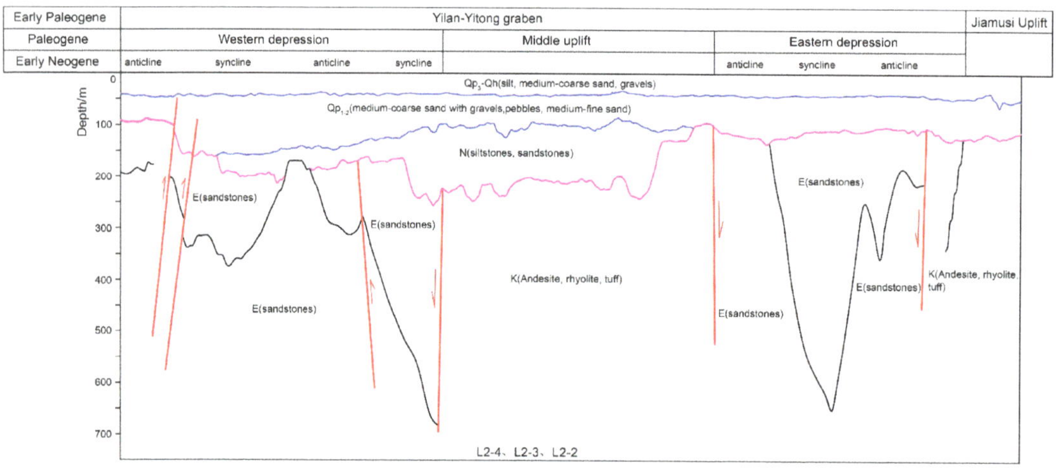

Figure 15. Tectonic framework across the TFD revealed by the seismic reflection data.

The seismic time sections suggest that the Neogene strata were distributed locally, thinly in the west and thickly in the east, which is consistent with the drilling data. However, the spatial distribution difference in the thickness of the Neogene strata indicate that the Yilan-Yitong graben formed huge thick deposits in the Paleogene, and then the Neogene appeared to have different spatial subsidence and rise characteristics. To the west, the graben, i.e., the rising section, resulted in sporadic deposition of Neogene strata, while the Neogene deposits were widespread, and the thickness increased significantly to the east. Only in the early Pleistocene did the whole graben and its surrounding area have a large range of subsidence.

6. Conclusions

(1) North of Tangyuan County, the YYF is composed of two main branches, of which the western YYF branch is a late-Pleistocene active fault and the eastern one is a middle-to-early-Pleistocene fault.

(2) Field investigation and trench excavation suggested that the most recent paleoearthquake may have occurred before about 4.5 ka along the Tangyuan segment of the YYF. Compared with the seismic recurrence interval, the risk of a large earthquake in Tangyuan is relatively low, but this requires more evidence in order to be verified.

(3) The seismic reflection data and sedimentary sequence reveal that the YYF north of Tangyuan is composed of three tectonic belts: a western depression, a central bulge, and an eastern depression. Each tectonic belt is composed of several small folds formed from the end of Paleogene to the beginning of Neogene. After the Neogene, differing subsidence and uplift arose in various parts of the YYF, and after the early Pleistocene, it showed consistent subsidence.

In this paper, we provided the shallow geometry of the Tangyuan segment of the YYF and preliminarily estimated the times of paleoearthquake occurrence. In the future, we suggest that focus be placed on deep structural and detailed paleoearthquake sequences to discuss tectonic origins and seismic hazard risks.

Author Contributions: Conceptualization, Q.W. and G.H.; methodology, W.M.; formal analysis, W.M. and G.H.; investigation, Q.W., G.H. and M.Z.; writing—original draft preparation, Q.W.; writing—review and editing, Q.W. and G.H.; visualization, G.H. and M.Z.; funding acquisition, W.M. All authors have read and agreed to the published version of the manuscript.

Funding: This research was funded by the National Key Research and Development Project of China, grant number 2022YFC3003501.

Institutional Review Board Statement: Not applicable.

Informed Consent Statement: Not applicable.

Data Availability Statement: No new data were created or analyzed in this study.

Conflicts of Interest: The authors declare no conflicts of interest.

References

1. Bai, Z.; Zhang, Z.; Wang, Y. Crustal structure across the Dabie–Sulu orogenic belt revealed by seismic velocity profiles. *J. Geophys. Eng.* **2007**, *4*, 436–442. [CrossRef]
2. Deng, Y.; Fan, W.; Zhang, Z.; Badal, J. Geophysical evidence on segmentation of the Tancheng-Lujiang fault and its implications on the lithosphere evolution in East China. *J. Asian Earth Sci.* **2013**, *78*, 263–276. [CrossRef]
3. Lin, A.; Miyata, T.; Wan, T. Tectonic characteristics of the central segment of the Tancheng–Lujiang fault zone, Shandong Peninsula, eastern China. *Tectonophysics* **1998**, *293*, 85–104. [CrossRef]
4. Wang, W.; Jin, Q.; Ma, Z. Meso-Cenozoic Evolution of the Tanlu Fault and Formation of Sedimentary Basins. *Acta Geol. Sin.-Engl. Ed.* **1998**, *72*, 350–362.
5. Yuan, X. On continental basal structure in China. *Chin. J. Geophys.* **1995**, *38*, 448–459. (In Chinese with English Abstract)
6. Chang, E.Z. Collisional orogene between north and south China and its eastern extension in the Korean Peninsula. *J. Southeast Asian Earth Sci.* **1996**, *13*, 267–277. [CrossRef]
7. Gu, C.; Zhu, G.; Zhai, M.; Lin, S.; Song, L.; Liu, B. Features and origin time of Mesozoic strike-slip structures in the Yilan-Yitong Fault Zone. *Sci. China Earth Sci.* **2016**, *59*, 2389–2410. [CrossRef]
8. Kamali, Z.; Nazari, H.; Rashidi, A.; Heyhat, M.R.; Khatib, M.M.; Derakhshani, R. Seismotectonics, Geomorphology and Paleoseismology of the Doroud Fault, a Source of Seismic Hazard in Zagros. *Appl. Sci.* **2023**, *13*, 3747. [CrossRef]
9. Kang, D.; Shen, W.; Ning, J.; Ritzwoller, M.H. Seismic evidence for lithospheric modification associated with intracontinental volcanism in Northeastern China. *Geophys. J. Int.* **2016**, *204*, 215–235. [CrossRef]
10. Min, W.; Liu, Y.; Jiao, D.; Shen, J.; Pan, X. Evidence for Holocene activity of the Yilan-Yitong fault, northeastern section of the Tan-Lu fault zone in Northeast China. *J. Asian Earth Sci.* **2013**, *67–68*, 207–216. [CrossRef]
11. Tian, Z.; Du, Y. Formation and evolution of the Yilan-Yitong graben. *Tectonophysics* **1987**, *133*, 165–173. [CrossRef]
12. Wang, E.; Meng, Q.; Burchfiel, B.C.; Zhang, G. Mesozoic large-scale lateral extrusion, rotation, and uplift of the Tongbai–Dabie Shan belt in east China. *Geology* **2003**, *31*, 307–310. [CrossRef]

13. Xu, M.; Li, Y.; Hou, H.; Wang, C.; Gao, R.; Wang, H.; Han, Z.; Zhou, A. Structural characteristics of the Yilan–Yitong and Dunhua–Mishan faults as northern extensions of the Tancheng–Lujiang Fault Zone: New deep seismic reflection results. *Tectonophysics* **2017**, *706–707*, 35–45. [CrossRef]
14. Yu, Z.; Yin, N.; Shu, P.; Li, J.; Wei, Q.; Min, W.; Zhang, P. Late Quaternary paleoseismicity and seismic potential of the Yilan-Yitong Fault Zone in NE China. *J. Asian Earth Sci.* **2018**, *151*, 197–225. [CrossRef]
15. Zhao, T.; Zhu, G.; Lin, S.; Wang, H. Indentation-induced tearing of a subducting continent: Evidence from the Tan–Lu Fault Zone, East China. *Earth-Sci. Rev.* **2016**, *152*, 14–36. [CrossRef]
16. Zhu, G.; Liu, G.S.; Niu, M.L.; Xie, C.L.; Wang, Y.S.; Xiang, B. Syn-collisional transform faulting of the Tan-Lu fault zone, East China. *Int. J. Earth Sci.* **2009**, *98*, 135–155. [CrossRef]
17. Zhang, L.; Barnden, J.A.; Hendley, R.J.; Lee, M.G.; Wallington, A.; Wen, Z. Affect detection and metaphor in e-drama. *Int. J. Contin. Eng. Educ. Life Long Learn.* **2008**, *18*, 234–252. [CrossRef]
18. Deng, Q.; Ran, Y.; Yang, X.; Min, W.; Chu, Q. *Active Tectonics Map of China*; Seismology: Beijing, Chian, 2007. (In Chinese)
19. Gu, C.; Zhu, G.; Zhang, S.; Liu, C.; Li, Y.; Lin, S.; Wang, W. Cenozoic evolution of the Yilan–Yitong Graben in NE China: An example of graben formation controlled by pre-existing structures. *J. Asian Earth Sci.* **2017**, *146*, 168–184. [CrossRef]
20. Li, Z.; Jia, D.; Chen, W.; Zhang, Y.; Wang, M.; Li, Y.; Li, H.; Li, S.; Zhou, X.; Wu, L.; et al. Structural deformation and evolution of right-lateral strike-slip tectonics of the Liaohe western depression during the early Cenozoic. *J. Asian Earth Sci.* **2013**, *77*, 1–11. [CrossRef]
21. Tang, D.-Q.; He, S.; Chen, H.-H. Fault system's characteristics of Yitong Basin and its evolution. *J. Jilin Univ. Earth Sci. Ed.* **2009**, *39*, 386–395.
22. Tang, D.; Chen, H.; Sun, J.; Zhang, H.; Chen, L. Cenozoic tectonic evolution of the Yitong part of the Tan-lu fault zone and its control on Yitong basin. *Geotecton. Metallog.* **2010**, *34*, 340–348.
23. Shen, Y.; Liu, C.; Jiang, L.; Wang, S.; Wang, D. Fault characteristics and analysis of tectonic styles in Tangyuan Fault-depression. *J. Jilin Univ. (Earth Sci. Ed.)* **2006**, *36*, 104–108.
24. Yang, J.-G.; Liu, J.-L.; Wu, H.-Y.; Liu, C.-G. The characteristics and evolvement of the tectonic in Tangyuan rift in northern Heilongjiang. *J. Daqing Pet. Inst.* **2007**, *31*, 19.
25. Guo, S.-B.; Wang, H. Sequence stratigraphic characteristics and hydrocarbon prospect of Mesozoic Formation in Suibin Depression, Sanjiang Basin. *Pet. Explor. Dev.* **2008**, *35*, 44–51. [CrossRef]
26. Wang, D.; Jia, L. Study on Seismic Geological Conditions and Acquisition Techniques in Sanjiang Basin. *Geophys. Prospect. Pet.* **2006**, *45*, 423–426. (In Chinese)
27. Wu, W.; Huang, J.; Li, X.; Liu, Z.; Zhu, H. Crustal structure in the Dasanjiang basin and its adjacent areas. *Acta Seismol. Sin.* **2022**, *44*, 286–301. [CrossRef]
28. Zhang, F.-Q.; Chen, H.-L.; Yang, S.-F.; Feng, Z.-Q.; Wu, H.-Y.; Batt, G.E.; Zhao, X.-Q.; Sun, M.-D.; A, M.-N.; Wang, S.-H.; et al. Late Mesozoic–Cenozoic evolution of the Sanjiang Basin in NE China and its tectonic implications for the West Pacific continental margin. *J. Asian Earth Sci.* **2012**, *49*, 287–299. [CrossRef]
29. Ramsey, C.B. Bayesian analysis of radiocarbon dates. *Radiocarbon* **2009**, *51*, 337–360. [CrossRef]
30. Reimer, P.J.; Bard, E.; Bayliss, A.; Beck, J.W.; Blackwell, P.G.; Ramsey, C.B.; Buck, C.E.; Cheng, H.; Edwards, R.L.; Friedrich, M.; et al. IntCal13 and Marine13 radiocarbon age calibration curves 0–50,000 years cal BP. *Radiocarbon* **2013**, *55*, 1869–1887. [CrossRef]

sustainability

MDPI

Article

Machine Learning Assessment of Damage Grade for Post-Earthquake Buildings: A Three-Stage Approach Directly Handling Categorical Features

Yutao Li [1], Chuanguo Jia [1,2,*], Hong Chen [3,4], Hongchen Su [1], Jiahao Chen [1] and Duoduo Wang [1]

1 School of Civil Engineering, Chongqing University, Chongqing 400045, China
2 Key Laboratory of New Technology for Construction of Cities in Mountain Area (Chongqing University), Ministry of Education, Chongqing 400045, China
3 School of Computer Science and Engineering, Beihang University, Beijing 100191, China
4 State Key Lab of Software Development Environment, Beihang University, Beijing 100191, China
* Correspondence: jiachuanguo@cqu.edu.cn

Abstract: The rapid assessment of post-earthquake building damage for rescue and reconstruction is a crucial strategy to reduce the enormous number of human casualties and economic losses caused by earthquakes. Conventional machine learning (ML) approaches for this problem usually employ one-hot encoding to cope with categorical features, and their overall procedure is neither sufficient nor comprehensive. Therefore, this study proposed a three-stage approach, which can directly handle categorical features and enhance the entire methodology of ML applications. In stage I, an integrated data preprocessing framework involving subjective–objective feature selection was proposed and performed on a dataset of buildings after the 2015 Gorkha earthquake. In stage II, four machine learning models, KNN, XGBoost, CatBoost, and LightGBM, were trained and tested on the dataset. The best model was judged by comprehensive metrics, including the proposed risk coefficient. In stage III, the feature importance, the relationships between the features and the model's output, and the feature interaction effects were investigated by Shapley additive explanations. The results indicate that the LightGBM model has the best overall performance with the highest accuracy of 0.897, the lowest risk coefficient of 0.042, and the shortest training time of 12.68 s due to its relevant algorithms for directly tackling categorical features. As for its interpretability, the most important features are determined, and information on these features' impacts and interactions is obtained to improve the reliability of and promote practical engineering applications for the ML models. The proposed three-stage approach can provide a reference for the overall ML implementation process on raw datasets for similar problems.

Keywords: building damage assessment; earthquake disaster; categorical feature; machine learning; LightGBM; interpretability method; Shapley additive explanation

Citation: Li, Y.; Jia, C.; Chen, H.; Su, H.; Chen, J.; Wang, D. Machine Learning Assessment of Damage Grade for Post-Earthquake Buildings: A Three-Stage Approach Directly Handling Categorical Features. *Sustainability* **2023**, *15*, 13847. https://doi.org/10.3390/su151813847

Academic Editors: Chong Xu, Su Chen and Shuang Li

Received: 18 July 2023
Revised: 8 September 2023
Accepted: 14 September 2023
Published: 18 September 2023

1. Introduction

In recent years, earthquake disasters have resulted in an enormous number of casualties and economic losses [1–3]. In 2010, Haiti was severely impacted by an earthquake with a moment magnitude (M_W) of 7 which left 316,000 people dead or missing, millions of people homeless, and more than half of the buildings around the epicenter damaged [4]. In 2015, an earthquake with an M_W of 7.8 struck Gorkha, Nepal, resulting in over 30,000 people dead or injured, eight million people displaced, 500,000 houses destroyed, and another 250,000 houses damaged [5]. In 2023, a 7.8 M_W earthquake happened in Turkey and Syria, where there were around 55,000 fatalities, 130,000 injuries, and 50,000 destroyed or badly damaged structures as a result of the tragedy, which affected approximately 18 million people [6,7]. Given the above shocking and alarming facts, it would be useful to be able to rapidly identify buildings with severe and light damage after an earthquake to implement proper rescue and reconstruction, facilitating the sustainable development of cities.

Some conventional assessment techniques have been developed based on empirical principles. Ningthoujam and Nanda proposed a robust and straightforward approach to evaluating the safety of concrete buildings against earthquakes in India, considering features such as whether the buildings are soft story, their age, whether they have substantial overhang, and so on [8]. Khan et al. employed a rapid visual screening approach of FEMA P-154 to assess the seismic vulnerability of existing buildings in Malakand, a region with an elevated risk [9]. Diana et al. predicted displacement demand and building damage grades (DGs) more accurately by typological capacity curves and the modified N2 method for Nordic countries [10]. Ozer et al. developed a comprehensive framework to assess seismic damage efficiently and accurately in multi-story buildings. This framework operates in near real time and encompasses three distinct levels of analysis: an empirical formulation, the effective stiffness concept, and the finite element model [11]. Martínez-Cuevas et al. performed statistical modeling and found discrimination metrics based on various national criteria and a ranking of features to assess the habitability of houses after earthquakes [12].

Machine learning (ML) is another favorable approach to address the subjectivity underlying empirical formulas and consider the potential impact of a broader range of features. Chaurasia et al. performed predictions using a neural network and the random forest model, and the best-performing model's F1 score was 0.743. [13]. Chen and Zhang used a cloud model and Bayesian networks to predict the DG in three levels with an accuracy of 0.888 based on a dataset of 9920 buildings [14]. A year later, they performed a three-grade damage prediction by ensemble learning with an accuracy of 0.783 based on a dataset of 12,045 buildings [15]. K.C. et al. trained and tested four ML models to predict building damage in five grades for 549,251 buildings and found that XGBoost had the highest accuracy of 0.577 [16].

It is acknowledged that ML can achieve high-accuracy damage grade assessments; however, it is typically realized by complex models, many of which are black boxes that are challenging even for experts to interpret [17]. Moreover, erroneous decisions that arise from the models may have severe repercussions, necessitating interpretability methods to guarantee the reliability of the decisions. Feature importance offers a simplified and global interpretation by ranking the importance of the features [18]. Partial dependence plots and accumulated local effects plots can present the average impact of features on predictions, and the latter has a broader range of applications as it is free from the limitation of the feature independence assumption [19,20]. Shapely additive explanations stem from game theory and elucidate the importance of features by comparing the mean change in model output through the presence or absence of features after constructing combinations of different features [21,22].

In the above studies, conventional empirical assessment techniques incorporate empirical formulations that are constrained by the researchers' experience and that have limited regional applicability. In addition, the limited considered features make it challenging to capture the potential feature influence and gain deeper insight into the mechanism in complex scenarios. Consequently, ML assessment approaches are preferred. However, when the ML models were implemented in the above studies, there was inadequate attention given to data preprocessing, affecting the accuracy and efficiency of the ML models. Additionally, none of them directly dealt with categorical features but employed one-hot encoding, increasing the feature dimensions and computational cost. Moreover, there was not enough consideration of the reliability the models when applied to practical engineering projects.

To bridge this gap, this paper proposed a three-stage approach, which can directly handle categorical features and achieve high-accuracy damage grade assessments of post-earthquake buildings. Meanwhile, the proposed approach enhanced the entire problem-solving methodology, i.e., before, during, and after the implementation of the ML models, to improve the models' accuracy and reliability. The remaining sections of the paper are organized as follows. In Section 2, the fundamental concepts and principles underlying the utilized ML models, evaluation metrics, and interpretability method are reviewed. In Section 3, the implementation of the three-stage approach is described in detail, including data preprocessing, the development of the ML models, and the development of the interpretability method for the best-performing model. In Section 4, the primary findings and limitations of the study are provided, and potential avenues for further research are presented.

2. Overview of Machine Learning and Interpretability Methods

2.1. Machine Learning Methods

2.1.1. K-Nearest Neighbors (KNNs)

The KNN algorithm is a non-parametric and supervised model, classifying samples relying on their closest neighbor's majority vote [23]. KNNs are influenced by the predominant number of samples among the closest K points, which, in turn, determines the quantity of neighbors K in the final result [24,25]. The distance metrics can be determined by various functions, among which the Heterogeneous Euclidean-Overlap Metric (HEOM) is suggested, as presented in Equation (1) [26]:

$$d(X,Y) = \sqrt{\sum_{n=1}^{m} d_n(x_n - y_n)} \tag{1}$$

where $d_n(x_n - y_n)$ denotes the distance between two observed samples in the nth feature.

A KNN is easy to implement and adapts easily, but it does not scale well and is prone to overfitting. The KNN algorithm works on the following principles [27]:

(1) Determine the distance between each training and test set of data;
(2) Sort by the increasing distance;
(3) Choose the K points with the shortest distance;
(4) Determine the frequency of occurrence of the category in which the first K points belong;
(5) Provide the data class that appears the most frequently in the first K points as the classification result of the test data.

2.1.2. Extreme Gradient Boosting (XGBoost)

In ML techniques, extreme gradient boosting is an algorithm for ensemble learning, commonly called XGBoost [28]. It integrates statistical boosting approaches and contains classification and regression trees (CARTs) that are pretty straightforward. Boosting is a technique that improves a model's accuracy by building several trees rather than a single tree and then combining those trees into a single prediction framework. This helps refine the estimation precision by increasing the model's accuracy [29]. XGBoost constructs the tree by iteratively utilizing the residuals from the previous trees as contributions to the resulting tree. Consequently, the resulting tree evolves by capturing the accumulated errors of the previous trees, ultimately refining the overall prediction. The growth of the new tree ceases either upon reaching the predetermined maximum limit of trees or when the training error fails to improve over a specified quantity of successive trees. Including random sampling in gradient boosting significantly improves the accuracy of estimations and the speed of execution. This integrated technique, probabilistic boosting, enhances the algorithm's overall performance [30]. The objective function of XGBoost consists of a loss function and a penalty term, as presented in Equation (2) [31,32]:

$$Obj^{(t)} = \sum_i l\left(y_i, \hat{y}_i^{(t)}\right) + \sum_k \Omega(f_k) \tag{2}$$

where t represents the tth iteration of training; l quantifies the difference between the actual value and the predicted value; and $\Omega(f_k) = \gamma T + \frac{1}{2}\lambda\|\omega\|^2$ is the penalty term governing the complexity of the kth tree [33]. In the penalty term, γ and λ represent the complexity of the leaves and the penalty parameter. T and $||\omega||^2$ denote the leaf count and the output of each leaf node [34]. Further details of the XGBoost model can be found in the studies [35–37].

2.1.3. Categorical Boosting (CatBoost)

CatBoost, an unbiased gradient boosting algorithm, handles categorical features directly [38]. Its notable characteristics include the capability to deal with categorical features and the novel order-boosting method without predicting shifts. It offers distinct solutions for various categorical features. CatBoost optimizes its tree splitting process instead of preprocessing. The features contain a limited number of classes, so the classifier incorporates one-hot encoding to transform the categorical features into numeric representations based on their frequency. As for combined features, the classes are substituted with the mean target value. This feature transformation will signify the loss of information regarding the interplay between categorical features. Therefore, CatBoost considers the prior amalgamation of the characteristics of the present state alongside the remaining categorical features. CatBoost can decrease model overfitting using the feature conversion value, as presented in Equation (3) [39,40]:

$$\hat{x}_i^k = \frac{\sum_{j=1}^{n} \varphi\left(x_j^k = x_i^k\right) Y_j + \alpha p}{\sum_{j=1}^{n} \varphi\left(x_j^k = x_i^k\right) + \alpha} \tag{3}$$

where x_i^k and x_j^k are the kth feature vectors in the ith and jth sample groups, respectively, and i and j fall within the range of 1 to n (the number of the sample groups); φ, Y_j, α, and p are the indicator function, ith target value, prior weight, and prior value, respectively.

The Minimal Variance Sampling (MVS) training mechanism is also a highlight of CatBoost. MVS is a weighted sampling variant of the regularization sampling technology. CatBoost incorporates all the settings needed to build separate decision trees and set up the random forest. The literature [41–43] can be referenced for further information about CatBoost.

2.1.4. Light Gradient Boosting Machine (LightGBM)

LightGBM is an exceptionally efficient implementation of the gradient boosting decision tree (GBDT) algorithm [44]. Unlike the traditional approach of level-wise tree growth, LightGBM adopts a leaf-wise growth approach for its decision trees, which are displayed in Figure 1a,b, respectively. The level-wise growth approach splits leaves on the identical layer simultaneously, assisting in model complexity controlling and multi-threading optimization. It processes all leaves in the same layer without distinction, increasing superfluous calculations and inefficiency. The leaf-wise growth approach, however, only selects and splits the leaf with the highest information gain each time. The information gain is calculated as follows [45,46]:

$$IG(A, W) = En(A) - \sum_{w \in Values(W)} \frac{|A_w|}{A} En(A_w) \tag{4}$$

$$En(A) = \sum_{d=1}^{D} -p_d \log_2 p_d \tag{5}$$

where En (A) and En (A_w) are the information entropy of the collection A and its subset A_w whose feature value is w, respectively; w, D, and p_d are the value of feature W, the number of categories, and the ratio of A to category d, respectively. Though the leaf-wise approach can assist in reducing errors and achieving greater accuracy, it may lead to overfitting due to the growth of a pretty deep decision tree, necessitating a maximum depth constraint.

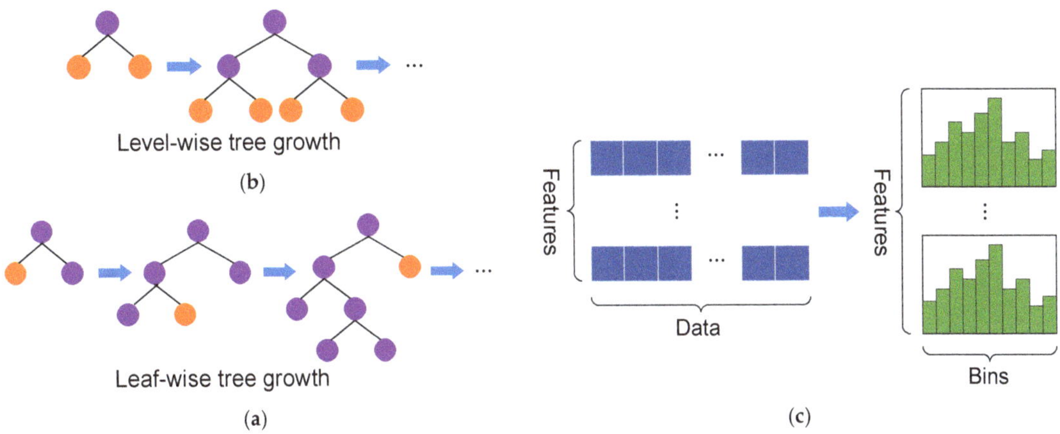

Figure 1. Schematic diagrams of some approaches related to LightGBM: (**a**) Conventional level-wise tree growth; (**b**) LightGBM leaf-wise tree growth 2; and (**c**) Histogram algorithm.

Figure 1c displays how the characteristics are binned, and the histogram approach is employed. Additionally, Gradient-based One-Side Sampling (GOSS) reduces training samples, while Exclusive Feature Bundling (EFB) merges features. The GOSS method assumes that samples with greater gradients affect information gain more than those with lower gradients. GOSS first sorts samples by their gradient absolute value in decreasing order. Next, a part of the top samples is chosen. GOSS also picks a fraction of the remaining data for sampling. EFB views characteristics as vertices and clashes as weighted edges, grouping them as graph colors. GOSS and EFB reduce computational complexity without affecting accuracy. LightGBM has a high accuracy, fast training speed, efficient handling of big data, and GPU-accelerated learning. Details of the LightGBM are provided in the literature [47–49].

2.2. Machine Learning Evaluation Metrics

Accuracy, precision, and recall are three common fundamental metrics. These metrics are formally defined by true positive (TP), false negative (FN), false positive (FP), and true negative (TN), as illustrated in Figure 2.

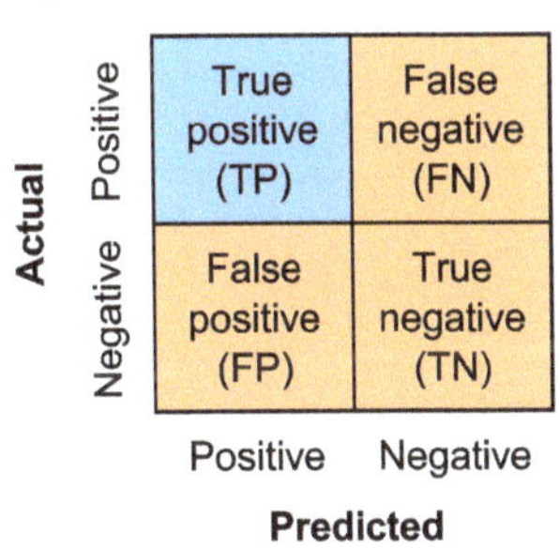

Figure 2. Schematic diagram of TP, FN, FP, and TN.

Accuracy, an overall metric that evaluates the performance of all categories, is defined by the proportion of correctly predicted samples, i.e., TP and TN, to the total number, as presented in Equation (6). The classification performance of a specific category can be evaluated by precision and recall. Precision is the truly predicted proportion of the predicted positive, and recall is the proportion that is successfully predicted as positive in the actual positive, which are defined in Equations (7) and (8).

$$\text{Accuracy} = \frac{\text{TP} + \text{TN}}{\text{TP} + \text{TN} + \text{FP} + \text{FN}} \tag{6}$$

$$\text{Precicsion} = \frac{\text{TP}}{\text{TP} + \text{FP}} \tag{7}$$

$$\text{Recall} = \frac{\text{TP}}{\text{TP} + \text{FN}} \tag{8}$$

The confusion matrix is a visualized approach to assessing the performance of ML classification models. The classification of the three DGs in this study is represented by a confusion matrix consisting of three rows and three columns, as demonstrated in Figure 3.

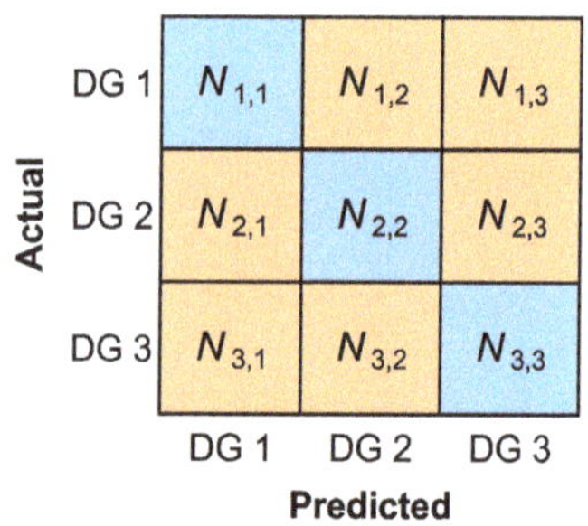

Figure 3. Confusion matrix for the classification of the three DGs.

From Figure 3, the element on the pth row and qth column is $N_{p,\,q}$, representing the number of samples that are predicted as the pth category and actually belong to the qth category. The elements on the main diagonal are true positives, and the depth of blue indicates their sizes. Other elements are misclassified samples, including true negatives, false positives, and false negatives, and the depth of orange indicates their sizes. The confusion matrix provides a clear view of how the categories are predicted and confused in the classification. Meanwhile, it can also be used to calculate accuracy, precision, and recall.

Reliability is essential for models that may be applied to practical engineering projects. In this problem, misclassifying a high DG as a low DG leads to considerable risk. To quantitatively assess this risk, it is supposed that damage resulting from the misclassification of DG as 3 to 2 and 2 to 1 is set to 1, and the damage caused by the misclassification of DG as 3 to 1 is set to 2. Then, a novel metric risk coefficient is proposed, which can be defined as follows:

$$\text{Risk coefficient} = \frac{N_{3,2} + N_{2,1} + 2N_{3,1}}{\sum_{p=1}^{3} \sum_{q=1}^{3} N_{p,q}} \tag{9}$$

The smaller the risk coefficient, the less potential damage the model may cause due to misclassification.

Moreover, the training time should also be taken into consideration. In practical applications, the number of samples involved in training can be numerous, while the time available for training is limited since the relevant people need to race against time to save victims and reinforce buildings after an earthquake. At this time, the shorter the training time, the better the model is. It can be implemented with the time library, which is a Python native library and does not require an additional download.

2.3. Interpretability Method: Shapely Additive Explanations

The Shapely Additive Explanations (SHAP) framework was introduced by Lundberg and Lee, drawing inspiration from game theory and local explanations. This methodology offers a promising approach to interpreting ML models [21,22]. As defined by SHAP, the explanation for a model with m input features is represented by a linear function g consisting of m binary features as follows:

$$g(z) = \varphi_0 + \sum_{i=1}^{m} \varphi_i z_i{'} \tag{10}$$

where $z_i{'}$ is 1 when a feature is observed; otherwise, it is 0. φ_0 represents the base value of the prediction, and φ_i denotes the contribution of the ith feature.

The calculation of the contribution of the ith feature, referred to as the SHAP value, is performed using the conditional output of subsets derived from game theory:

$$\varphi_i = \sum_{S \in M\{i\}} \frac{|S|!(M - S - 1)!}{M!} [f_x(S \cup \{i\}) - f_x(S)] \tag{11}$$

where M is the set of m features; S denotes all feature subsets that exclude the ith feature; the output of a tree when considering a specific feature subset S is $f_x(S) = [E(f(x)|x_s)]$; and $f_x(S \cup \{i\})$ and $f_x(i)$ are the expected output of models that include and exclude the ith feature, respectively. A comprehensive explanation of SHAP can be referred to in the literature [50].

3. Modeling of Damage Grade Assessment for Buildings

On 25 April 2015 at 11:56 AM NPT (UTC+5:45), an earthquake occurred in Gorkha, Nepal, as presented in Figure 4 [51]. The magnitude was 7.8. The location of the epicenter was 28.15° N and 84.71° E, and the depth was 15.0 km. Furthermore, there were three significant aftershocks with magnitudes of about 7. The first two occurred on the 25th and 26th of April, while the other happened on the 12th of May. More than 600,000 structures in Kathmandu and its surrounding areas experienced varying degrees of damage or destruction. Additionally, a considerable loss of human life occurred, with a reported death toll of approximately 9000 individuals. This seismic event was experienced throughout Nepal's central and eastern regions, a significant portion of the Ganges River valley in India's northern part, the northwestern area of Bangladesh, the southern plateau of Tibet, and the western region of Bhutan [51].

In 2016, Kathmandu Living Labs and the Central Bureau of Statistics conducted a comprehensive household survey utilizing mobile technology. The effects of the earthquake, the state of households, and demographic and economic data were all covered in this survey. The dataset utilized in this study was acquired from a comprehensive dataset and subsequently streamlined to align with the research goals of this paper, as documented by Driven Data [52]. The dataset consists of 214,839 rows and 39 columns. In this context, it can be observed that each row within the given dataset corresponds to a distinct building, while each column represents a specific feature associated with these buildings. Three distinct grades of damage exist, namely one, two, and three, which correspond to minimal damage, moderate damage, and near-total devastation, respectively.

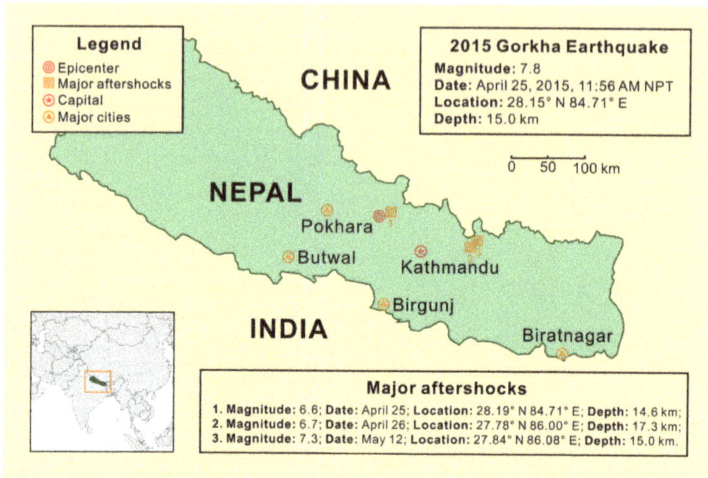

Figure 4. Location of the epicenter and major aftershocks of the 2015 Gorkha earthquake, as well as the capital and major cities.

3.1. *Stage I: Data Preprocessing*

First, an integrated data preprocessing framework was proposed, including missing value processing, outlier processing, duplicate value processing, normalization processing, feature selection, and variable encoding [53]. The whole process is shown in Figure 5.

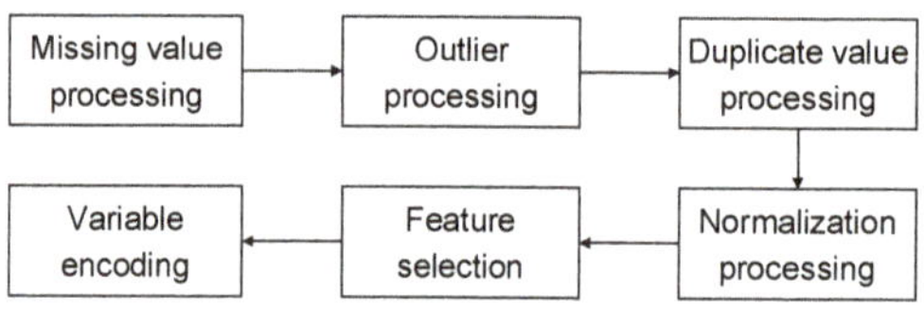

Figure 5. The flowchart of data preprocessing.

3.1.1. Missing Value Processing

Missing values can occur for several reasons, such as the unsuccessful acquisition of data, an accident while storing or transferring data, etc. First, for a building, if more than half of its 39 features have missing values, then this row of data should be discarded directly. If not, training after filling in the missing values may affect the model's accuracy. Then, owing to the categorical features in the dataset, linear interpolation and cubic spline interpolation filling methods are not appropriate. The missing value is filled with the nearest non-missing value in that column of data. If the two non-missing values closest to the missing value are equal in distance to it, then one is randomly chosen to fill the missing value.

3.1.2. Outlier Processing

Outliers can also occur in the dataset, perhaps due to the wrong information being obtained while acquiring, storing, or transferring data. It is noted that there are three types of data in the dataset: int, categorical, and binary, and the possible values are specified. Thus, if a value in the column is outside the specified range, it is considered an outlier. Here, the range is determined by the median absolute deviation algorithm [53]. Similarly, if more than half of a building's feature data are outliers, the building's data should be discarded. The replacement of outliers was performed using a method similar to the missing value filling method, i.e., replacing an outlier with the nearest non-outlier in that column.

3.1.3. Duplicate Value Processing

Duplicate values appearing in the variable building_id need to be handled. Because the values of the building_id variable represent the unique identification of a building, two buildings cannot have the same value. The building_id variable is first tabulated and sorted in descending order. For building_id variables with occurrences greater than 1, the row with fewer missing values and outliers is kept; otherwise, the front row is kept.

3.1.4. Normalization Processing

Since the dimensions and specific ranges of features are different, normalization is required to speed up the training of ML models and may improve accuracy. The normalization method is min–max normalization realized by the MinMaxScaler function of the sklearn library [54]. It should be noted that only int-type variables are processed here, binary variables are not processed, and categorical variables are processed in subsequent steps.

3.1.5. Feature Selection

There are 39 features in the dataset, which is quite a lot. Thus, the features need to be selected to accelerate the training and prediction of the model. Before that, it is noted that the variables associated with has_superstructure and has_secondary_use are all binary variables, which can be turned into categorical variables to reduce the number of features and compact the information expression. Afterwards, 19 features are able to express the information expressed by the original 39 features. Furthermore, the building_id feature is removed because it only serves to distinguish each building and does not substantially contribute to the judgment of DG. Information on the remaining 18 features is shown in Table 1.

Table 1. The building features after preprocessing for ML modeling.

Feature	Identifier	Type	Description	Possible Values
geo_level_1_id	N_{g1}		from broadest (level 1) to narrowest	0–30
geo_level_2_id	N_{g2}		(level 3) in terms of the building's	0–1427
geo_level_3_id	N_{g3}		geographical location	0–12,567
count_floors_pre_eq	N_{fl}	int	post-earthquake number of floors of the building	no exact value range
age	N_a		number of years the building has been in existence	no exact value range
count_families	N_{fa}		number of family members residing in the building	0–9
area_percentage	P_a		normalized area of the building footprint	no exact value range
height_percentage	P_h		normalized height of the building footprint	no exact value range
land_surface_condition	C_{ls}		the surface condition of the land for the location of the building	n, o, t.
foundation_type	T_f		foundation type of the building	h, i, r, u, w.
roof_type	T_r		roof type of the building	n, q, x.
ground_floor_type	T_g		ground floor type of the building	f, m, v, x, z.
other_floor_type	T_o	categorical	type of constructions utilized for stories above ground level (other than the roof)	j, q, s, x.
position	P_b		position of the building	j, o, s, t.
plan_configuration	C_p		building plan configuration	a, c, d, f, m, n, o, q, s, u.
has_superstructure	C_{ss}		the material of the superstructure	am, b, cmb, cms, mmb, mms, o, rce, rcne, sf, t
legal_ownership_status	C_{los}		legal ownership status of the land for the location of the building	a, r, v, w.
has_secondary_use	C_{su}		whether the building is being used secondarily and what its specific purpose is	a, go, h, hp, ind, ins, no, o, r, s, up

In Table 1, the specific meanings of the classes for C_{ss} are am (adobe/mud), b (bamboo), cmb (cement mortar–brick), cms (cement mortar–stone), mmb (mud mortar–brick), mms (mud mortar–stone), o (other materials), rce (engineered reinforced concrete), rcne (non-engineered reinforced concrete), sf (stone), and t (timber). Additionally, the specific meanings of the classes for C_{su} are a (agricultural purposes), go (government office), h (hotel), hp (health post), ind (industrial purposes), ins (institution), no (no secondary use), o (other purposes), r (rental purposes), s (school), and up (police station). Further details about the features and classes are provided in the literature [55].

During the feature selection, a combination of objective and subjective methods was employed. The variance thresholding method was used for the objective feature selection method, which is effective only for int-type features. The variance along with the min, max, and standard deviation of the features for reference are presented in Table 2.

Table 2. The statistical description of int-type features.

Feature	N_{g1}	N_{g2}	N_{g3}	N_{fl}	N_a	P_a	P_h	N_{fa}
Min	0	0	0	1	0	1	2	0
Max	30	1427	12,567	9	995	100	32	9
Standard deviation	8.07	413.05	3661.10	0.73	75.65	4.52	1.95	0.42
Variance	65.08	170,606.27	13,403,651.08	**0.54**	5723.67	20.47	3.82	**0.18**

The bolded ones are the variance values less than 1, and the corresponding features are considered for elimination.

For categorical features, since the variance cannot be calculated, according to a similar idea of variance thresholding, the dominant class proportion, i.e., the proportion of the class with the largest number to the total number (214,839) in the features, was used as the criterion. The calculation results are shown in Table 3.

Table 3. The proportion of the class with the largest number to the total number in the features.

Feature	C_{ls}	T_f	T_r	T_g	T_o	P_b	C_p	C_{ss}	C_{los}	C_{su}
Proportion	**0.83**	**0.83**	0.69	**0.80**	0.63	0.77	**0.96**	0.72	**0.96**	**0.89**

Similarly, the values greater than 0.8 are bolded, and their corresponding features are given consideration for elimination.

For the subjective feature selection method, manual scoring was conducted here, primarily considering their potential contribution to the DG in terms of the physical significance of the features. The scoring range is from 0 to 1, and the results are displayed in Table 4.

Table 4. Manual scoring considering the contribution of features based on physical significance.

Feature	N_{g1}	N_{g2}	N_{g3}	N_{fl}	N_a	P_a	P_h	N_{fa}	C_{ls}
Score	1	1	1	0.8	0.8	0.7	0.7	**0.3**	0.8
Feature	T_f	T_r	T_g	T_o	P_b	C_p	C_{ss}	C_{los}	C_{su}
Score	0.7	0.5	0.7	**0.4**	0.6	0.6	0.7	**0.4**	**0.3**

The bolded values are less than 0.5, and the corresponding features are deemed to be eliminated.

For a comprehensive evaluation, the information in the above three tables was considered simultaneously. The features with bolded values for their variance (or proportion) and score, i.e., N_{fa}, T_o, C_{los}, and C_{su}, were directly eliminated. For T_f, T_g, and C_p, although their scores are greater than 0.5, they are all less than or equal to 0.7. Furthermore, they provide less information because their proportion is over 0.8, so they were also eliminated. For N_{fl} and C_{ls} with bolded values for their variance (or proportion), their scores are both 0.8, revealing that they are essential and should be retained.

Therefore, the retained features are N_{g1}, N_{g2}, N_{g3}, N_{fl}, N_a, P_a, P_h, C_{ls}, T_r, P_b, and C_{ss}.

3.1.6. Categorical Variable Encoding

Among the selected models, KNN and XGBoost do not support categorical features, while CatBoost and LightGBM do. For the data to be trained in KNN and XGBoost, one-hot encoding is required, which can be implemented by the get_dummies function of the pandas library [56]. The process of one-hot encoding is illustrated in Figure 6. A categorical feature with N_c specific categories can be transformed into N_c binary variables with 0 and 1 representing whether it belongs to a category or not. For a sample, only one of these binary variables is 1, and the rest are 0. The binary variables are guaranteed to be equidistant for each category, unlike directly mapping specific categories to integers $0, 1, \ldots, N_c-1$, which are non-equidistant. However, this method results in a rise in dimensions, which can be particularly significant when N_c is large, thus greatly increasing the training and prediction time of the model.

Sample	Tr
S1	n
S2	q
S3	x

one-hot encoding →

Sample	Tr_n	Tr_q	Tr_x
S1	1	0	0
S2	0	1	0
S3	0	0	1

Figure 6. The process of one-hot encoding (taking feature T_r as an example).

Similarly, apart from features in the KNN and XGBoost, the training labels also need to be encoded, but one-hot encoding cannot be employed here. Instead, they are directly mapped to integers from 0 to N_L-1, where N_L-1 is the number of categories of the label, which is named label encoding.

3.2. *Stage II: Development of the Machine Learning Models*

First, the data were visualized to exhibit the distribution of DG on the relationship plot between the features. Since there are categorical features, the pairplot function of the seaborn library is not appropriate. Hence, the function scatterplot was utilized [57]. The plots were completed in the following order: N_{g1}, N_{g2}, N_{g3}, N_{fl}, N_a, P_a, P_h, C_{ls}, T_r, P_b, and C_{ss}. For a certain feature, its relationships with the next-to-last features were plotted to ensure orderliness and non-repetition. For the visualization of DG, the damage severity increases from 1 to 3 and is represented in orange, red, and dark red with different shapes, respectively, indicating the increasing severity. Due to the excessive amount of data, 500 samples were randomly selected for plotting, and the relationships between the features with DGs are displayed in Figure 7.

From Figure 7, abundant information can be obtained. For instance, from the third-to-last subfigure in Figure 7, it is found that for (Tr, Pb), the DG tends to be one when it is (x, o), (x, t), or (x, s). When it is (n, o), (n, s), (q, j), (q, t), or (x, j), the DG is generally two. When it is (n, j), (n, t), or (q, s), the DG is more prone to be three. Preliminarily understanding the influence of the relationship between the features on the DG can help establish a general and comprehensive view of the data.

Subsequently, four ML models, KNN, XGBoost, CatBoost, and LightGBM, were selected and developed. The KNN was selected because of its straightforward concept and few hyperparameters. Then, three ensemble learning algorithms, XGBoost, CatBoost, and LightGBM, were chosen due to their typically superior performance. These four algorithms can be divided into two groups for a subsequent comparison and discussion: the KNN and XGBoost cannot directly handle categorical features, whereas CatBoost and LightGBM can. The environment is Anaconda 3 and python 3.10.4. The primarily utilized libraries are sklearn 1.1.2, lightgbm 3.3.2, xgboost 1.6.2, catboost 1.2, shap 0.41.0, numpy 1.23.1, and pandas 1.4.3. For splitting the dataset, seventy percent was allocated as the training dataset, while the remaining thirty percent was designated as the test dataset. Preliminary training was performed using the default hyperparameters of the model as a baseline, and then hyperparameters tuning was performed using GridSearchCV, setting a five-fold cross-validation [54]. The model's accuracy with the default and optimal hyperparameters are provided in Table 5.

From Table 5, with the assistance of GridSearchCV, the optimal hyperparameters of the model were found, and the accuracy of the training and test datasets of all the models except the KNN was improved considerably.

After obtaining the optimal hyperparameters, the models were trained and tested, and their performance could be intuitively grasped and analyzed through the confusion matrices. The elements on the main diagonal or other positions are true positives or misclassified samples, whose sizes are indicated by the depth of blue or orange, respectively. The confusion matrices of the predicted DGs for the models of the training dataset are displayed in Figure 8.

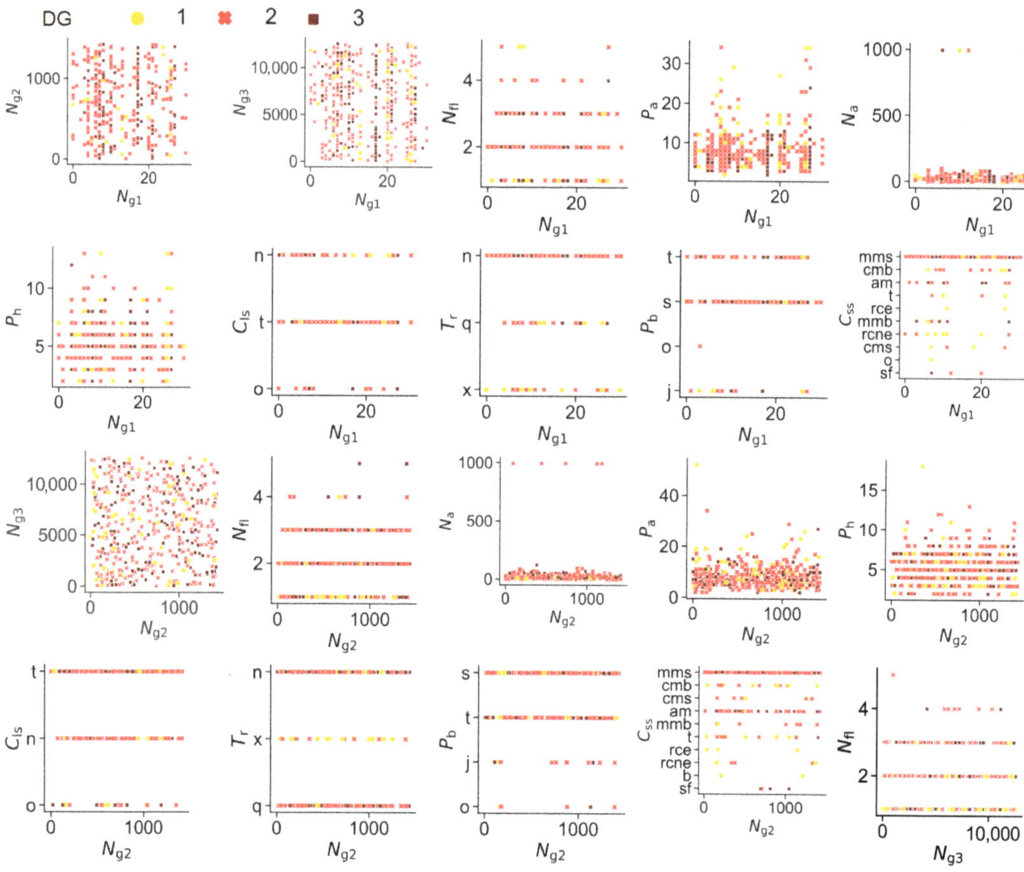

Figure 7. *Cont.*

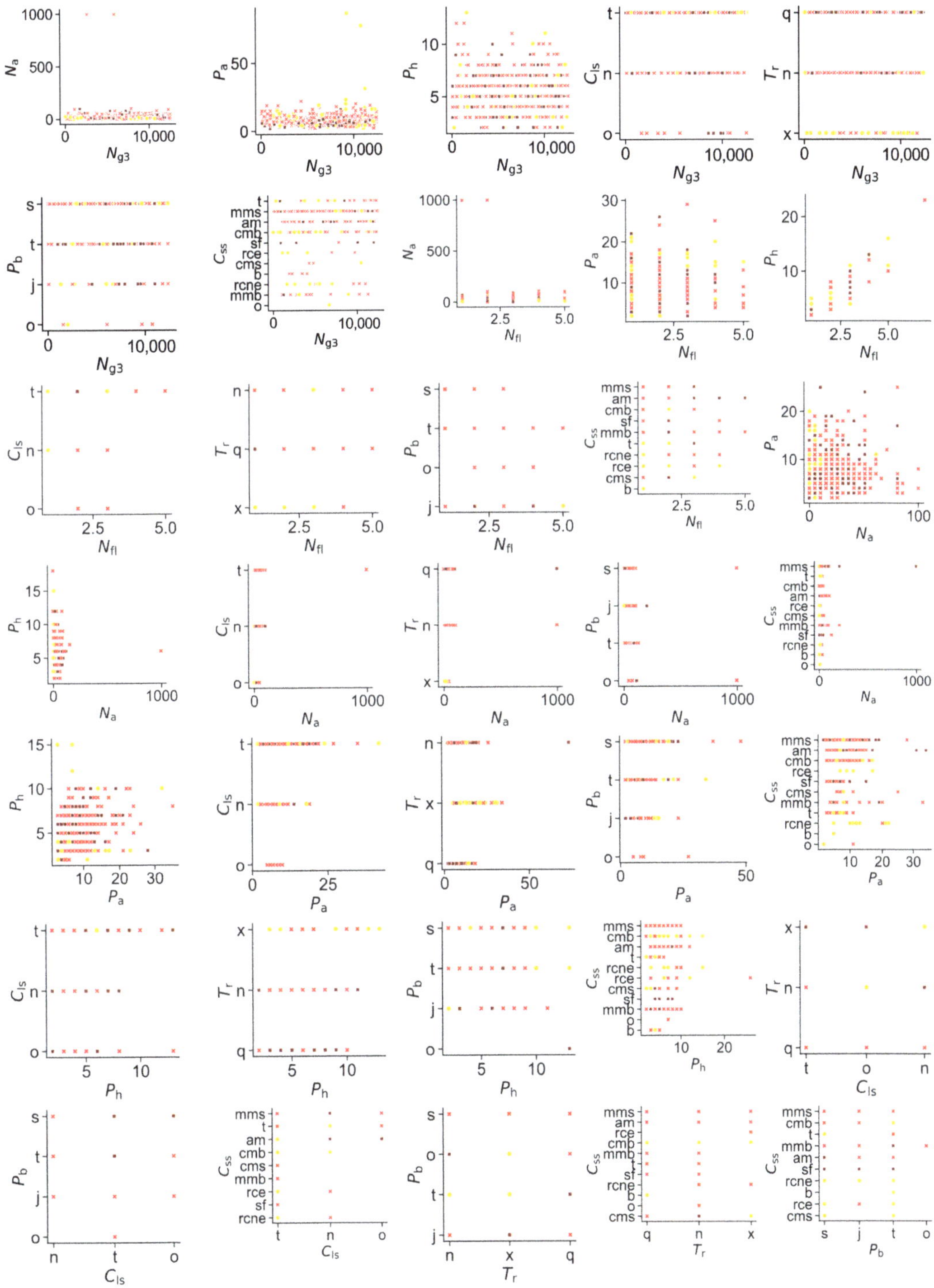

Figure 7. Visualization of relationships between features with DGs (500 samples).

Table 5. Model accuracy with default and optimal hyperparameters.

Model	Accuracy				Detailed Optimal Hyperparameters
	Default Hyperparameters		Optimal Hyperparameters		
	Training Set	Test Set	Training Set	Test Set	
KNN	0.841	0.770	0.876	0.770	n_neighbors = 3
XGBoost	0.877	0.861	0.969	0.898	learning_rate = 0.1, n_estimators = 1200, min_child_weight = 1, gamma = 0, max_depth = 8
CatBoost	0.858	0.849	0.908	0.874	depth = 8, learning_rate = 0.15, iterations = 1200
LightGBM	0.840	0.835	0.975	0.897	learning_rate = 0.1, n_estimators = 1200, num_leaves = 100, max_depth = 8, colsample_bytree = 0.8, subsample = 0.8

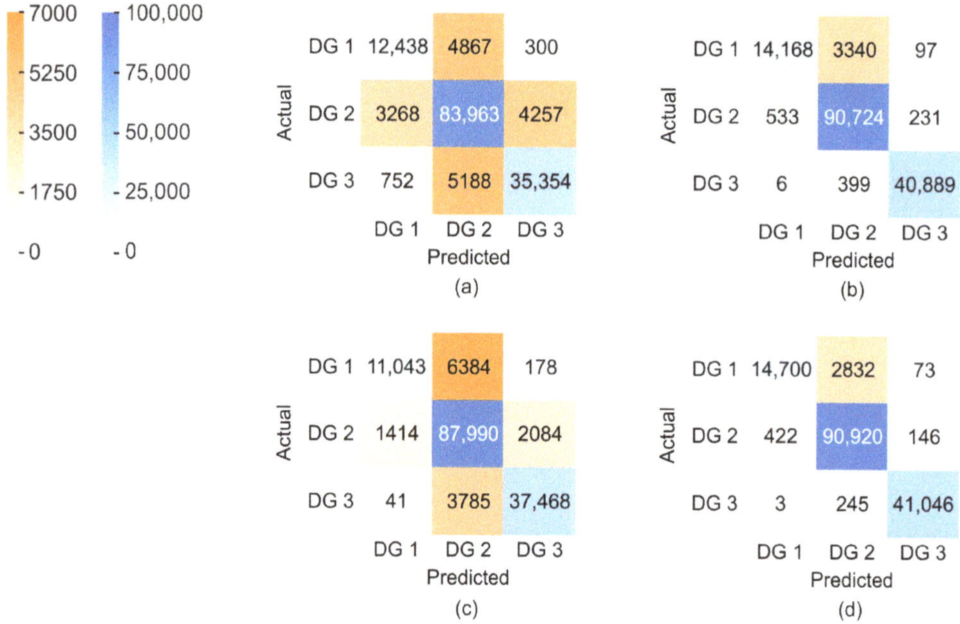

Figure 8. Confusion matrices of predicted DGs on the training dataset by different models: (**a**) KNN; (**b**) XGBoost; (**c**) CatBoost; and (**d**) LightGBM.

Based on the findings presented in Figure 8, it can be discerned that the KNN model exhibited the least favorable performance for the training set, whereas the LightGBM model demonstrated the most optimal performance. The XGBoost model exhibits a marginally inferior performance compared to the LightGBM model, yet it remains highly advantageous. The performance of the CatBoost model is moderate, particularly when evaluating the 6384 samples with a DG of 1 as 2. This misjudgment stands out as the most significant among the four models.

Figure 9 presents the confusion matrices depicting the predicted DGs of the models for the test dataset.

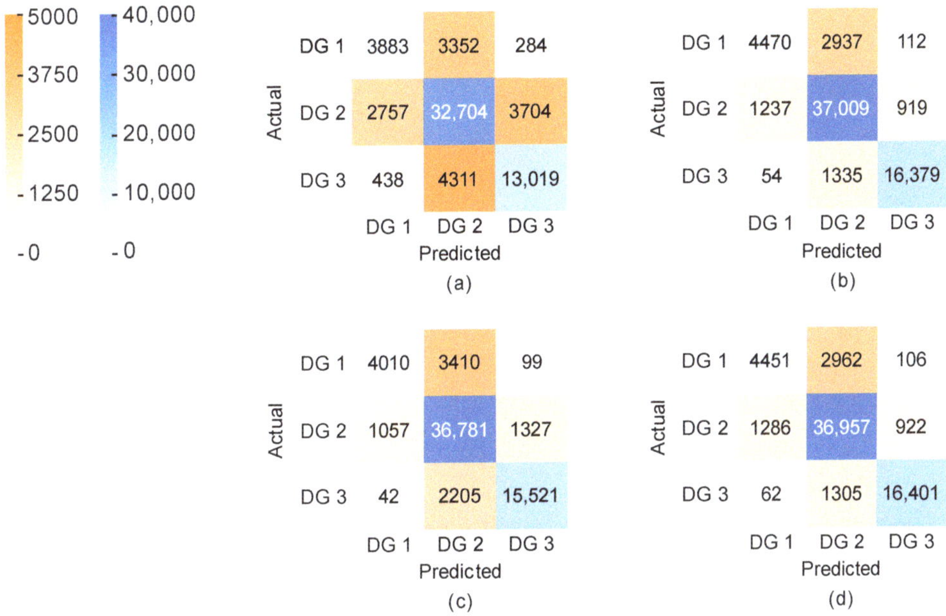

Figure 9. Confusion matrices of predicted DGs for the test dataset by different models: (**a**) KNN; (**b**) XGBoost; (**c**) CatBoost; (**d**) and LightGBM.

From Figure 9, it is indicated that for the test set, the KNN model still performed the worst, while the LightGBM and XGBoost models performed comparably, both being the best. For the CatBoost model, the number of samples misclassifying the DG as 1 instead of 2 is roughly the same as that of the KNN model, revealing that the misclassification has improved, though the overall performance remains moderate.

Next, the models with optimal hyperparameters were evaluated by comprehensive metrics, including accuracy, precision, recall, the proposed risk coefficient, and the training time, as presented in Table 6 and Figure 10.

Table 6. Performance metrics of the models with optimal hyperparameters on the test dataset (except training time on the training dataset).

Model	Accuracy	Precision			Recall			Risk Coefficient	Training Time (s)
		DG 1	DG 2	DG 3	DG 1	DG 2	DG 3		
KNN	0.770	0.549	0.810	0.766	0.516	0.835	0.733	0.123	**0.03**
XGBoost	**0.898**	**0.776**	**0.897**	**0.941**	**0.594**	**0.945**	**0.922**	**0.042**	108.19
CatBoost	0.874	**0.785**	0.868	0.916	0.533	0.939	0.874	0.052	169.51
LightGBM	**0.897**	0.768	**0.896**	**0.941**	**0.592**	**0.944**	**0.923**	**0.042**	**12.68**

The top two values of each metric are bolded to emphasize the two best performances [58].

As displayed in Table 6, the training time of the KNN model is the shortest, only 0.03 s, but none of its other metrics are in the top two. Thus, this model may be applied to a specific scenario with limited training time. For the CatBoost model, the precision of DG 1 reaches a high level. However, its general performance in other aspects makes it less appropriate for DG assessments. The XGBoost and LightGBM models performed fairly well, with eight out of nine metrics in the top two. To further compare the discrepancy between their parameters, the training time of the LightGBM model is 12.68 s, which is considerably smaller than the

108.19 s of the XGBoost model. Moreover, the precision of DG 1 of the LightGBM model is 0.768, slightly smaller than that of the XGBoost model (0.776).

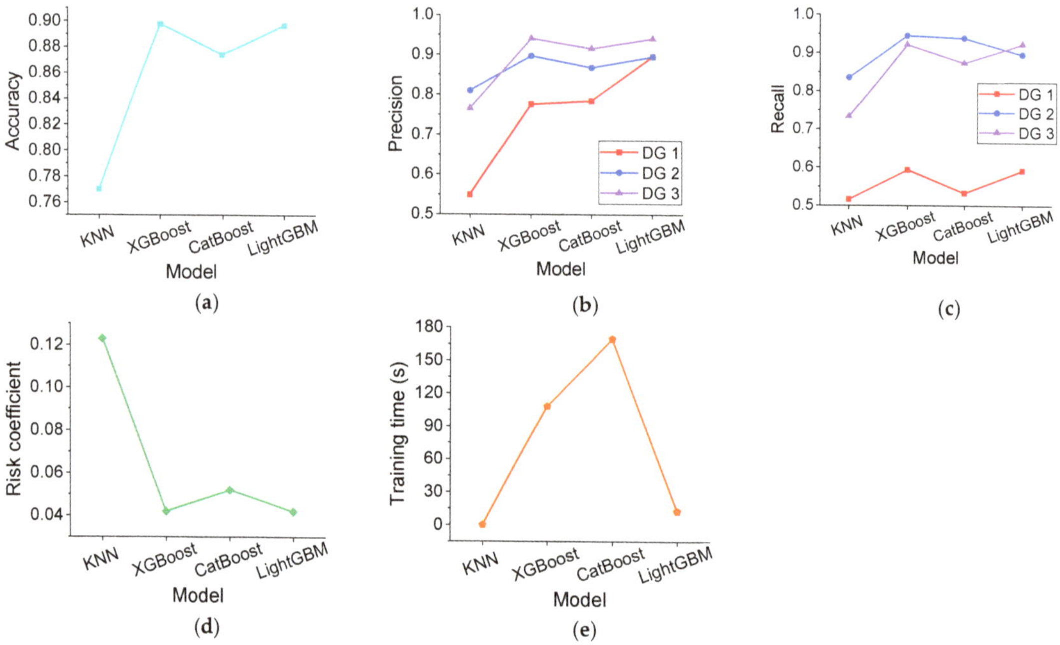

Figure 10. Performance metrics of the models with optimal hyperparameters on the test dataset (except training time on the training dataset): (**a**) Accuracy; (**b**) Precision; (**c**) Recall; (**d**) Risk coefficient; and (**e**) Training time.

Therefore, from Table 6 and Figure 10, the LightGBM model is regarded as the best-performing model for DG assessments in this paper after a comprehensive evaluation. The accuracy is 0.897, which is satisfactory compared to the studies of Chen and Zhang [14,15], with reported accuracies of 0.888 and 0.783.

3.3. Stage III: Development of the Interpretability Method for the Best-Performing Model

To enhance the comprehension of the model operation mechanism and ensure the safety of the application for practical engineering, the best-performing model, i.e., the LightGBM model, was interpreted by its application programming interface (API), as displayed in Figure 11.

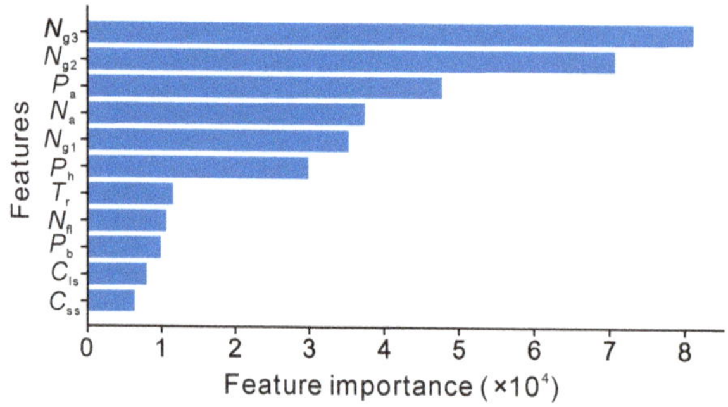

Figure 11. Feature importance of the LightGBM model by its API.

From Figure 11, the five most important features are N_{g3}, N_{g2}, P_a, N_a, and N_{g1}, indicating that the DG is closely connected with the buildings' locations, ages, and normalized footprint areas.

Then, the model was interpreted by SHAP. A global comprehension of the feature importance for different DGs can be implemented by visualizing the mean absolute SHAP value of the features, which is illustrated in Figure 12.

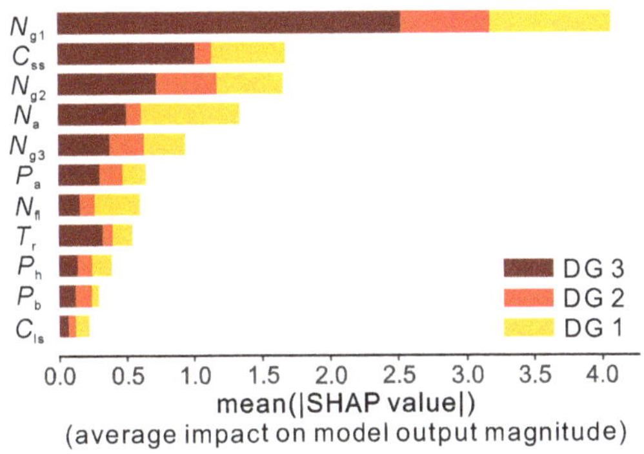

Figure 12. Feature importance of the LightGBM model by SHAP.

From Figure 12, the features N_{g1}, N_{g2}, and N_{g3}, which are related to the geographical locations, are essential, occupying three of the five most essential features. C_{ss} indicates the materials of the superstructures. The strength and stiffness of construction materials vary greatly, resulting in different levels of resistance to earthquakes, so C_{ss} is also crucial. N_a, the age of the buildings in years, is also vital. The aging of the buildings over time leads to a decrease in the load-bearing capacity of their components, thus increasing seismic damage. For DG 1, the three most prominent features are N_{g1}, N_a, and C_{ss}. For DG 2, they are N_{g1}, N_{g2}, and N_{g3}. For DG 3, they are N_{g1}, C_{ss}, and N_{g2}. They are generally similar but locally different.

Compared with Figure 11, both emphasize the importance of location and age. Summarizing the perspectives of the model API and SHAP, the six most important metrics are N_{g2}, N_{g1}, N_{g3}, N_a, P_a, and C_{ss}. Other features also have undeniable contributions to the model, but not as much as these previous features.

To investigate how the features impact the output of the LightGBM model, Figure 13 provides an information-dense summary. From Figure 13a, the SHAP values increase as N_a, N_{fl}, and P_h decrease, revealing a positive impact on DG 1 judgments. The effect becomes compounded for N_{g1}, N_{g2}, N_{g3}, and P_a, and their increase may have a positive or negative impact on DG 1's judgments. It is comprehensible since the features N_{g1}, N_{g2}, and N_{g3}, which are associated with geographic locations, are artificially defined, and their magnitudes do not have a high physical significance. As C_{ss}, T_r, C_{ls}, and P_b are categorical variables, no size difference exists between specific categories. Hence, they are presented in gray and are not discussed here.

From Figure 13b, with the increase in P_a and N_{fl}, the SHAP value rises, thus positively impacting DG 2 judgments. Similarly, the effects of N_{g1}, N_{g2}, and N_{g3} on DG 2 judgments are also compounded. Interestingly, for P_h and N_a, their reduction brings the SHAP value to nearly 0, revealing slight effects. In contrast, an elevation in P_h and N_a levels would lead to an SHAP value that is either positive or negative, indicating a potentially heightened positive or negative influence. By conducting a comparable analysis, as shown in Figure 13c, one can also deduce the influence of the variation in features on the judgment of DG 3.

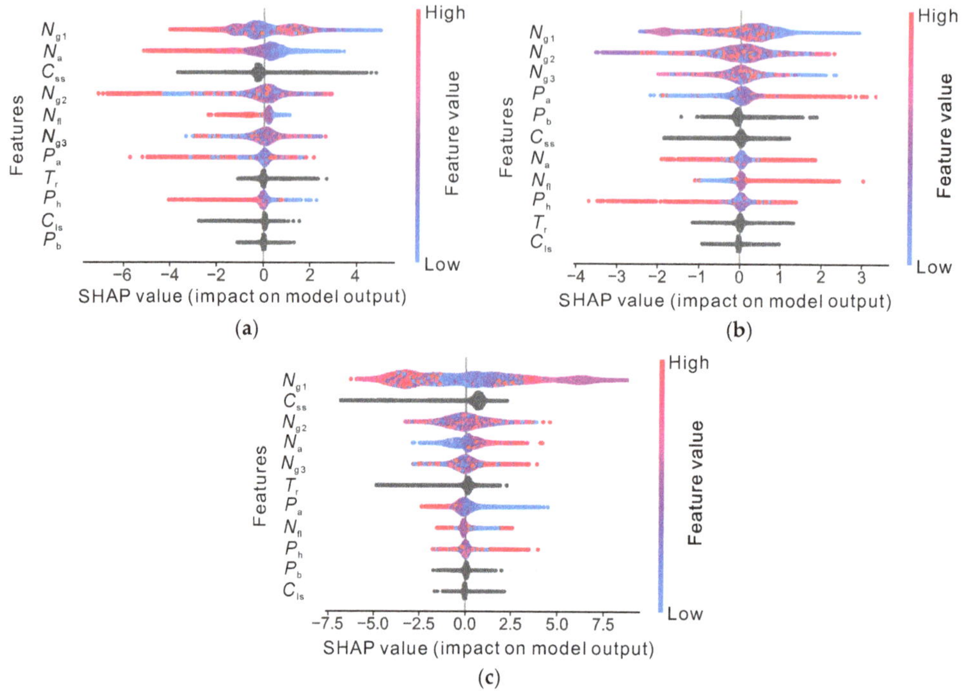

Figure 13. Beeswarm plots of SHAP values of features for the predicted results (the LightGBM model): (**a**) DG 1; (**b**) DG 2; and (**c**) DG 3.

Combining the information from the three figures yields the integrated impact of the features on the three DG judgments. Taking N_a as an example, Figure 13a,c illustrate that as N_a rises, the trends for the SHAP value are exactly the opposite. In Figure 13b, when the values of N_a are low, the SHAP values are close to 0. Furthermore, the increase in N_a gives rise to both positive and negative SHAP values. Therefore, it can be concluded that a newly constructed building tends to be judged as DG 1. When a building is older, it may be more likely to be judged as DG 2 or 3.

Furthermore, apart from illustrating the impact of the variation in a singular feature on the model output in Figure 13, SHAP can also demonstrate the interaction effects of the features, as depicted in Figure 14. Taking Figure 14a as an example, the SHAP values for the same value of N_{g1} are distinct, revealing nonlinear interaction effects between the two features of N_{g1} and N_{fl}. Otherwise, there will be no vertical dispersion in the figure, and the scatter points will ideally obey the line determined by the dependence_plot function. Figure 14c shows that the interaction effects between N_a and N_{fl} are particularly prominent. When the values of N_a are close to 0, the scatter points are predominantly blue, suggesting that the current values of N_{fl} are low, and the corresponding SHAP values are more than 0, indicating a positive impact on the model's output. However, most scatter points become red as N_a rises from 0 to 3. At this time, the values of N_{fl} are larger, and the corresponding SHAP values are nearly less than 0, reflecting a negative impact. Figure 14i shows that when T_r is n or x, with the increase in N_{fl}, the SHAP value gradually changes from negative to 0 and then to positive, signifying that the impact on the model's output progressively varies from negative to positive. When T_r equals q, the values of N_{fl} are typically greater, and the SHAP value may be positive or negative, implying a weak interaction.

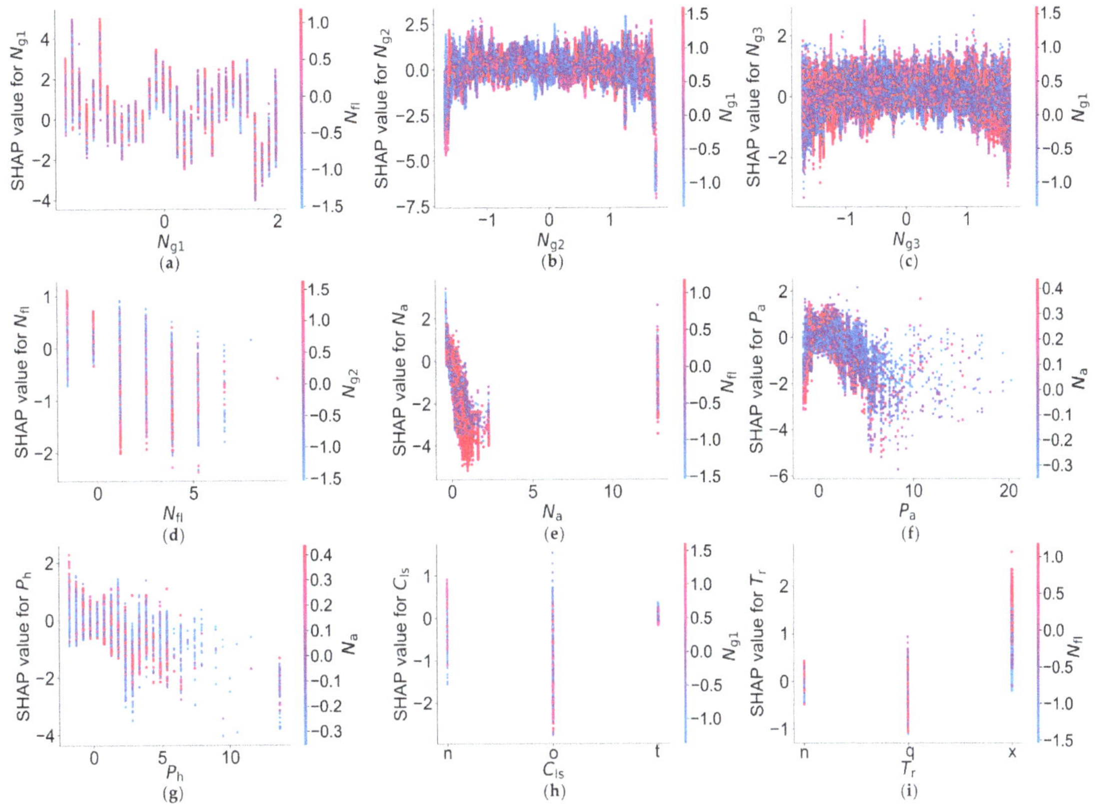

Figure 14. SHAP partial dependence scatter plots for the LightGBM model to present interaction effects of the features (selected features in DG 1): (**a**) N_{g1} and N_{fl}; (**b**) N_{g2} and N_{g1}; (**c**) N_{g3} and N_{g1}; (**d**) N_{fl} and N_{g2}; (**e**) N_a and N_{fl}; (**f**) P_a and N_a; (**g**) P_h and N_a; (**h**) C_{ls} and N_{g1}; and (**i**) T_r and N_{fl}.

The results and discussion in this stage provide a more informative and compact interpretation of the best-performing model compared to previous similar studies [15,16,59]. By directly coping with categorical features rather than employing one-hot encoding, information fragmentation attributed to transforming a categorical feature into several binary features is prevented. The proposed approach can comprehend the impact of features on the model's output and reflect the interaction effects between features involving categorical features, thereby enhancing the model's reliability. Simultaneously, a greater understanding of these models can also guide data acquisition to determine the features to be considered.

4. Conclusions

This paper employed a three-stage framework including four ML models, a series of evaluation metrics, and an interpretability method. In stage I, data preprocessing and feature selection were conducted on a dataset of the buildings after the 2015 Gorkha earthquake. In stage II, four ML models were trained and tested on the selected dataset to determine the best-performing model. In stage III, the best-performing model was interpreted to exhibit the feature importance and the relationships between the input and output variables. The conclusions can be summarized as follows.

(1) An integrated data preprocessing framework combined with subjective–objective feature selection was proposed to accelerate the model training and prediction without losing an excessive amount of accuracy, providing a reference for similar problems.
(2) The LightGBM and CatBoost models directly handle categorical features with training times of 12.68 s and 169.51 s, while the XGBoost model tackles features by one-hot encoding with a training time of 108.19 s. Given the guarantee of accuracy, the algorithms in LightGBM for handling categorical features can significantly accelerate training, while those in CatBoost instead increase the training time for this problem.
(3) The LightGBM and XGBoost models have the highest accuracies of 0.897 and 0.898, followed by the CatBoost model with 0.874. The KNN model may be less appropriate for this problem, with an accuracy of only 0.770.
(4) The risk coefficient was proposed to evaluate the safety of the model classification quantitatively, and the risk coefficients of the KNN, XGBoost, CatBoost, and LightGBM models were 0.123, 0.042, 0.052, and 0.042. The safety of the XGBoost and LightGBM models was the highest.
(5) Comparing the accuracy, precision, recall, risk coefficient, and training time of the four models, the LightGBM model has the best overall performance.
(6) To visualize the mechanism of the LightGBM model by its API and SHAP, the six most essential features are N_{g2}, N_{g1}, N_{g3}, N_{a}, P_{a}, and C_{ss}. The impact of N_{g1}, N_{g2}, and N_{g3} variations on the model output are all compound, while that of other features is related to the specific damage grade, presenting a monotonic or compound influence.

Nevertheless, this research is subject to some limitations. It only focused on features regarding buildings and their environments. Meanwhile, current interpretability approaches are inadequate for categorical features. For instance, beeswarm plots of SHAP values of features are incompatible with categorical features.

Therefore, future work can consider the features of seismic waves or other aspects to improve the accuracy and expand the range of potential applications. Moreover, more advanced interpretability techniques for categorical features should be developed and utilized to ensure the reliability and robustness of the models for practical engineering applications.

Author Contributions: Conceptualization, Y.L., H.C. and H.S.; methodology, Y.L. and H.C.; software, Y.L. and H.C.; validation, Y.L. and H.C.; formal analysis, Y.L. and H.C.; investigation, Y.L.; resources, Y.L., C.J. and H.C.; data curation, Y.L.; writing—original draft preparation, Y.L., H.C., and J.C.; writing—review and editing, Y.L., C.J., H.C., H.S. and J.C.; visualization, Y.L., J.C. and D.W.; supervision, C.J.; project administration, C.J.; funding acquisition, C.J. All authors have read and agreed to the published version of the manuscript.

Funding: This research was funded by National Natural Science Foundation of China, Grant No. 52278481 and Project Supported by Graduate Scientific Research and Innovation Foundation of Chongqing, China, Grant No. CYS23107.

Institutional Review Board Statement: Not applicable.

Informed Consent Statement: Not applicable.

Data Availability Statement: Some or all the data, models, or code generated or used during the study are available in the GitHub repository: https://github.com/Yu-tao-Li/Building-Damage-Grade-Assessment.

Conflicts of Interest: The authors declare no conflict of interest.

References

1. Cariolet, J.-M.; Vuillet, M.; Diab, Y. Mapping Urban Resilience to Disasters—A Review. *Sustain. Cities Soc.* **2019**, *51*, 101746. [CrossRef]
2. Han, L.; Ma, Q.; Zhang, F.; Zhang, Y.; Zhang, J.; Bao, Y.; Zhao, J. Risk Assessment of An Earthquake-Collapse-Landslide Disaster Chain by Bayesian Network and Newmark Models. *Int. J. Environ. Res. Public. Health* **2019**, *16*, 3330. [CrossRef] [PubMed]

3. Qiang, Y.; Huang, Q.; Xu, J. Observing Community Resilience from Space: Using Nighttime Lights to Model Economic Disturbance and Recovery Pattern in Natural Disaster. *Sustain. Cities Soc.* **2020**, *57*, 102115. [CrossRef]
4. DesRoches, R.; Comerio, M.; Eberhard, M.; Mooney, W.; Rix, G.J. Overview of the 2010 Haiti Earthquake. *Earthq. Spectra* **2011**, *27*, 1–21. [CrossRef]
5. Chaulagain, H.; Gautam, D.; Rodrigues, H. Chapter 1—Revisiting Major Historical Earthquakes in Nepal: Overview of 1833, 1934, 1980, 1988, 2011, and 2015 Seismic Events. In *Impacts and Insights of the Gorkha Earthquake*; Gautam, D., Rodrigues, H., Eds.; Elsevier: Amsterdam, The Netherlands, 2018; pp. 1–17. ISBN 978-0-12-812808-4.
6. Chen, W.; Rao, G.; Kang, D.; Wan, Z.; Wang, D. Early Report of the Source Characteristics, Ground Motions, and Casualty Estimates of the 2023 Mw 7.8 and 7.5 Turkey Earthquakes. *J. Earth Sci.* **2023**, *34*, 297–303. [CrossRef]
7. Omer, S. 2023 Turkey and Syria Earthquake: Facts, FAQs, and How to Help. Available online: https://www.worldvision.org/disaster-relief-news-stories/2023-turkey-and-syria-earthquake-faqs (accessed on 6 June 2023).
8. Ningthoujam, M.C.; Nanda, R.P. A GIS System Integrated with Earthquake Vulnerability Assessment of RC Building. *Structures* **2018**, *15*, 329–340. [CrossRef]
9. Khan, S.U.; Qureshi, M.I.; Rana, I.A.; Maqsoom, A. Seismic Vulnerability Assessment of Building Stock of Malakand (Pakistan) Using FEMA P-154 Method. *SN Appl. Sci.* **2019**, *1*, 1625. [CrossRef]
10. Diana, L.; Lestuzzi, P.; Podestà, S.; Luchini, C. Improved Urban Seismic Vulnerability Assessment Using Typological Curves and Accurate Displacement Demand Prediction. *J. Earthq. Eng.* **2021**, *25*, 1709–1731. [CrossRef]
11. Ozer, E.; Özcebe, A.G.; Negulescu, C.; Kharazian, A.; Borzi, B.; Bozzoni, F.; Molina, S.; Peloso, S.; Tubaldi, E. Vibration-Based and Near Real-Time Seismic Damage Assessment Adaptive to Building Knowledge Level. *Buildings* **2022**, *12*, 416. [CrossRef]
12. Martínez-Cuevas, S.; Morillo Balsera, M.C.; Benito, B.; Torres, Y.; Gaspar-Escribano, J.; Staller, A.; García-Aranda, C. Assessing Building Habitability after an Earthquake Using Building Typology and Damage Grade. Application in Lorca, Spain. *J. Earthq. Eng.* **2022**, *26*, 3417–3439. [CrossRef]
13. Chaurasia, K.; Kanse, S.; Yewale, A.; Singh, V.K.; Sharma, B.; Dattu, B.R. Predicting Damage to Buildings Caused by Earthquakes Using Machine Learning Techniques. In Proceedings of the 2019 IEEE 9th International Conference on Advanced Computing (IACC 2019), Tiruchirappalli, India, 13–14 December 2019; IEEE: New York, NY, USA, 2019; pp. 81–86.
14. Chen, W.; Zhang, L. Predicting Building Damages in Mega-Disasters under Uncertainty: An Improved Bayesian Network Learning Approach. *Sustain. Cities Soc.* **2021**, *66*, 102689. [CrossRef]
15. Chen, W.; Zhang, L. Building Vulnerability Assessment in Seismic Areas Using Ensemble Learning: A Nepal Case Study. *J. Clean. Prod.* **2022**, *350*, 131418. [CrossRef]
16. Sajan, K.C.; Bhusal, A.; Gautam, D.; Rupakhety, R. Earthquake Damage and Rehabilitation Intervention Prediction Using Machine Learning. *Eng. Fail. Anal.* **2023**, *144*, 106949. [CrossRef]
17. Zio, E. Prognostics and Health Management (PHM): Where Are We and Where Do We (Need to) Go in Theory and Practice. *Reliab. Eng. Syst. Saf.* **2022**, *218*, 108119. [CrossRef]
18. Wei, P.; Lu, Z.; Song, J. Variable Importance Analysis: A Comprehensive Review. *Reliab. Eng. Syst. Saf.* **2015**, *142*, 399–432. [CrossRef]
19. Friedman, J.H. Greedy Function Approximation: A Gradient Boosting Machine. *Ann. Stat.* **2001**, *29*, 1189–1232. [CrossRef]
20. Apley, D.W.; Zhu, J. Visualizing the Effects of Predictor Variables in Black Box Supervised Learning Models. *arXiv* **2019**, arXiv:1612.08468. [CrossRef]
21. Lundberg, S.M.; Lee, S.-I. A Unified Approach to Interpreting Model Predictions. In Proceedings of the Advances in Neural Information Processing Systems, Long Beach, CA, USA, 4–9 December 2017; Curran Associates, Inc.: New York, NY, USA, 2017; Volume 30.
22. Lundberg, S.M.; Erion, G.G.; Lee, S.-I. Consistent Individualized Feature Attribution for Tree Ensembles. *arXiv* **2019**, arXiv:1802.03888.
23. Bangyal, W.H.; Qasim, R.; Rehman, N.U.; Ahmad, Z.; Dar, H.; Rukhsar, L.; Aman, Z.; Ahmad, J. Detection of Fake News Text Classification on COVID-19 Using Deep Learning Approaches. *Comput. Math. Methods Med.* **2021**, *2021*, 5514220. [CrossRef]
24. Zhang, S.; Li, X.; Zong, M.; Zhu, X.; Cheng, D. Learning k for KNN Classification. *ACM Trans. Intell. Syst. Technol.* **2017**, *8*, 1–19. [CrossRef]
25. Zhang, S.; Li, X.; Zong, M.; Zhu, X.; Wang, R. Efficient KNN Classification With Different Numbers of Nearest Neighbors. *IEEE Trans. Neural Netw. Learn. Syst.* **2018**, *29*, 1774–1785. [CrossRef] [PubMed]
26. Salvador–Meneses, J.; Ruiz–Chavez, Z.; Garcia–Rodriguez, J. Compressed KNN: K-Nearest Neighbors with Data Compression. *Entropy* **2019**, *21*, 234. [CrossRef] [PubMed]
27. Guo, G.; Wang, H.; Bell, D.; Bi, Y.; Greer, K. KNN Model-Based Approach in Classification. In Proceedings of the On The Move to Meaningful Internet Systems 2003: CoopIS, DOA, and ODBASE, Sicily, Italy, 3–7 November 2003; Meersman, R., Tari, Z., Schmidt, D.C., Eds.; Springer: Berlin/Heidelberg, Germany, 2003; pp. 986–996.
28. Chen, T.; Guestrin, C. XGBoost: A Scalable Tree Boosting System. In Proceedings of the 22nd ACM SIGKDD International Conference on Knowledge Discovery and Data Mining, San Francisco, CA, USA, 13 August 2016; Association for Computing Machinery: New York, NY, USA, 2016; pp. 785–794.
29. Elith, J.; Leathwick, J.R.; Hastie, T. A Working Guide to Boosted Regression Trees. *J. Anim. Ecol.* **2008**, *77*, 802–813. [CrossRef] [PubMed]
30. Friedman, J.H. Stochastic Gradient Boosting. *Comput. Stat. Data Anal.* **2002**, *38*, 367–378. [CrossRef]

31. Fatahi, R.; Nasiri, H.; Homafar, A.; Khosravi, R.; Siavoshi, H.; Chelgani, S. Modeling Operational Cement Rotary Kiln Variables with Explainable Artificial Intelligence Methods—A "Conscious Lab" Development. *Part. Sci. Technol.* **2022**, *41*, 715–724. [CrossRef]
32. Le, T.-T.-H.; Oktian, Y.; Kim, H. XGBoost for Imbalanced Multiclass Classification-Based Industrial Internet of Things Intrusion Detection Systems. *Sustainability* **2022**, *14*, 8707. [CrossRef]
33. Nasiri, H.; Alavi, S.A. A Novel Framework Based on Deep Learning and ANOVA Feature Selection Method for Diagnosis of COVID-19 Cases from Chest X-Ray Images. *Comput. Intell. Neurosci.* **2022**, *2022*, e4694567. [CrossRef]
34. Fatahi, R.; Nasiri, H.; Dadfar, E.; Chelgani, S. Modeling of Energy Consumption Factors for an Industrial Cement Vertical Roller Mill by SHAP-XGBoost: A "Conscious Lab" Approach. *Sci. Rep.* **2022**, *12*, 7543. [CrossRef]
35. Alajmi, M.S.; Almeshal, A.M. Predicting the Tool Wear of a Drilling Process Using Novel Machine Learning XGBoost-SDA. *Materials* **2020**, *13*, 4952. [CrossRef]
36. Wan, Z.; Xu, Y.; Šavija, B. On the Use of Machine Learning Models for Prediction of Compressive Strength of Concrete: Influence of Dimensionality Reduction on the Model Performance. *Materials* **2021**, *14*, 713. [CrossRef]
37. Xu, B.; Tan, Y.; Sun, W.; Ma, T.; Liu, H.; Wang, D. Study on the Prediction of the Uniaxial Compressive Strength of Rock Based on the SSA-XGBoost Model. *Sustainability* **2023**, *15*, 5201. [CrossRef]
38. Prokhorenkova, L.; Gusev, G.; Vorobev, A.; Dorogush, A.V.; Gulin, A. CatBoost: Unbiased Boosting with Categorical Features. In Proceedings of the Advances in Neural Information Processing Systems, Montréal, QC, Canada, 3–8 December 2018; Curran Associates, Inc.: New York, NY, USA, 2018; Volume 31.
39. Zhang, M.; Chen, W.; Yin, J.; Feng, T. Health Factor Extraction of Lithium-Ion Batteries Based on Discrete Wavelet Transform and SOH Prediction Based on CatBoost. *Energies* **2022**, *15*, 5331. [CrossRef]
40. Nasiri, H.; Tohry, A.; Heidari, H. Modeling Industrial Hydrocyclone Operational Variables by SHAP-CatBoost—A "Conscious Lab" Approach. *Powder Technol.* **2023**, *420*, 118416. [CrossRef]
41. Kim, B.; Lee, D.-E.; Hu, G.; Natarajan, Y.; Preethaa, S.; Rathinakumar, A.P. Ensemble Machine Learning-Based Approach for Predicting of FRP–Concrete Interfacial Bonding. *Mathematics* **2022**, *10*, 231. [CrossRef]
42. Yin, J.; Zhao, J.; Song, F.; Xu, X.; Lan, Y. Processing Optimization of Shear Thickening Fluid Assisted Micro-Ultrasonic Machining Method for Hemispherical Mold Based on Integrated CatBoost-GA Model. *Materials* **2023**, *16*, 2683. [CrossRef] [PubMed]
43. Asad, R.; Altaf, S.; Ahmad, S.; Shah Noor Mohamed, A.; Huda, S.; Iqbal, S. Achieving Personalized Precision Education Using the Catboost Model during the COVID-19 Lockdown Period in Pakistan. *Sustainability* **2023**, *15*, 2714. [CrossRef]
44. Ke, G.; Meng, Q.; Finley, T.; Wang, T.; Chen, W.; Ma, W.; Ye, Q.; Liu, T.-Y. LightGBM: A Highly Efficient Gradient Boosting Decision Tree. In Proceedings of the Advances in Neural Information Processing Systems, Long Beach, CA, USA, 4–9 December 2017; Curran Associates, Inc.: New York, NY, USA, 2017; Volume 30.
45. Liang, W.; Luo, S.; Zhao, G.; Wu, H. Predicting Hard Rock Pillar Stability Using GBDT, XGBoost, and LightGBM Algorithms. *Mathematics* **2020**, *8*, 765. [CrossRef]
46. Liu, Y.; Zhao, H.; Sun, J.; Tang, Y. Digital Inclusive Finance and Family Wealth: Evidence from LightGBM Approach. *Sustainability* **2022**, *14*, 15363. [CrossRef]
47. Chen, C.; Zhang, Q.; Ma, Q.; Yu, B. LightGBM-PPI: Predicting Protein-Protein Interactions through LightGBM with Multi-Information Fusion. *Chemom. Intell. Lab. Syst.* **2019**, *191*, 54–64. [CrossRef]
48. Daoud, E.A. Comparison between XGBoost, LightGBM and CatBoost Using a Home Credit Dataset. *Int. J. Comput. Inf. Eng.* **2019**, *13*, 6–10.
49. Hu, Y.; Sun, Z.; Han, Y.; Li, W.; Pei, L. Evaluate Pavement Skid Resistance Performance Based on Bayesian-LightGBM Using 3D Surface Macrotexture Data. *Materials* **2022**, *15*, 5275. [CrossRef] [PubMed]
50. Mangalathu, S.; Hwang, S.-H.; Jeon, J.-S. Failure Mode and Effects Analysis of RC Members Based on Machine-Learning-Based SHapley Additive ExPlanations (SHAP) Approach. *Eng. Struct.* **2020**, *219*, 110927. [CrossRef]
51. Rafferty, J.P. Nepal Earthquake of 2015. Available online: https://www.britannica.com/topic/Nepal-earthquake-of-2015 (accessed on 12 July 2023).
52. Bull, P.; Slavitt, I.; Lipstein, G. Harnessing the Power of the Crowd to Increase Capacity for Data Science in the Social Sector. *arXiv* **2016**, arXiv:1606.07781.
53. Li, Y.; Qin, Y.; Wang, H.; Xu, S.; Li, S. Study of Texture Indicators Applied to Pavement Wear Analysis Based on 3D Image Technology. *Sensors* **2022**, *22*, 4955. [CrossRef] [PubMed]
54. Pedregosa, F.; Varoquaux, G.; Gramfort, A.; Michel, V.; Thirion, B.; Grisel, O.; Blondel, M.; Prettenhofer, P.; Weiss, R.; Dubourg, V.; et al. Scikit-Learn: Machine Learning in Python. *J. Mach. Learn. Res.* **2011**, *12*, 2825–2830.
55. DrivenData Richter's Predictor: Modeling Earthquake Damage. Available online: https://www.drivendata.org/competitions/57/nepal-earthquake/page/136/ (accessed on 11 July 2023).
56. McKinney, W. Data Structures for Statistical Computing in Python. In Proceedings of the 9th Python in Science Conference, Austin, TX, USA, 28 June–3 July 2010; van der Walt, S., Millman, J., Eds.; 2010; pp. 56–61.
57. Waskom, M.L. Seaborn: Statistical Data Visualization. *J. Open Source Softw.* **2021**, *6*, 3021. [CrossRef]

58. Nguyen, H.D.; LaFave, J.M.; Lee, Y.-J.; Shin, M. Rapid Seismic Damage-State Assessment of Steel Moment Frames Using Machine Learning. *Eng. Struct.* **2022**, *252*, 113737. [CrossRef]
59. Ghimire, S.; Gueguen, P.; Giffard-Roisin, S.; Schorlemmer, D. Testing Machine Learning Models for Seismic Damage Prediction at a Regional Scale Using Building-Damage Dataset Compiled after the 2015 Gorkha Nepal Earthquake. *Earthq. Spectra* **2022**, *38*, 2970–2993. [CrossRef]

sustainability

MDPI

Article

Comparative Study on Shaking Table Tests for a Pile–Nuclear Island Structure under Different Soil Conditions

Yunlun Sun [1], Gang Wang [2,*], Yougang Wang [1], Jian Tu [1], Liping Jing [3] and Wenhao Qi [3]

[1] China Nuclear Energy Technology Co., Ltd., Beijing 100094, China; sunyunlun@126.com (Y.S.); ygwang@chinergy.com.cn (Y.W.); jiantu@chinergy.com.cn (J.T.)
[2] School of Civil Engineering, Xuzhou University of Technology, Xuzhou 221018, China
[3] Key Laboratory of Earthquake Engineering and Engineering Vibration, Institute of Engineering Mechanics, China Earthquake Administration, Harbin 150080, China; jlp_iem@163.com (L.J.); qwhtky@163.com (W.Q.)
* Correspondence: wanggang198903@126.com

Abstract: In this paper, the shaking table tests of a Seismic–Soil–Pile–Superstructure Interaction (SSPSI) in medium-soft and hard base soil were carried out. Silted clay with a unit weight of 1.70 g/cm^3 and a shear wave velocity of 175 m/s was adopted to simulate the medium-soft soil, while the composite soil obtained by adding 20% quicklime to silted clay with a unit weight of 1.75 g/cm^3 and a shear wave velocity of 300 m/s was adopted to simulate the hard soil in the tests. By inputting the artificial seismic motion time history with different amplitudes synthesized by the RG1.60 response spectrum commonly used in nuclear power engineering to the models, the dynamic interaction characteristics and seismic response laws of the soil–pile–nuclear island structure in the medium-soft and hard base soil were compared, the internal force and deformation distribution characteristics of the pile foundation under different ground conditions were analyzed, and the site conditions and mechanism of seismic failure of the pile group foundation were described. The research results can provide a reference for site selection and seismic design of a nuclear power plant.

Keywords: nuclear island structure; pile foundation; shaking table test; SSPSI

Citation: Sun, Y.; Wang, G.; Wang, Y.; Tu, J.; Jing, L.; Qi, W. Comparative Study on Shaking Table Tests for a Pile–Nuclear Island Structure under Different Soil Conditions. *Sustainability* **2023**, *15*, 11988. https://doi.org/10.3390/su151511988

Academic Editors: Chong Xu, Su Chen and Shuang Li

Received: 2 July 2023
Revised: 25 July 2023
Accepted: 1 August 2023
Published: 4 August 2023

1. Introduction

Nuclear energy has been recognized and widely used by all countries as a clean, economical and efficient energy. The construction of nuclear power plants was mostly located in coastal bedrock sites for reasons of cooling and seismic safety in the past. However, the available coastal bedrock sites have gradually decreased with the rapid development of nuclear power, and the construction of nuclear power plants will extend to non-rock sites inevitably. Although there are a great deal of advantages with nuclear power generation, nuclear material is strongly radiative. It will bring a catastrophic impact on humans and the surrounding environment once nuclear leakage occurs. Therefore, the primary consideration for the construction of nuclear power plants is safety. The superstructure of nuclear power engineering is designed according to the strictest standards at present, of which safety can be ensured in the event of malicious impacts by large commercial aircraft and strong earthquakes [1]. The bedrock site is favorable for seismic design, on which the construction of nuclear power plants can meet the seismic safety requirements. However, in the construction of nuclear power plants in non-rock sites, the base soil has the characteristics of large discreteness and strong nonlinearity compared with the upper engineering structure of the nuclear island plant, which has a significant impact on the floor response spectrum and the inertial force of the structure. Therefore, the subsoil and foundation are the key to the seismic design of the nuclear island on the non-rock site, and it can be said that the seismic safety of the nuclear island plant depends on properties of the non-rock site and the seismic design of the foundation. The foundation design of a nuclear island in a non-rock site needs further study due to the characteristics of nuclear power structures and the complexity of the ground [2].

Pile foundations are generally adopted for important projects on non-rock soil, especially on soft soil, with the advantages of large bearing capacity, good stability and small differential settlement, and is widely used in bridge structures, high-rise buildings, port wharfs and offshore platforms. Earthquake damage investigations show that pile foundations have good seismic performance, but there is still some earthquake damage. The earthquake damage to a pile foundation is often concealed, and it is difficult to repair. For this reason, many scholars have carried out experimental and theoretical studies on the seismic performance of a pile foundation [3–5].

The theoretical research on the performance of a pile foundation was mostly focused on the bearing capacity of single pile and pile groups in the early stage, and the experimental research was mostly carried out on static pile pressing tests in the field to determine the bearing capacity and dynamic tests to test the integrity of the piles. With the accumulation of the seismic damage phenomenon of pile foundations during earthquakes and the development of test technology for seismic simulation, the seismic response law and damage to pile foundations has received more and more attention by engineers and scholars. In particular, the numerical simulation method is adopted to study the dynamic interaction of soil–pile–superstructure interactions under a seismic load (Seismic-Soil–pile–Superstructure-Interaction, referred to as SSPSI). However, the results of numerical simulations need to be tested and verified since the interaction involves complex nonlinear problems of the soil, the soil–pile contact surface, etc. The large-scale shaking table test of earthquake simulations is the most widely used laboratory test method for the study of soil–pile–superstructure dynamic interaction at home and abroad. The main feature of the test is that it is carried out in the 1 g gravity field, so a high-stress field, like in a centrifuge test, is not available. But the shaking table is capable of motion input in two dimensions and three dimensions, and the test model is generally much larger than that in the centrifuge test. Moreover, the shaking table test is not affected by the Coriolis effect in the centrifuge test, so it has unique advantages. Its 1 g test environment is most suitable for the test of cohesive soil, where the undrained stress–strain relationship and the confining pressure are independent.

Meymand [6] designed and used a cylindrical flexible soil box to carry out a large-scale shaking table test on a dynamic soft clay–pile–superstructure interaction. The test results show that the horizontal dynamic stiffness of the pile head is a function of the load level, and the results calculated according to the theory of elasticity are larger. Wei Xiao [7] carried out a series of shaking table tests on a dynamic soil–pile–pier structure interaction by using a rigid soil box, including a single-column pile-pier model, a single-pier pile group model and a double-pier pile group model test. Wu Xiaoping [8] used a layered shear soil box to carry out the shaking table test of an SSPSI for the first time in China, and the test verified the rationality of this soil box to eliminate the 'model box effect'. After that, many domestic research institutions and scholars designed and manufactured this layered shear deformation soil box and achieved many research results [9–11]. Chau K T [12] found a possible failure form of a pile foundation via a shaking table test. This failure is caused by the impact between the pile foundation and the soil; that is, the failure is caused by inertial force, which may be one of the main forms of pile foundation failure in large earthquakes. Tokimatsua [13] conducted shaking table tests to study kinematic and inertial SSPSIs and found that if the natural period of the superstructure is greater than that of the soil, the phase of the kinematic and inertial interaction tends to be the same, and the internal force in the pile increases. The maximum value occurs when both the kinematic and inertial interactions reach their peak values in the same direction. Hokmabadi et al. [14] carried out shaking table tests with a layered shear box to investigate the effect of SSPSI on the dynamic response of superstructures with different heights. The experimental results indicate that the SSPSI increases the lateral displacement of the superstructure and that the higher the superstructure, the more obvious this effect is. In order to study the influence of SSPSI on the dynamic characteristics of a pile foundation system in liquefiable sand ground, Huang Zhanfang et al. [15] carried out the shaking table test of four cases: natural ground and ground reinforced by a pile foundation with a pile spacing of 3D, 3.5D, and 4D. The test results show that the liquefaction of base soil with reinforcement of the pile foundation lags behind that in natural ground, and the improvement of liquefaction resistance of reinforced foundation with pile spacing of 3D and 3.5D is more obvious. With the increase in pile spacing, the degree of improvement will be weakened. Li Xiaojun et al. [16] took the CAP1400 nuclear power plant structure as an example to conduct a shaking table test on a nuclear power plant structure in a non-rock site for the analysis of the applicability and seismic response characteristics of nuclear power plant structures under the non-rock site condition. The results show that the seismic motion in all directions is amplified in the model site, and the low frequency zone of the site response spectrum is greatly affected by the superstructure. Cracks appear in the site under the action of seismic motion lower than the reference, and the structure is separated from the soil under the action of seismic motion, meeting the reference. Jing Liping et al. [17] studied the dynamic response of a soil–pile foundation–nuclear island system by a shaking table test; analyzed the internal force distribution characteristics, deformation law and failure mode of a pile group foundation under a seismic load in a medium-soft soil site; and compared them with the numerical simulation results.

Compared to general building structures, nuclear islands have greater mass and stiffness. Therefore, the impact of nuclear islands on the seismic response of their pile foundations may be different from that of general building structures. Considering the importance of nuclear islands and the potential significant harm after damage, it is necessary to conduct research on their seismic performance. At present, due to the limitation of various factors and test conditions, there are relatively few large-scale shaking table test studies on the SSPSIs of nuclear power plants at home and abroad. The research results are more common in the studies of pile foundation theory and numerical simulation [18–20]. The stiffness ratio between soil and pile is a very important influence factor of the SSPSI The shaking table test of a single soil layer simulating medium-soft soil, medium-hard soil and hard soil is carried out so as to study the dynamic interaction law of a soil–pile–nuclear island containment structure and the failure mechanism and mode of the pile foundation in different subsoil layers under a horizontal seismic load. This paper presents a comparative analysis of the test results under medium-soft and hard soil site conditions in the series of tests, mainly analyzing the dynamic response law of the upper nuclear island structure as well as the deformation and internal force response law of the pile foundation. In addition, the seismic failure mechanism and mode of the pile group foundation are discussed.

2. The Facilities and Design of the Test

2.1. The Shaking Table System

The large-scale seismic simulation shaking table system (Yanjiao Park) of the Key Laboratory of Earthquake Engineering and Engineering Vibration, Institute of Engineering Mechanics, China Earthquake Administration has a table size of 5 m × 5 m, a bearing capacity of 30 t, a maximum displacement of ±0.5 m, a maximum velocity of 1.5 m/s and a maximum acceleration of 2 g.

2.2. The Layered Shear Soil Box

The soil container in the test is a self-developed three-dimensional laminated shear model box, as shown in Figure 1. The external size of the model box is 3.7 m in length, 2.4 m in width and 1.7 m in height. The box is composed of multi-layer square steel tube frames with equal spacing, and the spacing between the upper and lower frames is 20 mm. The natural frequency of the shear model soil box is 7.5 Hz, and the damping ratio is 0.226. The layered shear model box has completed many soil–structure dynamic interaction tests, and the results show that it has a good effect for boundary simulation [10].

Figure 1. Soil box for the test.

2.3. Soil Model in the Test

The silted clay in a project site and the composite soil mixed with 20% quicklime were selected as the test soil for the simulation of the medium-soft and hard soil sites, respectively. The actual filling depth in the model box is 1.5 m. In order to ensure the uniformity of soil density and water content, the soil was artificially compacted and evenly sprinkled with each 0.1 m thickness soil filling, and then the density test and resonance column test of soil samples were conducted. The average density measured in laboratory test of the medium-soft soil is 1.70 g/cm^3, while the hard soil is 1.75 g/cm^3, and the shear wave velocities of the medium-soft and hard soil are 175 m/s and 300 m/s according to the resonance column test, respectively.

2.4. Model of the Pile Foundation and the Containment

According to the size of the model soil box in the test, the pile foundation is composed of the lower pile group and the upper embedded pile cap, and the surface of the pile cap is on an even height with the surface of the soil. The pile group consists of five piles arranged in a cross shape. The pile head is chipped after the pile group has been embedded, and then the upper pile cap is poured to simulate the pile cap connection mode of the actual site. The layout of the pile group is shown in Figure 2. The unit of the numbers in Figure 2 is millimeters, and ①–⑤ are serial number of piles. The length of the pile is 1.35 m, the diameter of the pile is 0.1 m, and there are four HRB335 longitudinal reinforcements with a diameter of 8 mm. The circular stirrups are arranged in a 100 mm interval with a diameter of about 4 mm. The pile spacing is 600 mm (six times the pile diameter), and the 20 mm thick steel plate set at the bottom of the pile is welded with the bottom of the model box to simulate the actual socketed pile stress. The plane size of the pile cap model is 1.9 m $\times$ 1.9 m, and the thickness is 0.15 m. The spacing of the internal HRB335 reinforcing bars with a diameter of 8 mm is 150 mm, which is configured as a double-layer and two-way steel mesh. The vertical reinforcing bars with hooks are arranged between the upper and lower layers of the steel mesh to enhance the integrity of the pile cap steel mesh. The C30 fine aggregate concrete was adopted to form the piles and the pile cap. The actual measured axial compressive strength of the concrete is 35.6 MPa, and the elastic modulus is 2.2567 $\times$ 10^4 MPa.

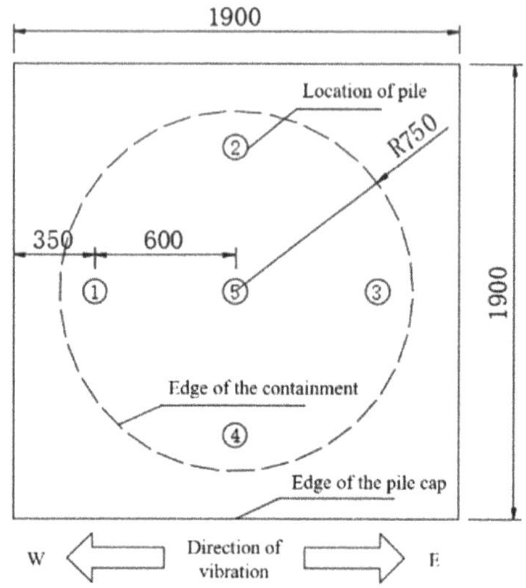

Figure 2. Schematic diagram of pile group layout.

The superstructure is composed of a nuclear island containment and a concrete baseplate, as shown in Figure 3. The unit of the numbers in Figure 3 is millimeters. The model is designed according to a 1/20 geometric scale and is made of steel concrete. The total height of the model is 1.769 m, and the height of the baseplate is 0.15 m. There is a barrier at every 0.4 m in the vertical direction of the model. The external diameter of the model is 1.5 m, while the internal diameter is 1.368 m. The inner and outer steel plates are pulled by bolts with a diameter of 4 mm and a spacing of 200 mm × 200 mm. For the convenience of subsequent analysis, the position of the upper surface on the pile cap in Figure 3 is defined as the first layer, and the positions of the three barriers from bottom to top are defined as the second, third, and fourth layer. The size and material of the model baseplate are consistent with those of the pile cap, and four high-strength bolts are used in the four corners of the baseplate to connect with the pile cap, as shown in Figure 4. The containment model is equipped with the clump weight to make its total mass reach about 10^4 kg.

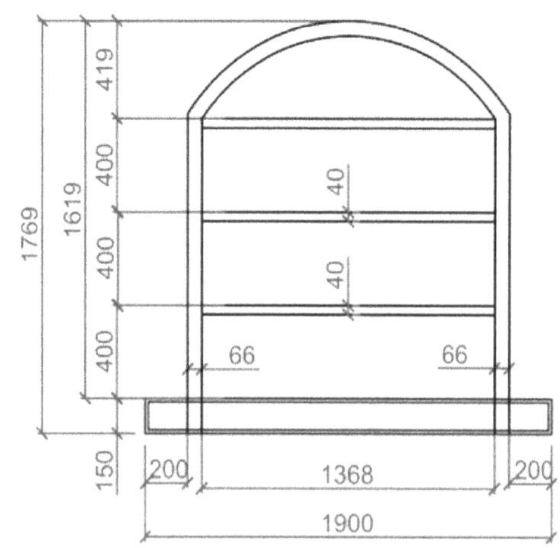

Figure 3. Schematic diagram of nuclear island plant model.

Figure 4. Bolted connection between the baseplate and the pile cap.

2.5. The Sensors Layout and the Data Acquisition System

The main purpose of the sensor arrangement scheme is monitoring the dynamic response of piles under horizontal load input so as to study the seismic dynamic response and failure mechanism of the pile foundation. As shown in Figure 5, the sensors arranged on the piles are draw-wire displacement sensors and strain gauges. There are two kinds of strain gauges in the test: resistance strain gauges and fiber Bragg grating strain gauges. The draw-wire displacement sensors are arranged on piles 1# and 2#, the resistance strain gauges are arranged on piles 2# and 5#, and the fiber Bragg grating strain gauges are arranged on pile 3#. Moreover, accelerometers are arranged at the bottom of the nuclear island containment and on the internal partition plates so as to study the influence of the dynamic interaction of the soil–pile–superstructure system on the dynamic response of the superstructure.

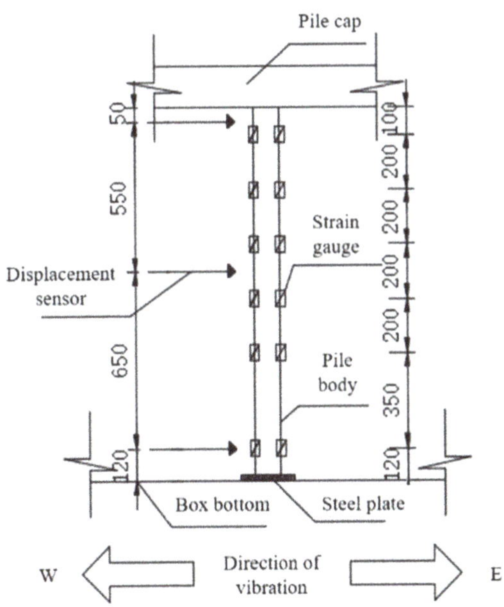

Figure 5. Schematic diagram of sensors layout.

2.6. The Seismic Motion Input

The seismic simulation shaking table test for the soil–structure dynamic interaction under the condition of constant gravitational acceleration can only give qualitative dynamic response characteristics and laws because it cannot strictly meet the similarity relation. Since the main attention should be paid to the influence of far-field large earthquakes in the process of nuclear power plant site selection, a seismic motion time history manually fitted according to the US RG1.60 design response spectrum and two natural seismic records are selected for the test. The amplitude of the seismic motions input was adjusted, while the duration was not adjusted.

Due to the limitation of the length of this paper, only the test results of RG1.60 artificial seismic motion input are discussed here. The time history and spectrum characteristics of the seismic motion are shown in Figure 6. The RG1.60 seismic motion has more high-frequency components (10–20 Hz) compared to general natural seismic motions. Amplitudes of the seismic motion are 0.1 g, 0.2 g and 0.4 g in the test.

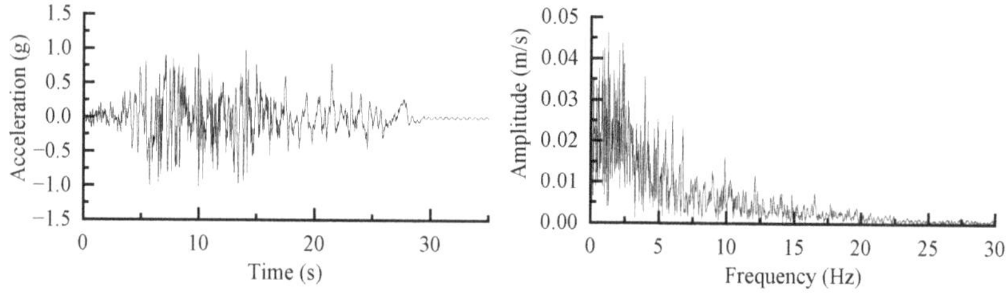

Figure 6. Time history and Fourier spectrum of the artificial seismic motion input in the test.

3. Macroscopic Phenomenon of Pile Foundation in the Test

The shaking table test of the soil–structure dynamic interaction system is different from that of the structural model because the structure is buried in the soil. During the test, the macroscopic test phenomenon of the pile foundation in soil cannot be observed intuitively. Although it can be determined indirectly whether the pile body is damaged by measurement results of the strain gauges arranged on the pile body during the test, the detailed and intuitive damage of the pile body can only be observed after all the test conditions and the excavation.

It can be found that the strain gauges on piles were unable to transfer data back when the input peak acceleration was 0.4 g in the test of the model with medium-soft soil. Based on this, it is determined that the pile body has been damaged, so the test is ceased. Then, the soil was excavated, and the macroscopic damage of piles was shown in Figure 7.

It can be seen from Figure 7 that damage occurs at the head of foundation piles, with failure modes of annular fracture, vertical and oblique cracks. Among them, the failure phenomenon at the head of piles 3# and 4# is obvious, part of concrete at the head of pile 2# has fallen off, and the vertical and oblique cracks appear at the heads of pile 1# and 5#, with obvious crack development trends. In addition, annular cracks appear at every 100 mm vertical interval of each pile body, and the position of the cracks is approximately in the middle of two stirrups. From the perspective of pile failure phenomenon, the damage degree from high to low is pile 3#, 4#, 2#, 5# and 1#. It is speculated that the damage of the pile foundation in the dynamic interaction of the system first occurs on the side stake on one side along the direction of motion, then passes to the side stakes in both sides normal to the direction of motion and the center stake, and finally passes the side stake in the other side along the direction of motion.

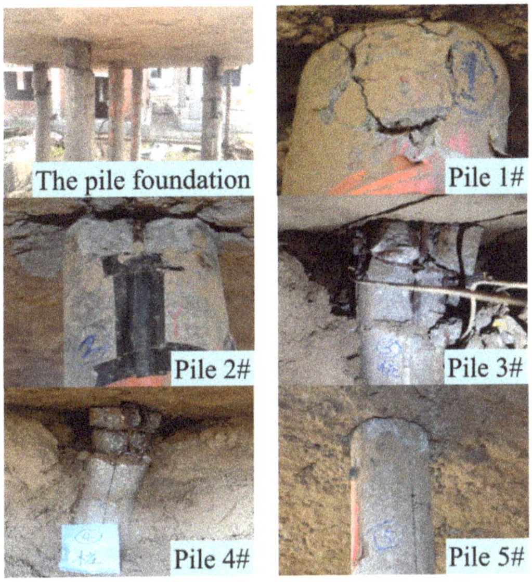

Figure 7. Failure of pile group in medium-soft soil site.

The strain gauges on piles can still work normally after seismic input with an amplitude of 0.4 g in the test of the model with hard soil, so the amplitude of the input motion was increased to 0.8 g gradually. Although the test results of the strain gauges on piles showed that the deformation of piles was still small, a vibration phenomenon with a large amplitude of the upper containment model was observed. The test was ceased for safety reasons. The macroscopic phenomenon of the piles after excavation is shown in Figure 8.

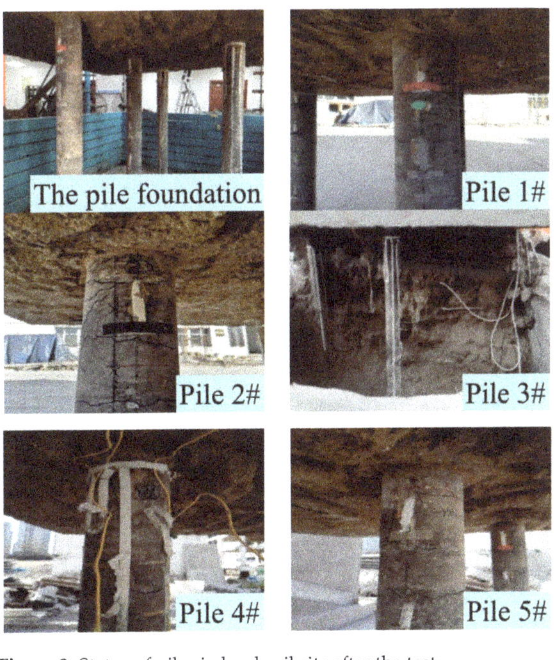

Figure 8. Status of piles in hard soil site after the test.

During the process of excavation, it was found that the base soil is hard and dense, and there is almost no separation between the base soil and the pile foundation. This is because the hard soil ground is a mixture composed of silted clay and quicklime, and the maintenance time from model forming to test conducting further increases the stiffness of the composite soil. There is no damage at the connection between the piles and the cap. After excavation, it is found that the pile body is basically intact, with small annular concrete cracks in the surface of the pile body.

From the macroscopic phenomenon of the pile foundation in the test, it can be seen that the stiffness of the base soil has a great influence on the damage degree of the pile foundation under the seismic load, and the damage of the pile foundation in the medium-soft soil site is serious.

4. Analysis of Test Results

Although the maximum acceleration amplitude in the hard subsoil test reached 0.8 g, only the law of system response in the artificial RG1.60 seismic motion input condition with amplitudes of 0.1 g, 0.2 g and 0.4 g was analyzed for comparison with the medium-soft subsoil test.

4.1. Acceleration Response of Nuclear Island Containment

Figure 9 shows the typical acceleration response time histories of the nuclear island model collected in the test. It can be seen intuitively from the diagram that high frequency components in the acceleration time history response of the superstructure are more abundant in the hard soil site than in the medium-soft soil site.

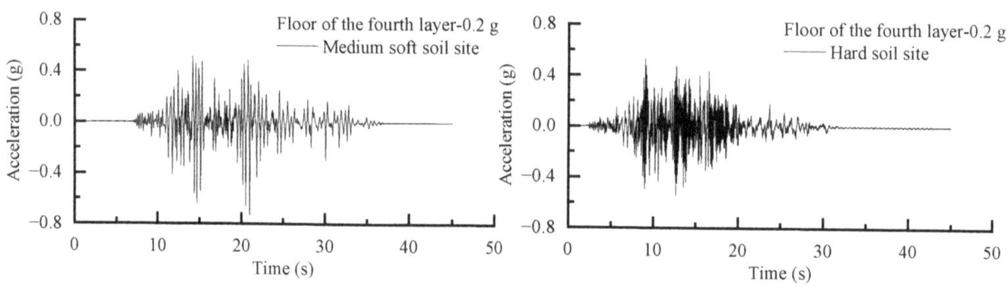

Figure 9. Typical acceleration time history collected in the test.

Figure 10 shows the amplification factor of the peak acceleration response at each layer in the nuclear island containment model relative to the input acceleration peak of the shaking table in conditions of two subsoils. It can be seen from the figure that the peak acceleration amplification factor of the nuclear island model on the medium-soft soil site is greater than that on the hard soil site when the amplitude of the seismic motion input is less than 0.2 g, indicating that the base soil is approximately in the elastic state when the amplitude of the input seismic motions is small and the response of the nuclear island structure on the site with smaller soil stiffness is larger. When amplitude of the seismic motion input is 0.4 g, however, the soil in medium-soft site reaches the strong nonlinear state, which has a greater dissipative effect on the energy of the input seismic motion, and the pile foundation is also destroyed, while the soil in the hard site is in the quasi-elastic state, which can transfer more energy of seismic motion, so the peak acceleration amplification factor of the nuclear island structure on the hard soil site is larger.

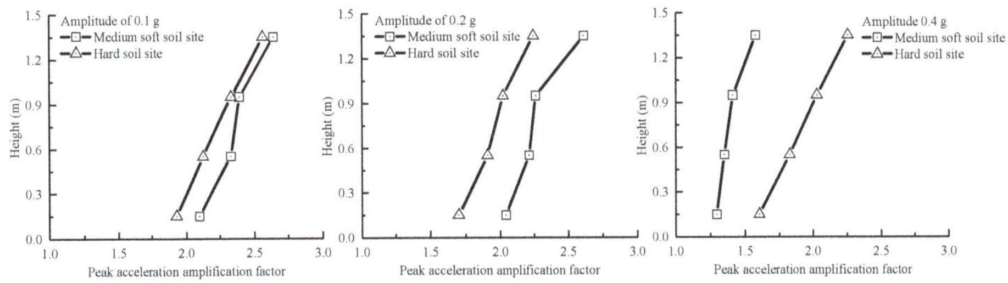

Figure 10. Comparison diagram of peak acceleration amplification factor of nuclear island model.

Figure 11 shows a comparison of the floor spectrum in the nuclear island containment model on two different base soils. It can be seen from the figure that the predominant period of the floor spectrum in the nuclear island structure is very close to that of the input seismic motion in the test, which is about 0.4 s, and the structure may have a resonance response when the input amplitude is 0.1 g and 0.2 g, while the predominant period of the floor spectrum is about 0.6 s, indicating that the soil has been destroyed, which makes the spectrum period of the superstructure longer when the input amplitude is 0.4 g in the soft soil site. In the case of 0.1 g, 0.2 g and 0.4 g inputs in the hard soil site, the period of the floor spectrum is basically about 0.1 s, far away from the predominant period of the input seismic motion. From the floor spectrum of the nuclear island structure, it can be found that properties of the soil around the pile foundation have a significant impact on the response characteristics of the superstructure. In addition, it also shows that the reason for macroscopic failure of the pile foundation in the test is mainly the small stiffness of the medium-soft soil and the weak constraint on the pile foundation. Furthermore, the natural period of the system is close to the predominant period of the input seismic motion, which may produce a resonance effect and cause serious damage to the pile foundation. Nevertheless, in the hard soil site, the stiffness of the soil is large, and the constraint effect on the pile foundation is strong. Moreover, the natural period of the system is quite different from the predominant period of the input seismic motion, and no resonance phenomenon occurs in the test with the hard soil model. Therefore, the pile foundation in the hard soil site remains intact.

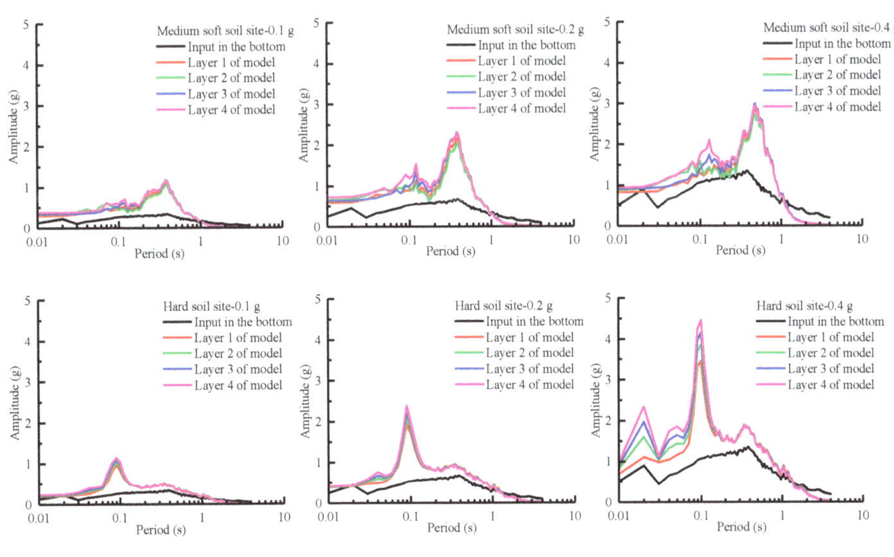

Figure 11. Comparison diagram of floor response spectrum in nuclear island model.

4.2. Displacement Response of Pile Foundation

Figure 12 shows the comparison of the peak displacement amplification factors along the height of the two piles (pile 1# and pile 2#) measured by the draw-wire displacement sensors in the cases of the two different sites. It can be seen from the figure that the peak displacement amplification factors of the two piles in the medium-soft soil site condition are larger than those in the hard soil site condition regardless of the input seismic motion amplitude, the peak displacement of piles increases with the increase in the height, which is largest in the pile head, and the deformation of the piles is also great in the medium-soft soil site condition, while the peak displacements of piles change little with the height in the hard soil site condition. This indicates that the soil has little constraint on the deformation of the pile body when the site soil is soft. It is easy to show the separation phenomenon between pile body and soil so that the displacement and deformation of the pile body are large, and the pile foundation failure occurs.

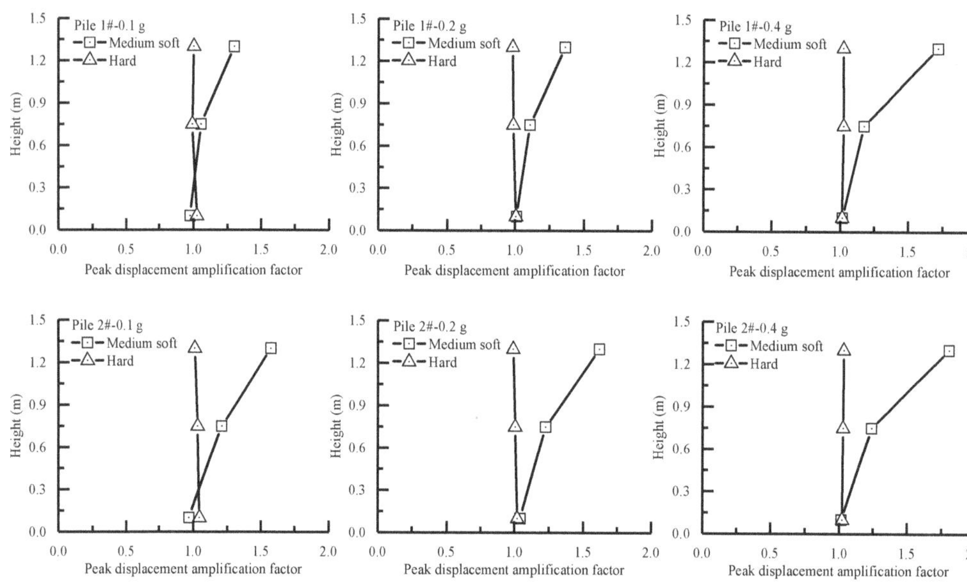

Figure 12. Comparison diagram of peak displacement amplification factor of piles.

4.3. Bending Moment Response of Pile Foundation

Figure 13 shows the comparison of the peak bending moment envelope diagram of three piles (pile 2#, pile 3# and pile 5#) in two site soil conditions. The bending moment of piles is calculated according to Formula (1) with the strain measured by the strain gauges on piles. Since the pile foundation has been damaged in the medium-soft soil site condition when the input seismic motion amplitude reaches 0.4 g, only the peak bending moment in conditions of 0.1 g and 0.2 g amplitude seismic motion input is compared in Figure 13.

$$M = \frac{\varepsilon_1 - \varepsilon_2}{D} \times EI \quad (1)$$

where M is the bending moment of the pile, ε_1 and ε_2 are bending strains on both sides of the pile body, D is the pile diameter, and EI is the bending stiffness of the pile section.

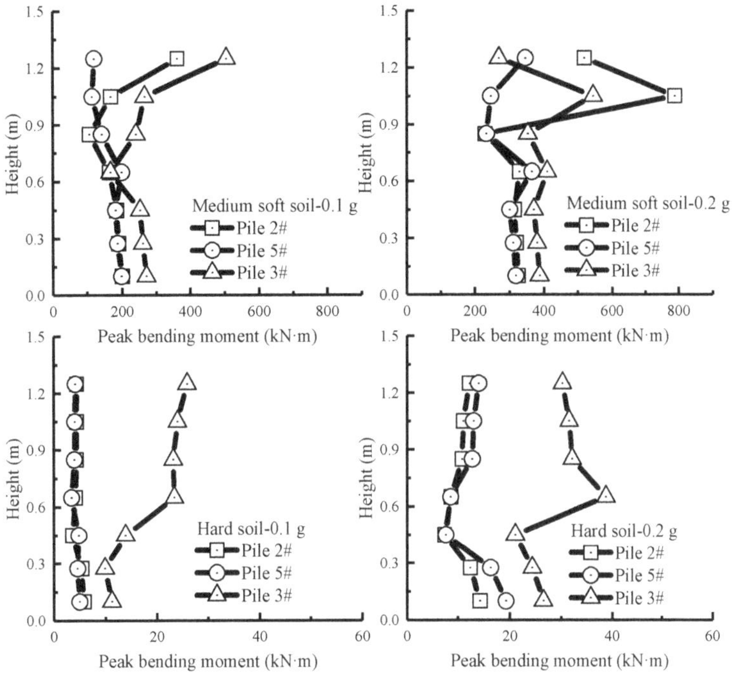

Figure 13. Comparison diagram of envelope curves of pile bending moment.

The following can be seen from Figure 13: (1) In the condition of the medium-soft soil site, the maximum peak bending moment of piles 2# and 3# appears at the pile head, while that of pile 5# appears at the middle of the pile when the input amplitude is 0.1 g, and the maximum peak bending moment of piles 2# and 3# moves down to the upper part of the pile with a height of about 1.05 m, while that of pile 5# still appears in the middle of the pile, but the peak bending moment at the pile head also increases significantly when the input amplitude is 0.2 g. (2) In the condition of the hard soil site, the maximum peak bending moment of piles 2# and 5# appears at the lower part of the pile, with a height of 0.275 m and 0.45 m, respectively, while that of pile 3# appears at the middle part of the pile, with a height of 0.65 m when the input amplitude is 0.1 g, and the position of the maximum peak bending moment occurrence remains unchanged. Furthermore, the peak bending moment at the upper part of piles 2# and 5#, with a height of 0.85 m, increases significantly when the input amplitude is 0.2 g. (3) In the condition of the medium-soft soil site, the order of the peak bending moment value at the three piles' heads is pile 3# > pile 2# > pile 5# when the input amplitude is 0.1 g, while the order is pile 2# > pile 5# > pile 3# when the input amplitude is 0.2 g. In the condition of the hard soil site, the peak bending moment value of pile 3# is largest in the three piles regardless of the input magnitude. (4) With the same input seismic motion amplitude, the peak bending moment of each pile in the medium-soft soil site condition is greater than that in the hard soil site condition, which is consistent with the analysis results of the pile displacement above and shows the reliability of the test results of draw-wire displacement sensors used on piles.

From the above analysis, it can be seen that the peak bending moment of the end-bearing pile group foundation under the action of seismic motion is affected by the hardness of the site soil and the position of the pile in the pile group due to the soil–pile–superstructure dynamic interaction. Specifically, the following can be concluded: (1) In the medium-soft soil, the deformation of the pile body is large, so the peak bending moment of the pile body is large, too. The maximum bending moment of the pile body generally appears at the pile head, which is also proved by the serious damage phenomenon at the pile head after the test. Conversely, in the hard soil, both the deformation and the peak bending moment of the pile body are small. The peak bending moment generally appears in the middle or lower part of the pile body. (2) The stress of the side piles in the pile group (such as pile 3#) is the largest. When the side piles are damaged, the stress of the middle side pile (such as pile 2#) increases significantly. Therefore, the failure order of the piles is judged as follows: the side pile, the middle side pile and the central pile. This order is consistent with the analysis results of the macroscopic failure phenomenon of the pile foundation above.

5. Numerical Simulation

5.1. The Finite Element Model

For further analysis, a finite element model of the medium-soft soil–pile group–nuclear island system was established. The size and material of the finite element model are consistent with those of the shaking table test model. Except that the reinforcement in the pile foundation was built with two-node rod elements, the rest of the model was built with eight-node solid elements, as shown in Figure 14.

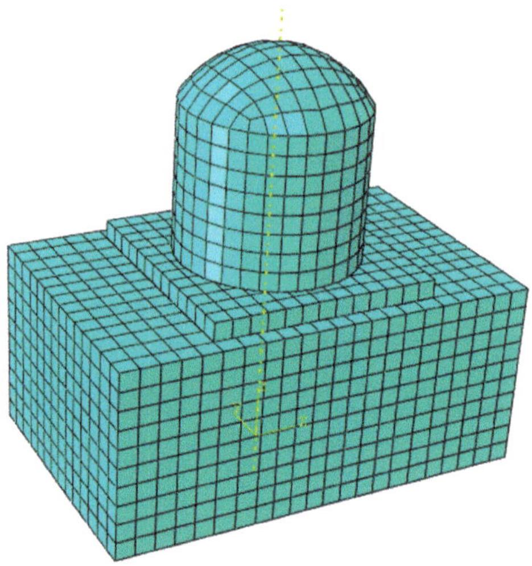

Figure 14. The finite element model.

The Mohr–Coulomb constitutive model was adopted for the soil, the plastic damage model was adopted for the pile concrete, and the ideal elastic–plastic constitutive model was adopted for the steel bars. Due to the high stiffness of the pile cap and nuclear island, and because the deformation during the test was very small, both were modeled using linear elastic constitutive models. The parameters of each material in the finite element model are shown in Tables 1–4.

Table 1. Parameters of the soil.

Item	Density (kg/m^3)	Elastic Modulus (kPa)	Poisson's Ratio	Cohesion (kPa)	Friction Angle (°)	Dilation Angle (°)
Soil surface	1700	1	0.35	0.01	20	0.1
Soil bottom	1700	15,000	0.35	25	20	0.1

When the parameter values in the surface and the bottom of soil are different, it indicates that the parameter values vary linearly in the vertical direction in the soil.

Table 2. Parameters of the concrete.

Item	Density (kg/m^3)	Elastic Modulus (Gpa)	Poisson's Ratio	Cohesion (°)	Eccentricity	f_{b0}/f_{c0}	K_c
Value	2400	22	0.2	35	0.1	1.16	0.6667

Table 3. Parameters of the reinforcement inside the pile.

Item	Density (kg/m^3)	Elastic Modulus (Gpa)	Poisson's Ratio	Yield Stress (Mpa)
Longitudinal reinforcement	7800	200	0.2	335
stirrup	7800	200	0.2	235

Table 4. Parameters of the pile cap and the nuclear island.

Item	Density (kg/m^3)	Elastic Modulus (Gpa)	Poisson's Ratio
Pile cap	2500	100	0.2
Baseplate of nuclear island	2500	100	0.2
Nuclear island containment	17,900	200	0.2

In the finite element model, the pile foundation reinforcement was placed inside the pile with embedded constraints. Binding contact was adopted between the pile and the pile cap. Normal hard contact and tangential penalty contact were adopted between the soil and the pile foundation. The friction coefficient was set to 0.577. Binding contact was set between the nuclear island containment and the baseplate and between the baseplate and the pile cap. The side boundary of the model adopted the degree of freedom binding boundary, and the bottom boundary adopted the acceleration input boundary. Then, the RG1.60 seismic motion with an amplitude of 0.4 g was input into the model.

5.2. Development Process of the Plastic Damage

Figure 15 shows the development of the nephogram of the tensile damage area on the pile under the RG seismic motion with an amplitude of 0.4 g. From Figure 15, it can be seen that the sequence of the tensile damage positions is as follows: the top of pile 3#; the top of piles 2#, 4# and 5# and the bottom of pile 3#; the top and bottom of pile 1#; the top and bottom of piles 2#, 4#, 5#; damage distributed at a certain distance along pile 1#; damage distributed at a certain distance along pile 3#; the middle area of piles 2#, 4# and 5#; the remaining areas of piles 1# and 3#.

Figure 16 shows the development of the nephogram of the pressure damage area on the pile under the RG seismic motion with an amplitude of 0.4 g. From Figure 16, it can be seen that the sequence of the pressure damage positions is as follows: the top of pile 3#; the top of piles 2#, 4# and 5#; the top of pile 1#; the bottom of all piles; and the middle upper and middle lower parts of all piles.

From Figures 15 and 16, it can be seen that the order of damage of the piles is generally as follows: side piles located in one direction of the vibration, side piles located in the middle position and the central pile and side piles located in the other direction of the vibration. This is consistent with the damage phenomenon of each pile in the shaking table test under medium-soft soil conditions. The order of damage locations on the pile in the numerical model is pile top, pile bottom, middle of the pile body. In the shaking table test of medium-soft soil, there was obvious damage to the top and body of the piles, but the degree of damage in the bottom of the piles was relatively mild. It is speculated that the addition of a 20 mm thick steel plate at the bottom of the piles in the shaking table test reduced the stress concentration phenomenon at the bottom of the piles.

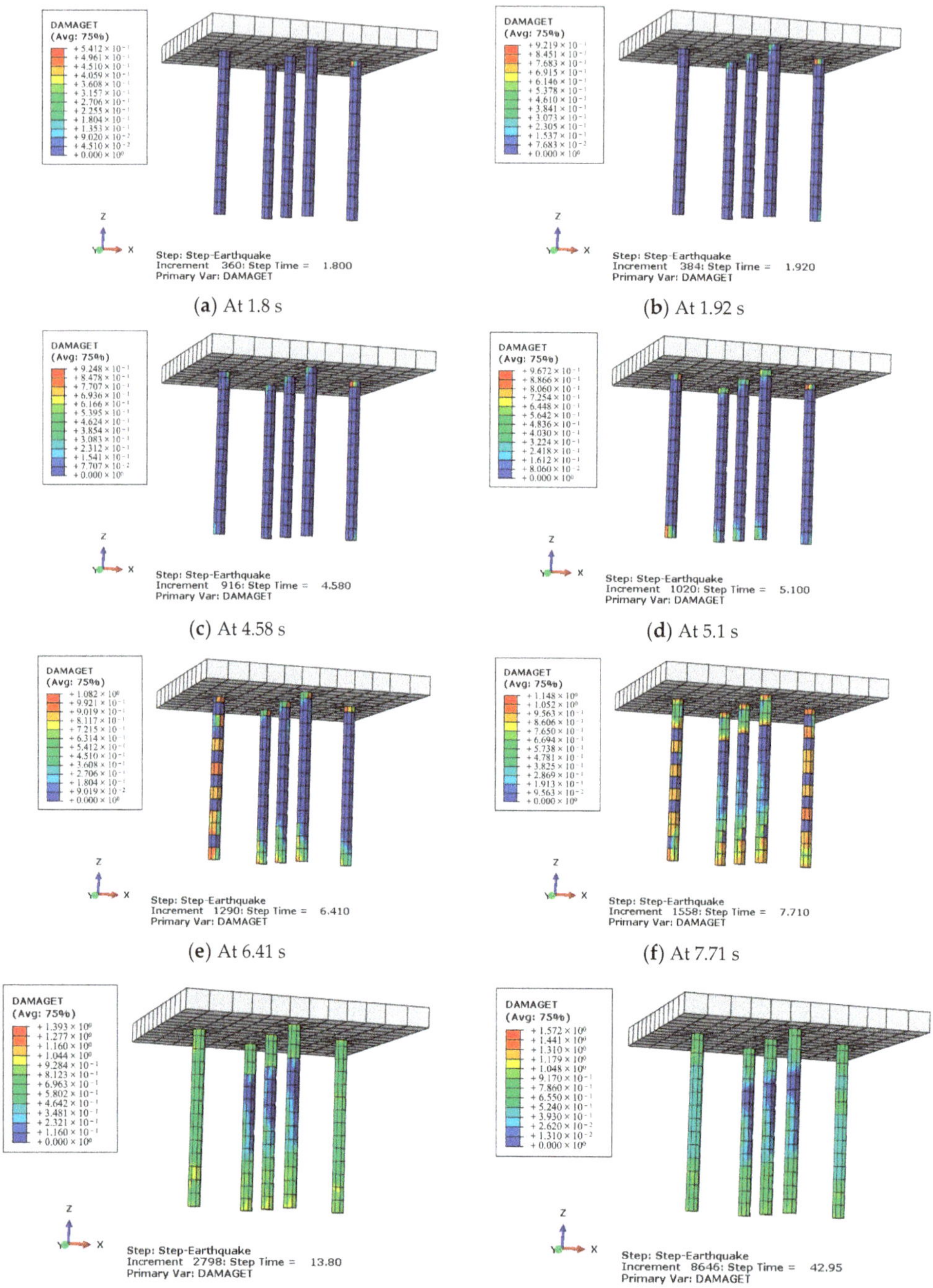

(**a**) At 1.8 s (**b**) At 1.92 s

(**c**) At 4.58 s (**d**) At 5.1 s

(**e**) At 6.41 s (**f**) At 7.71 s

(**g**) At 13.8 s (**h**) At 42.95 s

Figure 15. Development of pile tensile damage nephogram.

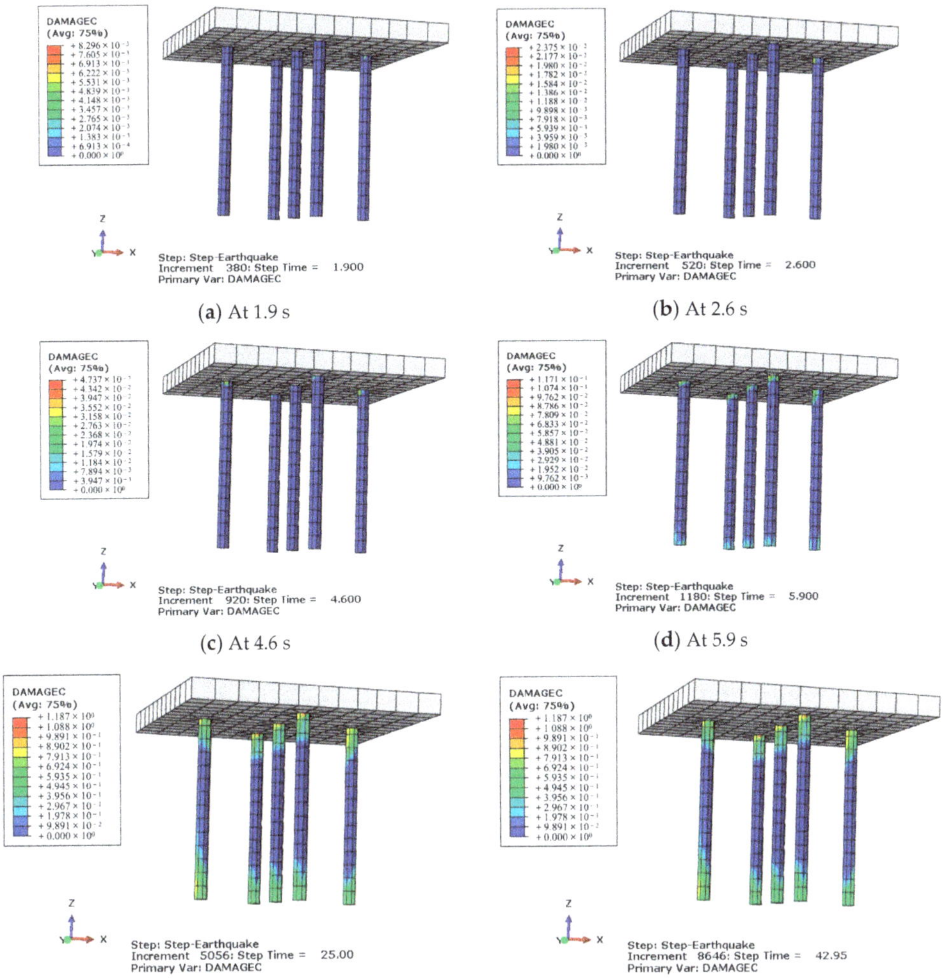

(**a**) At 1.9 s (**b**) At 2.6 s

(**c**) At 4.6 s (**d**) At 5.9 s

(**e**) At 25 s (**f**) At 42.95 s

Figure 16. Development of pile pressure damage nephogram.

6. Conclusions

In this paper, by comparing the shaking table test results of a soil–pile–nuclear island structure dynamic interaction model in conditions of simulating medium-soft soil and hard soil sites, the acceleration response law of the nuclear island model and the deformation and internal force response law of the pile foundation are analyzed. Moreover, the seismic failure mechanism and mode of the pile group foundation are discussed. The following conclusions can be drawn:

(1) When the amplitude of the input seismic motion is small (less than 0.2 g in this paper), the peak acceleration amplification factor of the nuclear island plant in the medium-soft soil site is larger than that in the hard soil site. However, when the amplitude of the input seismic motion is large (0.4 g in this paper), the peak acceleration amplification factor of the upper nuclear island structure is smaller than that in the hard soil site because the medium-soft soil has entered a strong nonlinear stage, and the pile foundation has been damaged.

(2) The properties of the soil around the pile foundation have a significant effect on the response characteristics of the superstructure. The predominant period value of the floor spectrum of the nuclear island plant in the soft soil site is greater than that in the hard soil site. Resonance occurs when the predominant period value of the floor spectrum is similar to that of the far-field seismic motion response spectrum, which aggravates the vibration of the plant structure and its pile foundation, resulting in great damage to the structural system.
(3) The properties of site soil have a great influence on the seismic damage of piles. The deformation constraint effect of soil on piles is small under the action of the seismic motion when the site soil is soft, and it is easy for the separation phenomenon between piles and soil to occur, resulting in the large displacement and deformation of piles, which may cause damage to the pile foundation.
(4) The internal force of piles is also related to the position of the pile in the pile group. The piles at the edge along the direction of the seismic motion input of the pile group have a large internal force and the most serious damage. Therefore, attention should be paid to strengthening the protection of the side piles in the seismic design of the nuclear island containment structure.
(5) The pile failure is most likely to occur at the head, which is damaged by the combined action of bending, shear and pressure with the increase in seismic motion amplitude when the site soil is soft, followed by the possible bending failure of the pile body.

The test results show that the hardness grade of the site soil has a great influence on the damage to the pile foundation of a nuclear island plant under the action of seismic motion, which should be taken into consideration during the site selection of a nuclear island plant structure and the seismic design of the pile foundation.

Author Contributions: Conceptualization, Y.S. and L.J.; methodology, Y.S. and L.J.; software, G.W.; validation, Y.W. and J.T.; formal analysis, Y.S.; data curation, L.J.; writing—original draft preparation, G.W.; writing—review and editing, G.W.; visualization, Y.W.; project administration, W.Q.; funding acquisition, W.Q. All authors have read and agreed to the published version of the manuscript.

Funding: This research was funded by the National Science and Technology Major Project of the Ministry of Science and Technology of China (Grant No. 2018ZX06902016) and the Scientific Research Fund of Institute of Engineering Mechanics, China Earthquake Administration (Grant No. 2019B10).

Institutional Review Board Statement: Not applicable.

Informed Consent Statement: Not applicable.

Data Availability Statement: Data are available in a publicly accessible repository.

Conflicts of Interest: The authors declare no conflict of interest.

References

1. Yang, Y.; Wu, H.; Fang, Q. Numerical simulation analyses of damage and vibration response of a nuclear power plant under aircraft impact. *J. Vib. Shock* **2023**, *42*, 312–324.
2. Yang, J.; Ma, Y.; Cai, L.; Shen, L. Research on the seismic response analysis of nuclear island buildings with pile foundation. *Ind. Constr.* **2021**, *51*, 31–36.
3. Tian, Z.; Li, P.; Zhu, S.; Zhang, Y.; Li, Y.; Liu, Y.; Zhou, K.; Yan, L. Shaking table tests on negative friction of piles in soft soils under strong earthquake motion. *Chin. J. Geotech. Eng.* **2022**, *44*, 550–559.
4. Zhuang, H.; Zhao, C.; Yu, X.; Chen, G. Earthquake responses of piles-soil dynamic interaction system for base-isolated structure system based on shaking table tests. *Chin. J. Geotech. Eng.* **2022**, *44*, 979–987.
5. Liang, F.; Liang, X.; Zhang, H. Simplified analysis of dynamic response of pile-supported bridge under local scour and verification by centrifugal shaking tests. *Chin. J. Geotech. Eng.* **2021**, *43*, 1771–1780.
6. Meymand, P. *Shaking Table Scale Model Tests of Nonlinear Soil-Pile-Superstructure Interaction in Soft Clay*; University of California: Berkeley, CA, USA, 1998.
7. Wei, X. *Shake Table Tests of Soil-Pile-Structure Interaction and Its Theoretical Analysis*; Tongji University: Shanghai, China, 1999.
8. Wu, X. *Shake Table Tests of Sand-Pile-Structure Interaction*; Tongji University: Shanghai, China, 2002.
9. Chen, G.; Wang, Z.; Zuo, X.; Du, X.; Han, X. Development of laminar shear soil container for shaking table tests. *Chin. J. Geotech. Eng.* **2010**, *32*, 89–97.

10. Sun, H.; Jing, L.; Meng, X.; Wang, N. A three-dimensional laminar shear soil container for shaking table test. *J. Vib. Shock* **2012**, *31*, 26–32.
11. Ni, K. *Shaking Table Test of Pile and Composite Foundations' Dynamic Behavior in Layered Soils Subjected to Earthquake Excitation*; China Academy of Building Research: Beijing, China, 2013.
12. Chau, K.; Shen, C.; Guo, X. Nonlinear seismic soil–pile–structure interactions: Shaking table tests and FEM analyses. *Soil Dyn. Earthq. Eng.* **2008**, *29*, 300–310. [CrossRef]
13. Tokimatsu, K.; Suzuki, H.; Sato, M. Effects of inertial and kinematic interaction on seismic behavior of pile with embedded foundation. *Soil Dynamatics Earthq. Eng.* **2005**, *25*, 753–762. [CrossRef]
14. Hokmabadi, A.S.; Fatahi, B.; Samali, B. Physical Modeling of Seismic Soil-Pile-Structure Interaction for Buildings on Soft Soils. *Int. J. Geomech.* **2015**, *15*, 04014046. [CrossRef]
15. Huang, Z.; Bai, X. Shaking table model test for seismic response of a pile group foundation with liquefiable sandy soil. *J. Vib. Shock* **2013**, *32*, 153–158.
16. Li, X.; Wang, X.; He, Q.; Liu, A. Shaking Table Test for Evaluation of Seismic Behavior of Nuclear Power Plants on Non-Rock Site. *Nucl. Power Eng.* **2017**, *38*, 31–35.
17. Jing, L.; Wang, G.; Li, J.; Sun, Y.; Zhou, Z.; Qi, W. Shaking Table Test and Numerical Simulation of Soil-Pile-Nuclear Island System Dynamic Interaction. *Chin. J. Geotech. Eng.* **2022**, *44*, 163–172.
18. Hu, C.; Yang, X. Analysis of seismic response on pile in liquefiable site. *J. Vib. Shock* **2007**, *106*, 133–137.
19. Zhang, Y.; Li, J. Numerical solution of pile-soil interaction problem in ocean. *J. Vib. Shock* **2009**, *28*, 160–163.
20. Luo, C.; Zhan, C.; Lou, Y.; Jin, X.; Ding, Z. Effects of pile-soil nonlinear coupling actions on dynamic response features of pile foundation under earthquake. *J. Vib. Shock* **2017**, *36*, 20–26.

sustainability

Review

A Review on the Modelling Techniques of Liquid Storage Tanks Considering Fluid–Structure–Soil Interaction Effects with a Focus on the Mitigation of Seismic Effects through Base Isolation Techniques

M. Chaithra, A. Krishnamoorthy and A. R. Avinash *

Department of Civil Engineering, Manipal Institute of Technology, Manipal Academy of Higher Education, Manipal 576104, India
* Correspondence: avinash.ar@manipal.edu

Abstract: Globally, tanks play a major part in the provision of access to clean drinking water to the human population. Beyond aiding in the supply of fresh water, tanks are also essential for ensuring good sanitary conditions for people and for livestock. Many countries have realized that a robust water supply and a robust sanitation infrastructure are necessary for sustainable growth. Therefore, there is large demand for the construction of storage tanks. Further, liquid storage tanks are crucial structures which must continue to be operational even after a catastrophic natural event, such as an earthquake, to support rehabilitation efforts. From an engineering point of view, the various forces acting on the tanks and the behaviour of the tanks under various loads are important issues which need to be addressed for a safe design. Analyses of the tanks are challenging due to the interaction between the fluid and tank wall. Thus, researchers have conducted several investigations to understand the performance of storage tanks subjected to earthquakes by considering this interaction. This paper discusses the historical development of various modelling techniques of storage tanks. The interaction with the soil also influences the behaviour of the tanks, and hence, in this paper, various modelling approaches for soil structure interaction are also reviewed. Further, a brief history of various systems of base isolation and modelling approaches of base-isolated structures are also discussed in this article.

Keywords: liquid storage tank; dynamic analysis; base isolation; fluid–structure interaction; soil–structure interaction

Citation: Chaithra, M.; Krishnamoorthy, A.; Avinash, A.R. A Review on the Modelling Techniques of Liquid Storage Tanks Considering Fluid–Structure–Soil Interaction Effects with a Focus on the Mitigation of Seismic Effects through Base Isolation Techniques. *Sustainability* **2023**, *15*, 11040. https://doi.org/10.3390/su151411040

Academic Editors: Chong Xu, Su Chen and Shuang Li

Received: 29 May 2023
Revised: 1 July 2023
Accepted: 3 July 2023
Published: 14 July 2023

1. Introduction

Liquid storage tanks play a crucial role in day-to-day activities, as they are used for storing not only essential supplies such as water and oil but also to safely store hazardous chemicals. For the sustainable development of any country, it is necessary to provide access to fresh water. For this, a vast network of storage tanks is essential. Additionally, any major sanitary work involves the use of storage containers to temporarily store the sewage water before it is sent for further treatment. Storage tanks cater to the massive water demand from the construction industry and are thus essential for any country to build a robust infrastructure. One of the sustainable development goals (Goal 6) as set by the United Nations, is to provide access to drinking water, sanitization, and hygiene globally by 2030 [1]. This target indicates that there will be a huge demand for storage tanks in the near future. As many countries are working towards this goal, they are building massive infrastructure which involves a vast network of tanks. Due to industrialization, massive amounts of hazardous chemicals are stored in storage containers. Be it a water storage tank or hazardous material storage container, any damage to these structures during a natural hazard, such as an earthquake, can lead to disastrous consequences. Thus, it is essential that these structures are resilient to natural disasters such as earthquakes, cyclones and tsunamis to ensure the sustainable growth of any country. Thus, engineers must have the knowledge and tools to design long-lasting tanks capable of resisting natural disasters.

Further, understanding the structural behaviour of the tanks helps the designer to achieve safety and economy in their design, which ensures sustainability in the long run. In the event of a natural disaster, there is always a risk of damage to tanks. Any such failure of a water tank may hamper the rehabilitation work as a continuous water supply is necessary to manage the possible subsequent fires and also to avoid any disease outbreak post disaster. Further, natural disasters can lead to the spillage of combustible products from the damaged tanks and may lead to fire hazards, potentially harming the environment [2]. The tanks are prone to earthquakes mainly due to their large mass. In the event of strong ground motions, and unlike other structures, such as buildings and bridges, tanks are also subjected to large hydrodynamic forces, making them critical structures, vulnerable to earthquake-induced vibrations. A lack of careful evaluation of these forces can lead to inadequate design, resulting in failure during strong ground motion. Several storage tanks have failed in the past due to strong earthquakes [3]. Over 30 tanks failed during the Turkey earthquake in 1999 [4], and 167 failed during the 2011 Tohoku earthquake [5]. Depending on the overturning moment and fluid pressure variation activated through seismic motion, the tank may undergo elastic–plastic or elastic buckling failure [6].

In structures such as buildings, interaction typically happens with the connected solid elements, such as between frames and infills or between columns and slabs, which are solid-to-solid interactions. However, tanks are unique in this sense; in tanks, the interaction happens between wall structure and liquid, which is a solid and fluid interaction. This fluid–structure interaction (FSI) is fairly complicated to model due to the interaction between two highly incompatible materials. Before the 1960′s, and owing to the difficulty involved with this modelling, tank walls were idealized as rigid and the effects of the interaction between the liquid and the wall of the tank were ignored [7]. However, the failure of such tanks during earthquakes between 1960 and 1970 showed that the storage tanks have much complex behaviour during ground motion, and that the flexibility of the tank cannot be ignored [2]. Thus, researchers nowadays consider the FSI effects in the modelling for a realistic analysis of tanks. Due to the evolution of various numerical techniques and advancements in computer technology, many researchers have been able to develop new modelling techniques for tanks, such as boundary element methods (BEM) and finite element methods (FEM), to investigate the response of fluid-filled storage containers subjected to seismic excitations [8].

Usually, the foundations of tanks are idealized as rigid units in conventional analysis. However, being made of different materials, there will be an interaction effect between tank and supporting soil on the tank behaviour. This soil–structure interaction (SSI) has been investigated by numerous researchers and it was found that ignoring the SSI effects by assuming the foundation to be rigid leads to unrealistic results [9–12]. It has been found that the soil stiffness significantly influences the behaviour of tanks subjected to ground motion [13]. Thus, when the SSI effect is considered, the tank's overall stiffness changes, which considerably alters its time period. The realistic analysis of a tank therefore depends on both FSI and SSI modelling and therefore these effects need to be considered during the analysis to accurately represent tank behaviour.

Due to the high significance attached to the fluid tanks, many researchers have focused on developing techniques to make the tanks earthquake resistant. Base isolation is one such promising and widely used technique for aseismic design [14]. In a base isolation system, a layer of relatively lesser stiffness is incorporated between the structure and the ground. This low-stiffness layer offsets the transfer of harmful frequencies to the tanks. To date, researchers have developed a wide variety of base isolation bearings which can be used as a standalone devices or can be used in a hybrid configuration with other types of earthquake resistant devices. A brief discussion on base isolation techniques is carried out later, in Section 4. Readers can refer to a number of review articles on various base isolation techniques and systems in [14–21].

Several review papers on the historical development of tank modelling techniques [14,22,23] are currently available. The scope of these papers is typically limited to either fixed-base tanks [22] or base-isolated tanks. This means there is a lack of review articles which address these different tank types together. Additionally, a specific and detailed discussion on the SSI effect, which is one of the major influencing factors on tank behaviour, is absent in these papers. Therefore, the authors of this paper felt the need to address the lack of a recent comprehensive review paper which collectively discusses the various aspects of modelling for both fixed and base-isolated tanks. In this regard, the developments in the various modelling techniques of fluid tanks, including the effects of FSI and SSI, are discussed in detail in the present paper. Further, Section 4.3 of this article is dedicated to discussions on the modelling of base-isolated tanks. Thus, this article aims to bridge the gap between discussions on the advancements in the area of fixed tanks and base-isolated tanks by collectively addressing the developments in these areas. The main contribution of this article would be to review various aspects of tank modelling, including the effects of SSI and FSI, with an emphasis on base isolation techniques used for the seismic resistance of tanks.

Some researchers have reported that elevated tanks are more vulnerable when compared with tanks that are supported on the ground. This is mainly due to the larger time period of the elevated tanks when compared with the ground-supported tanks. This higher time period is attributed to the flexibility provided by the tank staging. Though the discussions in this article are directed towards ground-rested tanks, the modelling concepts are also applicable to elevated tanks. The main distinction between the analysis of ground-rested and elevated tanks is the additional modelling requirements of tank staging. Therefore, very limited discussion has been carried out regarding elevated tanks.

2. Modelling Techniques for Storage Tanks Considering FSI Effects

For studies before the 1960s, researchers generally used a simplified technique developed by Westergaard [24] for tank analysis. This technique was initially developed to evaluate the fluid pressure on rigid dams subjected to harmonic excitation and later adapted for tank analysis [25]. Since then, researchers have proposed various modelling techniques exclusive to tanks and these are discussed in the subsequent sections.

2.1. *Equivalent Mechanical Models*

The equivalent mechanical model was initially proposed by Housner [7,26] for analysing ground supported and elevated tanks filled with fluid. In this model, the tank walls are idealized as rigid units. The equivalent mechanical model was derived from the models of Biot [27] and Housner et al. [28], which were used earlier for a response spectrum analysis. As shown in Figure 1, in this model, two masses are considered, one for the convective mass (M_C) and the other for the non-sloshing part (M_R). The convective mass denotes the sloshing part of the fluid, and the rigid mass represents the non-sloshing part. These fluid masses M_C and M_R are responsible for the forces developed on the tank walls. The expressions to calculate these masses and stiffness k for a cylindrical tank with radius R and depth of water h are as follows:

$$M_R = M\frac{\tanh(1.7^R/_h)}{1.7^R/_h} \tag{1}$$

$$M_C = 0.6M\frac{\tanh(1.8^R/_h)}{1.8^R/_h} \tag{2}$$

$$k = 5.4\frac{M_C^2}{M}\frac{gh}{R^2} \tag{3}$$

where M is the total fluid mass. The heights of these masses h_R and h_C can be obtained by using the expressions as follows:

$$h_R = \frac{3}{8}h\left\{1 + 1.33\left[\frac{M}{M_C}\left(\frac{R}{h}\right)^2 - 1\right]\right\} \tag{4}$$

$$h_C = h\left[1 - 0.185\frac{M}{M_C}\left(\frac{R}{h}\right)^2 - 1.12\frac{R}{h}\sqrt{\left(\frac{MR}{3M_C h}\right)^2 - 1}\right] \tag{5}$$

Similarly, the expressions related to rectangular tanks of width 2L are as follows:

$$M_R = M\frac{\tanh(1.7L/h)}{1.7L/h} \tag{6}$$

$$M_C = 0.83\,M\frac{\tanh(1.6L/h)}{1.6L/h} \tag{7}$$

$$k = 3\frac{M_C^2}{M}\frac{gh}{L^2} \tag{8}$$

$$h_R = \frac{3}{8}h\left\{1 + 1.33\left[\frac{M}{M_C}\left(\frac{L}{h}\right)^2 - 1\right]\right\} \tag{9}$$

$$h_C = h\left[1 - \frac{1}{3}\frac{M}{M_C}\left(\frac{L}{h}\right)^2 - 1.26\frac{L}{h}\sqrt{0.28\left(\frac{ML}{M_C h}\right)^2 - 1}\right] \tag{10}$$

In addition, the author suggested that this approach is applicable to an elevated tank as well.

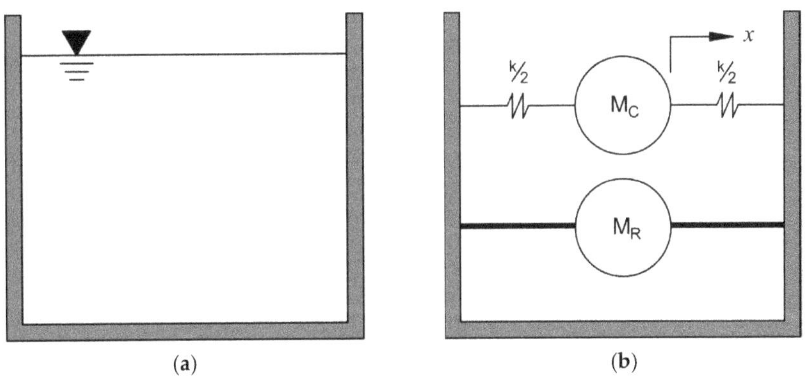

Figure 1. (**a**) Tank with partially filled water; (**b**) equivalent two-mass model, as adapted from [7].

The equivalent model proposed by Housner [7,26] was widely used for some time and even incorporated in the codes and standards of practice of some countries [29,30]. However, many storage tanks which were designed as per the codal provisions of the 1960's failed during earthquakes. These failures indicated that the codal provisions which were developed based on the assumption that tank walls are rigid could not realistically represent the forces on the tanks. Thus, the researcher Veletsos [31] proposed a tank model including the flexibility aspect of tank walls. In this study, it is only the effect of impulsive forces that is evaluated with the assumption of single degree of freedom (SDOF) behaviour of a tank–fluid system. The hydrodynamic forces are calculated using the method proposed by Chopra [32,33]. The equation of motion considered by Veletsos [31] is as follows:

$$\left(m^*_{I,S} + m^*_{I,L}\right)\ddot{w} + c^*\dot{w} + k^*w = -\left(m^*_{R,S} + m^*_{R,L}\right)\ddot{u}(t) \quad (11)$$

where, m*, k* and c* represent the effective mass, stiffness and damping coefficients of the system. The subscripts R and I represent the rigid and impulsive components, respectively, and the subscripts S and L indicate the contribution of masses from structure and liquid, respectively. The responses of the tank in terms of base moment and base shear are calculated. The model developed by this approach shows a significant effect on the seismic response of tanks compared with the seismic response of tanks modelled by assuming rigid tank walls. However, in this study, the convective component of the response is neglected based on the assumption that this response depends on the sloshing period, which is much higher than the tank period. Later, Haroun and Housner [34,35] presented another model in which the fluid mass is represented with three lumped masses. These masses comprise three parts representing the rigid (M_R), convective (M_C), and impulsive (M_I) masses of the fluid. These masses are calculated from the numerical study conducted by Haroun [2]. Figure 2 shows the mechanical model as proposed by the researchers Haroun and Housner [34,35]. Several researchers later used this model as a reference in their studies and validated their results through experimental investigations [36–40].

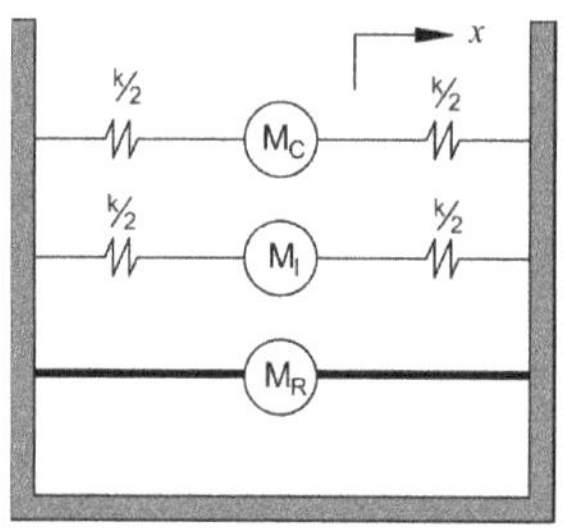

Figure 2. Equivalent three mass model, adapted from [34].

Malhotra et al. [41] proposed another model to evaluate the tank's response using sloshing and impulsive masses to model the fluid mass. As shown in Figure 3, a dashpot system is used in the model to represent the damping effects. The convective and impulsive natural period for a system with tank height H and radius r is calculated using the expression as follows:

$$T_C = C_C\sqrt{r} \quad (12)$$

$$T_I = C_I\frac{H\sqrt{\rho}}{\sqrt{t/r} \times \sqrt{E}} \quad (13)$$

Here, t is the tank wall thickness, ρ is the liquid mass density, and E is the modulus of elasticity of the tank material. C_I and C_C are impulsive and convective coefficients which depend on the aspect ratio of the tank. These coefficients are derived from the study conducted by Veletsos and co-workers [42–44]. Interested readers may refer to the table given by Malhotra et al. [41] to obtain the values of C_I and C_R and other parameters required for the seismic design of storage tanks. The model has been found to be efficient, especially for the analysis of cylindrical containers. The design procedure is also incorporated in the European code of practice [41].

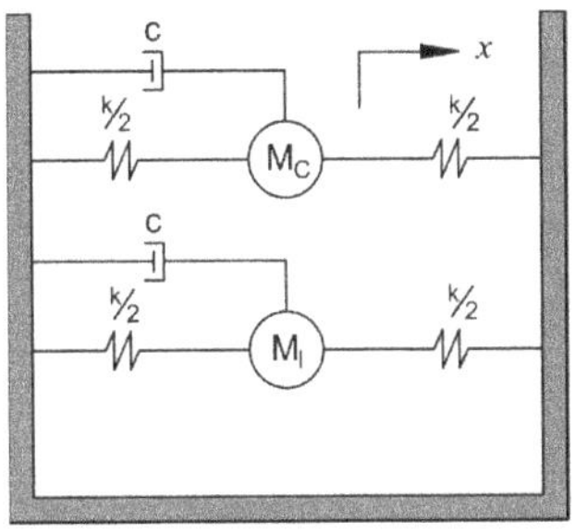

Figure 3. Mechanical model, as adapted from Malhotra et al. [41].

The simplified models discussed in this section are useful in studying the global aspects of tank response, such as base moment or base shear. However, numerical methods are recommended to conduct a detailed analysis of tanks that include the response even at the local level [45]. The following section is dedicated to the discussion of different numerical methods for tank modelling and analysis.

2.2. Direct Modelling Approaches

The advancement in computer technologies has enabled researchers to develop various numerical approaches, such as FEM and BEM, which can be used for the direct modelling of tanks. Unlike the mechanical models, which are limited only to tanks with regular geometries, numerical methods can also be used for tanks of arbitrary geometries [46]. The researcher Edwards [8] proposed one of the earlier numerical techniques to study a short (aspect ratio of less than 1) cylindrical tank. For this study, Edwards [8] used a refined shell theory to evaluate the displacements and stresses in a tank. Several researchers used this approach later with minor modifications to study the various aspects of storage tanks [47–49].

The FEM-based methods used to solve FSI problems are classified as added mass [24], Lagrangian (displacement based), and Eulerian (pressure based) formulations [50]. In the added mass technique, a part of the fluid mass is added to the structure's wall at the interface of the wall and fluid elements. This added mass (M_A) moves along with the wall during the seismic vibrations. The equation of motion for tanks inclusive of added mass can be considered as follows:

$$M^*\ddot{u} + C\dot{u} + Ku = -M^*\ddot{u}_g \tag{14}$$

where, M* is a matrix representing the combination of tank mass (M) and added mass (M_A). K and C are the stiffness and damping matrices of the tank, respectively. Additionally, $\ddot{u}_g$ in Equation (6) represents the ground acceleration. From Equation (6), it can be observed that part of the fluid mass is also added to the tank's mass for the total mass calculation. However, only the tank walls are considered when calculating the damping matrix, ignoring the contribution of the fluid. The researchers found that this method leads to conservative results [51]; therefore, several investigators used this approach in their storage tank analysis [24,52–56]. Because the damping effect is ignored, this method generally leads to higher force calculation. Though conservative, the forces calculated are not realistic as the damping effect is a major aspect, one which significantly affects the tank behaviour under the action of earthquakes.

In the generalized Lagrangian method used earlier for the dams, the liquid is discretized as a solid element with a negligible shear modulus [57,58]. The general form of equation of motion applicable to this approach is as follows:

$$M\ddot{u} + C\dot{u} + Ku + f_I + f_0 = 0 \tag{15}$$

where f_I and f_0 are the interface forces between the tank wall and fluid. This method yields symmetric matrices, resulting in narrow bandwidth matrices during the analysis [59]. This reduction in bandwidth is one of the major benefits this method offers, as it significantly reduces the computational time when used in computers. Due to this benefit of reduced computational effort, many researchers later proposed different forms of finite elements to model a fluid suitable for the Lagrangian approach [51,60–63]. The Lagrangian formulation since then has been used by several researchers to study storage tanks [64–66]. Though the Lagrangian approach leads to the reduction of computational efforts, some researchers have suggested that it leads to less accurate results [67,68].

The generalized Eulerian approach is based on the pressure formulation method [69] in which the motion of a liquid is directed by the continuity relation and the Navier–Stokes equation [61]. The general form of the equation of motion used for this approach, which couples the tank–fluid system, is as follows:

$$\begin{bmatrix} K & S \\ 0 & H \end{bmatrix}\begin{bmatrix} u \\ p \end{bmatrix} + \begin{bmatrix} C & 0 \\ 0 & A \end{bmatrix}\begin{bmatrix} \dot{u} \\ \dot{p} \end{bmatrix} + \begin{bmatrix} M & 0 \\ S^T & E \end{bmatrix}\begin{bmatrix} \ddot{u} \\ \ddot{p} \end{bmatrix} = [f(t)] \tag{16}$$

where H, A, and E are matrices for the fluid domain, u and p are displacement and fluid pressure, respectively, f(t) is the external force on the system, and S is the interaction matrix for a tank–fluid system, and which can be obtained based on boundary conditions between them. Although this approach leads to an unsymmetrical matrix and subsequent increase in computational cost [70], the method has wide application due to the lesser number of variables when compared with the Lagrangian approach. Moreover, the method becomes computationally efficient when the compressibility effect of fluid is neglected [61]. Hence, several researchers favour the Eulerian approach when assessing the behaviour of storage containers [71–73].

Another approach to evaluate storage tank behaviour is with the use of BEM [74]. Unlike FEM where the entire fluid is discretised into different finite elements, in this approach the fluid is discretized only at the boundary using surface elements. This method is computationally efficient as it reduces the number of elements required considerably compared with completely FEM-based tank models. Moreover, BEM and FEM can be used together for a tank's analysis, here the liquid can be modelled using BEM, and the tank walls are modelled using FEM, thus providing the benefits of both approaches. This advantage has prompted several researchers to use this method to solve the FSI problem in dams [75–77] and tanks [78–82]. The governing equation of motion for a coupled tank–fluid system based on FEM–BEM is as follows:

$$\begin{bmatrix} M_{oo} & M_{oc} & 0 \\ M_{co} & M_{cc} + M_{cc}^A & M_{cf}^A \\ 0 & M_{fc}^A & M_{ff}^A \end{bmatrix}\begin{Bmatrix} \ddot{u}_o \\ \ddot{u}_c \\ \ddot{\xi}_e \end{Bmatrix} + \begin{bmatrix} K_{oo} & K_{oc} & 0 \\ K_{co} & K_{cc} & 0 \\ 0 & 0 & K_{ff} \end{bmatrix}\begin{Bmatrix} u_o \\ u_c \\ \xi_e \end{Bmatrix} = \{0\} \tag{17}$$

where M and K are mass and stiffness matrices respectively, u indicates the horizontal motion, ξ_e indicates the sloshing vector, subscript 'o' indicates the degrees of freedom of the tank which is not in contact with fluid, subscript 'c' indicates the degrees of freedom of the tank which is in contact with fluid and M^A denotes the added mass matrix of the fluid.

3. Modelling Approaches for Storage Tanks Considering Combined FSI and SSI Effects

In many of the earlier studies, the tanks resting on the ground were assumed to be supported by a rigid surface, and accordingly, the analyses were carried out by idealizing the tank foundation as fixed [43]. However, soil conditions significantly affect the behaviour of structures when subjected to ground vibrations, as the response of a structure varies with the stiffness of soil. The effect of support conditions on the response of a structure was first recognized by Martel [83], based on the observations of the failure of structures during many earthquakes in Japan [84]. Later, many researchers would go on to conduct numerical and experimental investigations on the SSI effect on the behaviour of building structures [11,85–92]. Since tanks are also supported on the ground, it is logical to assume the presence of interaction effects between tanks and the supporting soil. Therefore, for the analysis of tanks, both FSI and SSI effects need to be incorporated for a realistic analysis. The research carried out by Chopra et al. [93] represents one of the early attempts to address the combined effects of FSI and SSI on tank behaviour. These authors proposed a general technique to study the response of gravity dams considering the interaction effect between the water and dam wall. Further, several such investigations on gravity dams and bridges have been conducted by investigators considering the combined effects of FSI and SSI on dams [93–99]. Taking a cue from this research in the area of dams, Veletsos and Tang [100] proposed a simplified FSI–SSI combined model exclusively for tanks. As shown in Figure 4, the researchers used a two degree of freedom (2DOF) model in which the mass of the foundation, the tank, and the portion of the non-sloshing mass are represented together as $M_{0'}$ and sloshing fluid mass as M_1 for the investigation. In the model, the effect of SSI is considered by a spring and dashpot system. The governing differential equation considered by Veletsos and Tang [100] is as follows:

$$D\frac{\partial^4 w}{\partial z^4} + \frac{Eh}{a^2}w + \rho h\frac{\partial^2 w}{\partial t^2} = p(z,t) \tag{18}$$

where D is the flexural rigidity of the tank wall and p(z,t) is the hydrodynamic wall pressure which is a function of the axial component of the coordinate system (z) for the tank and time (t), a is the tank radius, and h is the wall thickness. This model was later to be used by many researchers in their studies [101,102].

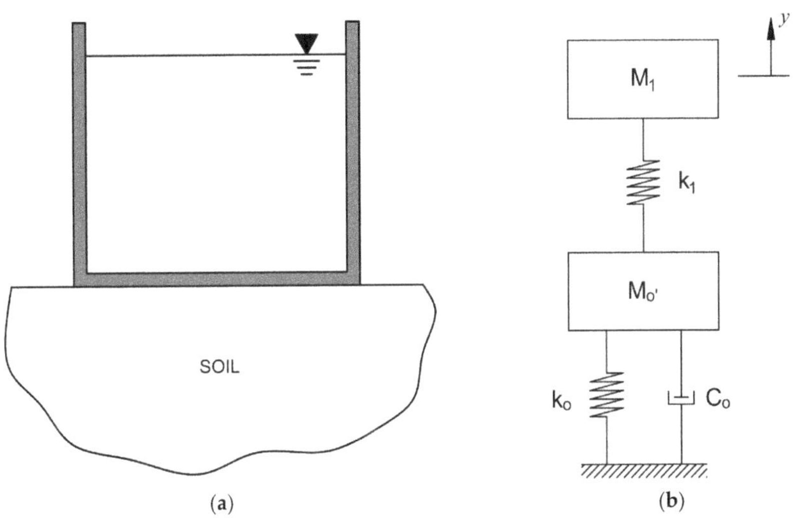

Figure 4. Model of tank considering soil flexibility: (**a**) actual tank; (**b**) mechanical model, as adapted from [100].

Although the method presented by Veletsos and Tang [100] was simple and easy to apply, for their study they considered only harmonic loading, and the effect of random vibration was not included. Seismic excitation belongs to the random vibration domain, and hence the response of the tanks to these vibrations is probabilistic in nature. Hence, Chatterjee and Basu [103] presented another method to estimate the SSI effect on the performance of a tank by applying random vibration theory. As shown in Figure 5, the tank and liquid are represented by the spring–mass dashpot system, and the foundation is modelled as a spring–mass system with m_i as the impulsive mass of the fluid and m_f as the foundation mass. These masses are connected with a spring of stiffness K and a viscous damper of coefficient C. The foundation mass is connected with a complex impedance function K_x. The governing equation of motion for the system considered is as follows:

$$m_i(\ddot{x}_i(t) + \ddot{x}(t)) + m_f\ddot{x}(t) + K_x(x(t) - x_g(t)) = 0 \tag{19}$$

Here, double overdot represents the second derivative of displacement with respect to time, and x_g is the ground motion. From the study, the authors concluded that the effect of SSI increases the response of tanks. Additionally, the SSI effect is found to be greater in slender tanks compared with broad tanks.

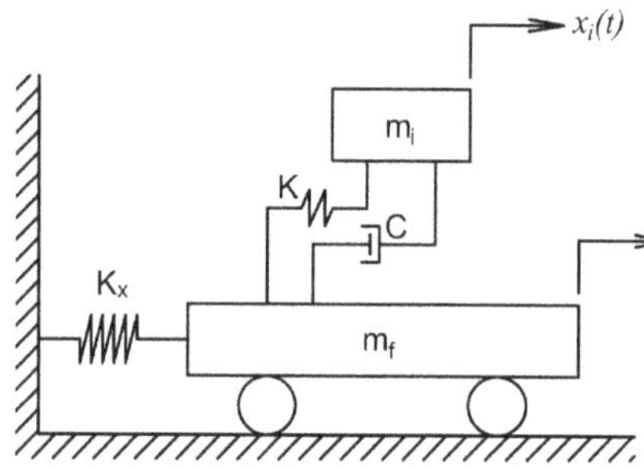

Figure 5. The mechanical model, as adapted from Chatterjee and Basu [103].

The researcher, Larkin [104] proposed a frequency domain approach to evaluate the seismic performance of storage tanks supported on the soil stratum. As shown in Figure 6, the system in the proposed model is assumed to have a single mass M with lateral stiffness of K and damping coefficient C. This model is capable of simulating both rocking and translational motions of the tanks subjected to ground motions. This study suggested that the SSI effect is prominent for slender tanks resting on soft soil. It is also shown in the study that the impulsive period of the tank increases with a decrease in soil stiffness.

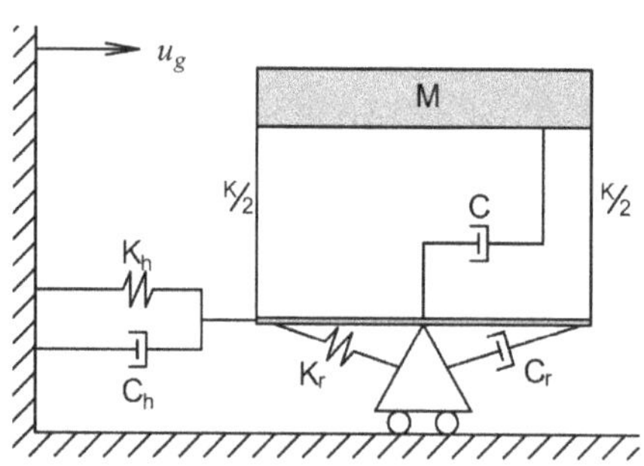

Figure 6. Model of the tank soil system, as adapted from [104].

Meng et al. [105] proposed a time domain approach to investigate the seismic behaviour of storage containers resting on elastic soil. As shown in Figure 7, a tank–liquid system is modelled using three masses—rigid, convective, and impulsive—to form a spring–mass system, as proposed by Haroun and Ellaithy [37], while the soil is modelled using a second-order lumped parameter model as proposed by Wu and Lee [106]. The governing equations of motion are obtained using kinetic energy T, potential energy V and the energy absorbed by dampers δW. These parameters are calculated as follows:

$$\begin{aligned} T = {} & \tfrac{1}{2}M_C\left(\dot{u}_C + \dot{u}_R + H_C\dot{u}_1^r + \dot{u}_g\right)^2 + \tfrac{1}{2}M_I\left(\dot{u}_I + \dot{u}_R + H_I\dot{u}_1^r + \dot{u}_g\right)^2 \\ & + \tfrac{1}{2}(M_R + M_B)\left(\dot{u}_R + H_R\dot{u}_1^r + \dot{u}_g\right)^2 + \tfrac{1}{2}(I_R + I_B)\left(\dot{u}_1^r\right)^2 \end{aligned} \tag{20}$$

$$\begin{aligned} V = {} & \tfrac{1}{2}k_C(u_C)^2 + \tfrac{1}{2}k_I(u_I)^2 + \tfrac{1}{2}k_1^h(u_R)^2 + \tfrac{1}{2}k_2^h\left(u_R - u_1^h\right)^2 + \tfrac{1}{2}k_1^r\left(u_1^r - u_2^r\right)^2 \\ & + \tfrac{1}{2}k_2^r\left(u_2^r\right)^2 + \tfrac{1}{2}k_3^r\left(u_2^r - u_3^r\right)^2 \end{aligned} \tag{21}$$

$$\begin{aligned} \delta W = {} & -C_C\dot{u}_C\delta u_C - C_I\dot{u}_I\delta u_I - C_1^h\dot{u}_R\delta u_R - C_2^h\dot{u}_1^h\delta u_1^h \\ & - C_1^r\left(\dot{u}_1^r - \dot{u}_2^r\right)\delta\left(u_1^r - u_2^r\right) - C_2^r\dot{u}_2^r\delta u_2^r - C_3^r\dot{u}_3^r\delta u_3^r \end{aligned} \tag{22}$$

In these equations, the superscripts h and r represent the horizontal and rocking degrees of freedom, respectively. From this study it has been observed that the SSI has a negligible effect on the convective component of responses; however, the effect is prominent for impulsive responses.

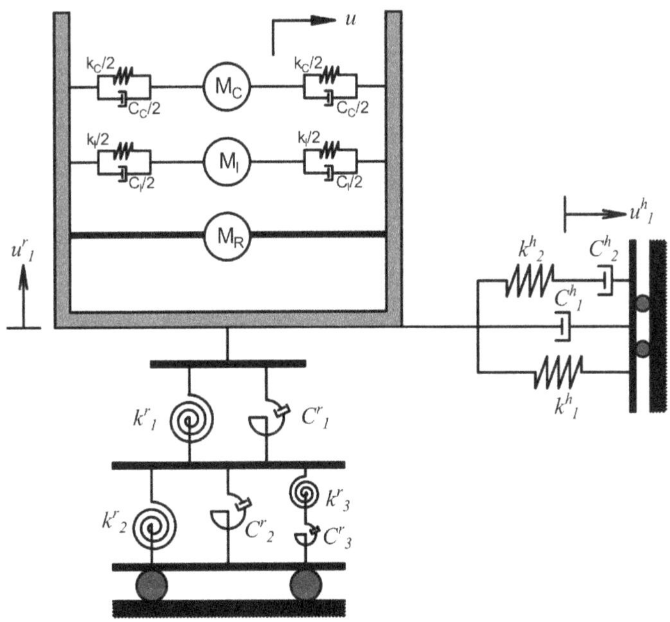

Figure 7. A combined mechanical model of a tank, fluid, and soil system, as adapted from [105].

Though the spring–mass model used by many investigators represents the effect of SSI effectively, some investigators realized that the simplified model tends to generate overly conservative results [107] and that a reasonable representation of the system can be obtained by using FEM [92]. In this regard, several researchers used FEM in their analysis of tanks [108–111]. Irrespective of the modelling approach used, it can be seen that the effect of SSI is prominent for tanks resting on soft soil.

4. Base Isolation Techniques for Storage Tanks

Due to the high importance attached to storage tanks as discussed in Section 1, it is necessary to mitigate the harmful effects of ground vibrations on tanks. Several earthquake resistant techniques have been proposed by researchers for the building of structures such as bracing systems, viscous fluid dampers and base isolation bearings. Out of these systems, base isolation systems are now widely used around the world for earthquake-resistant buildings. Apart from buildings, base isolation systems can also be used in tanks to improve earthquake resistant capabilities. In this section, an overview of various types of isolation systems and modelling techniques of isolated structures is presented.

4.1. Types of Base Isolation Systems

The primary purpose of base isolators is to decrease the frequency of structures to a value lesser than that of earthquakes, avoiding the transfer of harmful vibrations to the structure. Base isolation bearings also provide a means to mitigate seismic energy, reducing the force transferred to the structure. The isolation systems are broadly categorized as elastomer-based bearings and friction bearings. The elastomer bearings are manufactured using natural rubber as a main ingredient. Due to their elastic property, the bearings provide the restoring mechanism by default. Friction isolators work by the principle of sliding friction or rolling friction. In sliding isolators, the restoring capacities are provided either by changing the sliding surface geometry or by connecting to external restoring mechanisms. The usage of natural rubber blocks as base isolators in a school in Skopje in 1969 is regarded as one of the earlier known applications of elastomeric bearings [112]. Due to the low vertical stiffness, these rubber blocks later bulged sideways due to lack of vertical stiffness. This issue was later resolved by increasing the vertical stiffness through the usage of steel plates embedded within the rubber blocks [17]. The damping capacity of elastomer isolators can be controlled by adding some fillers, resins or oils to the natural rubber. The hysteretic behaviour of a damper plays a significant role in seismic energy dissipation. To enhance their energy dissipation, some researchers proposed elastomer bearings with lead cores [113] known as lead rubber bearings (LRB). As shown in Figure 8a, the lead core is inserted at the centre of an elastomer bearing. The behaviour of LRB is nonlinear, accordingly the hysteresis loop is idealised as a bilinear curve as shown in Figure 8b. Due to this arrangement, a single damper can provide vertical load carrying capacity, as well as restoring and damping capacities, improving their value [114].

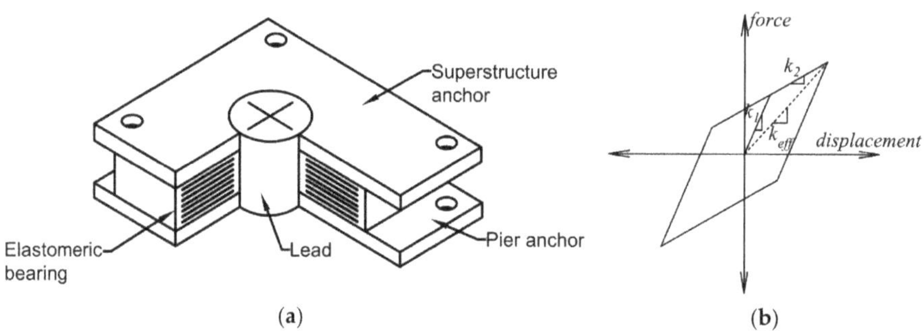

Figure 8. (**a**) Lead Rubber bearing (LRB), as adapted from [113]; (**b**) hysteresis loop for LRB.

Sliding bearings are another class of isolation systems used to dissipate energy through friction. The simplest form of sliding bearing system is the pure friction (PF) system which can be achieved by a layer of sand, Teflon bearings, or rollers between the foundation and superstructure [112]. Some researchers have also proposed a combination of PF systems with elastomeric bearings known as the "Electricité de France" (EDF) system [115]. Another type of system called resilient-friction base isolator (RFBI) was presented by Mostaghel and Khodaverdian [116] and consists of layers of sliding metal plates with a rubber core. The sliding elements provides base isolation, and the rubber core provides restoring capabilities. In PF isolation systems large residual displacements are typically observed; thus, some authors have proposed elliptical rolling rods at the base to overcome this issue [117]. Similarly, to address the issue of a lack of restoring mechanism in PF systems, Zayas et al. [118] proposed a curved geometry known as a friction pendulum (FP) isolator. As shown in Figure 9a, the curved geometry of the isolator helps restore the structure resting on the bearing to its original position after ground motion. The hysteresis loop of the FP system is shown in Figure 9b. k represents the post elastic stiffness and k_{eff} represents the effective stiffness of the isolator. The FP system is an extensively studied system and is commonly used, mainly due to its simplicity and the relative ease of its manufacture. Though FP bearings are efficient in a wide range of earthquakes, they show large sliding displacements under earthquakes with pulse-like frequencies.

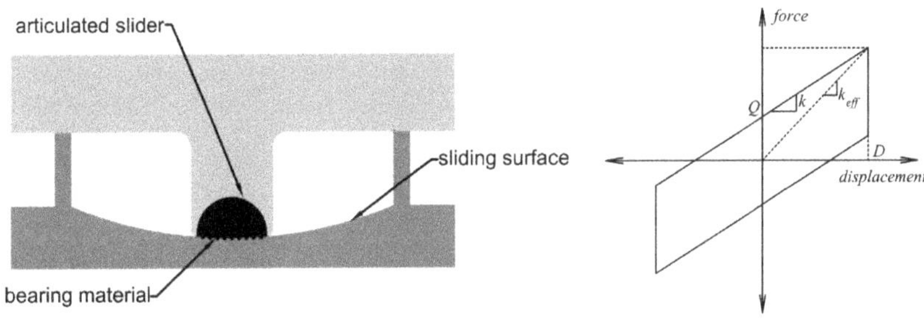

(**a**) (**b**)

Figure 9. (**a**) Typical section of a friction pendulum (FP) isolator; (**b**) hysteresis loop for FP isolator.

To address the drawback of FP bearings, researchers have proposed several solutions. Pranesh and Sinha [119] presented a bearing known as a variable frequency pendulum isolator (VFPI) that has a non-spherical (elliptical) surface. Due to the elliptical geometry, the frequency of the isolator varies along the sliding surface, avoiding the resonance issues which are observed in FP bearings. Another isolator system called the variable curvature friction pendulum system (VCFPS) was presented by Tsai et al. [120], wherein the radius of the curved surface is lengthened as the isolator displacement increases. The conical friction pendulum isolator (CFPI) is another system that has modified an FP bearing geometry [121]. Within a certain threshold region, the bearing behaves in a similar way to that of an FP bearing; however, beyond this threshold, the surface is flat and inclined. This isolator has shown a low residual displacement for near-fault earthquakes compared with FP. A further system was introduced by Krishnamoorthy [122], called a variable frequency and variable friction pendulum isolator (VFFPI), to solve the resonance issue in FP bearings. The shape of this isolator surface and its friction coefficients vary along a sliding surface. Due to the varied geometry the resonance issue is avoided in VFFPI. Further, the residual displacement of VFFPI has been found to be less than that of PF systems. Various additional isolation bearings with multiple sliding surfaces have been introduced in recent years. A detailed review of the advancements in isolation systems with sliding bearing can be seen in [19].

4.2. Base Isolated Tanks

The system of base isolation is effective not only for buildings and bridges but also for tanks. Several investigations have been conducted to assess the effectiveness of base isolation systems for tanks [123–125]. Studies conducted by the researcher Malhotra [126,127] on tanks isolated at base using elastomers show that the isolation scheme is effective when these tanks are subjected to the vertical component of ground motion. Similar investigations conducted later also found that the elastomer isolators are effective in reducing the seismic response of tanks [128–136]. In these investigations tanks were assumed to be full i.e., the liquid level in the tank remains fixed. However, in reality the weight of a tank varies due to the fluctuating liquid levels; as a result, the effectiveness of an elastomeric system varies, adversely affecting the reliability of isolation systems. Sliding isolation bearings, such as FP bearings, however, do not have the pitfalls of the elastomers as these bearings maintain a constant isolation period which is independent of a structure's weight. Thus, the liquid level in storage tanks does not affect the effectiveness of such bearings. In this regard, Wang et al. [137] proposed a procedure to evaluate the behaviour of the isolated tanks by considering FP bearings. Though the performance of FPS bearings is not affected by the level of water in the tank, due to their fixed isolation period, they do not perform satisfactorily under near-fault earthquakes as discussed in Section 4.1. Thus, in some studies, tanks isolated using a variable friction pendulum system (VFPS) were considered to investigate their performance under various near-fault ground motions [138,139]. It was concluded that the VFPS can be used effectively to control the response of fluid-filled containers under near-fault earthquakes. Similarly, many other researchers have considered different types of sliding isolation systems based on the modified geometry of FP bearings to evaluate the effect of base isolators in controlling the response of tanks subjected to ground motions [140–147]. In all of these studies, investigators found that isolators had been effective in regulating the storage tank responses.

Numerical methods have also been used recently to evaluate the behaviour of isolated tanks [81,82]. In these studies, the fluid and isolation system effects are modelled by coupling boundary and finite elements. Comparative studies conducted by Christovasilis and Whittaker [148] and Kalantari et al. [149], between FEM models and simplified mechanical models for isolated tanks, suggest that the simplified models could only be used for preliminary analysis and design but that FEM modelling is a better choice for detailed study. Krishnamoorthy [150] proposed a procedure for analysing a tank resting on FP bearings by using FEM. As shown in Figure 10 the fluid is discretized into four noded elements with one pressure degree of freedom, and the wall is modelled as frame elements with two translational degrees and one rotational degree of freedom at every node. The sliding–non sliding phases of the FP bearings are modelled with a fictitious spring as proposed by Yang et al. [151]. It was found that the FPS was efficient in considerably reducing the base shear and hydrodynamic pressure without greatly affecting the sloshing displacement. Several other studies have also been conducted to study the performance of isolated storage containers subjected to earthquakes considering various other numerical methods, interested readers may refer to [152–157].

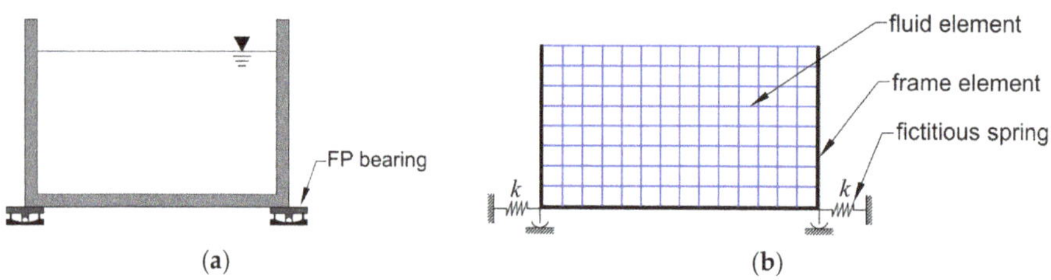

Figure 10. (**a**) Storage tank with FPS; (**b**) FE discretization of tank and liquid, as adapted from [142].

Kumar and Saha [158] investigated the effect of SSI on the seismic performance of isolated storage tanks. For this study, both ground-supported and elevated tanks are considered. Further, the effect of SSI on the effectiveness of laminated and lead rubber bearings is also discussed in this paper. The tank liquid is idealized as a three lumped mass approach [37] as discussed in Section 2.1, and the soil is discretized using finite elements. For the elevated tank, the staging is represented using a spring and dashpot arrangement. The researchers found that considering the SSI effect reduces the response of both ground-supported and elevated tanks fixed at the base. However, when isolated tanks are considered, the effect of SSI depends on the earthquake characteristics, the geometry of tank and soil flexibility.

Although the response of the tank reduces when it is isolated using elastomers or friction-based isolators, it was found that, when the isolation period is near to the sloshing period, a large sloshing of fluid is observed, leading to the failure of the tank [159]. Alternative methods, such as the use of baffles [160–162], which act as dampers and reduce the sloshing effect, have been proposed. Though flexible tank wall models are widely used, for tanks with baffles the wall flexibility has been found to be very low. Thus, several researchers have used rigid tank models in their study [160–162]. Some researchers have proposed rigid tank models resting on isolators to study the seismic behaviour of base-isolated tanks [161]. It has been found that the base isolation typically increases the sloshing height and therefore increases the damping ratios. However, studies conducted by Maleki and Ziyaeifar [163] have shown that the baffles were less effective in reducing the sloshing height in tanks with a lower aspect ratio. In recent years, several hybrid systems [159,164–167], wherein friction-based isolators or elastomers are combined with energy dissipating devices such as viscous fluid dampers or friction dampers, have been proposed and their effect on the reduction of sloshing has been investigated. These systems have been found to be effective in solving the sloshing resonance problem in isolated storage tanks. A summary of the models discussed so far is highlighted in Table 1.

4.3. Modeling Techniques of Base Isolated Structures

A discussion on various modelling techniques used for structures isolated at base is carried out in this section. Though the modelling techniques discussed in this section are developed for building structures, they also apply to tank structures. In one of the earliest studies on sliding friction, Hartog [168], developed a solution to the response of a sliding SDOF system with Coulomb damping at the interfaces between the base and structure. The proposed model was for an undamped system represented as a spring mass system. A similar study conducted by Hundal [169] provided a solution to an SDOF system which consists of viscous and Coulomb damping (Figure 11).

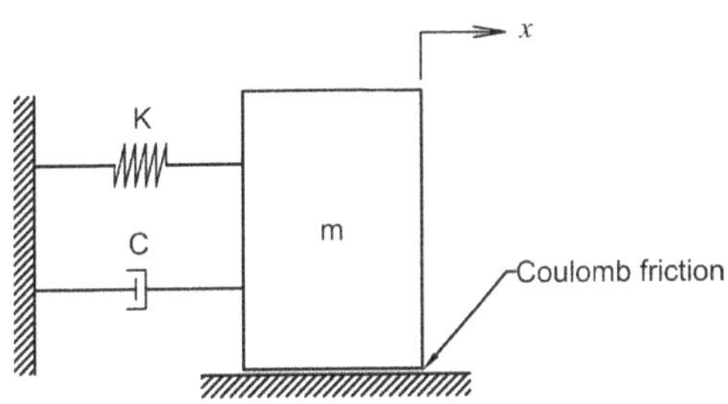

Figure 11. SDOF system with a spring, viscous damping and Coulomb friction [150].

A solution to the response of a 2DOF system with sliding surface was proposed by Qamaruddin et al. [170]. As shown in Figure 12, the structure is modelled as a spring–dashpot system with top mass (m_1) and a base mass (m_2). A constant friction coefficient is assumed between the sliding surfaces throughout the motion.

Yang et al. [171] proposed a base-isolated 2DOF model similar to the model of Qamaruddin et al. [170]. However, here the isolation system comprises lead rubber bearings instead of friction bearings (Figure 13).

Though several investigations have been conducted on isolated systems subjected to harmonic excitation [172–176], and random or seismic excitation [177–181], these studies idealize the structure as a 2DOF system. Yang et al. [151] presented a method to evaluate the response of a multi-degree-of-freedom system (MDOF) (Figure 14). The sliding device is idealized considering a fictitious spring which has no stiffness during the sliding stage and a large value during the static stage.

Table 1. Summary of the tank models discussed in the present article.

Reference	Type of Tank	Tank Wall Type (Rigid/Flexible)	SSI Effect Considered	Isolated Tank	Tank Model Description	Characteristics of Elements	Type of Dynamic Excitation	Type of Analysis
[7,26]	Ground supported, elevated	rigid	×	×	Two mass system	-	Seismic	-
[31]	Ground supported	flexible	×	×	SDOF system	-	Seismic	-
[34,35]	Ground supported	flexible	×	×	Three mass system derived from FE analysis	-	Seismic	-
[41]	Ground supported	flexible	×	×	Spring–dashpot system with impulsive and convective masses of fluid considering	-	Seismic	Linear
[52,53]	Ground supported	flexible	×	×	Liquid–tank system is modelled using added mass technique	-	Seismic	Linear, nonlinear
[54]	Ground supported	flexible	×	×	Liquid–tank system is modelled using added mass technique	-	Seismic	Nonlinear
[55]	Ground supported	flexible	×	×	Liquid–tank system is modelled using added mass technique	-	Seismic	Linear, nonlinear
[56]	Ground supported	flexible	×	×	Liquid–tank system is modelled using added mass technique	-	Seismic	Nonlinear
[64,65]	Ground supported	flexible	×	×	FE discretization with Lagrangian method for FSI effect	Linear	Seismic	Linear static and linear dynamic
[66]	Ground supported	flexible	×	×	FE discretization with Lagrangian method for FSI effect	Linear, four noded element with 3 DOF at all nodes	Seismic	Nonlinear
[71,72]	Ground supported	flexible	×	×	FE discretization with Eulerian approach for FSI effect	Two noded ring element with four DOF at each node for tank wall. Eight noded rectangular element for fluid	Seismic, harmonic	Linear

Table 1. *Cont.*

Reference	Type of Tank	Tank Wall Type (Rigid/Flexible)	SSI Effect Considered	Isolated Tank	Tank Model Description	Characteristics of Elements	Type of Dynamic Excitation	Type of Analysis
[73]	Ground supported	flexible	×	×	FE discretization with Eulerian approach for FSI effect	Eight-noded isoparametric element	Seismic	Linear
[78–81]	Ground supported	flexible	×	×	Coupled BE and FE discretization of tank–fluid system	Shell element for tank, BE for fluid	Seismic	Linear
[82]	Ground supported	flexible	×	✓	Coupled BE and FE discretization of tank–fluid system	Shell element for tank, linear boundary element for fluid, Bilinear isolator	Seismic	Nonlinear
[100]	Ground supported	flexible	✓	×	Two-DOF system model. Effect of SSI is modelled using spring–dashpot system	-	Harmonic	Linear
[103]	Ground supported	flexible	✓	×	Two-DOF linear mass–spring–dashpot system	-	Seismic	Linear
[104]	Ground supported	flexible	✓	×	SDOF linear mass–spring–dashpot for tank–fluid system which is attached to soil using two springs and two dampers	-	Harmonic, seismic	Linear
[105]	Ground supported	flexible	✓	×	Fluid is idealized as three mass system. Foundation and soil system is modelled using second-order lumped parameter model	-	Seismic	-

Table 1. *Cont.*

Reference	Type of Tank	Tank Wall Type (Rigid/Flexible)	SSI Effect Considered	Isolated Tank	Tank Model Description	Characteristics of Elements	Type of Dynamic Excitation	Type of Analysis
[108]	Ground supported	flexible	✓	✓	Coupled BE and FE discretization of tank–fluid system. Soil is represented by spring–dashpot system	Nine-noded shell element with 5 DOF at each node for the structure, BE for fluid, nonlinear base isolator	Seismic	Nonlinear
[109]	Ground supported	flexible	✓	×	FE discretization with Lagrangian method for FSI effect	Four-noded shell element with 6 DOF at each node.	Seismic	Linear
[110,111]	Ground supported	flexible	✓	×	FE discretization with Eulerian approach for FSI effect	Two-noded frame element with 3 DOF per node for tank wall and base, four-noded element with 2 DOF per node for foundation, four-noded element with SDOF for fluid.	Seismic	Linear
[126,127]	Ground supported	flexible	×	✓	Two-DOF system to model liquid–tank system isolated at base using elastomers	Nonlinear rubber bearing	Seismic	-
[137]	Ground supported	flexible	×	✓	MDOF system to model liquid–tank system isolated with FP isolator at base	-	Seismic	Nonlinear
[138,139]	Ground supported	flexible	×	✓	Three mass system to model liquid–tank system isolated at base using VFPS	-	Seismic	Linear
[141]	Ground supported	flexible	×	✓	Three mass system to model liquid–tank system isolated at base using double VFPS	-	Seismic	Nonlinear

Table 1. *Cont.*

Reference	Type of Tank	Tank Wall Type (Rigid/Flexible)	SSI Effect Considered	Isolated Tank	Tank Model Description	Characteristics of Elements	Type of Dynamic Excitation	Type of Analysis
[142]	Ground supported	flexible	×	✓	Three mass system to model liquid–tank system isolated at base using multiple FPS	-	Seismic	-
[150]	Ground supported	flexible	×	✓	FE discretization of tank–liquid system with FP isolator at base.	Two-noded frame element with 3 DOF per node for tank wall and base, four-noded element with SDOF for fluid.	Seismic	Linear
[158]	Ground supported, elevated	flexible	✓	✓	Three mass system to model liquid–tank system isolated at base using laminated and lead rubber bearings modelled using spring–dashpot system.	Linear and nonlinear isolation bearings	Seismic	-
[159]	Ground supported	rigid	×	✓	Three mass system to model liquid-tank system with hybrid isolation system.	-	Seismic	Linear
[160]	Ground supported	rigid	×	✓	BE discretization of isolated tank. Baffles are provided to reduce sloshing effect	Constant boundary elements	Seismic	Linear
[161–163]	Ground supported	rigid	×	✓	Lumped mass model for the isolated tank. Baffles are provided to reduce sloshing effect.	-	Seismic	Linear
[165–167]	Ground supported	Flexible, rigid	×	✓	Three mass system to model liquid-tank system with hybrid isolation system.	-	Seismic	Linear

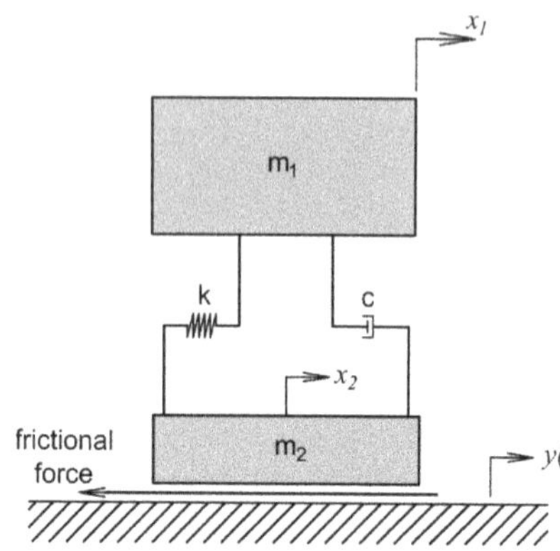

Figure 12. A 2DOF system with Coulomb friction, as adapted from [151].

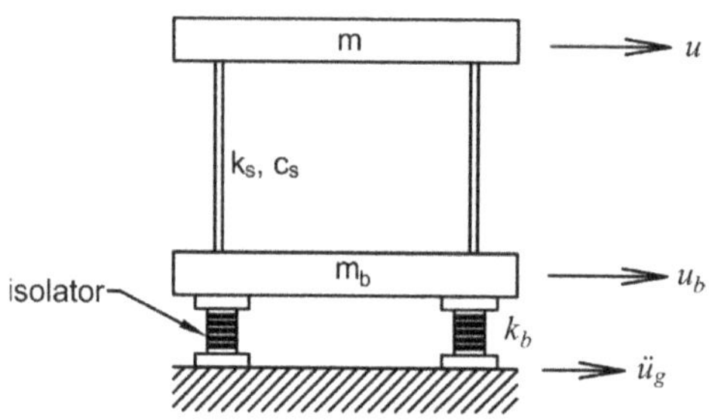

Figure 13. A 2DOF system with LRB, as adapted from [152].

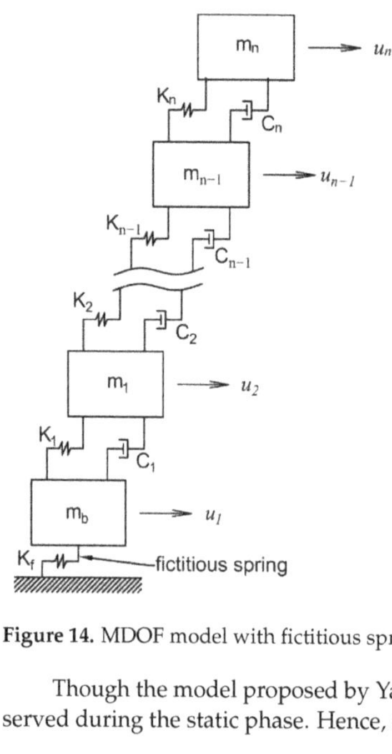

Figure 14. MDOF model with fictitious spring, as adapted from [163].

Though the model proposed by Yang et al. [151] is simple, a very low velocity was observed during the static phase. Hence, a modified model was proposed by Vafai et al. [182]

which assumed the stiffness during the static phase as infinity. Accordingly, these authors replaced the fictitious spring with a rigid plastic link (Figure 15).

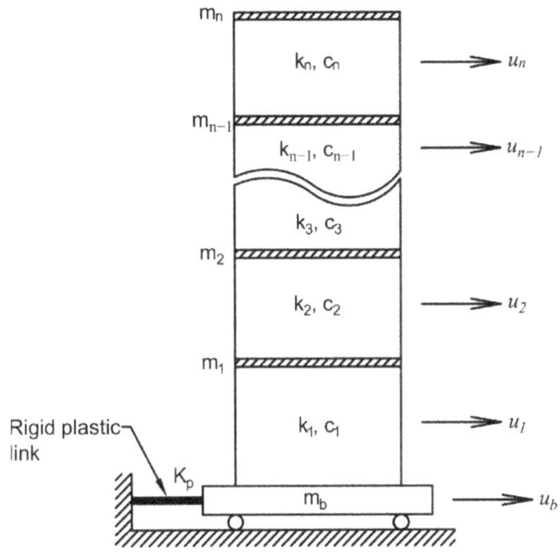

Figure 15. MDOF model with the rigid plastic link, as adapted from [164].

5. Forces and Factors Influencing the Seismic Behaviour of Tanks

In the past, researchers have observed many types of tank failures during earthquakes; some of the important failures that have been observed in steel tanks are "elephant foot" failure and diamond-shaped buckling [22,158]. Elephant footing failure is a result of the combined action of the axial force and the hoop tension reaching their critical stress values. In "elephant footing", walls fail by an elastoplastic buckling failure mode. In some cases, "diamond-shaped" failures in tank walls have been observed. This failure is a result of effects induced by the vertical acceleration of earthquakes on storage fluid. The fluid inside the tank provides a stabilizing effect against the axial buckling of tank walls. However, the vertical acceleration generated by an earthquake destabilizes the fluid and tank wall equilibrium, resulting in local buckling.

The typical failures observed in reinforced concrete (RC) tanks are different from the failure modes of steel tanks. As the walls of steel tanks are slender, they are prone to buckling failure. However, in the case of RC tanks the walls are quite thick and hence buckling failures are rare. The failures of RC tanks are mainly due to the bending and shear failure of staging beams, cracking of the connections, failure of shafts, and torsion failures of the staging system [183,184]. The toppling and sliding failures are global failures which are common in both steel and RC tanks. The failures of tanks are dependent on various parameters, such as the aspect ratio of tanks, types of supports, characteristics of seismic excitation and size of tanks. To avoid failures of tanks, it is therefore essential to identify the major factors which influence tank failures. In this regard, a brief discussion on the various parameters influencing a tank's behaviour is carried out in this section. The following are the factors which influence the seismic behaviour of tanks.

5.1. *Effect of Mass*

The mass of the fluid has a direct effect on the dynamic behaviour of tanks. As discussed in Section 2, the fluid in the tanks is idealized into different masses based on their contribution to hydrodynamic pressures. Further, the mass of the fluid in the tank varies due to the fluctuating fluid levels during its usage. The liquid in the tank may vary from empty to full with a partially filled condition in between. This effect of fluid fluctuation needs to be accounted for in the tank analysis for a realistic analysis. A tank's full and empty conditions can be easily modelled by idealizing the tank as SDOF, however modelling tanks for a partially full condition is quite challenging. Moreover, the partial full condition is the most realistic case in an actual scenario. Therefore, most of the researchers use partially full conditions in their study to account for the sloshing, convective and impulsive effects.

5.2. Effect of Base Shear

Base shear is the maximum horizontal force developed at the tank bottom during earthquake forces. If the base shear exceeds the frictional force at the base, there can be slippage between tanks and the foundation. The base shears should be evaluated accurately to ensure the safe performance of buildings. Many researchers have used the base shear as a basis for the evaluation of tank behaviour under seismic excitation; as an example, Haroun and Abou-Izzeddine [185] studied the effect of SSI on the base shear of tanks. Engineers try to reduce the base shear for the safe design of buildings by either using the fundamental structural analysis principles or using some innovative techniques. Base isolation systems, especially sliding systems, have been found to be very effective in reducing base shear. Many countries have recognized the importance of estimating the base shear and include simplified formulae for its calculation [186,187].

5.3. Effect of Overturning Moment

It has been found that tall tanks are much more prone to overturning than short tanks during a seismic event [158]. It is, therefore, essential to ensure that the tanks provide sufficient resistance to the rotation about the vertical axis. This resistance is measured in terms of base moments. The base overturning moments are induced due to the combined effects of sloshing, part of the liquid moving with the tank wall and fluid–wall interactions. It has been found that the impulsive portion of the liquid contributes significantly to the overturning moments [188]. The estimation of overturning moments is essential for the design of the connections of the anchored tanks. Furthermore, overturning forces lead to uplifting of the tanks, damaging the pipe connections. It has been found that the overturning moments are influenced by SSI effects and are reduced when the effect is considered in the analysis [158]. Liquid sloshing effects can impart significant damage to the freeboard section of the tank wall, and also have a tendency to amplify the overturning moment.

5.4. Effect of Stiffness

As discussed in Section 2, the tank walls had previously been assumed to be rigid in their analysis. However, investigations have shown that it is essential to consider the tank walls as flexible. The assumption of rigid walls in tanks led to an underestimation of seismic forces. This assumption led to the failure of many tanks in the past. Therefore, in all the recent studies the flexibility of the tank walls is taken into account. It has been found that higher stiffness results in reduced roof responses. At the same time, this additional stiffness leads to higher base shear demand.

6. Conclusions

This paper briefly discusses the historical development of various tank modelling techniques, various factors influencing the modelling techniques, and the benefits and limitations of different methods. Different modelling approaches of support conditions and their effect on the responses of tanks are also discussed. Further, this paper focuses on the brief history of different base isolation systems, modelling approaches, and their application in storage tanks. Based on the discussions so far, the main conclusions are:

(1) Analysis of liquid storage containers differs from that of typical structures such as buildings and bridges due to the additional interaction effects between the tank walls and liquid. Parameters such as the properties of the liquid, the shape of the tank and the flexibility of the tank walls, the soil properties, and the type of support condition, determine the seismic performance of tanks.

(2) In the past, many tanks had been analysed and designed with the assumption that their walls were rigid. However, the failure of tanks that had been designed on the basis of these analyses to withstand earthquakes motivated researchers to take the flexibility of the tank wall into consideration when analysing and designing future tanks.
(3) Many researchers have proposed various simplified models of tank–fluid systems to study the FSI effect on the performance of storage containers. However, most of these methods are limited to circular or rectangular tanks.
(4) The development of numerical methods, such as the added mass approach and the finite and boundary element methods, have enabled the analysis of tanks of irregular shape without significant loss in the accuracy of results.
(5) Along with FSI, the response of tanks also depends on the flexibility of the tank walls and on the soil. Thus, researchers have proposed several methods to model the SSI effects to evaluate the seismic performance of storage tanks subjected to ground motions.
(6) Though the simplified methods proposed using the mass–spring–dashpot system effectively represent the SSI effects, they have been found to produce overly conservative results. Hence, numerical methods can be used to model the soil–tank system for realistic responses.
(7) The base isolation technique is an innovative and practical technique to produce an earthquake-resistant tank.
(8) Base isolation systems generally tend to reduce the base shear demand when compared with the base shear response of fixed base tanks.
(9) Base isolation systems have also been found to reduce the top displacement demand as the tank moves as a rigid unit due to the high flexibility of the isolators in the horizontal direction.
(10) The SSI effect has a significant influence on the behaviour of base-isolated tanks. However, depending on the characteristics of the earthquake, the SSI effect could be either beneficial or detrimental to the tank response [158].
(11) Several studies have shown the effectiveness of elastomeric bearings for tanks with fixed liquid storage levels. However, some studies have shown that the effectiveness of the elastomer isolation system reduces with the fluctuation in the liquid levels inside tanks.
(12) Some researchers have favoured FP bearings over elastomer bearings as the period of isolation of FP bearings is independent of the structure's weight and stiffness.
(13) The resonance problem of FPS under pulse-like vibrations led investigators to develop alternative isolators with various geometrical surfaces, such as VFPI, VCFPS, VRFPS, CFPI, and VFFPI, which have non-fixed isolation periods.
(14) Analysis of isolated structures is quite challenging compared with conventional fixed base structures due to the presence of static and sliding phases. In this regard, models such as fictitious springs and rigid plastic links have been developed.
(15) For the analysis of isolated tanks due considerations are to be given to the combined effects of FSI and SSI. Thus, numerical methods have been found to be more appropriate for the accurate analysis of isolated tanks.
(16) Though base isolation systems are effective in reducing the seismic demand of storage tanks, it has been observed that the sloshing period has a tendency to match the isolation period, leading to the sloshing resonance problem.
(17) Several alternative techniques can be used to solve the sloshing resonance issue, such as using baffles which act as dampers or incorporating hybrid systems with a damper–isolator combination.

Author Contributions: Conceptualization, M.C., A.K. and A.R.A.; resources, M.C.; writing—original draft preparation, M.C.; writing—review and editing, A.K. and A.R.A.; visualization, M.C.; supervision, A.K.; project administration, A.K.; All authors have read and agreed to the published version of the manuscript.

Funding: This research received no external funding. The APC was funded by Manipal Academy of Higher Education, Manipal.

Institutional Review Board Statement: Not applicable.

Informed Consent Statement: Not applicable.

Data Availability Statement: No new data were created or analysed in this study. Data sharing is not applicable to this article.

Conflicts of Interest: The authors declare no conflict of interest.

References

1. THE 17 GOALS | Sustainable Development. Available online: https://sdgs.un.org/goals (accessed on 17 June 2023).
2. Haroun, M.A. *Dynamic Analyses of Liquid Storage Tanks*; Report No. EERL 80–4; Earthquake Engineering Research Laboratory, California Institute of Technology: Pasadena, CA, USA, 1980.
3. Razzaghi, M.S.; Eshghi, S. Behavior of Steel Oil Tanks Due to Near-Fault Ground Motion. In Proceedings of the 13th World Conference on Earthquake Engineering, Vancouver, BC, Canada, 1–6 August 2004.
4. Yazici, G.; Cili, F. Evaluation of the Liquid Storage Tank Failures in the 1999 Kocaeli Earthquake. In Proceedings of the 14th World Conference on Earthquake Engineering, Beijing, China, 12–17 October 2008.
5. Ibata, T.; Nakachi, I.; Ishida, K.; Yokozawa, J. Damage to Storage Tanks Caused by the 2011 Tohoku Earthquake and Tsunami and Proposal for Structural Assessment Method for Cylindrical Storage Tanks. In Proceedings of the 17th International Conference & Exhibition on Liquefied Natural Gas (LNG17), Houston, TX, USA, 16–19 April 2013.
6. Niwa, A.; Clough, R.W. Buckling of Cylindrical Liquid-Storage Tanks under Earthquake Loading. *Earthq. Eng. Struct. Dyn.* **1982**, *10*, 107–122. [CrossRef]
7. Housner, G.W. The Dynamic Behavior of Water Tanks. *Bull. Seismol. Soc. Am.* **1963**, *53*, 381–387. [CrossRef]
8. Edwards, N.W. *A Procedure for the Dynamic Analysis of Thin Walled Cylindrical Liquid Storage Tanks Subjected to Lateral Ground Motions*; University of Michigan: Ann Arbor, MI, USA, 1969.
9. Yasuo, S.; Yamaguchi, R. Vibration of a Building upon an Elastic Foundation. *Bull. Earthq. Res. Inst. Tokyo Univ.* **1957**, *35*, 545–565. [CrossRef]
10. Parmelee, R.A. Building-Foundation Interaction Effects. *J. Eng. Mech. Div.* **1967**, *93*, 131–152. [CrossRef]
11. Sarrazin-Arellano, M. *Soil-Structure Interaction in Earthquake Resistant Design*; Department of Civil Engineering, School of Engineering, Massachusetts Institute of Technology: Cambridge, MA, USA, 1970.
12. Scavuzzo, R.J.; Bailey, J.L.; Raftopoulos, D.D. Lateral Structure Interaction with Seismic Waves. *J. Appl. Mech.* **1971**, *38*, 125–134. [CrossRef]
13. Wolf, J.P. *Dynamic Soil-Structure Interaction*; Prentice-Hall: Englewood Cliffs, NJ, USA, 1985; ISBN 9780471486824/0471486825.
14. Panchal, V.R.; Soni, D.P. Seismic Behaviour of Isolated Fluid Storage Tanks: A-State-of-the-Art Review. *KSCE J. Civil. Eng.* **2014**, *18*, 1097–1104. [CrossRef]
15. Kelly, J.M. Aseismic Base Isolation: Review and Bibliography. *Soil. Dyn. Earthq. Eng.* **1986**, *5*, 202–216. [CrossRef]
16. Buckle, I.G.; Mayes, R.L. Seismic Isolation: History, Application, and Performance—A World View. *Earthq. Spectra* **1990**, *6*, 161–201. [CrossRef]
17. Jangid, R.S.; Datta, T.K. Seismic Behaviour of Base-Isolated Buildings: A State-of-the Art Review. *Proc. Inst. Civil. Eng. Struct. Build.* **1995**, *110*, 186–203. [CrossRef]
18. Girish, M.; Pranesh, M. Sliding Isolation Systems: State-of-the-Art Review. In Proceedings of the Second International Conference on Emerging Trends in Engineering (SICETE), Nagpur, India, 16–18 December 2013; pp. 30–35.
19. Calvi, P.M.; Calvi, G.M. Historical Development of Friction-Based Seismic Isolation Systems. *Soil. Dyn. Earthq. Eng.* **2018**, *106*, 14–30. [CrossRef]
20. De Luca, A.; Guidi, L.G. State of Art in the Worldwide Evolution of Base Isolation Design. *Soil. Dyn. Earthq. Eng.* **2019**, *125*, 105722. [CrossRef]
21. Avinash, A.R.; Krishnamoorthy, A.; Kamath, K.; Chaithra, M. Sliding Isolation Systems: Historical Review, Modeling Techniques, and the Contemporary Trends. *Buildings* **2022**, *12*, 1997. [CrossRef]
22. Rammerstorfer, F.G.; Scharf, K.; Fisher, F.D. Storage Tanks under Earthquake Loading. *Appl. Mech. Rev.* **1990**, *43*, 261–282. [CrossRef]
23. Saha, S.K.; Matsagar, V.A.; Jain, A.K. Reviewing Dynamic Analysis of Base-Isolated Cylindrical Liquid Storage Tanks under near-Fault Earthquakes. *IES J. Part. A Civil. Struct. Eng.* **2015**, *8*, 41–61. [CrossRef]
24. Westergaard, H.M. Water Pressures on Dams during Earthquakes. *Trans. Am. Soc. Civil. Eng.* **1933**, *98*, 418–433. [CrossRef]
25. Hoskins, L.M.; Jacobsen, L.S. Water Pressure in a Tank Caused by a Simulated Earthquake*. *Bull. Seismol. Soc. Am.* **1934**, *24*, 1–32. [CrossRef]
26. Housner, G.W. Dynamic Pressures on Accelerated Fluid Containers. *Bull. Seismol. Soc. Am.* **1957**, *47*, 15–35. [CrossRef]
27. Biot, M.A. Analytical And Experimental Methods in Engineering Seismology. *Trans. Am. Soc. Civil. Eng.* **1943**, *108*, 365–385. [CrossRef]

28. Housner, G.W.; Martel, R.R.; Alford, J.L. Spectrum Analysis of Strong-Motion Earthquakes*. *Bull. Seismol. Soc. Am.* **1953**, *43*, 97–119. [CrossRef]
29. United States Atomic Energy Commission. *Others Nuclear Reactors and Earthquakes*; United States Atomic Energy Commission: Washington, DC, USA, 1963.
30. Wozniak, R.S.; Mitchell, W.W. *Basis of Seismic Design Provisions for Welded Steel Oil Storage Tanks*; Chicago Bridge & Iron Company: Toronto, ON, Canada, 1978.
31. Veletsos, A.S. Seismic Effects in Flexible Liquid Storage Tanks. In Proceedings of the 5th World Conference on Earthquake Engineering, Rome, Italy, 1974; Volume 1, pp. 630–639.
32. Chopra, A.K. Reservoir-Dam Interaction during Earthquakes. *Bull. Seismol. Soc. Am.* **1967**, *57*, 675–687. [CrossRef]
33. Chopra, A.K. Earthquake Behavior of Reservoir-Dam Systems. *J. Eng. Mech. Div.* **1968**, *94*, 1475–1500. [CrossRef]
34. Haroun, M.A.; Housner, G.W. Seismic Design of Liquid Storage Tanks. *J. Tech. Counc. ASCE* **1981**, *107*, 191–207. [CrossRef]
35. Haroun, M.A.; Housner, G.W. Earthquake Response of Deformable Liquid Storage Tanks. *J. Appl. Mech.* **1981**, *48*, 411–418. [CrossRef]
36. Haroun, M.A.; Tayel, M.A. Dynamic Behavior of Cylindrical Liquid Storage Tanks under Vertical Earthquake Excitation. In Proceedings of the 8th World Conference on Earthquake Engineering, San Francisco, CA, USA, 21–28 July 1984; pp. 421–428.
37. Haroun, M.A.; Ellaithy, H.M. Seismically Induced Fluid Forces on Elevated Tanks. *J. Tech. Top. Civil. Eng.* **1985**, *111*, 1–15. [CrossRef]
38. Haroun, M.A.; Tayel, M.A. Axisymmetrical Vibrations, of Tanks—Analytical. *J. Eng. Mech.* **1985**, *111*, 346–358. [CrossRef]
39. Haroun, M.A.; Tayel, M.A. Axisymmetrical Vibrations of Tanks—Numerical. *J. Eng. Mech.* **1985**, *111*, 329–345. [CrossRef]
40. Haroun, M.A.; Tayel, M.A. Response of Tanks to Vertical Seismic Excitations. *Earthq. Eng. Struct. Dyn.* **1985**, *13*, 583–595. [CrossRef]
41. Malhotra, P.K.; Wenk, T.; Wieland, M. Simple Procedure for Seismic Analysis of Liquid-Storage Tanks. *Struct. Eng. Int.* **2000**, *10*, 197–201. [CrossRef]
42. Veletsos, A.S.; Yang, J. Earthquake Response of Liquid-Storage Tanks. In Proceedings of the Second Annual Engineering Mechanics Division Specialty Conference, North Carolina, CL, USA, 23–25 May 1977; pp. 1–24.
43. Veletsos, A.S. Seismic Response and Design of Liquid Storage Tanks. *Guidel. Seism. Des. Oil Gas Pipeline Syst.* **1984**, *102*, 255–370.
44. Veletsos, A.S.; Tang, Y. Soil-Structure Interaction Effects for Laterally Excited Liquid Storage Tanks. *Earthq. Eng. Struct. Dyn.* **1990**, *19*, 473–496. [CrossRef]
45. Tsipianitis, A.; Tsompanakis, Y.; Psarropoulos, P.N. Impact of Dynamic Soil–Structure Interaction on the Response of Liquid-Storage Tanks. *Front. Built Environ.* **2020**, *6*, 140. [CrossRef]
46. Ibrahim, R.A.; Pilipchuk, V.N.; Ikeda, T. Recent Advances in Liquid Sloshing Dynamics. *Appl. Mech. Rev.* **2001**, *54*, 133–199. [CrossRef]
47. Hsiung, H.C.H. *Dynamic Analysis of Hydroelastic Systems Using the Finite-Element Method*; University of Southern California: Los Angeles, CA, USA, 1973.
48. Shaaban, S.H.; Nash, W.A. *Finite Element Analysis of a Seismically Excited Cylindrical Storage Tank, Ground Supported, and Partially Filled with Liquid*; University of Massachusetts: Amherst, MA, USA, 1975.
49. Balendra, T.; Nash, W.A. *Earthquake Analysis of a Cylindrical Liquid Storage Tank with a Dome by Finite Element Method*; University of Massachusetts: Amherst, MA, USA, 1978.
50. Livaoğlu, R.; Doğangün, A. Simplified Seismic Analysis Procedures for Elevated Tanks Considering Fluid–Structure–Soil Interaction. *J. Fluids Struct.* **2006**, *22*, 421–439. [CrossRef]
51. Wilson, E.L.; Khalvati, M. Finite Elements for the Dynamic Analysis of Fluid-Solid Systems. *Int. J. Numer. Methods Eng.* **1983**, *19*, 1657–1668. [CrossRef]
52. Haroun, M.A.; Housner, G.W. Dynamic Characteristics of Liquid Storage Tanks. *J. Eng. Mech. Div.* **1982**, *108*, 783–800. [CrossRef]
53. Haroun, M.A.; Housner, G.W. Complications in Free Vibration Analysis of Tanks. *J. Eng. Mech. Div.* **1982**, *108*, 801–818. [CrossRef]
54. Barton, D.C.; Parker, J.V. Finite Element Analysis of the Seismic Response of Anchored and Unanchored Liquid Storage Tanks. *Earthq. Eng. Struct. Dyn.* **1987**, *15*, 299–322. [CrossRef]
55. Zeiny, A. Simplified Modeling of Liquid-Structure Interaction in the Seismic Analysis of Cylindrical Liquid Storage Tanks. In Proceedings of the 13th World Conference on Earthquake Engineering, Vancouver, BC, Canada, 1–6 August 2004; p. 1914.
56. Zhuang, Z.; Liu, H.; Li, Q.; Yamaguchi, S.; Toyoda, M. Development of a New Added Mass Model and Dynamic Analysis for Cylindrical Tank. In Proceedings of the ASME 2006 Pressure Vessels and Piping/ICPVT-11 Conference. Volume 8: Seismic Engineering, Vancouver, BC, Canada, July 2006; Volume 2006, pp. 239–248.
57. Shantaram, D.; Owen, D.R.J.; Zienkiewicz, O.C. Dynamic Transient Behaviour of Two- and Three-Dimensional Structures Including Plasticity, Large Deformation Effects and Fluid Interaction. *Earthq. Eng. Struct. Dyn.* **1976**, *4*, 561–578. [CrossRef]
58. Wilson, E.L. *Finite Elements for Foundations, Joints and Fluids*; Gudehus, G., Ed.; John Wiley & Sons Ltd.: Chichester, UK, 1977; pp. 319–350.
59. Olson, L.G.; Bathe, K.-J. A Study of Displacement-Based Fluid Finite Elements for Calculating Frequencies of Fluid and Fluid-Structure Systems. *Nucl. Eng. Des.* **1983**, *76*, 137–151. [CrossRef]
60. Chopra, A.K.; Wilson, E.L.; Farhoomand, I. Earthquake Analysis of Reservoir-Dam Systems. In Proceedings of the 4th World Conference on Earthquake Engineering, Santiago, Chile, 13–18 January 1969; pp. 1–10.

61. Zienkiewicz, O.C.; Bettess, P. Fluid-Structure Dynamic Interaction and Wave Forces. An Introduction to Numerical Treatment. *Int. J. Numer. Methods Eng.* **1978**, *13*, 1–16. [CrossRef]
62. Belytschko, T. Methods and Programs for Analysis of Fluid-Structure Systems. *Nucl. Eng. Des.* **1977**, *42*, 41–52. [CrossRef]
63. Bathe, K.-J.; Sonnad, V. On Effective Implicit Time Integration in Analysis of Fluid-Structure Problems. *Int. J. Numer. Methods Eng.* **1980**, *15*, 943–948. [CrossRef]
64. Doğangün, A.; Durmuş, A.; Ayvaz, Y. Static and Dynamic Analysis of Rectangular Tanks by Using the Lagrangian Fluid Finite Element. *Comput. Struct.* **1996**, *59*, 547–552. [CrossRef]
65. Dogangun, A.; Livaoglu, R. Hydrodynamic Pressures Acting on the Walls of Rectangular Fluid Containers. *Struct. Eng. Mech.* **2004**, *17*, 203–214. [CrossRef]
66. Farzin, A. *Nonlinear Dynamic Response of Seismically Excited Rectangular Concrete Liquid Filled Tanks*; Ryerson University Toronto: Toronto, ON, Canada, 2018.
67. Xu, Z.; Han, Z.; Qu, H. Comparison between Lagrangian and Eulerian Approaches for Prediction of Particle Deposition in Turbulent Flows. *Powder Technol.* **2020**, *360*, 141–150. [CrossRef]
68. Azarpira, M.; Zarrati, A.; Farrokhzad, P. Comparison between the Lagrangian and Eulerian Approach in Simulation of Free Surface Air-Core Vortices. *Water* **2021**, *13*, 726. [CrossRef]
69. Zienkiewicz, O.C.; Taylor, R.L. *The Finite Element Method; Volume 2: Solid. and Fluid. Mechanics Dynamics and Non-Linearity*; Wiley: New York, NY, USA, 1991.
70. MacNeal, R.H.; Citerley, R.; Chargin, M. A New Method for Analyzing Fluid-Structure Interaction Using MSC-NASTRAN. In Proceedings of the 5th International Conference on Structural Mechanics in Reactor Technology, Berlin, Germany, 13–17 August 1979; Paper B4/9. pp. 1–9.
71. Balendra, T.; Ang, K.K.; Paramasivam, P.; Lee, S.L. Free Vibration Analysis of Cylindrical Liquid Storage Tanks. *Int. J. Mech. Sci.* **1982**, *24*, 47–59. [CrossRef]
72. Babu, S.S.; Bhattacharyya, S.K. Finite Element Analysis of Fluid-Structure Interaction Effect on Liquid Retaining Structures due to Sloshing. *Comput. Struct.* **1996**, *59*, 1165–1171. [CrossRef]
73. Aslam, M. Finite Element Analysis of Earthquake-Induced Sloshing in Axisymmetric Tanks. *Int. J. Numer. Methods Eng.* **1981**, *17*, 159–170. [CrossRef]
74. Brebbia, C.A.; Dominguez, J. *Boundary Elements: An Introductory Course*; WIT Press: Southampton, UK; Computational Mechanics Publications: Southampton, UK, 1992; Volume 58, ISBN 978-1-85312-349-8.
75. Jablonski, A.M.; Humar, J.L. Three-Dimensional Boundary Element Reservoir Model for Seismic Analysis of Arch and Gravity Dams. *Earthq. Eng. Struct. Dyn.* **1990**, *19*, 359–376. [CrossRef]
76. Wepf, D.H.; Wolf, J.P.; Bachmann, H. Hydrodynamic-Stiffness Matrix Based on Boundary Elements for Time-Domain Dam-Reservoir-Soil Analysis. *Earthq. Eng. Struct. Dyn.* **1988**, *16*, 417–432. [CrossRef]
77. Humar, J.L.; Jablonski, A.M. Boundary Element Reservoir Model for Seismic Analysis of Gravity Dams. *Earthq. Eng. Struct. Dyn.* **1988**, *16*, 1129–1156. [CrossRef]
78. Park, J.-H.; Koh, H.M.; Kim, J. Fluid-Structure Interaction Analysis by a Coupled Boundary Element-Finite Element Method in Time Domain. In *Boundary Element Technology VII*; Springer Netherlands: Dordrecht, The Netherlands, 1992; pp. 227–243.
79. Lay, K.S. Seismic Coupled Modeling of Axisymmetric Tanks Containing Liquid. *J. Eng. Mech.* **1993**, *119*, 1747–1761. [CrossRef]
80. Koh, H.M.; Kim, J.K.; Park, J.-H. Fluid-Structure Interaction Analysis of 3-D Rectangular Tanks by a Variationally Coupled BEM-FEM and Comparison with Test Results. *Earthq. Eng. Struct. Dyn.* **1998**, *27*, 109–124. [CrossRef]
81. Kim, M.K.; Lim, Y.M.; Cho, S.Y.; Cho, K.H.; Lee, K.W. Seismic Analysis of Base-Isolated Liquid Storage Tanks Using the BE–FE–BE Coupling Technique. *Soil. Dyn. Earthq. Eng.* **2002**, *22*, 1151–1158. [CrossRef]
82. Shekari, M.R.; Khaji, N.; Ahmadi, M.T. A Coupled BE–FE Study for Evaluation of Seismically Isolated Cylindrical Liquid Storage Tanks Considering Fluid–Structure Interaction. *J. Fluids Struct.* **2009**, *25*, 567–585. [CrossRef]
83. Martel, R.R. Effect of Foundation on Earthquake Motion. *Civil. Eng. ASCE* **1940**, *10*, 7–10.
84. Singhal, A.; Tahiani, C. Influence of Soil-Structure Interaction on the Earthquake Response of Buildings. *Can. Geotech. J.* **1969**, *6*, 177–195. [CrossRef]
85. Ishizaki, H.; Hatakeyama, N. Experimental and Numerical Studies on Vibrations of Buildings. In Proceedings of the Second World Conference on Earthquake Engineering, Tokyo and Kyoto, Japan, 11–18 July 1960; pp. 1263–1284.
86. Merritt, R.G.; Housner, G.W. Effect of Foundation Compliance on Earthquake Stresses in Multistory Buildings. *Bull. Seismol. Soc. Am.* **1954**, *44*, 551–569. [CrossRef]
87. Lycan, D.L.; Newmark, N.M. Effect of Structure and Foundation Interaction. *J. Eng. Mech. Div.* **1961**, *87*, 1–31. [CrossRef]
88. Bielak, J. *Earthquake Response of Building-Foundation Systems*; Report EERL 71–04; Earthquake Engineering Research Laboratory, California Institute of Technology: Pasadena, CA, USA, 1971; Volume 71.
89. Meek, J.W.; Veletsos, A.S. *Dynamic Analysis and Behavior of Structure-Foundation Systems*; SRR Report 13; Department of Civil Engineering, Rice University: Houston, TX, USA, 1972.
90. Jennings, P.C.; Bielak, J. Dynamics of Building-Soil Interaction. *Bull. Seismol. Soc. Am.* **1973**, *63*, 9–48. [CrossRef]
91. Veletsos, A.S.; Meek, J.W. Dynamic Behaviour of Building-Foundation Systems. *Earthq. Eng. Struct. Dyn.* **1974**, *3*, 121–138. [CrossRef]

92. Krishnamoorthy, A.; Anitha, S. Seismic Effect of Soil Structure Interaction on Plane Frame Structure. *J. Mech. Civil. Eng.* **2014**, *2*, 55–60.
93. Chopra, A.K.; Chakrabarti, P.; Gupta, S. *Earthquake Response of Concrete Gravity Dams. Including Hydrodynamic and Foundation Interaction Effects*; Report No. UCB/EERC-80/01; Earthquake Engineering Research Center, University of California: Berkeley, CA, USA, 1980.
94. Chopra, A.K.; Gupta, S. Hydrodynamic and Foundation Interaction Effects in Earthquake Response of a Concrete Gravity Dam. *J. Struct. Div.* **1981**, *107*, 1399–1412. [CrossRef]
95. Chandrashaker, R.; Humar, J.L. Fluid–Foundation Interaction in the Seismic Response of Gravity Dams. *Earthq. Eng. Struct. Dyn.* **1993**, *22*, 1067–1084. [CrossRef]
96. Chen, M.; Penzien, J. *Nonlinear Soil-Structure Interaction of Skew Highway Bridges*; Technical Report UCB/EERC-77/24; Earthquake Engineering Research Center, University of California: Berkeley, CA, USA, 1977.
97. Crouse, C.B.; Hushmand, B.; Martin, G.R. Dynamic Soil–Structure Interaction of a Single-Span Bridge. *Earthq. Eng. Struct. Dyn.* **1987**, *15*, 711–729. [CrossRef]
98. Spyrakos, C.C. Assessment of SSI on the Longitudinal Seismic Response Short Span Bridges. *Constr. Build. Mater.* **1990**, *4*, 170–175. [CrossRef]
99. Spyrakos, C.C.; Vlassis, A.G. Effect of Soil-Structure Interaction on Seismically Isolated Bridges. *J. Earthq. Eng.* **2002**, *6*, 391–429. [CrossRef]
100. Veletsos, A.S.; Tang, Y. Dynamics of Vertically Excited Liquid Storage Tanks. *J. Struct. Eng.* **1986**, *112*, 1228–1246. [CrossRef]
101. Haroun, M.A.; Abou-Izzeddine, W. Parametric Study of Seismic Soil-Tank Interaction. II: Vertical Excitation. *J. Struct. Eng.* **1992**, *118*, 798–811. [CrossRef]
102. Malhotra, P.K. Seismic Response of Soil-Supported Unanchored Liquid-Storage Tanks. *J. Struct. Eng.* **1997**, *123*, 440–450. [CrossRef]
103. Chatterjee, P.; Basu, B. Non-Stationary Seismic Response of Tanks with Soil Interaction by Wavelets. *Earthq. Eng. Struct. Dyn.* **2001**, *30*, 1419–1437. [CrossRef]
104. Larkin, T. Seismic Response of Liquid Storage Tanks Incorporating Soil Structure Interaction. *J. Geotech. Geoenviron. Eng.* **2008**, *134*, 1804–1814. [CrossRef]
105. Meng, X.; Li, X.; Xu, X.; Zhang, J.; Zhou, W.; Zhou, D. Earthquake Response of Cylindrical Storage Tanks on an Elastic Soil. *J. Vib. Eng. Technol.* **2019**, *7*, 433–444. [CrossRef]
106. Wu, W.-H.; Lee, W.-H. Systematic Lumped-Parameter Models for Foundations Based on Polynomial-Fraction Approximation. *Earthq. Eng. Struct. Dyn.* **2002**, *31*, 1383–1412. [CrossRef]
107. Tang, Y.K.; Tang, H.T.; Kassawara, R.P.; Stepp, J.C. Validation of Soil-Structure Interaction Methods Using Earthquake Data in Lotung, Taiwan. In Proceedings of the 9th World Conference on Earthquake Engineering, Tokyo and Kyoto, Japan, 2–9 August 1988; Volume 8, p. 321.
108. Cho, K.H.; Kim, M.K.; Lim, Y.M.; Cho, S.Y. Seismic Response of Base-Isolated Liquid Storage Tanks Considering Fluid–Structure–Soil Interaction in Time Domain. *Soil. Dyn. Earthq. Eng.* **2004**, *24*, 839–852. [CrossRef]
109. Kianoush, M.R.; Ghaemmaghami, A.R. The Effect of Earthquake Frequency Content on the Seismic Behavior of Concrete Rectangular Liquid Tanks Using the Finite Element Method Incorporating Soil–Structure Interaction. *Eng. Struct.* **2011**, *33*, 2186–2200. [CrossRef]
110. Chaithra, M.; Krishnamoorthy, A.; Naurin Nafisa, P.M. Analysis of Soil—Structure Interaction on Response of Tanks Filled with Fluid. *Int. J. Civil. Eng. Technol.* **2017**, *8*, 813–819.
111. Chaithra, M.; Krishnamoorthy, A. Soil Structure Interaction Analysis of Tanks Filled with Fluid Subjected to Near-Fault Earthquakes. In *Lecture Notes in Civil Engineering*; Springer: Singapore, 2022; Volume 162, pp. 551–562.
112. Naeim, F.; Kelly, J.M. *Design of Seismic Isolated Structures: From Theory to Practice*; John Wiley & Sons, Inc.: Hoboken, NJ, USA, 1999; ISBN 9780470172742.
113. Robinson, W.H. Lead-Rubber Hysteretic Bearings Suitable for Protecting Structures during Earthquakes. *Earthq. Eng. Struct. Dyn.* **1982**, *10*, 593–604. [CrossRef]
114. Megget, L.M. Analysis and Design of a Base-Isolated Reinforced Concrete Frame Building. *Bull. N. Z. Soc. Earthq. Eng.* **1978**, *11*, 245–254. [CrossRef]
115. Guéraud, R.; Noël-Leroux, J.-P.; Livolant, M.; Michalopoulos, A.P. Seismic Isolation Using Sliding-Elastomer Bearing Pads. *Nucl. Eng. Des.* **1985**, *84*, 363–377. [CrossRef]
116. Mostaghel, N.; Khodaverdian, M. Dynamics of Resilient-Friction Base Isolator (R-FBI). *Earthq. Eng. Struct. Dyn.* **1987**, *15*, 379–390. [CrossRef]
117. Jangid, R.S.; Londhe, Y.B. Effectiveness of Elliptical Rolling Rods for Base Isolation. *J. Struct. Eng.* **1998**, *124*, 469–472. [CrossRef]
118. Zayas, V.A.; Low, S.S.; Mahin, S.A. A Simple Pendulum Technique for Achieving Seismic Isolation. *Earthq. Spectra* **1990**, *6*, 317–333. [CrossRef]
119. Pranesh, M.; Sinha, R. VFPI: An Isolation Device for Aseismic Design. *Earthq. Eng. Struct. Dyn.* **2000**, *29*, 603–627. [CrossRef]
120. Tsai, C.S.; Chiang, T.-C.; Chen, B.-J. Finite Element Formulations and Theoretical Study for Variable Curvature Friction Pendulum System. *Eng. Struct.* **2003**, *25*, 1719–1730. [CrossRef]

121. Lu, L.-Y.; Shih, M.-H.; Wu, C.-Y. Near-Fault Seismic Isolation Using Sliding Bearings with Variable Curvatures. In Proceedings of the 13th World Conference on Earthquake Engineering, Vancouver, BC, Canada, 1–6 August 2004; p. 3264.
122. Krishnamoorthy, A. Seismic Isolation of Bridges Using Variable Frequency and Variable Friction Pendulum Isolator System. *Struct. Eng. Int.* **2010**, *20*, 178–184. [CrossRef]
123. Kelly, T.E.; Mayes, R.L. Seismic Isolation of Storage Tanks. In *Seismic Engineering: Research and Practice*; ASCE: San Francisco, CA, USA, 1989; pp. 408–417.
124. Tajirian, F.F. Seismic Isolation of Critical Components and Tanks. In Proceedings of the ATC-17-1 Seminar on Seismic Isolation, Passive Energy Dissipation and Active Control, San Francisco, CA, USA, 11–12 March 1993; pp. 233–244.
125. Zayas, V.A.; Low, S.S. *Application of Seismic Isolation to Industrial Tanks*; American Society of Mechanical Engineers: New York, NY, USA, 1995.
126. Malhotra, P.K. New Method for Seismic Isolation of Liquid-Storage Tanks. *Earthq. Eng. Struct. Dyn.* **1997**, *26*, 839–847. [CrossRef]
127. Malhotra, P.K. Method for Seismic Base Isolation of Liquid-Storage Tanks. *J. Struct. Eng.* **1997**, *123*, 113–116. [CrossRef]
128. Chalhoub, M.S.; Kelly, J.M. *Theoretical and Experimental Studies of Cylindrical Water Tanks in Base Isolated Structures*; University of California at Berkeley: Berkeley, CA, USA, 1988.
129. Chalhoub, M.S.; Kelly, J.M. Shake Table Test of Cylindrical Water Tanks in Base-Isolated Structures. *J. Eng. Mech.* **1990**, *116*, 1451–1472. [CrossRef]
130. Shrimali, M.K.; Jangid, R.S. Non-Linear Seismic Response of Base-Isolated Liquid Storage Tanks to Bi-Directional Excitation. *Nucl. Eng. Des.* **2002**, *217*, 1–20. [CrossRef]
131. Liang, B.; Tang, J. Vibration Studies of Base-Isolated Liquid Storage Tanks. *Comput. Struct.* **1994**, *52*, 1051–1059. [CrossRef]
132. Jadhav, M.B.; Jangid, R.S. Response of Base-Isolated Liquid Storage Tanks. *Shock. Vib.* **2004**, *11*, 33–45. [CrossRef]
133. Jadhav, M.B.; Jangid, R.S. Response of Base-Isolated Liquid Storage Tanks to near-Fault Motions. *Struct. Eng. Mech.* **2006**, *23*, 615–634. [CrossRef]
134. Hashemi, S.; Aghashiri, M.H. Seismic Responses of Base-Isolated Flexible Rectangular Fluid Containers under Horizontal Ground Motion. *Soil. Dyn. Earthq. Eng.* **2017**, *100*, 159–168. [CrossRef]
135. Saha, S.K.; Sepahvand, K.; Matsagar, V.A.; Jain, A.K.; Marburg, S. Stochastic Analysis of Base-Isolated Liquid Storage Tanks with Uncertain Isolator Parameters under Random Excitation. *Eng. Struct.* **2013**, *57*, 465–474. [CrossRef]
136. Safari, S.; Tarinejad, R. Parametric Study of Stochastic Seismic Responses of Base-Isolated Liquid Storage Tanks under near-Fault and Far-Fault Ground Motions. *J. Vib. Control* **2018**, *24*, 5747–5764. [CrossRef]
137. Wang, Y.-P.; Teng, M.-C.; Chung, K.-W. Seismic Isolation of Rigid Cylindrical Tanks Using Friction Pendulum Bearings. *Earthq. Eng. Struct. Dyn.* **2001**, *30*, 1083–1099. [CrossRef]
138. Panchal, V.R.; Jangid, R.S. Variable Friction Pendulum System for Seismic Isolation of Liquid Storage Tanks. *Nucl. Eng. Des.* **2008**, *238*, 1304–1315. [CrossRef]
139. Panchal, V.R.; Jangid, R.S. Seismic Response of Liquid Storage Steel Tanks with Variable Frequency Pendulum Isolator. *KSCE J. Civil. Eng.* **2011**, *15*, 1041–1055. [CrossRef]
140. Abalı, E.; Uçkan, E. Parametric Analysis of Liquid Storage Tanks Base Isolated by Curved Surface Sliding Bearings. *Soil. Dyn. Earthq. Eng.* **2010**, *30*, 21–31. [CrossRef]
141. Soni, D.P.; Mistry, B.B.; Panchal, V.R. Double Variable Frequency Pendulum Isolator for Seismic Isolation of Liquid Storage Tanks. *Nucl. Eng. Des.* **2011**, *241*, 700–713. [CrossRef]
142. Weng, D.; Zhang, R.; Ge, Q.; Liu, S. Effect Investigation of Combination Isolation System for Liquid Storage Tank in Different Seismic Levels. In Proceedings of the 15th World Conference on Earthquake Engineering (15WCEE), Lisbon, Portugal, 24–28 September 2012.
143. Saha, S.K.; Matsagar, V.A.; Jain, A.K. Earthquake Response of Base-Isolated Liquid Storage Tanks for Different Isolator Models. *J. Earthq. Tsunami* **2014**, *8*, 1450013. [CrossRef]
144. Saha, S.K.; Matsagar, V.A.; Jain, A.K. Seismic Fragility of Base-Isolated Water Storage Tanks under Non-Stationary Earthquakes. *Bull. Earthq. Eng.* **2016**, *14*, 1153–1175. [CrossRef]
145. Bagheri, S.; Farajian, M. The Effects of Input Earthquake Characteristics on the Nonlinear Dynamic Behavior of FPS Isolated Liquid Storage Tanks. *J. Vib. Control.* **2018**, *24*, 1264–1282. [CrossRef]
146. Bagheri, S.; Raad, H.H. Seismic Performance Assessment of FPS Isolated Liquid Storage Tanks at Various Intensity Levels. In Proceedings of the Fourth Conference on Smart Monitoring, Assessment and Rehabilitation of Civil Structures, Zurich, Switzerland, 13–15 September 2017.
147. Uckan, E.; Umut, Ö.; Sisman, F.N.; Karimzadeh, S.; Askan, A. Seismic Response of Base Isolated Liquid Storage Tanks to Real and Simulated near Fault Pulse Type Ground Motions. *Soil. Dyn. Earthq. Eng.* **2018**, *112*, 58–68. [CrossRef]
148. Christovasilis, I.P.; Whittaker, A.S. Seismic Analysis of Conventional and Isolated LNG Tanks Using Mechanical Analogs. *Earthq. Spectra* **2008**, *24*, 599–616. [CrossRef]
149. Kalantari, A.; Nikoomanesh, M.R.; Goudarzi, M.A. Applicability of Mass-Spring Models for Seismically Isolated Liquid Storage Tanks. *J. Earthq. Tsunami* **2019**, *13*, 1950002. [CrossRef]
150. Krishnamoorthy, A. Finite Element Method of Analysis for Liquid Storage Tank Isolated with Friction Pendulum System. *J. Earthq. Eng.* **2021**, *25*, 82–92. [CrossRef]

151. Yang, Y.-B.; Lee, T.-Y.; Tsai, I.-C. Response of Multi-Degree-of-Freedom Structures with Sliding Supports. *Earthq. Eng. Struct. Dyn.* **1990**, *19*, 739–752. [CrossRef]
152. Moslemi, M.; Farzin, A.; Kianoush, M.R. Nonlinear Sloshing Response of Liquid-Filled Rectangular Concrete Tanks under Seismic Excitation. *Eng. Struct.* **2019**, *188*, 564–577. [CrossRef]
153. Rawat, A.; Matsagar, V.A.; Nagpal, A.K. Numerical Study of Base-Isolated Cylindrical Liquid Storage Tanks Using Coupled Acoustic-Structural Approach. *Soil. Dyn. Earthq. Eng.* **2019**, *119*, 196–219. [CrossRef]
154. Rawat, A.; Mittal, V.; Chakraborty, T.; Matsagar, V. Earthquake Induced Sloshing and Hydrodynamic Pressures in Rigid Liquid Storage Tanks Analyzed by Coupled Acoustic-Structural and Euler-Lagrange Methods. *Thin-Walled Struct.* **2019**, *134*, 333–346. [CrossRef]
155. Shekari, M.R.; Hekmatzadeh, A.A.; Amiri, S.M. On the Nonlinear Dynamic Analysis of Base-Isolated Three-Dimensional Rectangular Thin-Walled Steel Tanks Equipped with Vertical Baffle. *Thin-Walled Struct.* **2019**, *138*, 79–94. [CrossRef]
156. Vern, S.; Shrimali, M.K.; Bharti, S.D.; Datta, T.K. Response and Damage Evaluation of Base-Isolated Concrete Liquid Storage Tank under Seismic Excitations. *Eng. Res. Express* **2021**, *3*, 045002. [CrossRef]
157. Rawat, A.; Matsagar, V. Seismic Analysis of Liquid Storage Tank Using Oblate Spheroid Base Isolation System Based on Rolling Friction. *Int. J. Non Linear Mech.* **2022**, *147*, 104186. [CrossRef]
158. Kumar, H.; Saha, S.K. Effects of Soil-Structure Interaction on Seismic Response of Fixed Base and Base Isolated Liquid Storage Tanks. *J. Earthq. Eng.* **2021**, *26*, 6148–6171. [CrossRef]
159. Zhang, R.; Zhao, Z.; Pan, C. Influence of Mechanical Layout of Inerter Systems on Seismic Mitigation of Storage Tanks. *Soil. Dyn. Earthq. Eng.* **2018**, *114*, 639–649. [CrossRef]
160. Gedikli, A.; Ergüven, M.E. Seismic Analysis of a Liquid Storage Tank with a Baffle. *J. Sound. Vib.* **1999**, *223*, 141–155. [CrossRef]
161. Maleki, A.; Ziyaeifar, M. Damping Enhancement of Seismic Isolated Cylindrical Liquid Storage Tanks Using Baffles. *Eng. Struct.* **2007**, *29*, 3227–3240. [CrossRef]
162. De Angelis, M.; Giannini, R.; Paolacci, F. Experimental Investigation on the Seismic Response of a Steel Liquid Storage Tank Equipped with Floating Roof by Shaking Table Tests. *Earthq. Eng. Struct. Dyn.* **2010**, *39*, 377–396. [CrossRef]
163. Maleki, A.; Ziyaeifar, M. Sloshing Damping in Cylindrical Liquid Storage Tanks with Baffles. *J. Sound. Vib.* **2008**, *311*, 372–385. [CrossRef]
164. Krishnamoorthy, A. Variable Curvature Pendulum Isolator and Viscous Fluid Damper for Seismic Isolation of Structures. *J. Vib. Control.* **2011**, *17*, 1779–1790. [CrossRef]
165. Tsipianitis, A.; Tsompanakis, Y. Improving the Seismic Performance of Base-Isolated Liquid Storage Tanks with Supplemental Linear Viscous Dampers. *Earthq. Eng. Eng. Vib.* **2022**, *21*, 269–282. [CrossRef]
166. Chang, S.-P.; Makris, N.; Whittaker, A.S.; Thompson, A.C.T. Experimental and Analytical Studies on the Performance of Hybrid Isolation Systems. *Earthq. Eng. Struct. Dyn.* **2002**, *31*, 421–443. [CrossRef]
167. Luo, H.; Zhang, R.; Weng, D. Mitigation of Liquid Sloshing in Storage Tanks by Using a Hybrid Control Method. *Soil. Dyn. Earthq. Eng.* **2016**, *90*, 183–195. [CrossRef]
168. Hartog, J.P.D. LXXIII. Forced Vibrations with Combined Viscous and Coulomb Damping. *Lond. Edinb. Dublin Philos. Mag. J. Sci.* **1930**, *9*, 801–817. [CrossRef]
169. Hundal, M.S. Response of a Base Excited System with Coulomb and Viscous Friction. *J. Sound. Vib.* **1979**, *64*, 371–378. [CrossRef]
170. Qamaruddin, M.; Rasheeduzzafar; Arya, A.S.; Chandra, B. Seismic Response of Masonry Buildings with Sliding Substructure. *J. Struct. Eng.* **1986**, *112*, 2001–2011. [CrossRef]
171. Yang, T.Y.; Konstantinidis, D.; Kelly, J.M. The Influence of Isolator Hysteresis on Equipment Performance in Seismic Isolated Buildings. *Earthq. Spectra* **2010**, *26*, 275–293. [CrossRef]
172. Mostaghel, N.; Hejazi, M.; Tanbakuchi, J. Response of Sliding Structures to Harmonic Support Motion. *Earthq. Eng. Struct. Dyn.* **1983**, *11*, 355–366. [CrossRef]
173. Westermo, B.; Udwadia, F. Periodic Response of a Sliding Oscillator System to Harmonic Excitation. *Earthq. Eng. Struct. Dyn.* **1983**, *11*, 135–146. [CrossRef]
174. Li, Z.; Rossow, E.C.; Shah, S.P. Sinusoidal Forced Vibration of Sliding Masonry System. *J. Struct. Eng.* **1989**, *115*, 1741–1755. [CrossRef]
175. Jangid, R.S. Response of Pure-Friction Sliding Structures to Bi-Directional Harmonic Ground Motion. *Eng. Struct.* **1997**, *19*, 97–104. [CrossRef]
176. Younis, C.J.; Tadjbakhsh, I.G. Response of Sliding Rigid Structure to Base Excitation. *J. Eng. Mech.* **1984**, *110*, 417–432. [CrossRef]
177. Crandall, S.H.; Lee, S.S.; Williams, J.H. Accumulated Slip of a Friction-Controlled Mass Excited by Earthquake Motions. *J. Appl. Mech.* **1974**, *41*, 1094–1098. [CrossRef]
178. Mostaghel, N.; Tanbakuchi, J. Response of Sliding Structures to Earthquake Support Motion. *Earthq. Eng. Struct. Dyn.* **1983**, *11*, 729–748. [CrossRef]
179. Ahmadi, G. Stochastic Earthquake Response of Structures on Sliding Foundation. *Int. J. Eng. Sci.* **1983**, *21*, 93–102. [CrossRef]
180. Constantinou, M.C.; Tcidjbakhsh, I.G. Response of a Sliding Structure to Filtered Random Excitation. *J. Struct. Mech.* **1984**, *12*, 401–418. [CrossRef]
181. Kulkarni, J.A.; Jangid, R.S. Effects of Superstructure Flexibility on the Response of Base-Isolated Structures. *Shock. Vib.* **2003**, *10*, 368693. [CrossRef]

182. Vafai, A.; Hamidi, M.; Ahmadi, G. Numerical Modeling of MDOF Structures with Sliding Supports Using Rigid-Plastic Link. *Earthq. Eng. Struct. Dyn.* **2001**, *30*, 27–42. [CrossRef]
183. Rai, D.C. Performance of Elevated Tanks InM w 7.7 Bhuj Earthquake of 26 January 2001. *J. Earth Syst. Sci.* **2003**, *112*, 421–429. [CrossRef]
184. Soroushnia, S.; Tafreshi, S.T.; Omidinasab, F.; Beheshtian, N.; Soroushnia, S. Seismic Performance of RC Elevated Water Tanks with Frame Staging and Exhibition Damage Pattern. *Procedia Eng.* **2011**, *14*, 3076–3087. [CrossRef]
185. Haroun, M.A.; Abou-Izzeddine, W. Parametric Study of Seismic Soil-Tank Interaction. I: Horizontal Excitation. *J. Struct. Eng.* **1992**, *118*, 783–797. [CrossRef]
186. Jain, S.K.; Sameer, S.U. A Review of Requirements in Indian Codes for Aseismic Design of Elevated Water Tanks. *Bridge Struct. Eng.* **1993**, *23*, 1–16.
187. Jaiswal, O.R.; Rai, D.C.; Jain, S.K. Review of Seismic Codes on Liquid-Containing Tanks. *Earthq. Spectra* **2007**, *23*, 239–260. [CrossRef]
188. Malhotra, P. Practical Nonlinear Seismic Analysis of Tanks. *Earthq. Spectra* **2000**, *16*, 473–492. [CrossRef]

sustainability

Article

Shaking Table Test for Seismic Response of Nuclear Power Plant on Non-Rock Site

Xinyu Lu [1], Liping Jing [1,2,*], Ying Ma [1,3], Jianhua Yang [3] and Wenhao Qi [1]

1 Key Laboratory of Earthquake Engineering and Engineering Vibration, Institute of Engineering Mechanics, China Earthquake Administration, Harbin 150080, China; luxinyu@nuaa.edu.cn (X.L.); maying@cnpe.cc (Y.M.); qwhtky@163.com (W.Q.)
2 Institute of Disaster Prevention, Sanhe 065201, China
3 China Nuclear Power Engineering Co., Ltd., Beijing 100840, China; yangjh@cnpe.cc
* Correspondence: jing_liping@126.com

Abstract: In order to compare and analyze the seismic response characteristics of a safety-related nuclear structure on a non-rock site in the condition of raft and pile group foundations under unidirectional and multidirectional seismic motion input, a large-scale shaking table test of the soil-nuclear structure system was carried out in this paper. In the test, the soil was uniform silted clay, and the shear wave velocity was 213 m/s. Considering the similarity of the superstructure natural frenquency, the actual nuclear power structure was simplified to a three-story frame shear wall structure model. The annular laminated shear model box was used to take the boundary effect of soil into consideration; the seismic motions = were input in only one horizontal direction or three directions at the same time for the shaking table test, and the results were analyzed. The results of the test show that the acceleration response of the safety-related nuclear plant is affected by the directions of input seismic motion and the forms of the foundation. When the seismic motion is input simultaneously in three directions, the acceleration responses of the horizontal motion and vertical rocking of the safety-related plant are larger than those of the single-direction input. The acceleration response of the horizontal motion and vertical rocking of the safety-related structure with the pile group foundation is smaller than that with the raft foundation. The values of most frequency bands in the horizontal acceleration Fourier amplitude spectrum at the top of the pile-foundation structure are smaller than that at the top of the raft-foundation structure, while the displacement is basically the same as that of the raft-foundation structure. This is related to the relation between the frequency component of input seismic motion and the natural frequency of the structure system. Therefore, it is more reasonable to use three-dimensional seismic input in the seismic response analysis of nuclear power plants. The seismic performance of nuclear power plants can be enhanced by using pile group foundations.

Keywords: three-dimensional seismic motion; raft foundation; pile group foundation; shaking table test; safety-related nuclear structure

Citation: Lu, X.; Jing, L.; Ma, Y.; Yang, J.; Qi, W. Shaking Table Test for Seismic Response of Nuclear Power Plant on Non-Rock Site. *Sustainability* **2023**, *15*, 10366. https://doi.org/10.3390/su151310366

Academic Editors: Chong Xu, Su Chen and Shuang Li

Received: 30 May 2023
Revised: 24 June 2023
Accepted: 25 June 2023
Published: 30 June 2023

1. Introduction

Clean and efficient nuclear power is an important way to solve the global energy problem. However, it is the safety performance of the plant structure that must be given adequate consideration while vigorously developing the nuclear power industry. At present, most of the nuclear power plants built or under construction in China are located in coastal hard bedrock areas, according to the design specifications of nuclear power plants. In view of the decreasing number of sites that meet the requirements of standard design, it is inevitable to turn to inland non-rock sites when selecting bases for new nuclear power plants in the future. Although there are many examples of non-rock nuclear power plant construction abroad, according to statistics, 60% of nuclear power plants in France and 50% in the United States are located in soft rock or sandy soil [1]. The construction of nuclear power plants in China started later, and the theory and specifications are relatively imperfect. Compared with the structural dynamic analysis of nuclear power plants with high-quality bedrock site conditions, the seismic response analysis of safety-related nuclear structures on non-rock sites is more complicated and has more uncertainties. The influence of many factors involved has not been recognized [2].

Nuclear power plant structures have the characteristics of large mass, large stiffness, and high displacement control requirements, which are different from general civil buildings [3]. Currently, some scholars have completed partial research work on the seismic response of nuclear structures on non-rock sites. Ying Jianian et al. [4] calculated the impedance function of a pile foundation under a nuclear power plant with different methods, then suggested that the impedance of the pile foundation should be selected carefully for specific projects and the field test results should be used to assist in the determination of impedance for major projects; Mohamed A. Sayed et al. [5] calculated the dynamic response of a simplified nuclear power plant model; Li Weixin and Wang Guixuan [6,7] established a quasi-three-dimensional finite element model of the pile-soil-CPR1000 nuclear island plant structure and compared the seismic reaction of the plant before and after using a pile group foundation on a soil base; Luo Chuan et al. [8] compared the consequences of nuclear structures with and without piles in horizontally layered sites under earthquakes; Zou Degao et al. [9] analyzed the response of the three-dimensional pile foundation nuclear island model under earthquakes beyond the design basis; Yang Jianhua et al. [10] discussed the key influencing factors in the numerical calculation of the dynamic performance of a nuclear plant with a pile foundation; Zhu Shengdong et al. [11] completed a nonlinear seismic behavior analysis of the piled raft foundation AP1000 nuclear island structure; Chen Shaolin et al. [12] proposed a partition analysis method to improve the efficiency of time-domain soil–structure interaction analysis and applied it to seismic response analysis of nuclear structures.

The numerical simulation method adopted in the above research can reflect the dynamic response characteristics of nuclear power structures to a certain extent. However, the model test method is more intuitive and real, which can provide a reference for evaluating the reliability of various theoretical and numerical models. Jeong Gon Ha et al. [13] compared the results of field tests and centrifuge tests (horizontal unidirectional seismic input) of the seismic reaction of a nuclear power plant containment on a soil site in Hualien and advanced that centrifuge tests can be employed to simulate the soil–foundation–structure interaction of the actual nuclear power plant containment. Li Xiaojun et al. [14,15] conducted a shaking table test of the soil–structure system under three-dimensional seismic motion input with the research object of the CAP1400 nuclear power plant; Peng Lingyun et al. [16] tested the seismic performance of a plant structure with rigid foundations near the bottom plate edge in the CAP1400 nuclear power plant by shaking table test and numerical simulation; Zhang Xueming et al. [17] verified the applicability of the method for establishing a uniform hazard spectrum in actual nuclear power structures by a shaking table test of a nuclear power plant with rigid foundations; Zhou Zhiguang et al. [18,19] completed the shaking table test of the AP1000 nuclear power plant with and without isolation on rigid and soil base; Gao Yongwu et al. [20] studied the floor response spectrum of a nuclear power plant on a soil site under horizontal earthquakes.

This paper is a part of the project "Research on the dynamic response of nuclear island plants on soft soil bases", which is carried out by a nuclear power company for further analysis of the site adaptability of the safety-related nuclear structure of the third-generation nuclear power plant on a soil base. The shaking table test of the safety-related plant under unidirectional horizontal and multi-directional earthquakes was completed using a simplified model. By comparing the dynamic response of the safety-related plant under unidirectional and multi-directional seismic motion, the influence of multi-directional seismic motion on the horizontal seismic performance of the plant is analyzed, and by comparing the dynamic response of the safety-related plant with pile group and raft foundations, the seismic response characteristics of the plant with different foundation forms on a non-rock site are discussed. The research results in this paper can provide a reference for the seismic design of nuclear power plants on non-rock sites.

2. Overview of the Test

2.1. Background of the Project

The study subject in this paper is a third-generation nuclear island plant, whose layout is shown in Figure 1, including the reactor plant, fuel plant, electrical plant, safety-related plant, etc. All the nuclear plants above share a raft foundation. This study mainly takes safety-related plant B as the research object, which consists of walls and plates with 10 stories and 40 m in height. The base characteristics refer to the typical soil site in China, and the pile foundation contains rock-socketed piles with a diameter of 1.5 m and a length of 40 m.

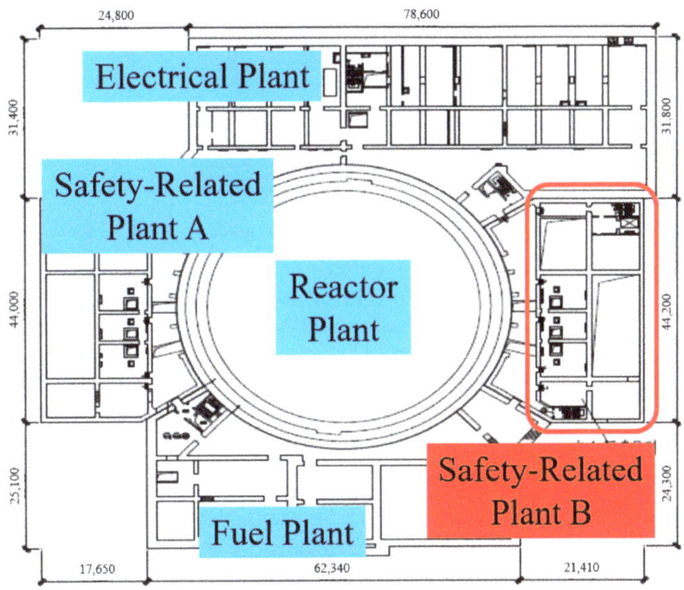

Figure 1. Diagram of nuclear power plant.

2.2. The Shaking Table

The shaking table system used in this experiment is a 5 m × 5 m three-direction, six-degree-of-freedom large-scale seismic simulation shaking table system of the Institute of Engineering Mechanics, China Earthquake Administration. The maximum acceleration input is 2 g, the maximum distance is 0.5 m, and the effective bearing capacity is 30 t.

2.3. Annular Layered Shear Box

A self-developed cylindrical laminated shear model box was adopted in the test. The model box is 2.5 m in height and 2.8 m in inner diameter. It is composed of several H-shaped frames. Three-way movable support elements are set between the frames, which can produce relative motion and simulate the shear deformation of the soil during vibration, as shown in Figure 2.

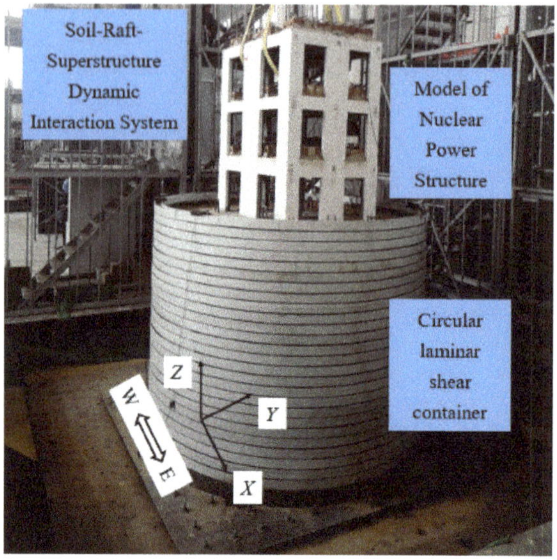

Figure 2. Model of soil–raft superstructure system.

2.4. Design and Modeling of Superstructure

In order to reflect the dynamic characteristics of the prototype structure quantitatively in the dynamic test, the model should meet the similarity relation. The soil–structure interaction model of the large-scale seismic simulation shaking table test includes two parts: the soil mass model and the scaled structure model. In principle, the scaled structure model can approximately satisfy the experimental similarity relation by appropriately applying artificial mass and selecting the fabrication material. However, since the soil is a three-phase dispersion with strong nonlinearity, the shaking table test of the soil is difficult to meet the similarity relation of physical quantities such as modulus and force at the same time under the condition of constant Gravitational acceleration. Therefore, it is impossible to find strict similarity relations in the large-scale shaking table test of soil structure [21]. It is feasible to take the similarity between the safety-related nuclear plant with a pile or raft foundation and the prototype structure into reasonable consideration in the design of the model. However, due to the difficulty in meeting the similarity relation of the soil, the practical significance of this approach is limited. Based on this, various factors are taken into comprehensive consideration to design the experimental model, such as the size and upper load limit of the shaking table, the size of the soil container, and the relation between the natural frequencies of the safety-related nuclear plant model and the prototype (the ratio of the two is preliminarily determined to be 4:1), and the final determination is that the length of the structure model is 1.6 m, the width is 1.1 m, and the total height is 1.85 m. It is designed as a three-layer building; the height of the first layer is 0.65 m, and both the second and third layers are 0.60 m. The model is made of microconcrete. The concrete strength grades of the model structure and the floor are C30 and C40, respectively. The thickness of the walls and floors is 40 mm and 30 mm, respectively. The reinforcement is made of two-direction double-row galvanized iron wire mesh with a diameter of 2 mm and a grid spacing of 10 mm.

In order to facilitate the connection between the superstructure and the pile cap, two test models were designed in total for the shaking table test of the raft and pile group foundation. The baseplate thickness of the superstructure with the raft foundation model is 300 mm, which is also used as the raft foundation. The baseplate thickness of the superstructure with the pile group foundation model is 150 mm, which is connected with the pile cap (with a thickness of 150 mm) by high-strength screws. The length and width of the superstructure baseplates in both test models are 2.0 m and 1.5 m, respectively, and the only difference is the thickness of the baseplates. The natural frequencies of the safety-related plant B and two experimental models are shown in Table 1. The ratios of natural frequencies between the models and the prototype in the X and Y directions are close to 4:1, which meets the design requirements. A model of the nuclear power safety-related plant with a raft foundation is shown in Figure 2.

Table 1. The natural frequencies of structural models and the prototype.

	The Prototype	The Structural Model with Raft Foundation	The Structural Model with Pile Group Foundation
In X direction	5.20	20.20	20.67
In Y direction	3.91	16.55	17.75

2.5. Model of Soil Mass

According to the difference in shear wave velocity, the base soil is classified in the Code for Seismic Design of Buildings (GB50011-2010) [22]. The shear wave velocity ranges of soft soil, medium soft soil, medium hard soil, and hard soil are less than 150 m/s, 150 m/s~250 m/s, 250 m/s~500 m/s, and greater than 500 m/s, respectively. In addition, the Code for Seismic Design of Nuclear Power Plants (GB50267-2019) [23] holds that soil mass with a shear wave velocity of less than 300 m/s can be regarded as a relatively weak base. Therefore, the classification limit for medium hard soil and hard soil bases of nuclear structures can be changed from 500 m/s to 300 m/s. Considering that the pile in the shaking table test model is made of fine aggregate concrete, the measured elastic modulus is about 22 GPa, while the elastic modulus of ordinary C30 concrete is about 30 GPa, and the ratio of the two is about 0.7, it can be calculated that the ratio of shear wave velocity is about 0.8 when the density and Poisson ratio are close. As a result, in the shaking table test of this paper, the shear wave velocity range of the aforementioned different types of soil needs to be reduced by 0.8 times to obtain the shear wave velocity range for the classification of nuclear structure subsoil that can match the pile group foundation in the model. The shear wave velocity ranges of soft soil, medium soft soil, medium-hard soil, and hard soil base after reduction are less than 120 m/s, 120~200 m/s, 200~240 m/s, and greater than 240 m/s, respectively.

The ordinary silted clay and medium-fine sand were mixed according to a mass ratio of 2:1 and then formed in the soil container when preparing the soil mass model in this experiment. The filling height was 2.3 m, and manual compaction was performed after each 100 mm filling to ensure the uniformity of the soil. The average density of the soil measured by the cutting ring method is 1.80 g/cm^3, and the shear wave velocity of the soil measured by the resonance column test method is 213 m/s. It can be judged that this is a medium-hard soil base according to the shear wave velocity range of different types of subsoil discussed above.

2.6. Model of Pile Group Foundation

The 3 $\times$ 3 arrangement of 9 foundation piles is adopted, and the net distance of foundation piles in the east–west and north–south directions is 0.6 m and 0.4 m, respectively. The foundation pile is a circular pile with a diameter of 100 mm and a length of 2 m, of which size and reinforcement are shown in Figure 3a,b, and the bottom steel plate is welded with the bottom of the soil container to simulate the rock-socketed end bearing pile, as shown in Figure 3c. The pile head and cap are poured integrally. The diagram of the pile group arrangement is shown in Figure 3d, where the circles with numbers represent the piles.

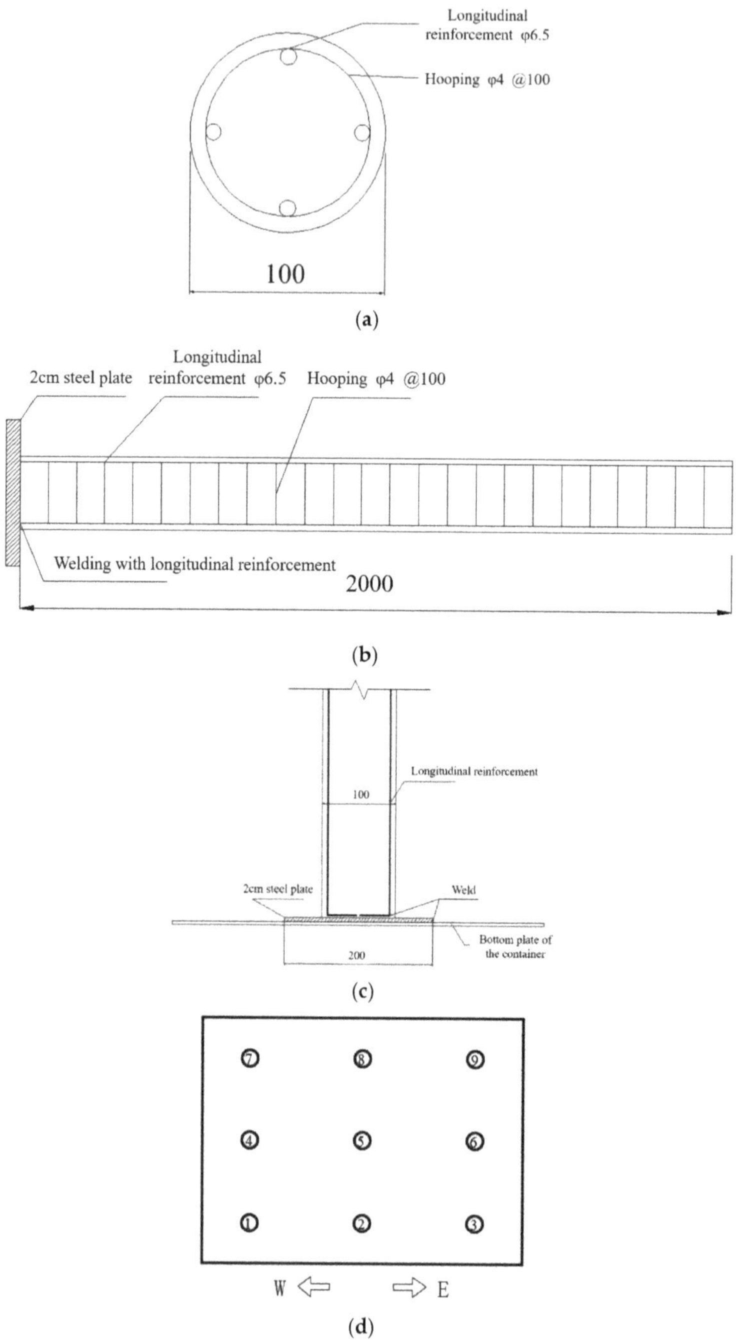

Figure 3. Diagram of pile group foundation. (**a**) Cross-section of single pile reinforcement. (**b**) Vertical section of single pile reinforcement. (**c**) Diagram of connection between piles and soil container. (**d**) Layout of pile group.

2.7. Instrumentation

Since it is the dynamic response of the upper safety-related plant structure in different seismic motion input directions and different foundation conditions that is compared in this paper, only the monitoring sites of three-direction acceleration sensors and draw-wire displacement sensors arranged in each layer of the superstructure are explained. The layout of the sensors is shown in Figure 4, where the red circle and the white rectangle represent the acceleration and displacement monitoring sites, respectively. Acceleration sensors are arranged at the center of the upper surface of the floor in each layer of the structure, and the bottom-up numbers are A0~A2. For the analysis of the plant structure's rocking motion, acceleration sensors are arranged at the midpoints of the eastern, western, southern, and northern edges of the structure, numbered A3-EM, A3-WM, A3-SM, and A3-NM, respectively. Monitoring sites of draw-wire displacement sensors are arranged at the midpoint of the western and northern edges of the top of each layer in the structure, and the bottom-up numbers are DW1~DW3 (western side) and DN1~DN3 (northern side).

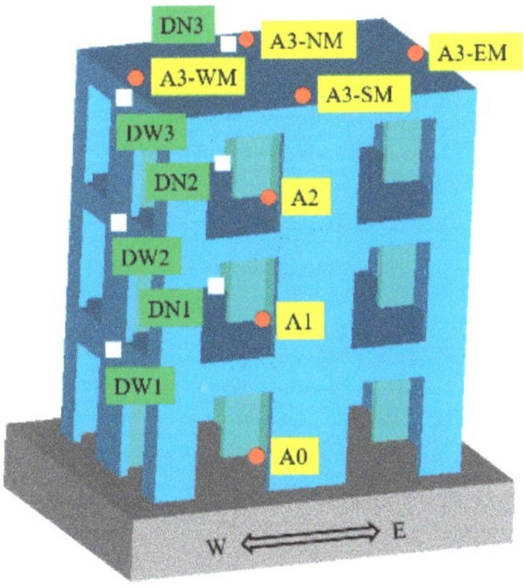

Figure 4. Layout of sensors.

2.8. Scheme of Dynamic Load Input

In the shaking table test, artificial acceleration time histories and actual strong earthquake records are generally selected as input motions for the test model. So, the artificially synthesized seismic motion according to the RG1.60 response spectrum commonly used in the United States nuclear power design (hereinafter referred to as RG1.60 seismic motion) and the two natural seismic motions, Landers and Chichi, are applied in the test. The normalized acceleration time history and Fourier amplitude spectrum of input seismic motions are shown in Figure 5.

After amplitude modulation of the selected seismic motions to 0.05 g, 0.10 g, and 0.20 g, they are input into the test model along only one direction (X) and three directions at the same time (X, Y, and Z, see Figure 2). When the motions were input in three directions simultaneously, the amplitude ratio of acceleration in the X, Y, and Z directions is 1:0.85:0.65 [22]. The graded loading increases with the amplitude mentioned above in the test. In addition, to get the natural vibration characteristics of the model at different stages of the test, white noise is input before and after all levels of loading.

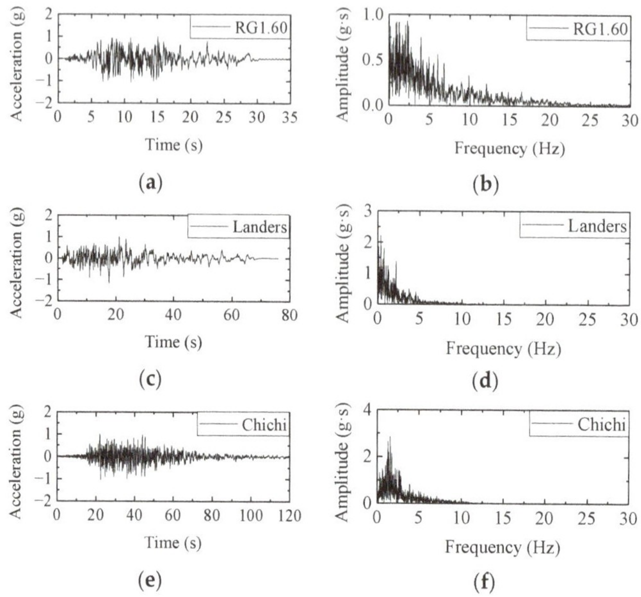

Figure 5. Acceleration time history and Fourier amplitude spectrum of input seismic motion. (**a**) Time history of RG1.60. (**b**) Fourier spectrum of RG1.60. (**c**) Time history of Landers. (**d**) Fourier spectrum of Landers. (**e**) Time history of Chichi. (**f**) Fourier spectrum of Chichi.

3. Analysis of Test Results

3.1. Experimental Phenomena and Structural System's Natural Frequency

In the process of most seismic input, the vibration amplitude of the two foundation-structure integral model systems is generally small, the soil container has no obvious relative deformation, and the structural baseplate has good contact with the surrounding soil layer. Only when the amplitude of input is 0.20 g is the structural baseplate separated from the surrounding soil, and cracks appear on the surface of the soil (Figure 6). Except for the obvious vertical rocking motion of the superstructure when the RG1.60 seismic motion is input into the raft foundation model with an amplitude of 0.20 g, the motion of the superstructure in other test cases can be regarded as approximate translation. It can be preliminarily judged that the two test models are intact and were not damaged during the test. The white noise method is used to analyze the frequency of the whole model. The results are shown in Table 2. It can be seen that the natural frequency of the models does not change before and after the test, which is consistent with the experimental phenomenon.

Figure 6. Cracks on surface of soil around the foundations after 0.20 g seismic motion input. (**a**) Test of pile group foundation model. (**b**) Test of raft foundation model.

Table 2. Results of frequency using white noise method (Hz).

Conditions	Pile Group Foundation Model	Raft Foundation Model
Before the test	16.10	14.14
After the test	16.14	13.92

3.2. Test Results Comparison of Seismic Motion Unidirectional and Three Directional Input

For the analysis of the energy transfer law in the process of seismic wave propagation, the Arias seismic motion intensity index I_a [24] is introduced, and its expression is

$$I_a = \frac{\pi}{2g}\int_0^{T_d} a(t)^2 dt \tag{1}$$

where $a(t)$ is the acceleration time history of the monitoring sites, whose unit is m/s^2, T_d is the duration of the seismic motion, and g is the gravity acceleration (m/s^2). The Arias intensity amplification factor is defined as the ratio of Arias intensity in the roof of each layer to the baseplate of the superstructure. Under the action of a unidirectional earthquake, the amplification factor is obtained from the unidirectional acceleration response of each monitoring site. Under the action of a multi-directional earthquake, the horizontal total acceleration response of each monitoring site is obtained along the synthetic direction of horizontal earthquake input (the angle with the X axis is 40°), and the amplification factor is acquired.

The Arias intensity amplification factor curve of the nuclear power plant on the pile group foundation is shown in Figure 7. It can be seen from Figure 7: (1) The Arias intensity of each floor increases with height under unidirectional or multi-directional earthquakes. (2) The amplification factor of each floor decreases with the increase in the input seismic motion amplitude, but the change is small. This is because the safety-related structure is not significantly damaged and is in the nonlinear elastic stage; (3) When the same seismic motion input is unidirectional and three-directional, the amplification factor of each layer of the structure in the unidirectional input case is slightly greater than that in the three-directional input case. The amplification factors at the top of RG1.60 seismic motion along one-way and three-way input are 2.42~3.07 and 2.22~2.85, respectively, of Landers seismic motion are 1.82~1.85 and 1.67~1.76, and of Chichi seismic motion are 1.79~1.93 and 1.65~1.86; (4) The amplification factor of the safety-related plant structure in the case of RG1.60 input is greater than that in the case of Landers and Chichi input. From Figure 5, it can be seen that this is because the main frequency components of Landers and Chichi are below 10 Hz, while RG1.60 still has more components from 10~20 Hz, which is located near the natural frequency of the test models and has a more obvious amplification effect.

The values of Arias intensity at the top of the safety-related structure under unidirectional and multi-directional earthquakes are compared in Table 3. Although the amplification factor of the safety-related structure under multi-directional seismic motion input is smaller than that under unidirectional input, the values of Arias intensity at each layer of the safety-related plant in multi-directional input cases are basically larger than those in horizontal unidirectional input cases. In the condition of three-way input, the Arias intensity values at the top of the superstructure are 0.98–1.58 times those of the same seismic input in a single direction.

Table 3. Arias intensity values at the top of structure model with pile group foundation (m/s).

Type of Seismic Motion	Unidirectional Input			Multi-Directional Input		
	0.05 g	0.10 g	0.20 g	0.05 g	0.10 g	0.20 g
RG1.60	1.172	4.005	12.139	1.546	5.493	13.666
Landers	0.317	1.402	7.472	0.499	2.084	11.342
Chichi	1.345	4.351	19.160	1.316	6.262	23.674

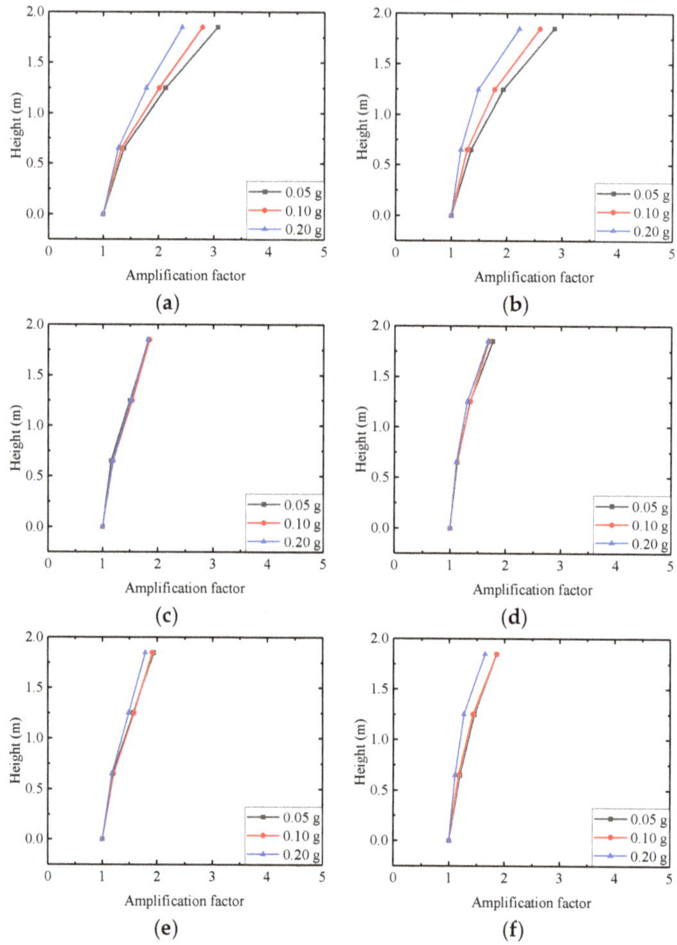

Figure 7. Arias intensity amplification factor curve of structure model with pile group foundation. (**a**) RG1.60 unidirectional input. (**b**) RG1.60 multi-directional input. (**c**) Landers unidirectional input. (**d**) Landers multi-directional input. (**e**) Chichi unidirectional input. (**f**) Chichi multi-directional input.

Figure 8 shows the floor spectra of the safety-related plant structure with a pile group foundation under a 0.05 g amplitude input case. In cases of unidirectional input, the floor spectra are generated by the one-way acceleration response of each monitoring site. In cases of multi-directional input, the floor spectra are generated by the horizontal total acceleration response along the synthetic direction of horizontal earthquake input (the angle with the X axis is 40°) of each monitoring site. It can be seen from Figure 8 that (1) in the case of any seismic motion input, the shape of the floor spectrum of each floor of the structure is basically the same; only the amplitude is different, increasing with the height; and (2) while the same seismic motion is input along a single direction or a multi-direction, the shape of the floor spectra of the safety-related plant is relatively consistent, and the peak period is generally unchanged.

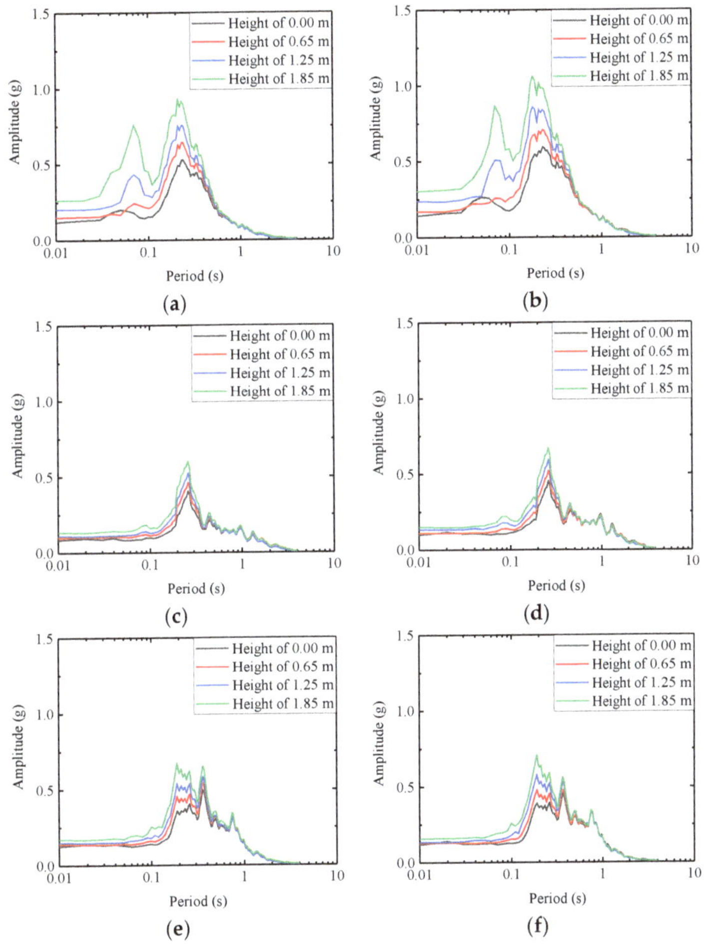

Figure 8. Floor spectra of structure model with pile group foundation. (**a**) RG1.60 unidirectional input. (**b**) RG1.60 multi-directional input. (**c**) Landers unidirectional input. (**d**) Landers multi-directional input. (**e**) Chichi unidirectional input. (**f**) Chichi multi-directional input.

Figure 9 shows the variation of the Arias intensity values of the rocking motion at the top of the safety-related plant with the input seismic motion amplitude. The solution of the intensity values is as follows: in the condition that the seismic motion is input in the X direction, subtract the average values of the records in the Z direction (as shown in Figure 2) of A4-EM and A4-WM—acceleration sensors on the eastern and western sides at the top of the structure—from the two records separately to obtain two new time histories and calculate the Arias intensity values of the new time histories. Then, the average value of Arias intensity values obtained from new time histories is acquired to represent the intensity of rocking motion. The calculation method for the seismic motion three-direction input condition is similar, except that the average value recorded by four sensors at the top is subtracted. The intensity values in the east–west and north–south directions are calculated separately using the above method and then summed up for the total intensity value.

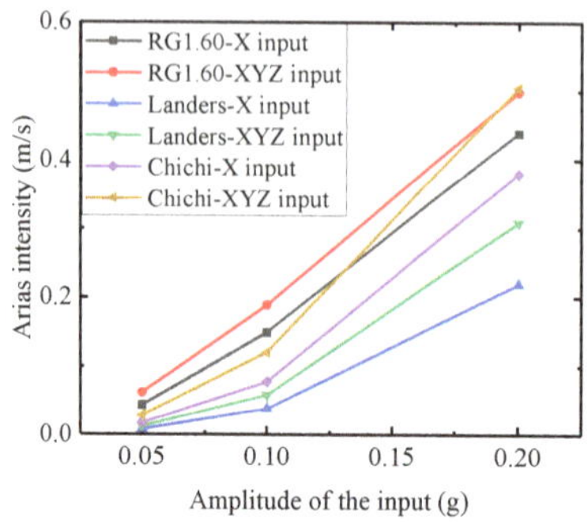

Figure 9. Arias intensity values of rocking motion at the top of superstructure with pile group foundation.

It can be drawn from Figure 9 that when the selected three seismic motions are input into the test model with different amplitudes, the rocking motion at the top of the plant model, when the seismic motion is input from three directions, is more violent than that when the seismic motion is input from the X direction, and the Arias intensity value is about 1.14~1.62 times that when the seismic motion is input from the X direction. According to Table 2, the Arias intensity value of the rocking motion of the pile-foundation structure model accounts for about 1.3% to 4.0% of the horizontal motion at the same position. In addition, though the duration of the RG1.60 seismic motion is the shortest of the three input motions, the Arias intensity value of the rocking motion at the top of the structure is the largest, which indicates that the safety-related plant is more sensitive to high-frequency seismic motion input.

Figure 10 is used to illustrate the displacement calculation basis of the three-way input condition of seismic motion in this experiment. When the test model generates displacement along the direction of the actual movement in Figure 10, the actual displacement ① at the monitoring site of the draw-wire displacement sensor is larger than the measured displacement ② in this direction. However, since the actual displacement ③ in the other horizontal direction is much smaller than the length of the wire of the displacement sensor, it can be approximately considered that the angle between the wire before and after the motion is very small, and the actual variation of the wire length (obtained by the initial length of the wire minus ②) is basically the same as the variation of the same seismic motion input along the single direction (obtained by the initial length of the wire minus ①). It can be found that ① is basically the same as ②, and the measured ② can be used to replace the real ① in the experiment. Because, in fact, ① is greater than ②, this practice will lead to a smaller measured value of structural displacement in the three-dimensional seismic motion input condition. Figure 11 shows the displacement amplitude at different heights of the nuclear power plant when the amplitude of input seismic motion is 0.05 g. Due to the large stiffness of the plant model, the displacement amplitude at different heights in the same test condition is generally the same; the interlayer displacement of the structure is small and can be approximately regarded as a translation of a rigid body. When the seismic motion is input from three directions, the horizontal displacement amplitude is about 1.07~1.19 times that of X unidirectional input. As a result of the fact that the displacement measured in the three-way input condition is less than the true value, the difference between the actual displacement value of the three-dimensional input and the one-dimensional input condition is greater.

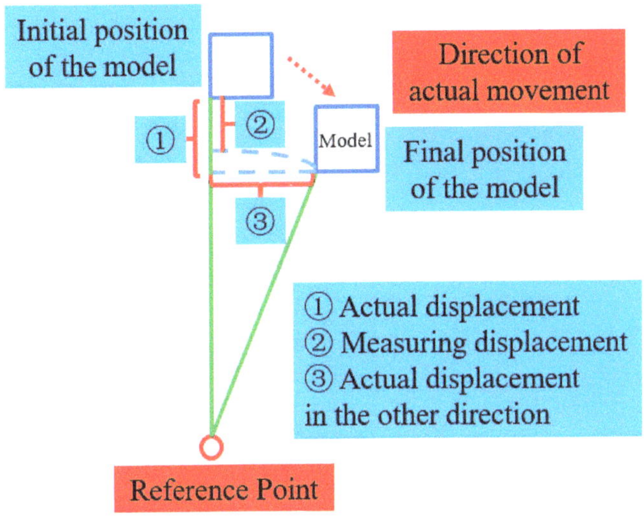

Figure 10. Diagram of displacement calculation in condition of multi-directional seismic motion input.

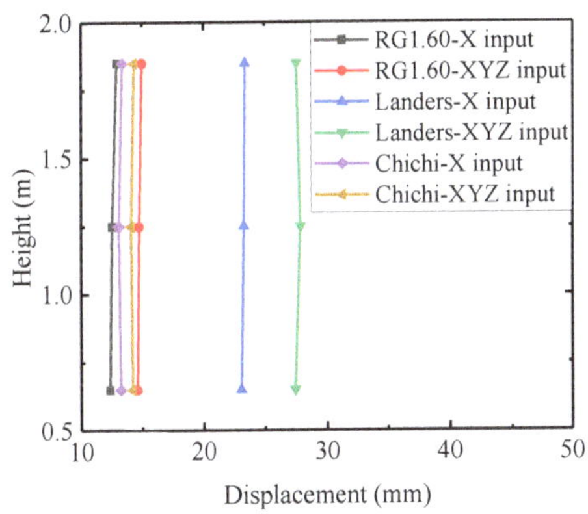

Figure 11. Maximum displacement of superstructure with pile group foundation.

3.3. *Comparison of Structures with Different Foundations Test Results*

Figures 12 and 13 are the comparison results of the Arias intensity values at each height and the Fourier amplitude spectrum at the top of the nuclear safety-related plant with different foundation forms under 0.05 g amplitude seismic motion input. The maximum Arias intensity appears at the top of the structure model, regardless of the foundation forms. When the pile group foundation is applied, the Arias intensity values at each height of the structure model are smaller than those for raft foundation conditions. While RG1.60, Landers, and Chichi seismic motion input, the Arias intensity values at the top of the structure on pile group foundation are reduced by 41%, 32%, and 48%, respectively, compared with the structure on raft foundation. The shapes of the acceleration Fourier spectra at the top of the plant structure with different foundation forms are similar. However, the amplitude of Fourier spectra at the top of the plant on the raft foundation is larger than that of the plant on the pile group foundation in most frequency bands. This is because the natural frequencies of both the two model systems of plants with different foundation forms are high, and components of the input seismic motions are mainly in the relatively low-frequency zone. The natural frequency of the model system in the case of a pile group foundation is higher than that in the case of a raft foundation. So the difference between most frequency components in the input seismic motions and the natural frequency of the test model is greater in the pile foundation case than in the raft foundation case, and the amplification effect is weakened.

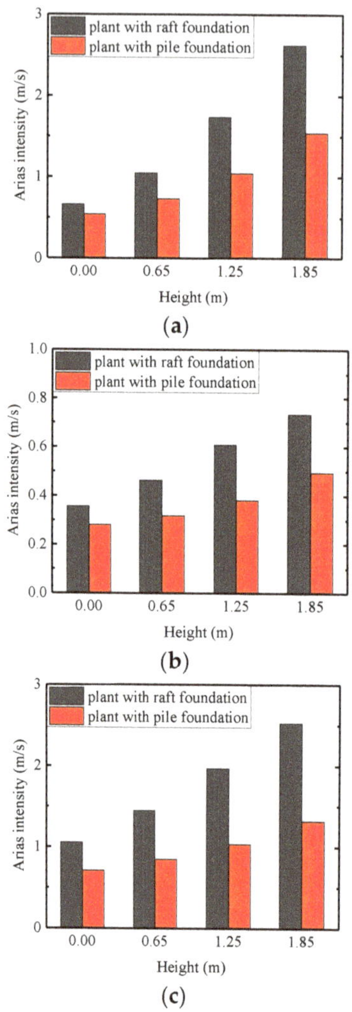

Figure 12. Comparison of Arias intensity values of superstructure with different foundation forms. (**a**) RG1.60 input. (**b**) Landers input. (**c**) Chichi input.

Figure 14 shows the comparison results of Arias intensity values of rocking motion at the top of the plant with raft and pile group foundations. The rocking motion of the raft-foundation structure caused by the three seismic motions selected in the test is greater than that of the pile-foundation structure, which indicates that the pile group foundation is beneficial to reduce the vertical rocking motion of the nuclear safety-related plant. The rocking motion of the plant caused by RG1.60 seismic motion is the most intense.

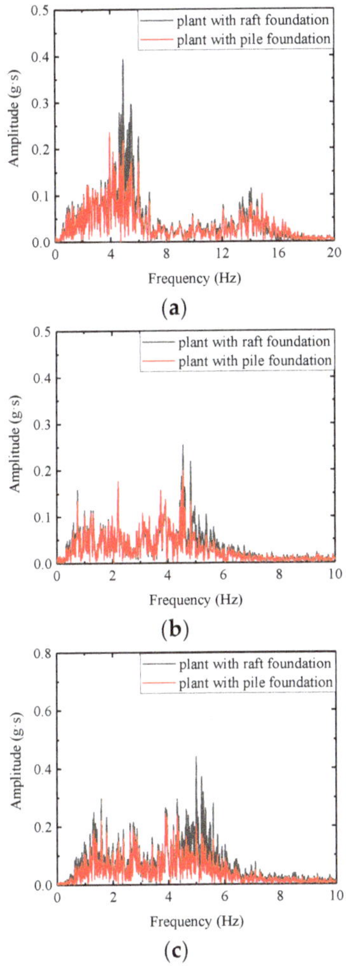

Figure 13. Comparison of spectrum of superstructure with different foundation forms. (**a**) RG1.60 input. (**b**) Landers input. (**c**) Chichi input.

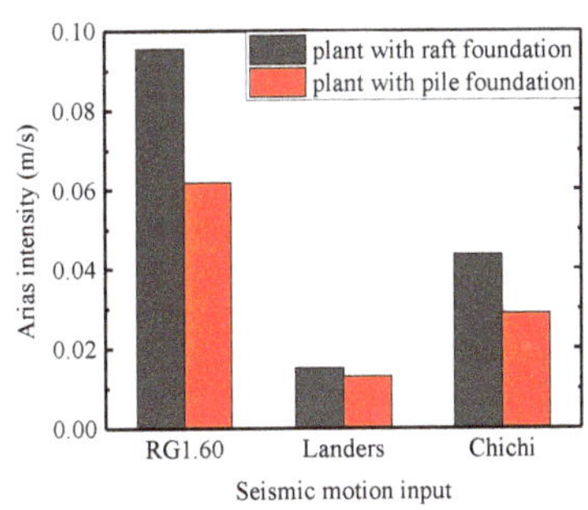

Figure 14. Comparison of Arias intensity of rocking motion of superstructure with different foundation forms.

Due to the large stiffness of the nuclear power plant model, the displacement at each height is basically the same, and only the displacement amplitude at the top of the plant with raft and pile group foundation is given, as shown in Figure 15. It can be seen from Figure 15 that the displacement of the nuclear structure model with two foundation forms is generally the same. This is because the components of displacement response are mainly in the low-frequency zone, which differs significantly from the high natural frequencies of nuclear power structure systems.

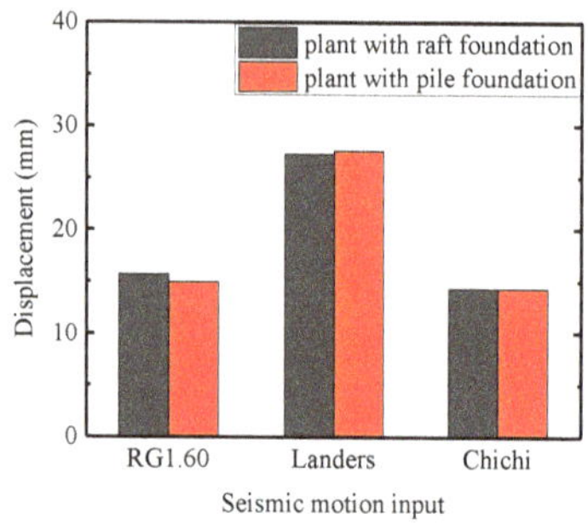

Figure 15. Comparison of displacement of superstructure with different foundation forms.

4. Conclusions

In this paper, based on the shaking table test of a subsoil-safety-related nuclear structure system on a non-rock site, the acceleration and displacement responses of a safety-related nuclear structure with a pile group and raft foundation under horizontal unidirectional and multi-directional seismic motion input are compared. The following three conclusions can be drawn:

(1) The Arias intensity amplification factor at different heights of the safety-related plant is related to the frequency components of seismic motion input, which is larger under RG1.60 seismic motion input than that of the other two natural seismic motion input cases. That is because RG1.60 seismic motion has more components near the natural frequency of the soil–structure system model. In addition, the high-frequency components of the RG1.60 seismic motion cause a more severe vertical rocking motion of the nuclear power plant than the other two natural seismic motions.

(2) When the seismic motion is input in three directions simultaneously, the Arias intensity value, displacement, and degree of vertical rocking motion of the safety-related plant are largerthan in unidirectional input cases. The shape of the floor spectrum is basically the same as that of the unidirectional input, and the peak period changes little.

(3) The Arias intensity value and rocking motion amplitude of the pile group-foundation safety-related plant are less than those of the raft-foundation safety-related plant. In most frequency bands, the value of the Fourier amplitude spectrum at the top of the raft-foundation structure is greater than that of the pile group-foundation structure. The reason for this is that the frequency of the input seismic motion is lower than the natural frequency of the test model system, and the natural frequency of the pile-foundation model system is higher than that of the raft-foundation model system, so the amplification effect is weakened. The displacement of the safety-related plant on different foundation forms is basically the same because the frequency of the input displacement is low and far away from the natural frequency of the model systems.

Based on the above conclusions, it is recommended to use three-dimensional seismic motion input in the seismic response analysis of nuclear power plants on non-rock sites to make the results less underestimated and more reasonable. In addition, the seismic performance of a nuclear power plant on a non-rock site can be enhanced by setting pile group foundations.

Author Contributions: Conceptualization, X.L. and L.J.; Methodology, X.L.; Writing—original draft, X.L.; Visualization, Y.M.; Project administration, W.Q. and J.Y. All authors have read and agreed to the published version of the manuscript.

Funding: This research was funded by Scientific Research Fund of Institute of Engineering Mechanics, China Earthquake Administration (Grant No. 2019B10).

Institutional Review Board Statement: Not applicable.

Informed Consent Statement: Not applicable.

Data Availability Statement: Data available in a publicly accessible repository.

Conflicts of Interest: The authors declare no conflict of interest.

References

1. Du, N. Exploration and practice of quality assurance for haixing nuclear power plant soft foundation treatment. In Proceedings of the Progress Report on China Nuclear Science and Technology (Roll VI)-Proceedings of the 2019 Annual Academic Meeting of the Chinese Nuclear Society Volume 10 (Nuclear Safety Subsection and Nuclear Security Subsection), Baotou, China, 20–23 August 2019; pp. 70–77.
2. Wang, F. *Study on the Effect of Non-Rock Site on Seismic Behavior of Safety-Related Nuclear Structure*; Institute of Geophysics, China Earthquake Administration: Beijing, China, 2021.
3. Chen, J.; Ma, J.; Jiang, D. Dynamic analysis for isolation structures of nuclear power plants. *Ind. Constr.* **2016**, *46*, 105–110.
4. Ying, J.; Wang, R.; Yu, Z. Earthquake response analysis of soil-pile-nuclear power station auxiliary workshop structure interaction system. *Earthq. Eng. Eng. Dyn.* **1995**, *1*, 44–52.
5. Sayed, M.; Kim, D.; Cho, S.G. *Seismic Analysis of Base Isolated Nuclear Power Plant Considering Nonlinear Pile-Soil Interaction*; Transactions, SMiRT-22: San Francisco, CA, USA, 2013; pp. 18–23.
6. Li, W.; Yin, X.; Wang, G. Seismic response analysis of nuclear power plant on soil foundation considering pile-soil-structure dynamic interaction. *J. Seismol. Res.* **2016**, *39*, 40–45+180.
7. Wang, G.; Zhang, R.; Zhao, J. Analysis on seismic bearing capacity of natural and pile foundations about a coastal nuclear power plant in soft ground. *J. Seismol. Res.* **2016**, *39*, 15–21+179–180.
8. Luo, C.; Yang, X.; Zhan, C.; Jin, X.; Ding, Z. Nonlinear 3D finite element analysis of soil-pile-structure interaction system subjected to horizontal earthquake excitation. *Soil Dyn. Earthq. Eng.* **2016**, *84*, 145–156. [CrossRef]
9. Zou, D.G.; Sui, Y.; Chen, K. Plastic damage analysis of pile foundation of nuclear power plants under beyond-design basis earthquake excitation. *Soil Dyn. Earthq. Eng.* **2020**, *136*, 106179. [CrossRef]
10. Yang, J.; Ma, Y.; Cai, J. Research on the seismic response analysis of nuclear island buildings with pile foundation. *Ind. Constr.* **2021**, *51*, 31–36.
11. Zhu, S.; Chen, G.; Jiang, P.; Chen, W.; Gao, W. 3D nonlinear response characteristics of the pile-raft-supported AP1000 nuclear island building in soft deposits subjected to strong ground motions. *Eng. Mech.* **2021**, *38*, 129–142.
12. Chen, S.L.; Lv, H.; Zhou, G.L. Partitioned analysis of soil-structure interaction for nuclear island building. *Earthq. Eng. Struct. Dyn.* **2022**, *51*, 2220–2247. [CrossRef]
13. Ha, J.G.; Lee, S.H.; Kim, D.S.; Choo, Y.W. Simulation of soil–foundation–structure interaction of Hualien large-scale seismic test using dynamic centrifuge test. *Soil Dyn. Earthq. Eng.* **2014**, *61–62*, 176–187. [CrossRef]
14. Li, X.; Wang, X.; He, Q. Shaking table tests for evaluation of seismic behavior of nuclear power plants on non-rock site. *Nucl. Power Eng.* **2017**, *38*, 31–35.
15. Wang, X.H.; Li, X.J.; Liu, A.W.; He, Q.; Hou, C. Seismic analysis of soil-structure system of nuclear power plant on non-rock site via shaking table test. *Soil Dyn. Earthq. Eng.* **2020**, *136*, 106209.
16. Peng, L.Y.; Kang, Y.J.; Tang, Z.Y.; Chen, H.T. Seismic Performance of CAP1400 Nuclear Power Station considering Foundation Uplift. *Shock. Vib.* **2018**, *2018*, 16. [CrossRef]
17. Zhang, X.; Yan, W.; Sun, Y.; Chen, S.; He, H. Shaking table test of nuclear power plant considering uniform hazard spectrum. *Nucl. Power Eng.* **2018**, *39*, 80–84.
18. Zhou, Z.; Guo, J.; Wei, X.; Zhang, J. Research of Coupling effect of horizontal and vertical seismic response of nuclear structures considering SSI effect. *Struct. Eng.* **2018**, *34*, 82–87.
19. Zhou, Z.; Zhou, L.; Zhao, J. Shaking table test study of AP1000 base-isolated nuclear power plant. *Earthq. Eng. Eng. Dyn.* **2020**, *40*, 64–72.
20. Gao, Y.; Wang, T.; Dai, J. Experimental research on seismic responses of a new type of nuclear power plant under different site conditions. *Eng. Mech.* **2020**, *37*, 116–124.
21. Jing, L.; Wang, G.; Li, J.; Sun, Y.; Zhou, Z.; Qi, W. Shaking table tests and numerical simulations of dynamic interaction of soil-pile-nuclear island system. *Chin. J. Geotech. Eng.* **2022**, *44*, 163–172+207–208.
22. *GB50011-2010*; Ministry of Housing and Urban-Rural Development of the People's Republic of China. Code for Seismic Design of Buildings. China Architecture Press: Beijing, China, 2010.

23. *GB50267-2019*; Ministry of Housing and Urban-Rural Development of the People's Republic of China. Code for Seismic Design of Nuclear Power Plants. China Planning Press: Beijing, China, 2019.
24. Arias, A. *A Measure of Earthquake Intensity, in Seismic Design for Nuclear Power Plants*; MIT Press: Cambridge, UK, 1970.

sustainability

Article

Influence of Borehole Casing on Received Signals in Downhole Method

Shi Yan [1], Yan Yu [2,*], Wenjun Zheng [1], Jie Su [3,4] and Zhenghua Zhou [4]

1 Guangdong Provincial Key Laboratory of Geodynamics and Geohazards, School of Earth Sciences and Engineering, Sun Yat-sen University, Guangzhou 510275, China
2 National Institute of Natural Hazards, Ministry of Emergency Management of China, Beijing 100085, China
3 School of Civil Engineering and Architecture, Wuyi University, Jiangmen 529030, China
4 Institute of Geotechnical Engineering, Nanjing Tech University, Nanjing 210009, China
* Correspondence: yanyu@ninhm.ac.cn

Abstract: Borehole shrinkage and collapse are likely to occur when downhole testing is conducted in soft or loose sandy soils, resulting in testing interruption. To prevent this situation from occurring, installing casing in the borehole is a common approach. However, in actual testing, the quality of the signal obtained from measuring points within the depth of the casing is often not ideal, and there is still no clear and unified justification for the causes of interference generated by the casing. Therefore, the team attempt to investigate and elucidate the impact of casing through on-site experiments and numerical simulations. Firstly, on-site tests on the impact of different materials of casing on the wave velocity test utilizing the downhole method are conducted, the waveform characteristics of the measurement points inside the PVC casing and steel casing boreholes are analyzed, and the usability of the test results are evaluated. Next, the contact state between shallow soil and casing is changed, and its impact on the waveform characteristics of signal at different depth measurement points is analyzed. Then, the ABAQUS finite element software is utilized to establish a three-dimensional finite element model for wave velocity testing using the casing method, and the dynamic response of the measuring points on the casing wall inside the hole under surface excitation is solved. By numerically simulating different casing materials, the contact state between the casing and the hole wall, and the presence of low wave velocity filling soil around the casing, the variation patterns of the affected measurement point signals in the time and frequency domains are investigated. Furthermore, combined with the measured data, the impact characteristics of the casing on the results of the wave velocity testing using the downhole method are systematically explored. This research can provide some insights for the application and data interpretation of signals in the downhole methods of cased wells.

Keywords: downhole method; numerical simulation; casing effect; shear wave velocity; travel time; in-situ experiment

Citation: Yan, S.; Yu, Y.; Zheng, W.; Su, J.; Zhou, Z. Influence of Borehole Casing on Received Signals in Downhole Method. *Sustainability* **2023**, *15*, 9805. https://doi.org/10.3390/su15129805

Academic Editor: Claudia Casapulla

Received: 1 March 2023
Revised: 4 May 2023
Accepted: 9 May 2023
Published: 20 June 2023

1. Introduction

The in-situ shear wave velocity profile of the site is a crucial geotechnical parameter for engineering sites [1–5], and the downhole method of shear wave testing is one of the most commonly used in-situ wave velocity testing methods for obtaining shear wave velocity under small strain on the site, which has advantages such as low cost, simple principle, and convenient operation [6–9]. At present, research on the downhole method testing mainly focuses on travel time calculation [10–12] and propagation distance calculation [13], mostly neglecting the analysis of the reliability of obtaining signals.

Many factors can affect the quality of signal acquisition in on-site testing. Currently, there is a consensus that the influencing factors include the contact state between the sleeper and the surface, the distance between the sleeper and the borehole, and etc. [14–16]. However, there is relatively little research on the impact of casing on the acquisition of signals through the downhole method of wave velocity testing. When conducting downhole testing on weak sites or sites with loose sandy soil layers, it is easy to experience borehole shrinkage and collapse, which can engender testing interruption. To prevent this situation, lowering the casing during drilling is a typical practice. However, in actual testing, the quality of the signal acquired from the measuring points within the depth of the casing is often not ideal. Larkin and Taylor [17] compared the signals collected by drilling holes before and after casing placement under forward and reverse percussion and found that the phase of the signals gained by shallow and mid layer measurement points after casing placement was significantly disturbed. The interference components were defined as the torsional and bending waves generated by percussion propagating along the casing, which attenuated rapidly with depth. Crice [18] compared the signals obtained from drilling with casing in the presence and absence of well fluid at measurement points and found that the waveform of the signals obtained in the wet hole state was substantially disturbed. Crice believed that the interference component was a pressure pulse formed by the coupling of soil, casing, and liquid, which had the characteristics of fast wave velocity and slow attenuation along depth. Therefore, in deeper measurement points, it could be mistakenly identified as shear waves and cause interference. In fact, there is still no clear and unified justification for the reason of interference caused by the casing. Moreover, due to the expansion of the hole during the drilling process, the loose sand layer in the hole falling off, and the bridge plug phenomenon during the soil subsidence process, the contact state between the casing and the hole wall soil is complex, which can easily produce a cavity situation and impact the acquisition of signals. The analysis of the contact state between the casing and the hole wall on the testing site is time-consuming and labor-intensive; therefore, it is generally believed that the signal obtained in the depth of the casing in the downhole method of wave velocity testing is not reliable, and the test signal of the measurement point in the depth of the casing is discarded to avoid the error generated by the casing. However, with the extensive construction of infrastructure in China's delta plain areas and cross river and sea engineering projects, there are more and more opportunities for encountering soft soil sites in in-situ wave velocity testing. Improving drilling stability through casing has become a norm, and the casing is also getting deeper. Many technical personnel in on-site wave velocity testing choose to abandon testing the wave velocity at shallow positions with casing. However, estimating the corresponding soil layer wave velocity referring to the wave velocity data of the nearby boreholes will result in more errors. It is meaningful to clarify the impact caused by the casing during the drilling method testing and find ways to eliminate the impact.

This study mainly focused on the influence of casing on the wave velocity testing signal in the downhole method. Field experiments on the influence of different materials of casing on the wave velocity test of downhole method were first conducted, and the signal wave characteristics and the usability of test results in the borehole of PVC casing and that of steel casing were analyzed. Then, the contact state between shallow soil and casing was altered to analyze its effect on the signal waveform characteristics of measurement points at different depths. In geotechnical engineering, there is a lot of available software, such as FLAC, UDEC or ABAQUS. Considering that our research group has purchased the genuine ABAQUS software, then in this study, the geotechnical simulation software is used to analyze the displacement contour of the roadway excavation [19]; the ABAQUS finite element software was harnessed to establish a three-dimensional finite element model for wave velocity testing with the casing downhole method and simulate the dynamic response of the measurement points on the casing wall in the hole under surface excitation. Through numerically simulating different casing materials, casing and borehole wall contact state, casing surrounded by a low-velocity soil layer, and other working conditions, the affected measurement point signal changes in the time domain and frequency domain were analyzed, and combined with the measured data, the characteristics of the influence of casing on the downhole method wave velocity test were systematically explored.

2. Field Experiment Preparation and Experimental Model

2.1. Experimental Site and Test Apparatus

The experimental site was located on the North Campus of the College of Disaster Prevention and Technology in Sanhe City, situated in the pre-Yanshan plain area, mainly alluvial from the Chaobai River and the Ji Canal. The main body of the site is fluvial deposits with sound stratigraphic conditions. A sketch of the stratigraphic section is shown in Figure 1a [20]. Two boreholes with 117 mm diameter and about 10 m apart were arranged in the site for placing the casing of different materials. The casing length of the PVC cased borehole exceeded 30 m, and that of the steel cased borehole exceeded 24 m. The acquisition system selected for the wave velocity test field experiment of the downhole method with cased borehole was the SE2404PLUS series integrated engineering detector (see Figure 1b) produced by Beijing Jopeng Group, equipped with a single three-component airbag probe (see Figure 1c). The power system excited shear energy by manually striking the ground sleeper horizontally (see Figure 1d,e). To increase the coupling between the sleepers and the ground, the ground surface was scraped and iron blocks were placed on top of the sleepers.

(a)

Layer number	Layer thickness(m)	Lithology characterization
1	1	backfill
2	3.5	Silty sand, light yellowish brown color, with black stripes buried in the near horizontal layer
3	6	Clay, yellow-gray to gray color
4	4	Gray fine sand with thin layer of clay
5	3.5	Silty clay, gray color
6	4	Silty clay ,with calcareous nodules
7	3.5	Earthy yellow clay mixed with silt, with lots of rust spots
8	11.5	Grey silty fine sand to fine sand, with a small amount of clay
9	1.5	Gravel layer, with particle size of 5-20 mm, consisting of muddy gravel and quartz gravel, moderately sorted and rounded
10	6	Silty sand, gray color, locally grayish green stripes and rusty clay
11	6	Clay, brownish red color, with copper green and rust stripes at the bottom
12	3	Silty sand, brownish yellow color, massive rust spots, calcareous nodules
13	2	Clay, brownish yellow brown color, with large rust spots
14	3.5	Clay, dark gray color, with disturbed structure and calcareous nodules

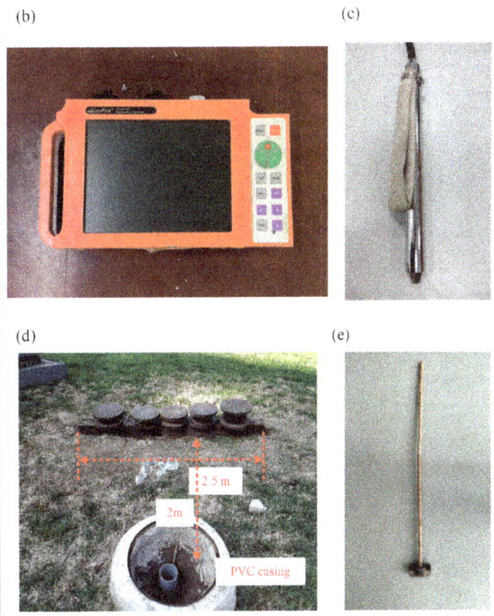

Figure 1. (**a**) Schematic diagram of soil profile of in-situ downhole method, (**b**–**e**) in-situ downhole method equipment.

2.2. Experimental Models and Test Method

2.2.1. Experimental Models of Casing Borehole with Different Materials

In the field experiment, PVC casing and steel casing were installed in two boreholes to develop two experimental models of casing, and the casing was in close contact with the hole wall at this time. In order to strengthen the connection between the casing and the borehole wall, the diameter of the borehole should be slightly smaller than that of the casing during the excavation process. After the casing was established, sand was poured into the void between the surface casing and the hole wall. A three-component probe was placed in the borehole with casing and the signal of each measurement point was recorded by pulling the probe from the bottom up. The distance between the vertical center of the percussive sleeper and the center of the casing hole was 2 m. The test depths of the two casing holes with casing were 4 m to 22 m, and the spacing between the measuring points was 2 m. In an effort to enhance the reliability of signal acquisition, the same measuring point was struck several times and the signal observed. The sampling interval was 0.05 ms, and no filtering was set for the acquisition process.

2.2.2. Experimental Models with Different Contact States

By excavating the soil around the casing and injecting fluid around the casing in the two cased boreholes in Section 2.2.1, two other experimental models were developed, and combined with the experimental model in Section 2.2.1, three experimental models have been constructed: "contact" model, "free casing" model, and "casing immersed in fluid" model (abbreviated as "fluid immersion" model). For the sake of experimental observation and practical operation, the excavation depth around the casing was 2.2 m, which means that only the first 2.2 m of the casing was falling off from the hole wall in the free casing model. The same is true for the fluid immersion model. Figure 2a–d presents the free casing model and fluid immersion model for PVC cased borehole and steel cased borehole, respectively. Considering the depth of the location where the casing contact state in the borehole changed, 4.0 m was established as the test depth. The distance between measurement points was shortened to 0.5 m. The distance of the percussion sleeper from the borehole remained unchanged, and striking was performed at the same measurement point several times with the signal recorded. 0.05 ms was positioned as the sampling interval, with no filtering set during the acquisition process.

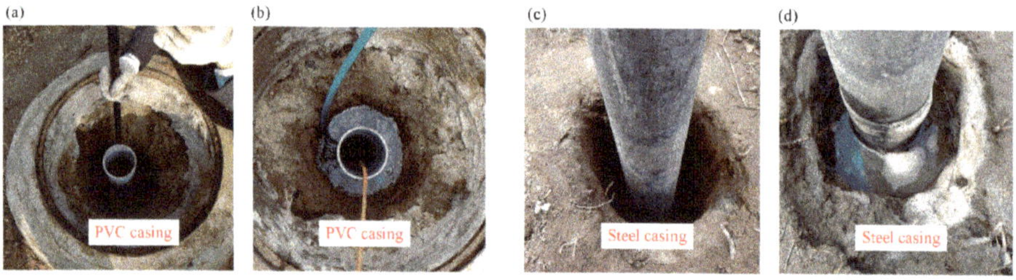

Figure 2. Different casing models: (**a**) free casing model of PVC cased borehole, (**b**) casing immersed in fluid model of PVC cased borehole, (**c**) free casing model of steel cased borehole, and (**d**) casing immersed in fluid model of steel cased borehole.

3. Numerical Simulation of Wave Velocity Test with Casing Downhole Method

The ABAQUS finite element analysis platform was leveraged to establish a three-dimensional analysis model to simulate and calculate the characteristics of the influence of the cased borehole on the signal obtained through the downhole method wave velocity test. As detailed in Figure 3, the size of the soil in the model was 20 m × 20 m × 30 m, the diameter of the borehole was 0.1 m, and the borehole ran through the whole soil model. The depth of the casing in the borehole from the surface to the bottom of the casing was measured at 20 m. The X–Y plane was shown as the horizontal plane of the actual test, and the center of the borehole was the center of the X–Y plane, which was set as the origin (marked by "O" with co-ordinates (0, 0, 0)); thus, the co-ordinates of the observation point on the casing in the X–Y plane should be (−0.05, 0, 0). In the model, the Z-axis was upward, and the negative Z-axis was the direction of depth increase. ABAQUS infinite element transmission boundaries [21,22] were used around the model and on the bottom surface to mitigate the interference of reflected waves from the model boundaries to the near-field region. The material properties of the infinite element unit were linear elastomer, and the rest of the material parameters was consistent with those of the connected soil unit. An approximate impulse force along the Y-axis was applied to the surface of the model to simulate the excitation source of the sleeper strike in the actual test, and the point of the load action was 2.0 m (marked by "S" with co-ordinates (−2.05, 0, 0)). The observation points on the surface of the casing and the surface of the soil borehole wall in contact with the casing were set along the negative direction of the Z-axis, and the observation points were located in the X–Y plane; hence, the horizontal Y direction of the observation points was parallel to the direction of the load action.

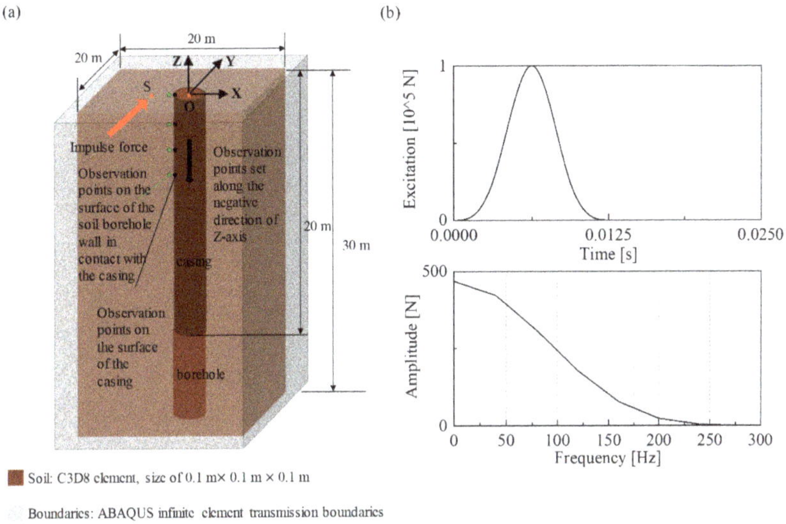

Figure 3. (**a**) Schematic diagram of 3D finite element analysis model, (**b**) approximate impulse concentrated force to simulate surface strike in downhole test.

According to Campanella and Stewart [6], and Ishihara [23], the strain level of the soil in the wave velocity test of the downhole method is in a small range; therefore, the soil was considered as a homogeneous, isotropic linear elastic half-space medium in the simulation of this paper. To minimize the influence of other factors, the soil as well as the casing was set as a single material. Different casing materials were simulated by varying the dielectric shear wave (S-wave) velocity and density of the casing, and by varying the unit contact between the casing and the soil, the contact state in practice was simulated. Given that the soil was in the small strain range under the pulse load, a sound contact state between the casing and the borehole wall was simulated by a common nodal contact between the casing and the soil unit, and a nodal no-contact state was simulated by the casing.

Figure 4 provides a schematic diagram of the different material casing models and the different contact models involved in the simulation. Figure 4a is used to examine the effects of different casing materials on the acquired signals, and to simulate the velocity responses of the measurement points for three conditions: no casing, PVC casing, and steel casing, respectively. Figure 4b,c simulates the effect of free casing between the casing and the borehole wall. In the model of Figure 4b, the free casing occurs in the first 10 m, and the casing is still in contact with the soil at the common node in the last 10 m. Indicated in the model of Figure 4c, the free casing is set in the depth range of 10–15 m, and the casing is still in contact with the soil at the common node in the remaining depth range. Figure 4d presents the new model added to the numerical simulation, in which a soil layer with low dielectric wave velocity is added between the casing and the original soil body to surround the casing. The thickness of the low velocity surround layer is 0.15 m, to simulate the solid medium filling between the casing and the borehole wall, and the three discrete units are in common node contact with each other. The model parameters of the model in Figure 4 are shown in Table 1. Three-dimensional eight-node hexahedral solid units are used to discretize the soil body, and the casing is discretized as a shell unit because its thickness is much smaller than its length, and the degree of freedom in the rotation direction is constrained in the modeling process. The size of soil and casing units and the time step of numerical calculation are set according to the stability criterion of numerical integration of time domain dynamic display [24], and both the size of the solid unit for discretizing soil and that of the shell unit for discretizing casing are not larger than 0.1 m. The size of unit for discretizing the soil's low velocity layer is not larger than 0.02 m. The soil and casing around 1/4 of the borehole in the X–Y plane is divided into five units. The time step of all models is 1×10^{-6} s.

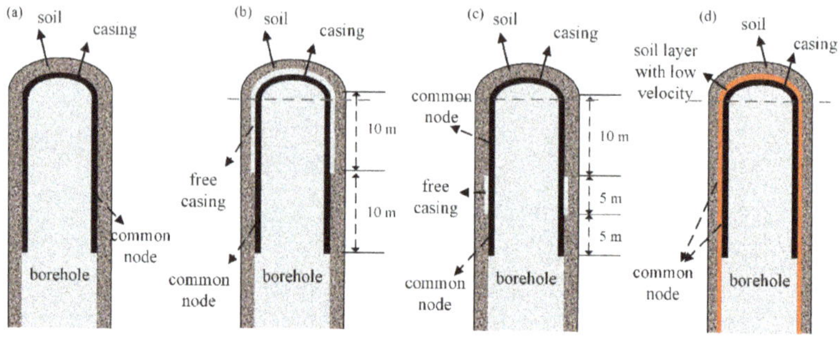

Figure 4. Schematic diagram of 4 touching condition model of cased borehole (**a**) cased borehole with different casing material, (**b**) free casing at the first 10 m depth, (**c**) free casing at 15–20 m depth, and (**d**) cased borehole surrounded by a soft soil layer with low shear wave velocity.

Table 1. Material parameters of models.

Model	Simulation Conditions	Soil			Casing			Contact Mode
		Shear Wave Velocity m/s	Poisson's Ratio	Density kg/m^3	Shear Wave Velocity m/s	Poisson's Ratio	Density kg/m^3	
Model 1	Condition 1: free casing	200	0.4	1700	-	-	-	-
	Condition 2: PVC casing	200	0.4	1700	1000	0.3	1400	common node
	Condition 3: steel casing	200	0.4	1700	3000	0.3	7300	common node
Model 2	Free casing in the depth range of 10 m	200	0.4	1700	3000	0.3	7300	first 10 m: common node post 10 m: free contact
Model 3	Free casing in the depth range of 10–15 m	200	0.4	1700	3000	0.3	7300	0–10 m: common node 10–15 m: free contact 15–20 m: common node
Model 4	Surrounded by a soil layer with low dielectric wave velocity	200/ 60	0.4/ 0.25	1700/ 1400	3000	0.3	7300	Both contact surfaces are in common node mode

4. Results and Discussion

4.1. Effect of Cased Material

Figure 5 displays the horizontal signals and the corresponding spectra acquired in the field experiment for PVC cased borehole and steel cased borehole within 22 m test depth each. In order to lessen the interference of noise, the signals of multiple strikes were recorded and then normalized and superimposed. Except for the signals from the 14 m and 16 m test points, clear peaks were observed at all depths of the PVC cased borehole, and more vibration cycles were noticed at the shallow test points than at the deeper test points (e.g., 18–22 m test points) as shown in Figure 5a. Through the Fourier variation of the signal obtained from the PVC cased borehole, the signal band was acquired, which ranged from 25 Hz to 100 Hz, similar to the shear band range recorded by other researchers in the downhole method [10,14], and did not change much with the depth of the measurement point. Notwithstanding, it can be seen that the main frequencies of the signals at 4 m, 10 m, and 12 m were higher than those at other depths. These depths were located in the sand layer, and it is assumed that the soil medium characteristics had a greater influence on the signal wave traces and spectral characteristics. The signals observed from the steel cased boreholes were of better quality than those from the PVC cased boreholes, and the signals from the 14 m and 16 m measurement points did not show signal delay, as indicated in Figure 5. Nevertheless, the signal from the shallow measurement points of the steel cased borehole also witnessed more cycles in the wave traces than the signal from the deep measurement points (e.g., 18–24 m measurement points). Shown in Figure 5d, the band range of the steel cased borehole signals was also within the range of 25–100 Hz, close to the band range of the signals obtained from the PVC cased borehole. However, the frequency band of the deeper measurement points obtained from the steel cased borehole was less attenuated around 75 Hz than that from the PVC cased borehole.

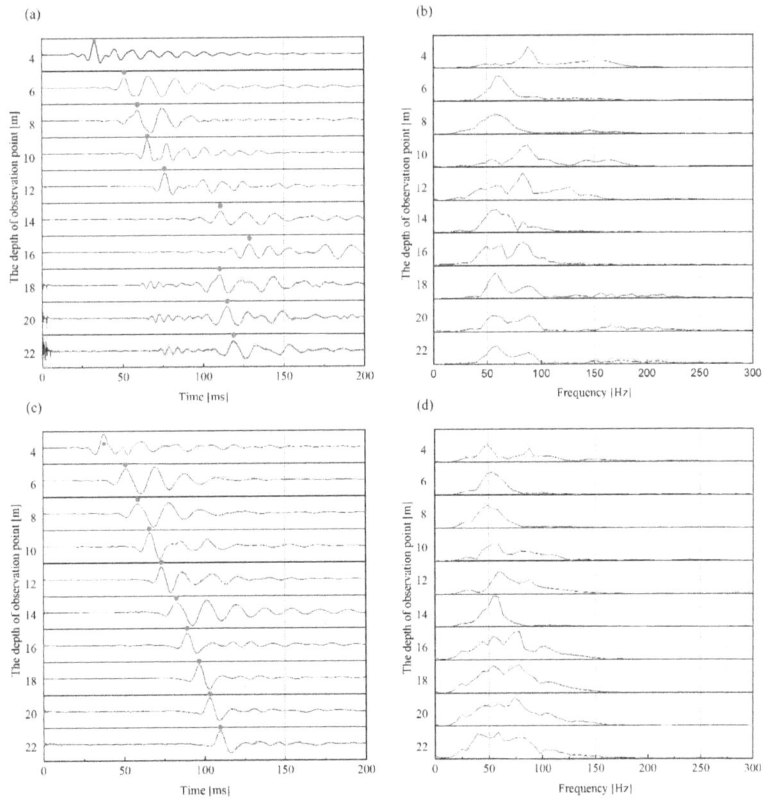

Figure 5. (**a**,**b**) Signal traces of horizontal direction of receivers and its corresponding FFT spectra at all testing depths in downhole test with PVC tube, (**c**,**d**) signal traces of horizontal direction of receivers and its corresponding FFT spectra at all testing depths in downhole test with steel tube.

Selecting and calculating the time difference of wave peaks between signals recorded at different depths is a regular approach to calculate shear wave brigades, also known as the peak-to-peak method [6,13]. Then, the shear wave travel distance was calculated based on the ray path, i.e., the shear wave velocity propagated in a given soil layer could be acquired based on the mathematical relationship between "distance–time–velocity" [6,13,25,26]. Figure 6 exhibits a comparison of the peak times of the signal wave traces recorded in the cased borehole tests of two materials and the shear wave velocities calculated from the signals of the adjacent measurement points. It can be found in Figure 6 that the signal peaks of the two types of cased boreholes were close to each other between 4 m and 12 m. Due to abnormal signal delay at 14 m and 16 m, the signal peaks of the PVC cased boreholes were larger than those of the steel cased boreholes from 14 m onwards. Since the vicinity of 14 m was exactly the partition interface of the site soil layer, and the nature of layer 4 and layer 5 soil was fine sand and powder sand, which could be easily disturbed, thereby leading to the collapse of the hole wall, it is presumed that the delay of PVC casing drilling signal was affected by the bad contact between casing and hole wall. When the depth of the measurement point continued to increase to enter the clay layer, the contact degree between PVC casing and hole wall increased at this time, and the signal gradually returned to normal. For a steel cased borehole with a casing medium wave velocity of up to 2000 m/s, the shear wave velocity was calculated from the recorded signal in the test depth of 4–22 m between 200 m/s and 300 m/s. Similar shear wave velocities were also observed in the PVC cased holes between 6 and 10 m. Due to abnormal signals obtained at 14 m and 16 m, a significant deviation was detected in the shear wave velocity calculated from the other tests of the PVC cased borehole.

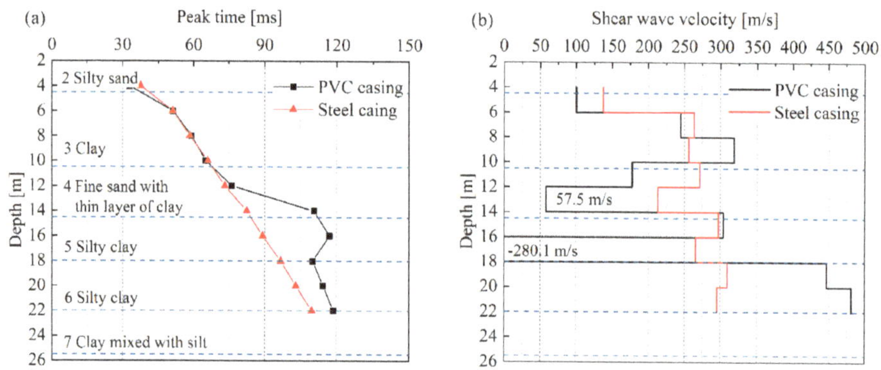

Figure 6. (**a**) Comparison of peak time of signals in horizontal direction received in PVC and steel casing downhole test, (**b**) comparison of shear wave velocity calculated through selected peak time based on peak-to-peak method.

As indicated in Figures 5 and 6, the recorded signals from the borehole with casing failed to identify the obvious wave components generated by the casing (possibly due to the small interference components generated by the casing), and the shear wave velocity calculated from the signals was not significantly increased by the casing, whether the shear wave velocity was 700 m/s for the PVC casing or 2500 m/s for the steel casing [7]. Therefore, it is assumed that the borehole with casing can record valuable signals normally, and the quality of the recorded signals in the borehole with casing is more likely to be deteriorated by the unsatisfactory contact between the casing and the borehole wall.

Figure 7 displays the numerical simulation results of the horizontal directional component response. It can be noted that, in the simulation, when the casing and soil are in common contact to simulate the close contact between the casing and the hole wall in the actual test, the amplitude and peak of the velocity response wave traces recorded at the same depth observation point without casing and that with two different media wave velocities are very close. Nonetheless, with the increase in the shear wave velocity of casing media, the amplitude of the response wave traces decreases and the position of the wave peak moves forward continuously. Through comparing the velocity response wave traces of the observation points at three depth positions, it is found that the amplitude of the response wave traces decreases with the increase in depth and the peak time decreases accordingly when the measurement points are located within the depth range of casing. The peak time difference between adjacent observation points was calculated based on the peak-to-peak method. Figure 7b demonstrates the effects of different casing materials on the calculated shear wave velocity in the numerical simulation. Based on the shear wave velocity test results of the free casing model, it can be observed that the relative error of shear wave velocity calculated by the model with a casing medium wave velocity of 1000 m/s is very small, basically less than 1%. For the casing medium wave velocity model with 3000 m/s, the calculation of shear wave velocity is not affected by the decrease in corresponding wave amplitude and the premise of wave peak time. For example, the relative error of 13–19 m depth is less than 1%. However, it should be noted that the relative error of shear wave velocity calculated by the casing medium wave velocity 3000 m/s model starts at 9 m depth and gradually increases with depth, and this feature still demands further investigation, but the final relative error is only 5%. It can be seen that the relative error of shear wave velocity for both casing models increases significantly near the depth of the bottom of the casing, even up to 13.3% (the model with casing medium wave velocity of 3000 m/s). This suggests that although the casing is in good contact with the soil, the difference in the response wave traces of the observation points between the transition zone of the casing and the soil in the borehole can still generate a major error in the calculated shear wave velocity, especially when the signals of the adjacent observation points are used for calculation.

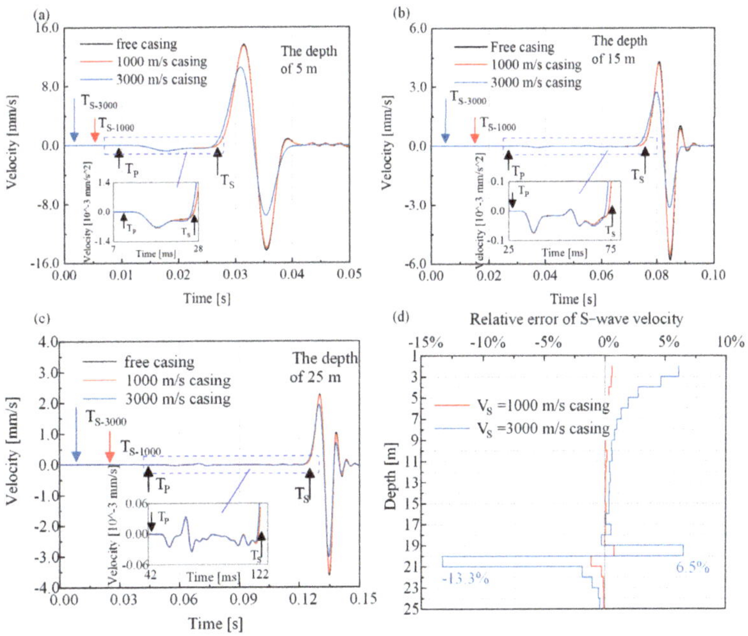

Figure 7. Results of numerical simulation of different casing materials models, (**a**) signal traces of horizontal Y-direction component of velocity response at the depth of 5 m, (**b**) signal traces of horizontal Y-direction component of velocity response at the depth of 15 m, (**c**) signal traces of horizontal Y-direction component of velocity response at the depth of 25 m, and (**d**) relative error of S-wave velocity between cased borehole model and non-cased borehole model with shear wave velocity calculation based on the adjacent signals of velocity response using peak-to-peak method.

Based on the numerical simulation results, it can be concluded that the presence of casing does cause errors in the calculated S-wave velocity, and the larger the shear wave velocity of the casing medium, the more significant the errors. However, the actual effect of the presence of the casing on the calculated shear wave velocity is very small, and the only thing that requires attention is the casing transition region. This further reflects that in the state of close contact between the casing and borehole wall, the presence of the casing, and the difference in casing material have little effect on the signal obtained in the borehole, and that the high-quality signal can still be recorded for the calculation of reliable shear wave velocity at this time. In contrast, the variation in the contact state seems to have a more pronounced effect on the recorded signals. For this reason, field experiments are conducted for three contact states and numerical simulations of the downhole method wave velocity tests performed for two of these contact states for better comparative studies.

4.2. *Effect of Contact State between Casing and Hole Wall*

As shown in Figure 2, a free casing model was formed after the soil around the casing was excavated and a "fluid immersion" model after filling the void with fluid. Due to the limitation of manual excavation, only a 2.2 m deep casing could be formed; therefore, the spacing of measurement points was changed to 0.5 m to record enough measurement signals for comparative analysis. Figure 8 presents the signals and spectral information of the 1.5 m measurement points recorded under the three models. By observing Figure 8a, it can be found that the measuring point could still record signals after PVC free casing. The signal arrival time was then delayed compared with the casing contact model, and the signal waveform attenuation slowed down, and there were still obvious up and down vibration wave traces after 50 ms. The frequency band of the signal recorded at the measurement point under the free casing model became significantly narrower, and the main frequency was above 100 Hz, significantly larger. As demonstrated in Figure 8c,d, the wave traces and spectral change characteristics appearing after free casing were especially obvious when the casing material was steel. The test was conducted again after a fluid injection in the void. At this time, the number of cycles of the signal wave traces was similar to that of the contact model, and the peak position of the signal wave traces is different from the delayed position of the casing after dehollowing, but close to the peak position of the contact model again, or even slightly ahead of the contact model. Observing the frequency spectrum of the signal recorded in the water injection model in Figure 8b,d, it was found that the frequency band range increased compared with that in the free casing model and was again close to that in the contact model, especially when the casing material was PVC.

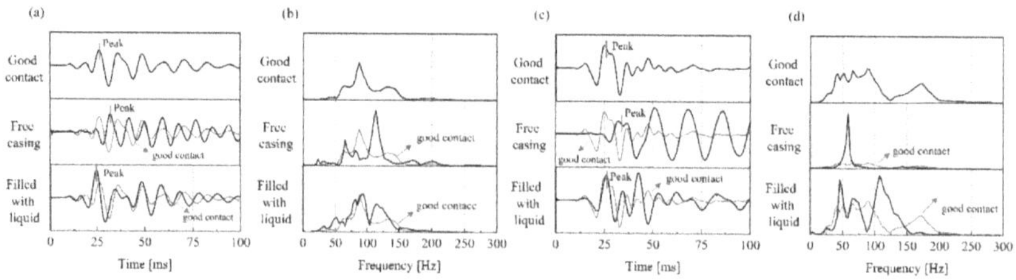

Figure 8. Signal traces of horizontal direction and corresponding FFT spectra of 1.5 m receiver in different casing model (**a**,**b**) of PVC cased borehole, (**c**,**d**) of steel cased borehole.

The results in Figure 8 suggest that when the lower part of the casing was still in favorable contact with the hole wall, even if the casing was dehollowed, the signal could still be recorded at the dehollowing location. Based on the signal delay, it can be inferred that the signal propagated upward along the casing through the dehollowing position and then was detected by the measuring point in the dehollowing position. When the casing was surrounded by water, the measured points in the casing could record signals similar to those when the casing was in good contact with the hole wall. This means that in addition to filling the gap between the casing and the hole wall with sand, pebbles, concrete, and solid objects, the gap filled with liquid could also solve the problem of unfavorable contact between the casing and the hole wall. The liquid could not transmit the shear wave, but since the cased borehole was located in three-dimensional space, shear waves generated by horizontal percussion traveled in a spherical shape, thereby causing the vibration of soil particles. Therefore, more energy was transferred around the casing as shear waves around the liquid and was transferred to the casing in the form of a compression wave, which is then captured by the geophone. Considering that the compressional wave velocity is larger than the shear wave, the signal recorded at the measurement point is advanced, which also explains the advancement of the signal wave crest of the water injection model in Figure 8a,c.

Figure 9 demonstrates the signal and spectrum information of all the measurement points within the test depth of the three contact state models of the PVC cased borehole and the steel cased borehole. It can be seen that in Figure 9a,c, the signal wave traces recorded in the dehollowing area (1.0–2.0 m) and the un-dehollowing area (2.5–4.0 m) differed greatly in the dehollowing model with the excavation surface (2.2 m) as the boundary. Especially when the steel casing was dehollowed, the signal was attenuated very slowly. At the same time, the signal arrival of the measurement point in the dehollowing area was delayed with the increasing shallow depth of the measurement point, shown in Figure 10. This indicates that the wave recorded at the measuring point in the casing dehollowing area propagated along the casing from bottom to top. The recorded signal frequency bands of the measuring points in the casing dehollowing area were narrowed, but still in the range of the soil shear wave (clearly observed in Figure 9c). This means that the signal propagation through the casing to the measurement point still retained the component of the excitation signal propagating through the soil, presumed to be amplified by the influence of the casing when it propagates in the casing, and therefore could be seen in the steel casing, made of more rigid material. Hence, it can be seen that the signal wave traces recorded in the steel casing with greater stiffness had a more obvious feature of increased circulation and a narrower signal band. The shape and spectral characteristics of the recorded signal wave traces in the fluid immersion model were very close to those of the casing contact model. As can be seen from Figure 10, the shear wave velocity between the measurement points calculated using the peak-to-peak method and the linear path of the adjacent measurement points was still different between the contact model and the fluid immersion model, but the relative error of the calculated shear waveform velocity was still within the acceptable range for engineering testing.

Numerical simulation was adopted to make up for the shortcoming of the shallow depth of manual excavation dehollowing in the field test by increasing the depth of the free casing to the first 10 m, and it was also leveraged to calculate the dehollowing condition in the middle region of the casing only, which could be difficult to realize in the field. The modeling process is shown in Section 3. The material parameters of the model were the same as those of the model with good casing contact (Model 1, working condition 3). The numerical results are displayed in Figure 11. For the first 10 m of the casing after dehollowing, the velocity response of the observation point in the dehollowing area showed the characteristics of delayed wave arrival time and increased wave cycle, and the signal arrival time shows an increase with the shallow depth of the observation point, while the spectrum changed from a wide band before dehollowing to two narrower spectral peaks. These changes were similar to the change pattern of the field test results (free casing model). When the dehollowing position occurs in the middle section of the casing, it can be observed from Figure 11c,d that the wave traces of the observation point response arrived earlier than when the casing was in good contact, and the wave peak position was delayed with the increase in the observation point depth, indicating that the wave propagation in the dehollowing area became top-down and propagated at a faster speed than the wave speed of the soil shear wave. The main frequency after the narrowing was larger than that in the first 10 m of the free casing model.

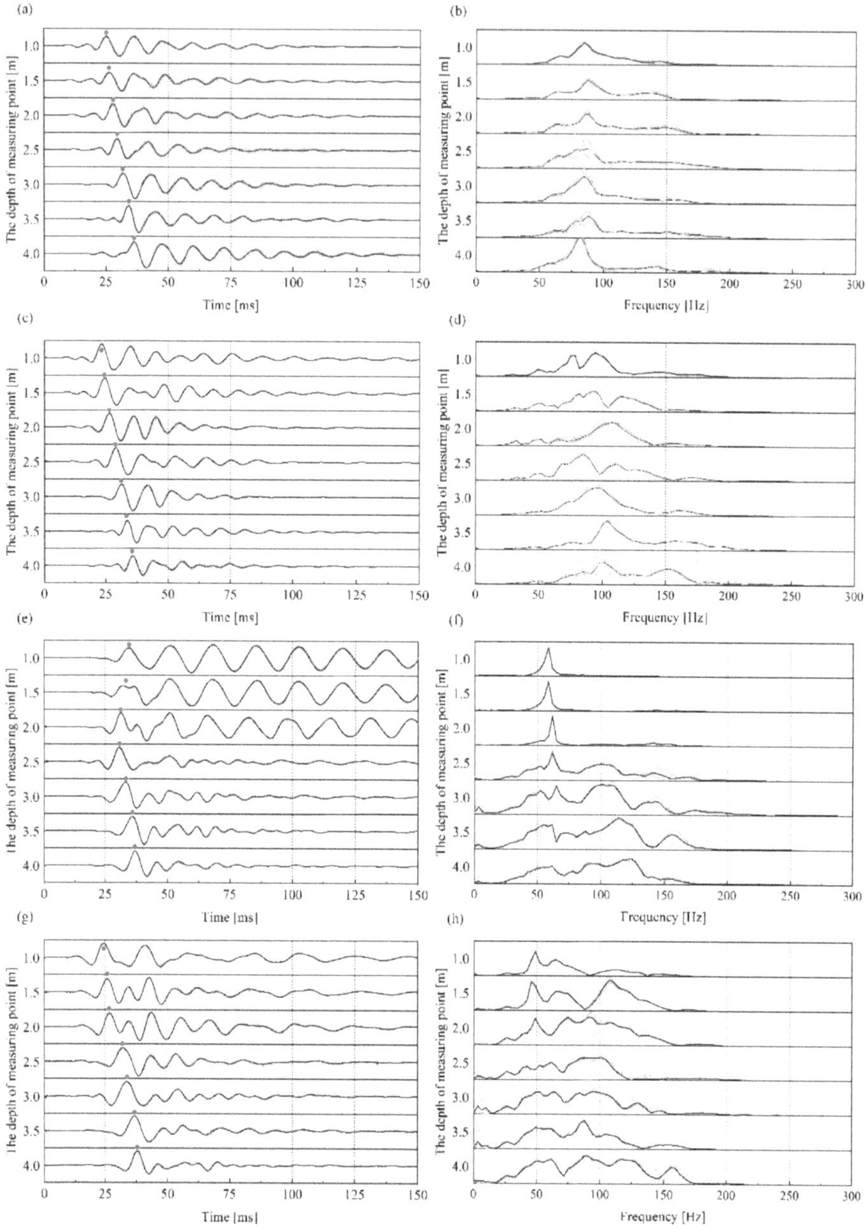

Figure 9. (**a**,**b**) Signal traces of horizontal direction and its corresponding FFT spectra of receivers at all testing depths, (**a**,**b**) in free casing model of PVC cased borehole, (**c**,**d**) in casing immersed in fluid model of PVC cased borehole, (**e**,**f**) in free casing model of steel cased borehole, (**g**,**h**) in casing immersed in fluid model of steel cased borehole.

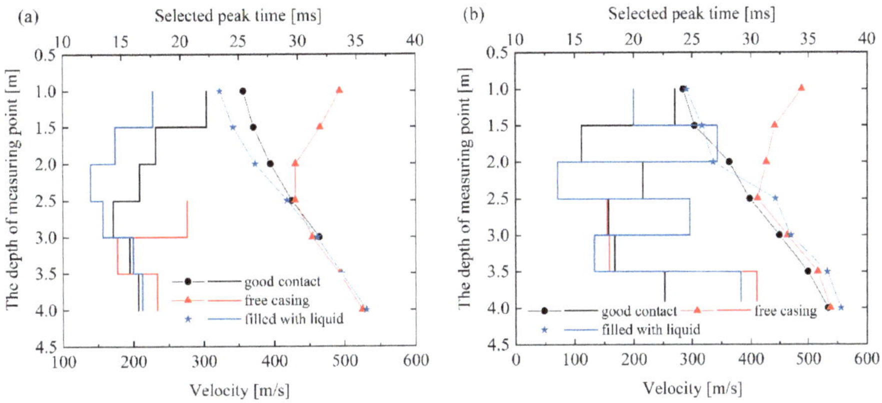

Figure 10. Comparison of peak time of signals and shear wave velocity calculated through selected peak time based on peak-to-peak method in three casing model, (**a**) PVC cased borehole, (**b**) steel cased borehole.

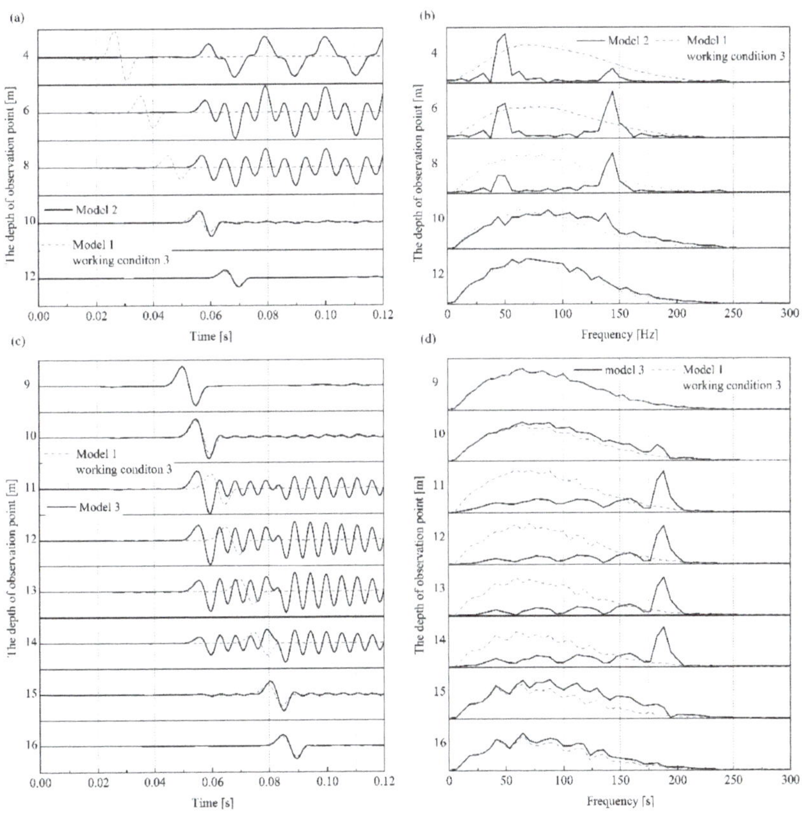

Figure 11. Signal traces of horizontal Y-direction component of velocity response and its corresponding FFT spectra from the numerical simulation of case borehole models with different touching conditions, (**a**,**b**) the model with a free casing at the depth of 0–10 m, (**c**,**d**) the model with a free casing at the depth of 10–15 m.

Figure 12 indicates the numerical simulation results of the signals obtained from the measurement points inside the casing after the casing is filled with soft soils with low dielectric wave velocity. Under the modeling conditions of Model 1 and working condition 3, a layer of soil layer with medium shear wave velocity of 60 m/s was added around the casing. The specific modeling parameters are shown in Table 1. As reflected in Figure 12, the newly added medium wave velocity layer caused a decrease in the amplitude and a slight delay in the peak time of the horizontal Y-direction component of the velocity response at the observation point, but the overall wave traces were close to those without the low velocity layer. The results are revealed in Figure 12d, which shows that the calculated shear velocities were similar to those without the low velocity layer at most depths, except for the top and bottom of the casing (19–20 m interval), where the shear velocities were inaccurate compared to those in the good contact model. It can be noted that when the casing was surrounded by a low-velocity soil layer, the signals obtained from the observation points in the borehole were basically the same as those when the casing was surrounded by a preset soil body. However, due to the presence of the low-velocity soil layer, the top of the casing was influenced by the near-field effect, and the error caused by the contact medium conversion of the probe at the bottom of the casing was also magnified.

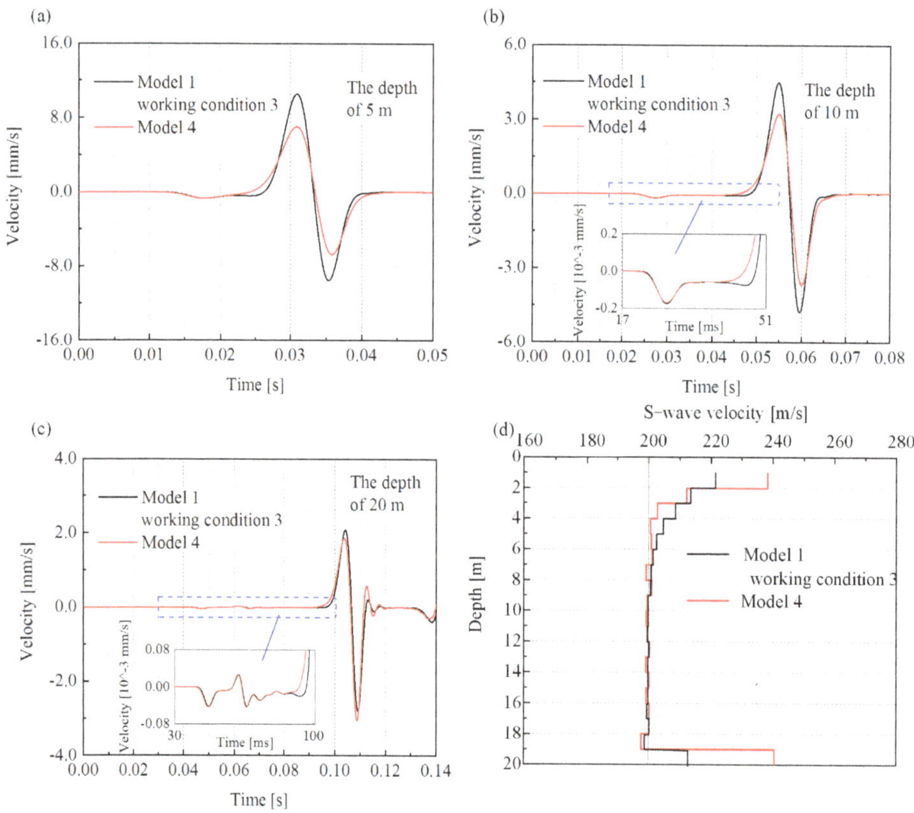

Figure 12. Results of numerical simulation of cased model with a casing that is surrounded by low-velocity soil layer, (**a**) signal traces of horizontal Y-direction component of velocity response at the depth of 5 m, (**b**) signal traces of horizontal Y-direction component of velocity response at the depth of 10 m, (**c**) signal traces of horizontal Y-direction component of velocity response at the depth of 20 m, (**d**) shear wave velocity calculation based on the adjacent signals of velocity response using peak-to-peak method.

5. Conclusions

In the current research, the wave traces characteristics and spectral characteristics of the signals obtained from the measurement points under different casing materials were compared with the results of 3D finite element numerical simulations of wave velocity tests under the downhole method with casing. For the change of contact form between the borehole and the casing, the influence of the presence of casing dehollowing and the liquid filling in the casing dehollowing area on the signal of the measurement points in the casing was also studied by field experiments and numerical simulations. The main research conclusions are as follows.

(1) Under the condition of desirable contact between the casing and the borehole wall, both PVC cased borehole and steel cased borehole can receive good quality signals in the casing. There are no casing-induced components in the signal wave traces of the measurement point. Both the signal wave traces of different casing materials and the frequency band characteristics are similar. However, due to the existence of the casing, the soil layer prone to collapse and the boundary between soil layers will more easily contribute to the abnormal signal recorded by the measurement points in the casing.
(2) On account of characteristics such as the change in arrival time, the slowdown in wave attenuation, and the narrowing of the main frequency, significant errors in the calculation of shear wave speed will arise. It should be noted that when the dehollowing occurs in a certain area of the casing it will be more complex to identify based on the signals recorded at the dehollowing area, and the shear wave speed calculation error is relatively significant at this time. While dehollowing, the greater the stiffness of the casing material, the greater the impact on the signal of the measurement point.
(3) Fluid injection between the casing and the hole wall can eliminate to a greater extent the error caused by the free casing, but at this time the measurement point signal still contains interference components generated by the casing. However, the calculated shear wave velocity error is within the acceptable range of engineering.
(4) Compared with the presence of casing, the contact state between the casing and the hole wall in the wave velocity test with the casing downhole method has a greater impact on the signal of the measurement point. With a view to acquire better signal of measuring points and eliminate the error of calculating wave velocity, it is an effective approach to, after drilling holes, fill the space between the hole wall and the casing with materials, such as grout and pebble sand.
(5) Two improvements can be made for the further study. One is to apply artificially built site models to accurately control the depth and characteristics of different layers of soil, and better comprehend the impact produced by soil layer differences. The other is to establish a more refined numerical simulation model to simulate the effects of unsaturated soil and the presence of groundwater, as a means to better simulate the actual situations.

Author Contributions: S.Y.: Writing—original draft, Visualization. Y.Y.: Conceptualization, Methodology, Supervision, Funding acquisition, Writing—original draft, Writing—review and editing. W.Z.: Methodology, Project administration, Writing—review and editing. J.S.: experiments and experimental data organization. Z.Z.: Methodology, Project administration, Writing—review and editing. All authors have read and agreed to the published version of the manuscript.

Funding: This study was supported by the National Key R&D Program of China (No. 2019YFE0115702).

Institutional Review Board Statement: Not applicable.

Informed Consent Statement: Not applicable.

Data Availability Statement: No new data were created.

Conflicts of Interest: The authors declare that they have no known competing financial interest or personal relationship that could have appeared to influence the work reported in this paper.

References

1. Wang, H.Y.; Jiang, W.P. In-situ shear-strain profile assessment in geotechnical array with ground motion data. *Soil Dyn. Earthq. Eng.* **2023**, *165*, 107667. [CrossRef]
2. Ansary, M.A.; Tasmiah, A.; Al Noman, M.A. Development of empirical correlations between shear wave velocity and in situ penetration test results for different types of soils in DMDP area, Bangladesh. *Arab. J. Geosci.* **2023**, *16*, 280. [CrossRef]
3. Chandrasekaran, S.; Santibanez, F.; Tripathi, B.B.; DeRuiter, R.; Bruegge, R.V.; Pinton, G. In situ ultrasound imaging of shear shock waves in the porcine brain. *J. Biomech.* **2022**, *134*, 110913. [CrossRef] [PubMed]
4. Thornley, J.; Dutta, U.; Fahringer, P.; Yang, Z.J. In situ shear-wave velocity measurements at the Delaney Park downhole array, Anchorage, Alaska. *Seismol. Res. Lett.* **2019**, *90*, 395–400. [CrossRef]
5. Mital, U.; Kawamoto, R.; Andrade, J.E. Effect of fabric on shear wave velocity in granular soils. *Acta Geotech.* **2020**, *15*, 1189–1203. [CrossRef]
6. Campanella, R.G.; Stewart, W.P. Seismic cone analysis using digital signal processing for dynamic site characterization. *Can. Geotech. J.* **1992**, *29*, 477–486. [CrossRef]
7. Seylabi, E.; Hallal, M.M.; Cox, B.R. Site characterization at Treasure Island and Delaney Park downhole arrays by heterogeneous data assimilation. *Earthq. Spectra* **2022**, *38*, 2398–2421. [CrossRef]
8. Sahadewa, A.; Zekkos, D.; Stokoe, K.H.; Woods, R.D. Field Testing Method for Evaluating the Small-Strain Shear Modulus and Shear Modulus Nonlinearity of Solid Waste. *Geotech. Test. J.* **2015**, *38*, 427–441. [CrossRef]
9. Di Fiore, V.; Cavuoto, G.; Tarallo, D.; Punzo, M.; Evangelista, L. Multichannel Analysis of Surface Waves and Down-Hole Tests in the Archeological "Palatine Hill" Area (Rome, Italy): Evaluation and Influence of 2D Effects on the Shear Wave Velocity. *Surv. Geophys.* **2016**, *37*, 625–642. [CrossRef]
10. Sully, J.P.; Campanella, R.G. Evaluation of in situ anisotropy from crosshole and downhole shear wave velocity measurements. *Geotechnique* **1995**, *45*, 267–282. [CrossRef]
11. Hou, X.; Yang, X.; Liao, Z.; Bo, J. An optimized approach for single-hole method of shear wave velocity measurement based on correlation functions. *Rock Soil Mech.* **2006**, *2006*, 1161–1165. (In Chinese)
12. Akram, J.; Eaton, D.W. A review and appraisal of arrival-time picking methods for downhole microseismic data. *Geophysics* **2016**, *81*, KS71–KS91. [CrossRef]
13. Kim, D.S.; Bang, E.S.; Kim, W.C. Evaluation of various downhole data reduction methods for obtaining reliable Vs profiles. *Geotech. Test. J.* **2004**, *27*, 585–597.
14. Beeston, H.E.; McEvilly, T.V. Shear wave velocities from down-hole measurements. *Earthq. Eng. Struct. Dyn.* **1977**, *5*, 181–190. [CrossRef]
15. Hoar, R.J.; Stokoe, K.H. *Generation and Measurement of Shear Waves In Situ/Dynamic Geotechnical Testing*; ASTM International: West Conshohocken, PN, USA, 1978.
16. Hunter, J.A.; Benjumea, B.; Harris, J.B.; Miller, R.D.; Pullan, S.E.; Burns, R.A.; Good, R.L. Surface and downhole shear wave seismic methods for thick soil site investigations. *Soil Dyn. Earthq. Eng.* **2002**, *22*, 931–941. [CrossRef]
17. Larkin, T.J.; Taylor, P.W. Comparison of down-hole and laboratory shear wave velocities. *Can. Geotech. J.* **1979**, *16*, 152–162. [CrossRef]
18. Crice, D. Borehole Shear-Wave Surveys for Engineering Site Investigations. Geostuff. 2002. Available online: www.georadar.com/geostuff (accessed on 10 May 2021).
19. Chen, J.; Liu, L.; Zeng, B.; Tao, K.; Zhang, C.; Zhao, H.; Li, D.; Zhang, J. A Constitutive Model to Reveal the Anchorage Mechanism of Fully Bonded Bolts. *Rock Mech. Rock Eng.* **2023**, *56*, 1739–1757. [CrossRef]
20. Deng, M. *Study on the Geometry Distribution and Paleoearthquake of the Xiadian Buried Fault in the Area of Beijing Plain*; Institute of Disaster Prevention: Langfang, China, 2018. (In Chinese)
21. Fei, K. *Application of ABAQUS in Geotechnical Engineering*; China Water & Power Press: Beijing, China, 2010. (In Chinese)
22. *Abaqus Theory Manual*; ABAQUS Inc.: Palo Alto, CA, USA, 2010.
23. Ishihara, K. *Soil Behaviour in Earthquake Geotechnics*; Oxford University Press: Oxford, UK, 1996.
24. Liao, Z. *Introduction to Wave Motion Theories in Engineering*; Science Press: Beijing, China, 1996. (In Chinese)
25. Stewart, W.P.; Campanella, R.G. Practical aspects of in situ measurements of material damping with the seismic cone penetration test. *Can. Geotech. J.* **1993**, *30*, 211–219. [CrossRef]
26. ASTM. *Standard Test Methods for Downhole Seismic Testing*; Annual book of ASTM standard; ASTM: West Conshohocken, PA, USA, 2007; Volume 4, p. D7400.

sustainability MDPI

Article

Discussion on Adjustment Method of the Characteristic Period of Site Response Spectrum with Soft Soil Layer

Yuandong Li, Bing Hao, Zhen Chen, Zhenghua Zhou *, Zhu Bian, Yi Han and Cheng Peng

College of Transportation Engineering, Nanjing Tech University, Nanjing 210009, China; wddd2017@njtech.edu.cn (Y.L.)
* Correspondence: bjsmoc@njtech.edu.cn

Abstract: Twelve site models were established based on the analysis of the influence of site conditions on earthquake damage and the influence of the soft soil layer on-site seismic response. The equivalent linearization site seismic response analysis is carried out at different input ground motion levels to discuss the influence of soft soil layer thickness and buried depth. The results show that the characteristic period of the response spectrum exhibits a gradual increase as the buried depth or thickness of the soft soil layer increases. Furthermore, the characteristic period of the response spectrum also increases with the rise in the input ground motion peak. Moreover, according to the influence characteristics of soft soil thickness, buried depth, and input ground motion intensity on the characteristic period of the site acceleration response spectrum, a method for adjusting the characteristic period of the site acceleration response spectrum with a soft soil layer is put forward.

Keywords: site conditions; ground motion; characteristic period; equivalent linearization; earthquake damage

Citation: Li, Y.; Hao, B.; Chen, Z.; Zhou, Z.; Bian, Z.; Han, Y.; Peng, C. Discussion on Adjustment Method of the Characteristic Period of Site Response Spectrum with Soft Soil Layer. *Sustainability* **2023**, *15*, 8837. https://doi.org/10.3390/su15118837

Academic Editors: Chong Xu, Su Chen and Shuang Li

Received: 22 March 2023
Revised: 5 May 2023
Accepted: 25 May 2023
Published: 31 May 2023

1. Introduction

Site conditions play a decisive role in the influence of ground motion [1,2]. In engineering seismic design, the engineering design and construction solutions that are compatible with the characteristics of the site conditions can effectively reduce the damage to the engineering structure from seismic effects, extend its safe service life, and contribute to the sustainability of the engineering structure. Site conditions are mainly considered in the following aspects: engineering geological conditions and hydrogeological conditions near the surface [3,4], local topographic effects [5,6], and fault site effects [7,8]. The engineering geological conditions near the surface can be investigated from three angles, such as geotechnical type, overburden thickness, and soil structure [9]. Generally speaking, the ground motion on the bedrock site is smaller, followed by the hard soil site, and the soft soil site is the largest under the same ground motion. Meanwhile, the existing analysis shows that the soil structure also has a significant influence on ground motion [10,11]. Specifically, it has been observed that as the burial depth of the hard interlayer increases, both the peak acceleration and the response spectrum of the ground surface increase. Moreover, as the thickness of the hard interlayer increases, the peak acceleration of the ground surface first decreases gradually, then increases gradually, while the response spectrum value increases [12–15]. On the other hand, increasing the burial depth or thickness of the soft interlayer leads to a decrease in both the peak acceleration and the peak of the response spectrum [16,17].

In recent years, people have paid attention to the influence of soft soil on the site's seismic response, and some scholars have carried out in-depth analyses from different perspectives. Xu et al. [18] conducted a study on the seismic damage mechanism of soft soil sites in Fuzhou. The findings indicated that soft soil increases the site excellence period to some extent, and the excellence period is closely related to the stiffness of the foundation soil. Yao et al. [19] suggested that the existence of a local soft interlayer can significantly amplify or attenuate ground motion, which might affect the lagged spatial consistency between spatially varying ground motions. Cao [20] analyzed the effect of the burial depth of the soft interlayer on the ground motion. The results revealed that the amplification and predominant frequency of the site decreased with the deepening of the soft interlayer location. In a study conducted by Wang et al. [21], the influence of the buried depth of a soft interlayer on ground motion parameters was investigated through site seismic response analysis using the equivalent linear method. The findings revealed that as the burial depth of the soft interlayer increased from shallow to deep, the peak acceleration, peak velocity, and peak response spectrum exhibited a decreasing trend. Additionally, the period corresponding to the maximum value of the characteristic period and response spectrum showed an increasing trend with an increase in the burial depth of the soft interlayer. Tian [22] proposed that under the same Class III site conditions, the presence of soft soil layers makes the ground motion parameter values vary greatly compared to those obtained for sites without soft soil layers. Furthermore, the different locations of the soft soil layers in the soil structure lead to large differences in the ground motion parameter values. Yan et al. [23] focus on the influence of soft interlayer and slope effects on the dynamic response of slope sites through acceleration amplification effects and seismic wave fluctuation mechanisms. They combine the traditional Fourier spectrum and Hilbert marginal spectrum methods to demonstrate the spectral variation characteristics of sites from the frequency domain perspective. Li and Xia et al. [24] calculated three profiles with thicknesses of 3 m, 5 m, and 9 m to analyze the effects of burial depth and thickness of soft soil interlayer on surface ground motion parameters under the condition of constant burial depth. Wang et al. [25] conducted research using an ideal site as the base model and varied the position of the soft soil layer to establish corresponding calculation models. Through soil response analysis, they investigated the effects of the soft interlayer at different locations on parameters such as peak surface acceleration, amplification coefficient, and equivalent shear wave velocity at the site. The study concluded that the influence of the soft interlayer on the site's peak acceleration exhibits an initial amplification followed by a reduction. Additionally, the propagation capacity of the four site types exhibits a certain range, and the equivalent shear wave velocity does not accurately reflect the soil layer structure.

In summary, the soft soil layer has a significant effect on the site's seismic response, especially in the form of a significant increase in the characteristic period. Additionally, compared with the Code [26], the characteristic period of the acceleration response spectrum of a site with a soft soil layer after the regulation is much larger than the value specified in the Code. The method of determining the characteristic period is the key technology in earthquake engineering, and there is little research on the adjustment method of the characteristic period of the site response spectrum. Although a few scholars have conducted relevant research [27–29], the currently available adjustment methods are not intended for sites with soft interlayers. Therefore, a new characteristic period calibration method applicable to soft soil sites has yet to be proposed (i.e., a new seismic engineering technology). In light of these findings, the present study aims to develop soft site models incorporating silt layers, building upon previous research. The influence of the soft soil layer on the seismic response of the site will be analyzed, and a method for adjusting the characteristic period of the response spectrum will be proposed. This research endeavor intends to provide a theoretical foundation for determining the characteristic period of the seismic response spectrum for sites with soft soil layers.

2. Ground Motion Input

The input ground motion for the seismic response analysis of a soft site with a silt layer is obtained by artificial synthesis [30–32]. The synthesized ground motions consist of peak accelerations of 50 gal, 100 gal, and 200 gal (1 gal = 1 cm/s^2), time intervals of 0.02 s, and discrete points of 2048. The acceleration time range is reduced by half in magnitude as the input ground motions are calculated for the one-dimensional soil seismic response analytical model. The input ground motion acceleration time range and acceleration response spectrum are drawn in Figure 1, and the characteristic periods of the response spectrum are 0.30 s, 0.35 s, and 0.40 s, respectively.

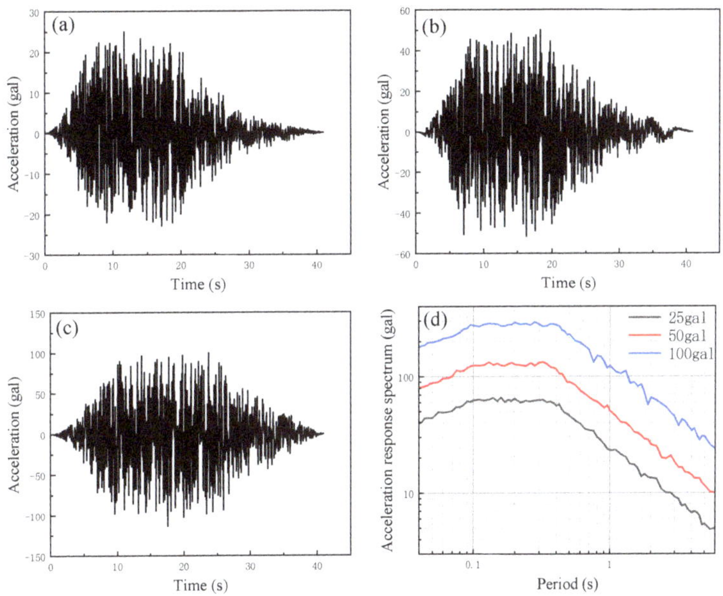

Figure 1. Input ground motion: (**a**) acceleration time range (PGA = 25 gal); (**b**) acceleration response spectrum (PGA = 50 gal); (**c**) acceleration time range (PGA = 100 gal); (**d**) acceleration response spectrum.

3. Site Seismic Response Analysis

3.1. Site Seismic Response Calculation Model and Determination of Dynamic Parameters

For analyzing the seismic response of the site with a soft soil layer, six analytical models are established, each representing a different burial depth of a single layer of silt. Additionally, six analytical models are created, representing different burial depths of two layers of silt. These models are developed based on the survey data and the specific engineering seismic conditions of the site. The objective is to analyze the influence of the characteristics of the thickness and burial depth of the soft soil layer on the site's seismic response.

To analyze the ground motion response of soil layers at the site, it is desirable to have detailed information about the soil profile, including the layered thickness and properties of each soil layer. Additionally, the mechanical properties of the soil also play a crucial role in analyzing the ground motion response. These properties include the shear wave velocity, density, and dynamic nonlinear characteristic parameter values of the soil. Based on the actual field investigation and experimental data and considering different silt layer thicknesses and burial depths, six analytical models with one layer and two layers of silt, each with different burial depths, have been established [33–35]. The profile and mechanical parameters of model 1 are summarized in Table 1. Models 2 to 6 are obtained on the basis of model 1 by varying the number of silty clay layers overlying the silt layer and gradually moving the silty clay from under the silt layer of model 1 to above the silty layer. For instance, model 2 consists of one layer of silty clay with soil class number 3. Model 3 includes two layers of silty clay with class numbers 3 and 4. Model 4 is composed of three layers of silty clay with class numbers 3, 4, and 5. Model 5 is constructed with four layers of silty clay with class numbers 3, 4, 5, and 6. Model 6 comprises five layers of silty clay with class numbers 3, 4, 5, 6, and 7. The profile and mechanical properties of model 7 are summarized in Table 2. Models 8 to 12 are also obtained by changing the number of layers of overlying silty clay based on model 7. This means that model 8 is overlaid with one layer of silty clay, classified as 3. Model 9 is overlaid with two layers of silty clay, classified as 3 and 4. Model 10 is overlaid with three layers of silty clay, classified as 3, 4, and 5. Model 11 is overlaid with four layers of silty clay, classified as 3, 4, 5, and 6. Lastly, model 12 is overlaid with five layers of silty clay, classified as 3, 4, 5, 6, and 7.

Table 1. Analysis model 1.

No.	Rock-Soil	Soil Class	Depth at the Bottom of Layer (m)	Layer Thickness (m)	Shear Wave Velocity (m/s)	Density (t/m^3)
1	Silt	1	5.0	5.0	112	1.58
2	silty clay	3	9.5	4.5	160	1.86
3	silty clay	4	13.0	3.5	165	1.87
4	silty clay	5	17.0	4.0	199	1.88
5	silty clay	6	21.0	4.0	212	1.96
6	silty clay	7	24.0	3.0	242	1.98
7	rounded gravel	8	27.0	3.0	258	2.20
8	fully weathered andesite	8	30.0	3.0	393	2.25
9	bedrock of model	9			516	2.65

Table 2. Analysis model 7.

No.	Rock-Soil	Soil Class	Depth at the Bottom of Layer (m)	Layer Thickness (m)	Shear Wave Velocity (m/s)	Density (t/m^3)
1	Silt	1	5.0	5.0	112	1.58
2	Silt	2	10.0	5.0	112	1.66
3	silty clay	3	14.5	4.5	160	1.86
4	silty clay	4	18.0	3.5	165	1.87
5	silty clay	5	22.0	4.0	199	1.88
6	silty clay	6	26.0	4.0	212	1.96
7	silty clay	7	29.0	3.0	242	1.98
8	rounded gravel	8	32.0	3.0	258	2.20
9	fully weathered andesite	8	35.0	3.0	393	2.25
10	bedrock of model	9			516	2.65

Table 3. Nonlinear parameters of dynamic shear of various soils at different shear strain levels.

Soil Class	Soil Layer	Modulus Ratio	Shear Strain (10^{-4})							
		Damping Ratio	0.05	0.1	0.5	1	5	10	50	100
1	silt	G/G_{max}	0.9902	0.98086	0.9105	0.8358	0.5045	0.3374	0.0923	0.0483
		ζ	0.0173	0.0244	0.0525	0.0711	0.1236	0.1429	0.1672	0.1712
2	silt	G/G_{max}	0.9913	0.9827	0.9189	0.8500	0.5313	0.3617	0.1018	0.0536
		ζ	0.0088	0.0135	0.0356	0.0525	0.1073	0.1303	0.1615	0.1669
3	silty clay	G/G_{max}	0.9918	0.9838	0.9241	0.8588	0.5489	0.3783	0.1085	0.0573
		ζ	0.0138	0.0199	0.0459	0.0641	0.1201	0.1428	0.1735	0.1788
4	silty clay	G/G_{max}	0.9925	0.9851	0.9296	0.8684	0.5689	0.3975	0.1166	0.0619
		ζ	0.0123	0.0176	0.0402	0.0561	0.1053	0.1258	0.1542	0.1592
5	silty clay	G/G_{max}	0.9939	0.9878	0.9419	0.8903	0.6187	0.4479	0.1396	0.0750
		ζ	0.0157	0.0218	0.0461	0.0626	0.1136	0.1356	0.1677	0.1736
6	silty clay	G/G_{max}	0.9943	0.9887	0.9460	0.8975	0.6365	0.4668	0.1490	0.0805
		ζ	0.0181	0.0249	0.0512	0.0688	0.1234	0.1473	0.1827	0.1894
7	silty clay	G/G_{max}	0.9950	0.9901	0.9524	0.9092	0.6669	0.5003	0.1668	0.0910
		ζ	0.0106	0.0152	0.0342	0.0478	0.0936	0.1154	0.1504	0.157
8	Rounded gravel and pebbles	G/G_{max}	0.990	0.970	0.900	0.850	0.700	0.550	0.320	0.200
		ζ	0.004	0.006	0.019	0.030	0.075	0.090	0.110	0.120
9	bedrock	G/G_{max}	1.000	1.000	1.000	1.000	1.000	1.000	1.000	1.000
		ζ	0.004	0.008	0.01	0.015	0.021	0.030	0.036	0.046

The dynamic nonlinear parameters and density values of each soil layer in different calculation site models are derived from the experimental results of the seismic safety evaluation project at the actual engineering site. The dynamic nonlinear parameters and density values of each soil layer are listed in Table 3, and the density values are listed in Tables 1 and 2.

3.2. Analysis of Calculation Results

When the equivalent linearization method [36,37] is adopted for the site seismic response calculation of each analytical model, the input ground motion acceleration time histories corresponding to three peak acceleration levels (25 gal, 50 gal, and 100 gal) are taken as the computational base incident ground motion of the one-dimensional soil response analysis model. The ground motion acceleration time histories and response spectrum values of the surface horizontal seismic response are obtained through the computational analysis of the horizontal seismic response of each analytical model.

The peak acceleration of surface horizontal seismic response under different ground motions for each analytical model is shown in Table 4, from which the dynamic amplification coefficients of surface horizontal seismic response for each analytical model are obtained and presented in Table 5.

Table 4. The peak acceleration of each analytical model.

Input Peak Acceleration/gal				Input Peak Acceleration/gal			
Surface Peak Acceleration/gal	**25**	**50**	**100**	**Surface Peak Acceleration/gal**	**25**	**50**	**100**
Analytical Model				**Analytical Model**			
1	53.1	97.1	193.2	7	48.9	96	174.2
2	43.8	86.5	152.3	8	38.7	62.6	110.1
3	41.2	72.4	118.7	9	33.7	55.2	102.8
4	38.7	58.2	109.9	10	31.1	53.9	83.9
5	36.4	52.6	100.6	11	30.4	48.2	69.4
6	30.6	45	87.4	12	29.6	46.4	64.6

Table 5. Dynamic amplification coefficient of the surface seismic response of each analytical model.

Input Peak Acceleration/gal				Input Peak Acceleration/gal			
Dynamic Amplification Coefficient	**25**	**50**	**100**	**Dynamic Amplification Coefficient**	**25**	**50**	**100**
Analytical Model				**Analytical Model**			
1	2.124	1.942	1.932	7	1.956	1.92	1.742
2	1.752	1.73	1.523	8	1.548	1.252	1.101
3	1.648	1.448	1.187	9	1.348	1.104	1.028
4	1.548	1.164	1.099	10	1.244	1.078	0.839
5	1.456	1.052	1.006	11	1.216	0.964	0.694
6	1.224	0.900	0.874	12	1.184	0.928	0.646

From Table 4, it can be seen that under the same input peak acceleration level, the thicker the soft soil layer, the smaller the surface peak acceleration; the deeper the soft soil layer is buried, the smaller the surface peak acceleration.

As can be seen from Table 5, at the same input peak acceleration level, the thicker the soft soil layer, the smaller the dynamic amplification coefficient of surface peak acceleration; the deeper the soft soil layer is buried, the smaller the dynamic amplification coefficient of surface peak acceleration; and the attenuation of the dynamic amplification coefficient is slower as the burial depth increases. Under the same thickness and burial depth of the soft soil layer, with the increase in input peak acceleration, the dynamic amplification coefficient of surface peak acceleration gradually decreases, which indicates that the site soil has significant nonlinearity.

The variation of peak ground acceleration with different burial depths of the soft soil layer is given according to Table 4, as depicted in Figure 2. Likewise, the variation of the peak ground acceleration dynamic amplification coefficient with different burial depths of the soft soil layer is given according to Table 5, illustrated in Figure 3.

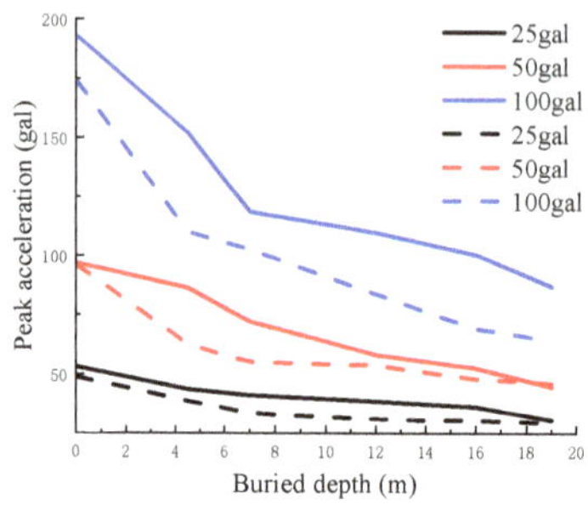

Figure 2. Variation characteristics of peak ground acceleration with different buried depths of soft soil layers (solid lines for models with one silt layer, dashed lines for models with two silt layers).

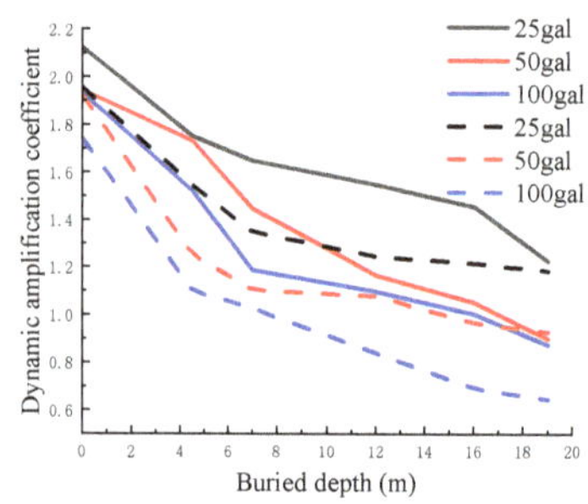

Figure 3. Variation characteristics of peak acceleration dynamic amplification coefficient with different buried depths of soft soil layer (solid lines for models with one silt layer, dashed lines for models with two silt layers).

From Figure 2, it can be seen that for a given input ground motion level, the peak ground acceleration decreases as the burial depth of the soft soil layer increases. The attenuation of the peak ground acceleration is more pronounced near the surface, and it becomes slower as the burial depth increases. Moreover, for different input ground motion levels, higher input peak accelerations result in faster attenuation of the peak ground acceleration. Similarly, the thickness of the soft soil layer also affects peak ground acceleration. A thicker layer of soft soil results in a smaller peak ground acceleration. The attenuation of the peak ground acceleration is faster near the shallow surface in thicker soil layers. This difference becomes more obvious as the input ground motion peak acceleration increases.

As seen in Figure 3, under the same input ground motion, the dynamic amplification coefficient of the peak surface acceleration is smaller when the burial depth of the soft soil layer increases. The attenuation of the dynamic amplification coefficient is faster near the surface and slower as the burial depth increases. Furthermore, for different input ground motion levels, the dynamic amplification coefficient attenuates faster as the input peak acceleration increases. Meanwhile, the thickness of different soft soil layers affects the peak surface acceleration. A thicker layer leads to a smaller dynamic amplification coefficient and faster attenuation near the shallow surface. The difference in the dynamic amplification coefficient becomes more obvious with an increase in the input ground motion peak acceleration.

The site-related response spectra for the damping ratio of 5% are also obtained in the seismic response analysis of each model, as shown in Figures 4 and 5.

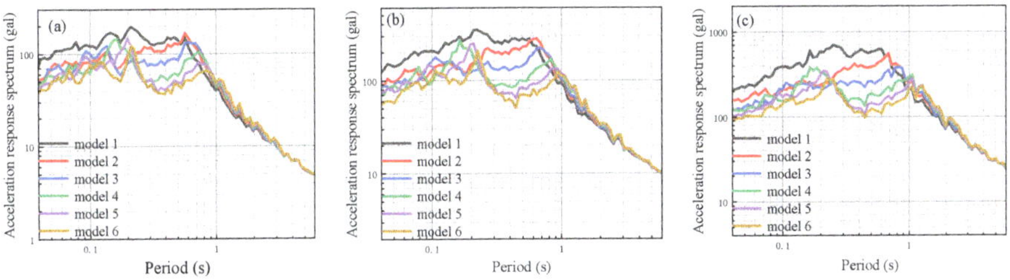

Figure 4. Site-related acceleration response spectra of each analysis model with a layer of silt at different input ground motions: (**a**) Input ground motion peak acceleration of 25 gal; (**b**) input ground motion peak acceleration of 50 gal; (**c**) input ground motion peak acceleration of 100 gal.

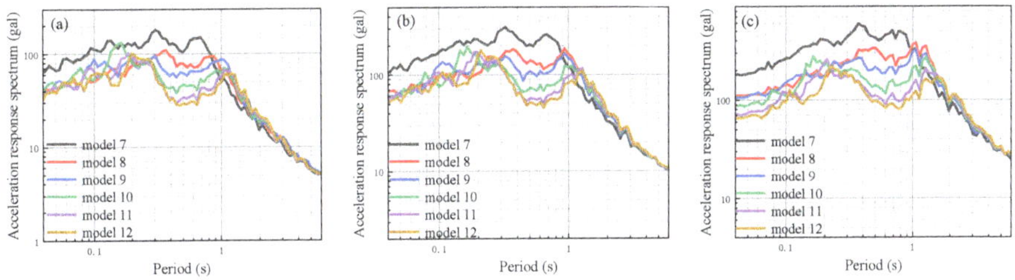

Figure 5. Site-related acceleration response spectra of each analysis model with two layers of silt at different input ground motions: (**a**) Input ground motion peak acceleration of 25 gal; (**b**) input ground motion peak acceleration of 50 gal; (**c**) input ground motion peak acceleration of 100 gal.

Figure 4 illustrates that in the model with one silt layer, at the same input ground motion peak acceleration level, the overall trend within the fluctuation band of the site-related response spectrum is that the acceleration response spectrum value decreases as the burial depth of the soft soil layer increases. Additionally, the initial frequency of the response spectrum attenuation section decreases, and the dominant frequency band of the response spectrum becomes wider. The variation trend of the response spectrum for each model is generally consistent at different input peak acceleration levels, while the greater the input peak acceleration, the larger the response spectrum value. The response spectra of models are close to each other for the periodic acceleration spectrum above 1 s. This suggests that the burial depth of the soft soil layer has less influence on the long-period ground motion.

Figure 5 demonstrates that in the model with two silt layers, at the same input ground motion peak acceleration level, the overall performance within the fluctuation band of the site-related response spectrum is that the acceleration response spectrum value decreases as the burial depth of the soft soil layer increases. Additionally, the initial frequency of the attenuation section decreases, and the dominant frequency band of the response spectrum becomes wider. The trend of variation in the response spectrum for each model remains generally consistent at different input peak acceleration levels. However, it is observed that as the input peak acceleration increases, the response spectrum values also increase. The response spectra of models are close to each other for the periodic acceleration spectrum above 1s, which indicates that the burial depth of the soft soil layer has less influence on the long-period ground motion.

Based on the free surface horizontal acceleration response spectrum of the engineering site, the site-related normalized response spectrum with a damping ratio of 5% is calculated and displayed in Figures 6 and 7.

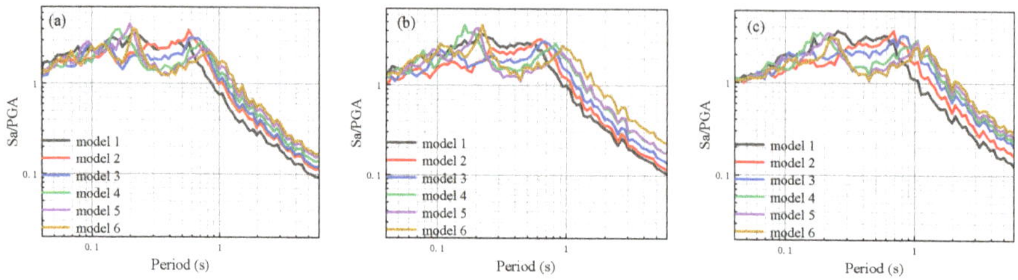

Figure 6. Site-related normalized spectrum of each calculation model with a layer of silt under different input ground motions: (**a**) Input ground motion peak acceleration of 25 gal; (**b**) input ground motion peak acceleration of 50 gal; (**c**) input ground motion peak acceleration of 100 gal.

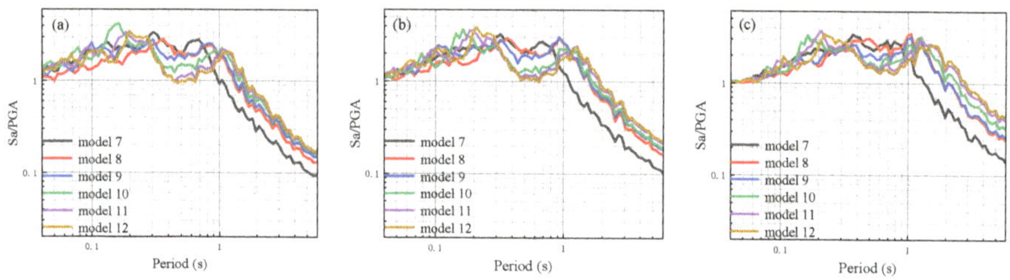

Figure 7. Site-related normalized spectrum of each calculation model with two layers of silt under different input ground motions: (**a**) Input ground motion peak acceleration of 25 gal; (**b**) input ground motion peak acceleration of 50 gal; (**c**) input ground motion peak acceleration of 100 gal.

As can be seen from Figure 6, for the calculation model containing one silt layer, the normalized spectrum of site correlation varies with the burial depth of the soft layer at the same input ground motion level. Within the normalized spectrum fluctuation band, the general performance is that the normalized spectrum value is lower the deeper the burial depth of the soft soil layer. Additionally, the initial frequency of the attenuation section of the normalized spectrum decreases, and the dominant band of the normalized spectrum becomes wider. The variation trend of the normalized spectrum for each model remains approximately the same under different input ground motion levels. However, it is noted that as the input peak acceleration increases, the normalized spectrum tends to decrease. In the normalized spectrum of 0.1 s and below, the results of the models are close to each other, indicating that the burial depth of the soft soil layer has less influence on the normalized spectrum of a short period. Meanwhile, in the period above 0.1 s, the burial depth of the soft soil layer has more influence on the normalized spectrum. The general performance is that the deeper the burial depth of the soft soil layer, the larger the normalized spectrum value and the difference between them becomes more noticeable.

It can also be seen from Figure 7 that for the calculation model containing two silt layers, the normalized spectrum of site correlation varies with the burial depth of the soft layer under the same input ground motion level. In the normalized spectrum fluctuation band, the general performance is that the normalized spectrum value decreases with increasing the burial depth of the soft soil layer. Additionally, the initial frequency of the attenuation section of the normalized spectrum decreases, and the dominant band of the normalized spectrum becomes wider as the burial depth of the soft soil layer increases. The variation trend of the normalized spectrum of the models remains generally consistent under different input ground motion levels. However, it is observed that the normalized spectrum tends to decrease as the input peak acceleration increases. In the normalized spectrum within the range of 0.1 s and below, the results of models show a similar trend, indicating that the burial depth of the soft soil layer has less influence on the normalized spectrum over a short period. However, in the period above 0.1 s, the burial depth of the soft soil layer has more influence on the normalized spectrum. The overall performance is that the deeper the burial depth of the soft soil layer, the larger the normalized spectrum value and the difference between the models becomes more pronounced.

4. A Method for Adjusting the Characteristic Period of Response Spectrum

According to the site category determination method [38] of the Code [26] in China, the site categories of each analytical model can be obtained, as summarized in Table 6.

Table 6. Site categories of analytical models.

Analytical Model	Overburden Thickness (m)	Equivalent Shear Wave Velocity (m/s)	Site Category	Analytical Model	Overburden Thickness (m)	Equivalent Shear Wave Velocity (m/s)	Site Category
1	30	156.0	II	7	35	134.5	III
2	30	156.0	II	8	35	134.5	III
3	30	156.0	II	9	35	134.5	III
4	30	156.0	II	10	35	142.0	III
5	30	161.3	II	11	35	161.3	II
6	30	182.4	II	12	35	182.4	II

From Table 6, it can be seen that the analyzed model site categories in this paper are II and III. Combined with Table 5.1.4-2 of the Code [26], the characteristic period of the site is 0.45 s (corresponding to Class II sites) or 0.65 s (corresponding to Class III sites) if considered following the third design seismic grouping.

The regularized response spectrum of the sites [39] is obtained following the format of the regularized response spectrum specified in the Code [26], which is presented in Figures 8–13. Additionally, the characteristic period of the response spectrum of each model is given in Table 7.

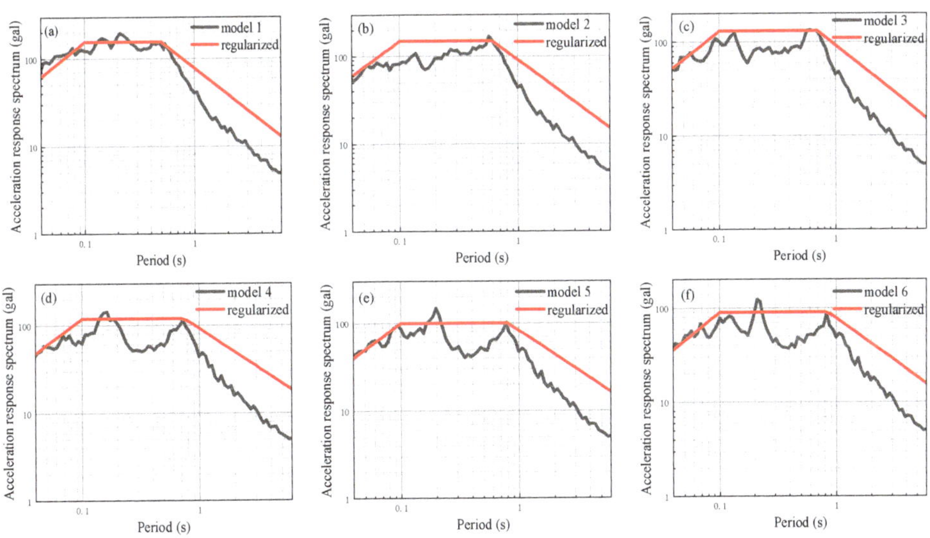

Figure 8. Site acceleration response spectrum of each analysis model with a layer of silt under input ground motion with a peak acceleration of 25 gal: (**a**–**f**) corresponding models 1 to 6.

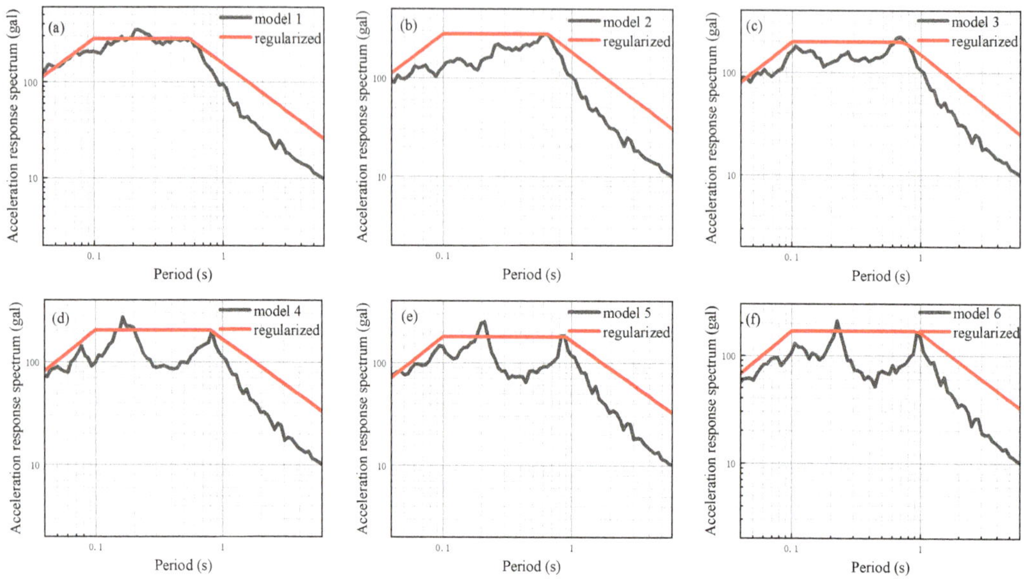

Figure 9. Site acceleration response spectrum of each analysis model with a layer of silt under input ground motion with a peak acceleration of 50 gal: (**a**–**f**) corresponding models 1 to 6.

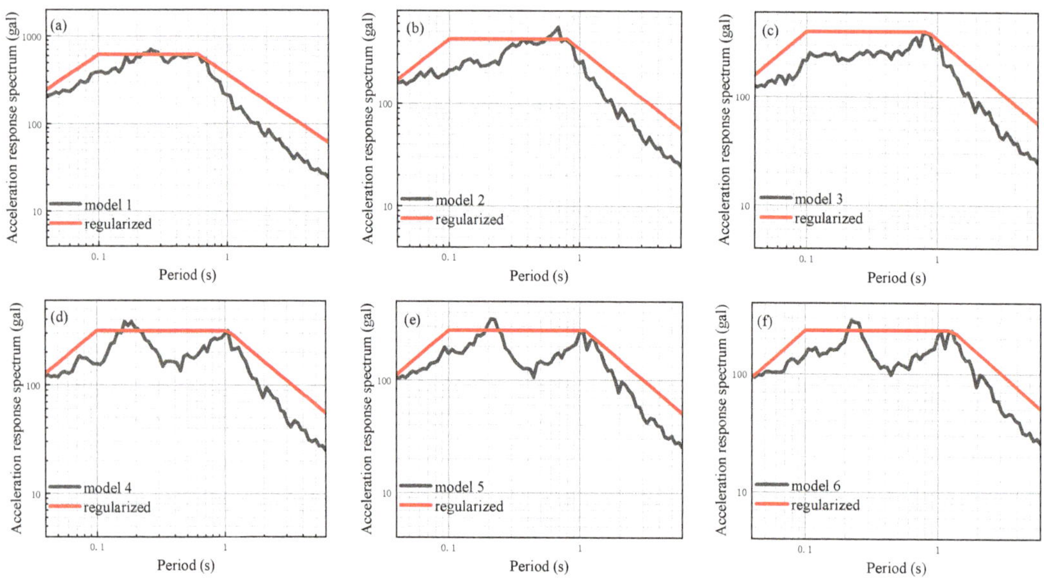

Figure 10. Site acceleration response spectrum of each analysis model with a layer of silt under input ground motion with a peak acceleration of 100 gal: (**a**–**f**) corresponding models 1 to 6.

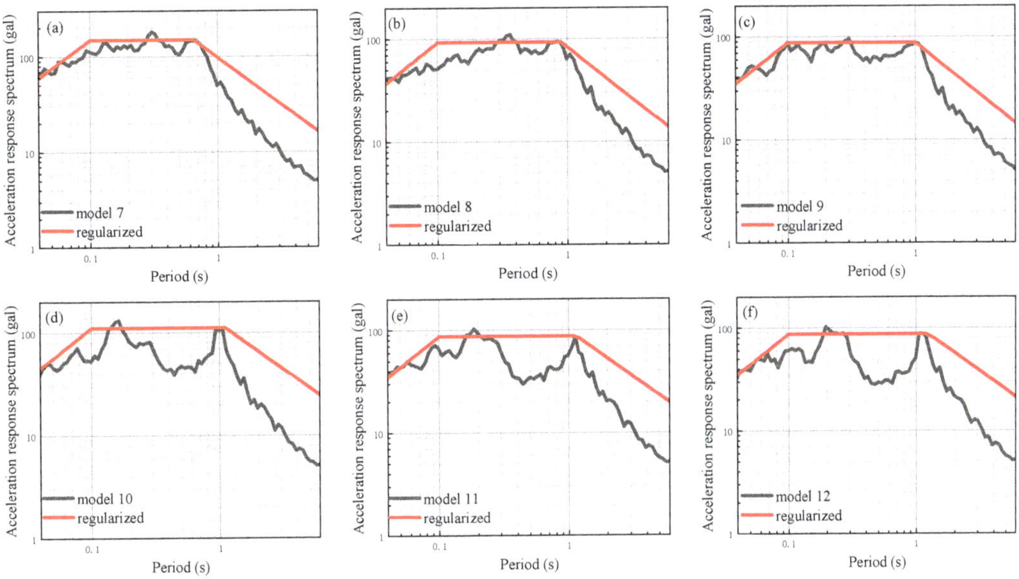

Figure 11. Site acceleration response spectrum of each analysis model containing two layers of silt under input ground motion with a peak acceleration of 25 gal: (**a**–**f**) corresponding models 7 to 12.

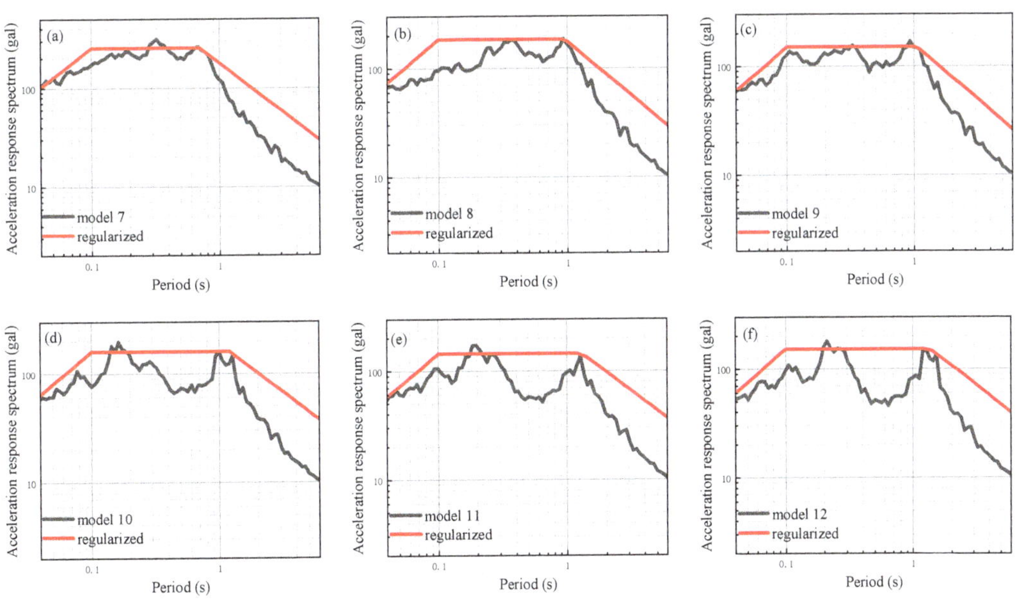

Figure 12. Site acceleration response spectrum of each analysis model containing two layers of silt under input ground motion with a peak acceleration of 50 gal: (**a**–**f**) corresponding models 7 to 12.

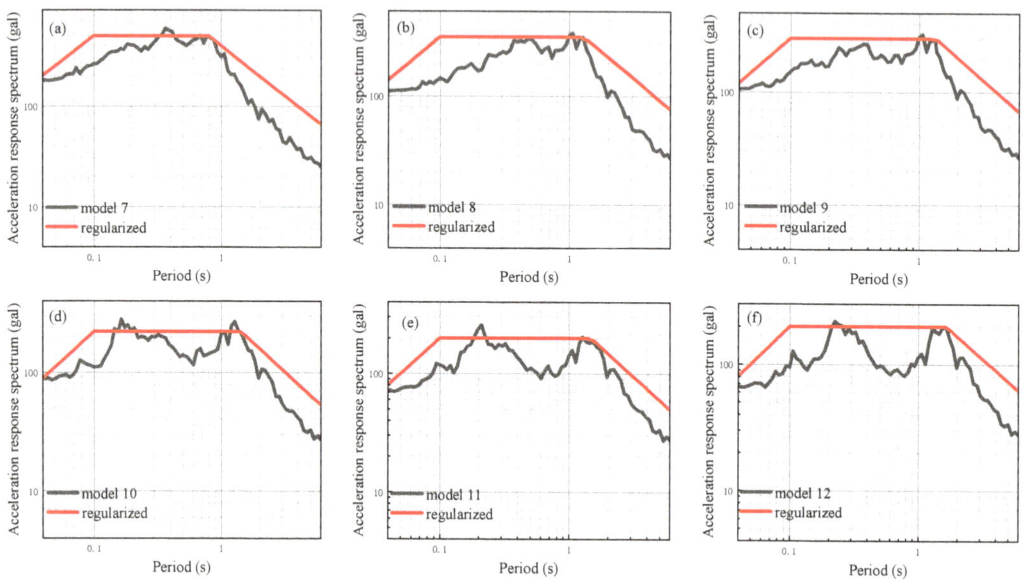

Figure 13. Site acceleration response spectrum of each analysis model containing two layers of silt under input ground motion with a peak acceleration of 100 gal: (**a**–**f**) corresponding models 7 to 12.

Table 7. The characteristic periodicity of the regularized site response spectrum of each analytical model.

Input Peak Acceleration/gal				**Input Peak Acceleration/gal**			
Characteristic Period/s	**25**	**50**	**100**	**Characteristic Period/s**	**25**	**50**	**100**
Analytical Model				**Analytical Model**			
1	0.5	0.55	0.6	7	0.7	0.75	0.95
2	0.6	0.65	0.8	8	0.9	0.95	1.2
3	0.7	0.75	0.85	9	1	1.05	1.35
4	0.75	0.8	1.05	10	1.1	1.2	1.45
5	0.8	0.9	1.1	11	1.15	1.3	1.5
6	0.85	0.95	1.2	12	1.25	1.35	1.65

From Figures 8–13 and Table 7, it can be seen that at the same input peak acceleration level, there is a trend where the response spectrum characteristic period increases with the thicker soft soil layer thickness. Additionally, the characteristic period of the response spectrum also increases with the deeper burial of the soft soil layer. Moreover, it can be observed that as the burial depth increases, the rate of increase in the response spectrum characteristic period gradually decreases, while the rate of increase near the shallow surface is faster. Additionally, the response spectrum characteristic period gradually increases with the increase in input peak acceleration while keeping the soft soil thickness and burial depth constant.

In order to further analyze the impact of different input peak acceleration levels on the seismic response of the site with a soft soil layer, additional seismic responses are calculated using input peak accelerations of 50 gal, 100 gal, 150 gal, 200 gal, and 300 gal. These calculations are performed by modulating the ground motion time history of 25 gal using amplitude modulation. The results of the one-dimensional site seismic response analysis of models 1 to 6 are presented in Figures 14–19, which show the site-related acceleration response spectra. On this basis, the calculated site-related acceleration response spectrum is regularized according to the aforementioned method, leading to the regularized response spectrum and related parameters. The regularized spectrum can be found in Figures 14–19. Additionally, the characteristic period of the regularized spectrum is summarized in Table 8.

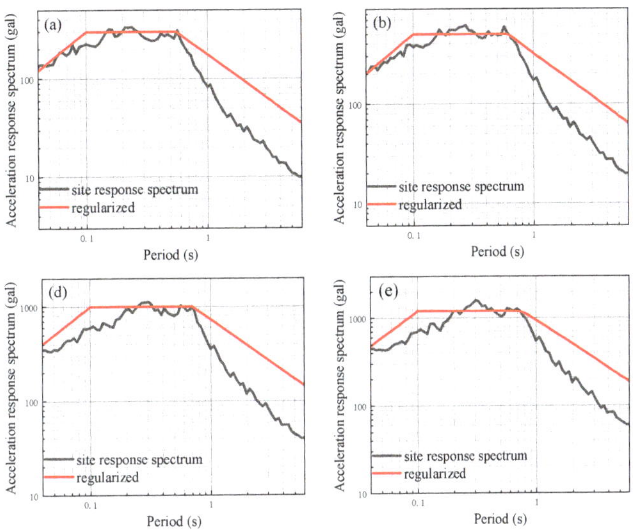

Figure 14. Site-related acceleration response spectrum of model 1 under different input levels: (**a**) Input ground motion peak acceleration of 50 gal; (**b**) input ground motion peak acceleration of 100 gal; (**c**) input ground motion peak acceleration of 150 gal; (**d**) input ground motion peak acceleration of 200 gal; (**e**) input ground motion peak acceleration of 300 gal.

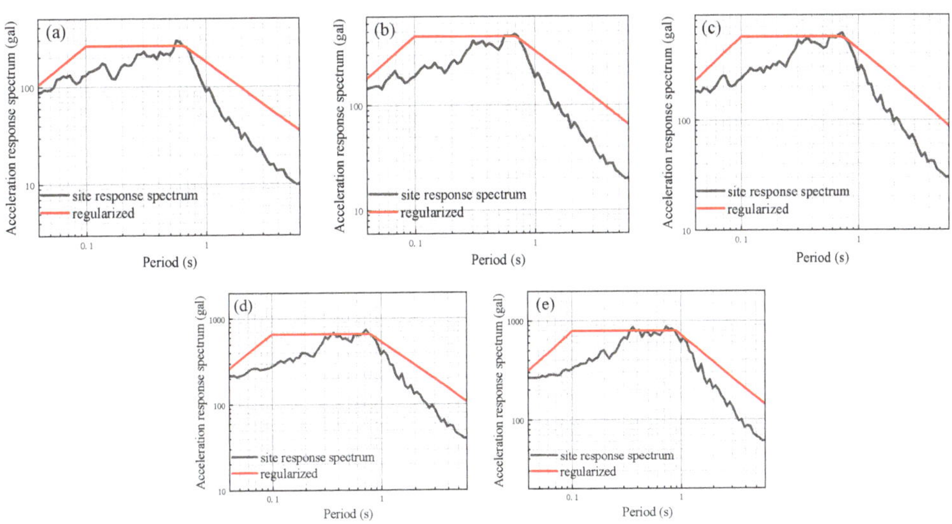

Figure 15. Site-related acceleration response spectrum of model 2 under different input levels: (**a**) Input ground motion peak acceleration of 50 gal; (**b**) input ground motion peak acceleration of 100 gal; (**c**) input ground motion peak acceleration of 150 gal; (**d**) input ground motion peak acceleration of 200 gal; (**e**) input ground motion peak acceleration of 300 gal.

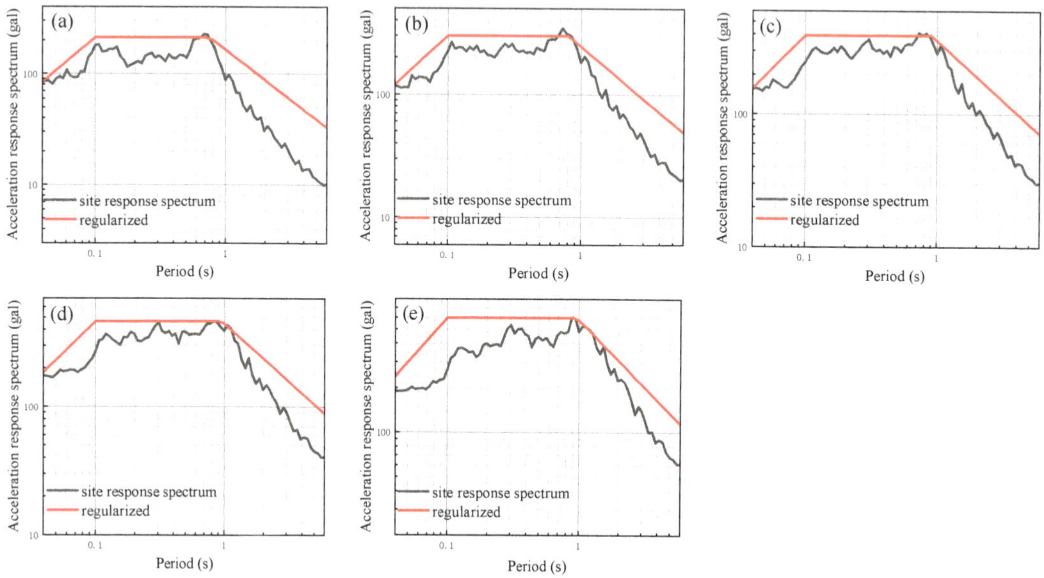

Figure 16. Site-related acceleration response spectrum of model 3 under different input levels: (**a**) Input ground motion peak acceleration of 50 gal; (**b**) input ground motion peak acceleration of 100 gal; (**c**) input ground motion peak acceleration of 150 gal; (**d**) input ground motion peak acceleration of 200 gal; (**e**) input ground motion peak acceleration of 300 gal.

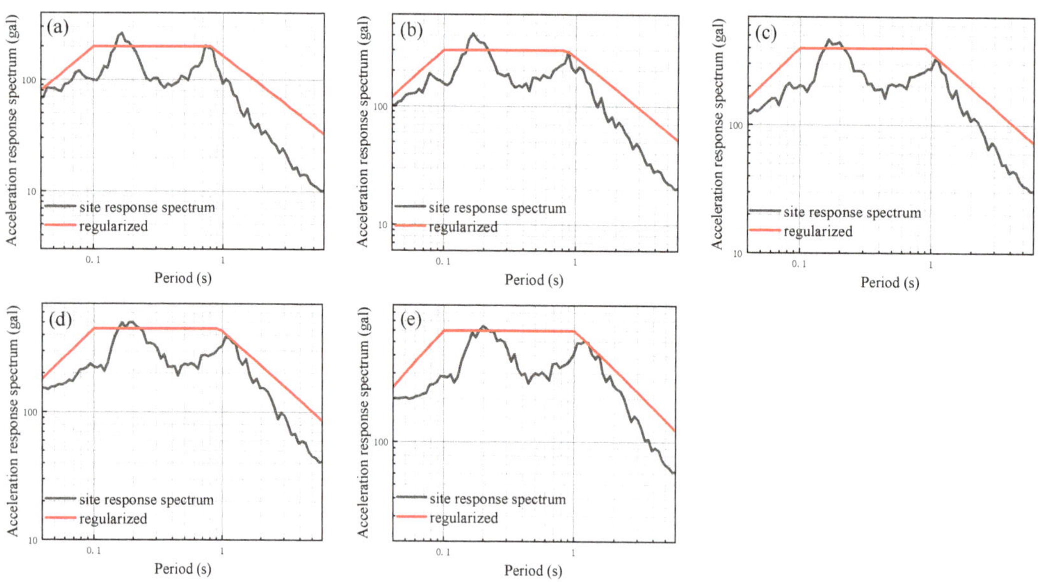

Figure 17. Site-related acceleration response spectrum of model 4 under different input levels: (**a**) Input ground motion peak acceleration of 50 gal; (**b**) input ground motion peak acceleration of 100 gal; (**c**) input ground motion peak acceleration of 150 gal; (**d**) input ground motion peak acceleration of 200 gal; (**e**) input ground motion peak acceleration of 300 gal.

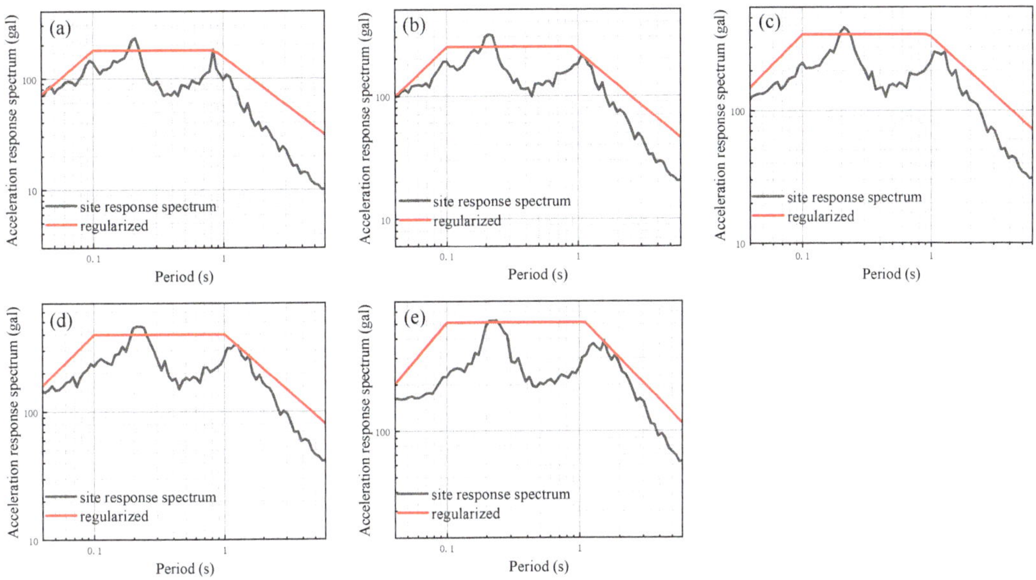

Figure 18. Site-related acceleration response spectrum of model 5 under different input levels: (**a**) Input ground motion peak acceleration of 50 gal; (**b**) input ground motion peak acceleration of 100 gal; (**c**) input ground motion peak acceleration of 150 gal; (**d**) input ground motion peak acceleration of 200 gal; (**e**) input ground motion peak acceleration of 300 gal.

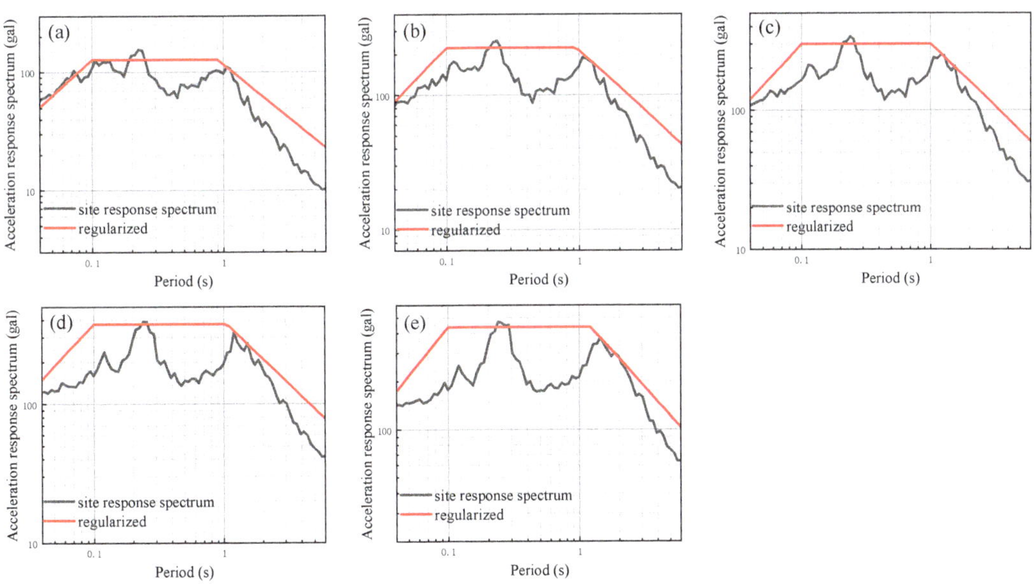

Figure 19. Site-related acceleration response spectrum of model 6 under different input levels: (**a**) Input ground motion peak acceleration of 50 gal; (**b**) input ground motion peak acceleration of 100 gal; (**c**) input ground motion peak acceleration of 150 gal; (**d**) input ground motion peak acceleration of 200 gal; (**e**) input ground motion peak acceleration of 300 gal.

Table 8. The characteristic periods of models 1–6 at different input levels.

Analytical Model						
Characteristic Period/s	**1**	**2**	**3**	**4**	**5**	**6**
Input Peak Acceleration/gal						
25 *	0.50	0.6	0.7	0.75	0.8	0.85
50	0.55	0.65	0.75	0.8	0.85	0.9
100	0.60	0.7	0.8	0.85	0.9	0.95
150	0.65	0.75	0.85	0.9	0.95	1.0
200	0.70	0.8	0.9	0.95	1.0	1.05
300	0.75	0.9	0.95	1.0	1.05	1.2

* The characteristic periods of each model for the input peak acceleration of 25 gal are derived from Table 7.

From Figures 14–19 and Table 8, it is shown that the response spectrum characteristic period of each analysis model increases with the increasing input ground motion level. The response spectrum characteristic period of model 1 increases gradually from 0.50 s to 0.75 s; model 2 increases from 0.6 s to 0.9 s; model 3 increases from 0.7 s to 0.95 s; model 4 increases from 0.75 s to 1.0 s; model 5 increases from 0.8 s to 1.05 s; and model 6 increases from 0.85 s to 1.2 s. It can be observed that the characteristic period of the response spectrum increases roughly linearly with the increase in the input ground motion level.

In summary, the soft soil layer has a significant influence on the characteristic period of the site acceleration response spectrum. Compared with the Code [26], the site response spectrum characteristic period with the soft soil layer after the regulation is much larger than the value specified in the code. In the subsequent analysis, we will utilize the previously established 12 site models as analytical models to propose a method for adjusting the characteristic period of the acceleration response spectrum for a site with a soft soil layer. This method will be based on the impact of the thickness and burial depth of the soft soil layer on the characteristic period of the acceleration response spectrum. The aim is to provide a theoretical basis for determining the characteristic period of the seismic response spectrum for sites with soft soil layers.

Taking a unit-area soil column of height h from a site with a soft soil layer, as shown in Figure 20a, the deformation of the column surface could be considered to be determined by the soft layer in the column. This assumes that the deformation of the upper and lower portions of the soil can be disregarded due to the low stiffness of the soft soil layer in comparison to its overlying and underlying soil layers. Based on this assumption, the soil column can be simplified to a spring–mass single degree of freedom system as depicted in Figure 20b, and the stiffness and mass of this system would be:

$$k = \frac{\rho_s \times {v_s}^2}{h_s} \quad (1)$$

$$m = h_s \rho_s + h_u \rho_u \quad (2)$$

where ρ_s, v_s, and h_s are the density, shear wave velocity, and thickness of the soft soil layer, respectively. ρ_u and h_u are the thickness and density of the overlying soil layer, respectively.

It is not difficult to obtain the natural period of this spring–mass single degree of freedom system as follows:

$$T_0 = 2\pi\sqrt{m/k} \quad (3)$$

By combining Equations (1)–(3), it is easily seen that as the thickness of the soft soil layer or the overlying soil layer increases, the stiffness of the spring–mass single degree of freedom system decreases. Additionally, the mass of the system increases, resulting in an increase in the natural period of the system. This inevitably affects the spectral characteristics of the site seismic response, which in turn increases the characteristic period of the site seismic response.

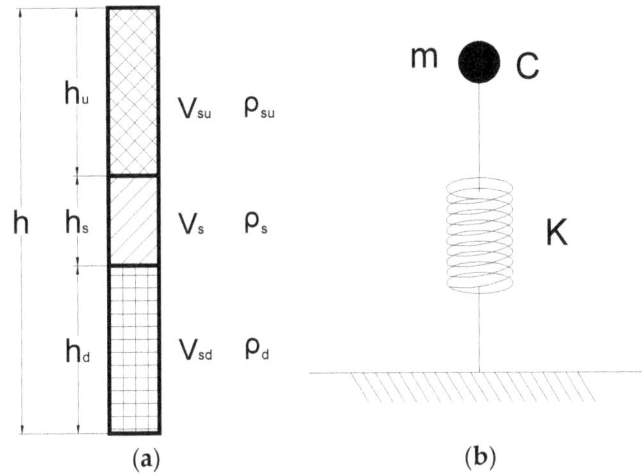

Figure 20. Analysis diagram: (**a**) Soil column; (**b**) spring–mass single degree of freedom system.

It is not difficult to explain the properties of the increase in the characteristic period of the site with the soft soil layer when the input intensity of ground motion increases. As the input ground motion intensity strengthens, several effects occur. Firstly, the strain level of the shallow surface soft soil layer increases. This leads to greater nonlinearity in the soft soil body. Secondly, the shear modulus ratio attenuates at a faster rate, resulting in a smaller ratio. Consequently, the shear modulus decreases, causing a decrease in the shear stiffness of the soft soil layer, This behavior is illustrated in Figure 20b for the spring–mass single degree of freedom system. As a result, the natural period of the system increases, subsequently leading to a greater characteristic period of the site with a soft soil layer. This qualitative interpretation also applies to the case where the difference between V_{su}, V_s, and V_{sd} is not very large. Normally, the results of seismic response analysis of sites with soft soil layers indicate that the strain maximum occurs in the soft layer. Furthermore, the shear modulus ratio of the soft soil attenuates faster, which leads to the shear modulus ratio being smaller than that of the overlying and underlying soil layers. As a consequence, the difference in shear modulus between the three layers becomes larger, and this difference becomes more significant with increasing input ground motion intensity. Therefore, the aforementioned hypothesis is valid.

5. Discussion

To further illustrate the reasonableness of the three wave velocities when their differences are not significant, we made modifications to the original model 2 (see Table 9). Specifically, we replaced the silt layer with an overlying and underlying silty clay, resulting in the creation of two analytical models. These models are documented in Tables 10 and 11. Subsequently, we calculated and obtained the corresponding acceleration response spectra, which are displayed in Figures 21 and 22.

Figures 21 and 22 clearly demonstrate that there is a noticeable difference in the characteristic period of the response spectra when the wave velocity of the silt layer is not significantly different from that of up-and-down soil layers. This difference becomes more obvious with increasing input ground motion intensity.

Table 9. The original analytical model.

No.	Rock-Soil	Soil Class	Depth at the Bottom of Layer (m)	Layer Thickness (m)	Shear Wave Velocity (m/s)	Density (t/m^3)
1	silty clay	3	4.5	4.5	160	1.86
2	silt	1	9.5	5.0	112	1.58
3	silty clay	4	13.0	3.5	165	1.87
4	silty clay	5	17.0	4.0	199	1.88
5	silty clay	6	21.0	4.0	212	1.96
6	silty clay	7	24.0	3.0	242	1.98
7	rounded gravel	8	27.0	3.0	258	2.20
8	fully weathered andesite	8	30.0	3.0	393	2.25
9	bedrock of model	9			516	2.65

Table 10. The supplementary analysis model 1.

No.	Rock-Soil	Soil Class	Depth at the Bottom of Layer (m)	Layer Thickness (m)	Shear Wave Velocity (m/s)	Density (t/m^3)
1	silty clay	3	4.5	4.5	160	1.86
2	silty clay	3	9.5	5.0	160	1.86
3	silty clay	4	13.0	3.5	165	1.87
4	silty clay	5	17.0	4.0	199	1.88
5	silty clay	6	21.0	4.0	212	1.96
6	silty clay	7	24.0	3.0	242	1.98
7	rounded gravel	8	27.0	3.0	258	2.20
8	fully weathered andesite	8	30.0	3.0	393	2.25
9	bedrock of model	9			516	2.65

Table 11. The supplementary analysis model 2.

No.	Rock-Soil	Soil Class	Depth at the Bottom of Layer (m)	Layer Thickness (m)	Shear Wave Velocity (m/s)	Density (t/m^3)
1	silty clay	3	4.5	4.5	160	1.86
2	silty clay	4	9.5	5.0	165	1.87
3	silty clay	4	13.0	3.5	165	1.87
4	silty clay	5	17.0	4.0	199	1.88
5	silty clay	6	21.0	4.0	212	1.96
6	silty clay	7	24.0	3.0	242	1.98
7	rounded gravel	8	27.0	3.0	258	2.20
8	fully weathered andesite	8	30.0	3.0	393	2.25
9	bedrock of model	9			516	2.65

A qualitative explanation for the increase in the characteristic period of seismic response of the site with soft soil is given in the previous discussion. To further quantitatively characterize the impact of the thickness and burial depth of the soft soil layer on the site characteristic period, it is proposed to establish the influence law of the thickness and burial depth of the soft soil layer on the site characteristic period via the statistical regression method. This method was based on the analysis data of the above models and the influence law of the input ground motion intensity on the site characteristic period of the soft soil layer.

$$T_g = a \times T_0 + T_g' + b \times T \quad (4)$$

Based on the qualitative analysis presented above, it is clear that the characteristic period of a site with a soft soil layer is influenced by the site's dominant period. Consequently, it is assumed that the characteristic period (T_g) of a site containing a soft soil layer is determined by the natural period (T_0) of the soft soil layer, the characteristic period of the input ground motion (T_g'), and the dominant period of the underlying soil (T).

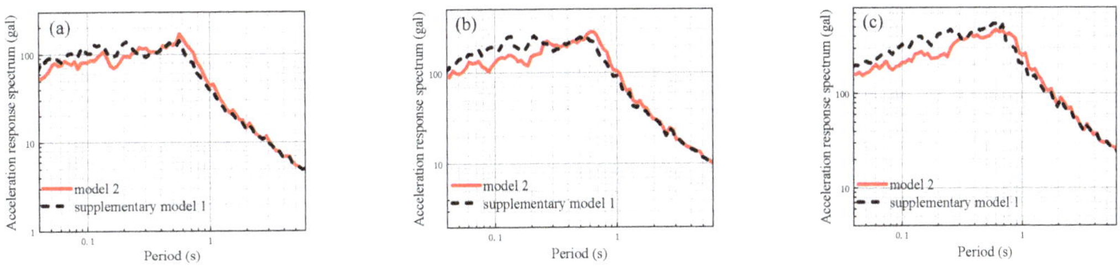

Figure 21. The site-related acceleration response spectra of supplementary model 1 and analysis model 2 under different input ground motions: (**a**) Input seismic peak acceleration 25 gal; (**b**) input seismic peak acceleration 50 gal; (**c**) input seismic peak acceleration 100 gal.

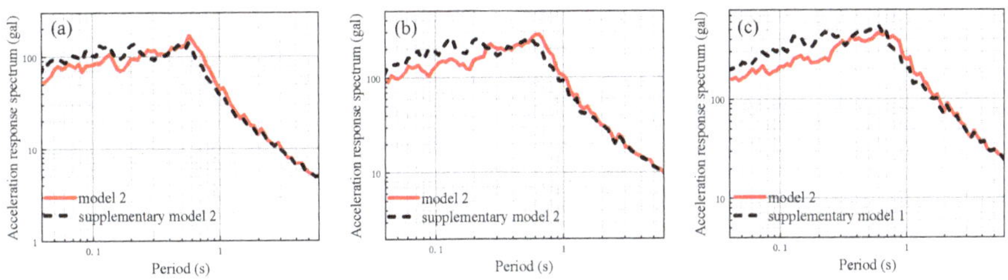

Figure 22. The site-related acceleration response spectra of supplementary model 2 and analysis model 2 under different input ground motions: (**a**) Input seismic peak acceleration 25 gal; (**b**) input seismic peak acceleration 50 gal; (**c**) input seismic peak acceleration 100 gal.

Where $T = \frac{4h_d}{v_{sd}}$, hypothetically, and v_{sd} is the equivalent shear wave velocity of the soil underlying the soft layer. h_d is the thickness of the soil underlying the soft layer; a and b are the regression coefficients.

From Equation (4), it is known that the site characteristic period takes into account the influence of the input ground motion spectrum characteristics, the overlying and underlying soils, and the soft soil layer. Based on the data in Tables 7 and 8, the regression coefficients a and b in Equation (4) are available by the least squares method, as listed in Table 12, and the regression results are presented in Figure 23.

Table 12. Summary of fitting results for a characteristic period of the site response spectrum under different input ground motion levels.

Models 1–6				Models 7–12			
Input Ground Acceleration (gal)	**a**	**b**	R^2	**Input Ground Acceleration (gal)**	**a**	**b**	R^2
25	0.838	−0.066	0.9998	25	0.972	−0.297	0.99912
50	0.941	−0.166	0.99905	50	1.050	−0.410	0.99964
100	1.231	−0.270	0.99972	100	1.246	−0.223	0.99938

When the variables are analyzed by linear regression in statistics, the parameters are estimated by the least squares method. R^2 is the ratio of the sum of squares of regression to the sum of squares of total deviation, which represents the proportion that can be explained by the sum of squares of regression. The larger the ratio, the more accurate the model and the more significant the regression effect. R^2 ranges from 0 to 1, where a value of 1 indicates a perfect fit of the model to the data. It is generally believed that the goodness of fit of models exceeding 0.8 is higher. The statistical relationship R^2 between the characteristic period of the site with a soft soil layer and the burial depth of the soft layer for the same thickness of soft soil layer at different input levels is about 1. This indicates that the correlation between the characteristic period of the site response spectrum and the burial depth of the soft soil layer is relatively good at different input ground motion levels. The variation curves of the characteristic period of the site response spectrum with the soft soil layer buried at different input ground motion levels for models 1–6 are approximately straight lines. Additionally, the variation curves of the characteristic period of the site response spectrum with the soft soil layer buried at different input ground motion levels for models 7–12 have greater curvature than those for models 1–6. Furthermore, it is observed that the contribution value of the natural period of the soft soil layer to the characteristic period increases as the thickness of the soft soil layer increases.

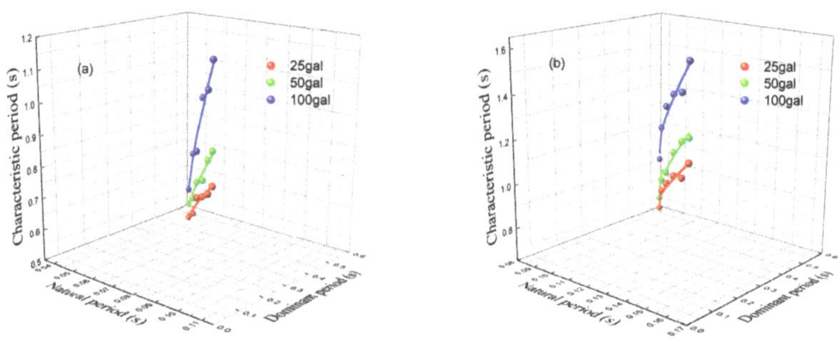

Figure 23. Relationship between the characteristic period of site response spectrum and thickness and buried depth of soft soil layer under different input ground motion levels: (**a**) Models 1–6; (**b**) Models 7–12.

In addition, the variation trend of the characteristic period T_g of the response spectrum, in relation to the input peak ground motion acceleration A, is obtained at the site with a specific thickness and depth of burial of soft soil layers. This information is derived from Table 8 and depicted by the scatter points in Figure 24. It is easy to see from Figure 24 that the characteristic period T_g of the site acceleration response spectra of different models is similar with an increasing peak acceleration of input ground motion. It is approximately linear with the input peak acceleration A. Therefore, the equation $T_g = \alpha + \beta \times A$ is proposed to fit the variation law of the characteristic period T_g of the site acceleration response spectrum with the input peak acceleration A. The regression coefficients α and β in the equation are obtained by linear regression based on the least squares method, as listed in Table 13, and the regression results are demonstrated as straight lines in Figure 24.

The regression analysis results, indicated by the high R^2 values close to 1, indicate a strong correlation between the characteristic period of the site response spectrum and various input peak ground motion accelerations. Moreover, the characteristic period of the response spectrum exhibits an approximately linear increase with the input peak acceleration for the site, which includes a specific thickness and burial depth of a soft soil layer. This finding suggests that as the input peak ground motion acceleration increases, the characteristic period of the site response spectrum also lengthens.

By considering the influence of soil structure on the characteristic period of the site seismic acceleration response spectrum, we propose a straightforward and physically meaningful adjustment formula. This formula considers not only the effects of soft soil layer thickness and burial depth but also the influence of different input ground motion levels. The proposed formula provides a clear and concise representation of these factors. The seismic response analysis conducted using a representative soil structure site model demonstrates the effectiveness of the adjustment method. The method accurately reflects the influence of the burial depth of the soft soil layer within the profile on the characteristic period of the site seismic acceleration response spectrum. Furthermore, the estimated value obtained from the adjustment method closely approximates the characteristic period of the response spectrum observed at the actual engineering site. This indicates that the adjustment method provides reliable and realistic results.

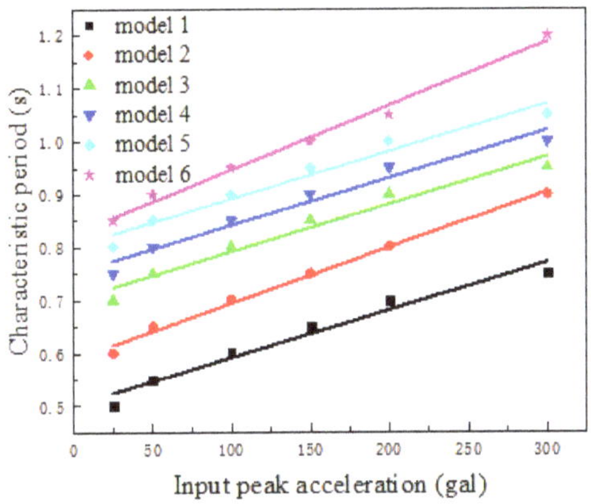

Figure 24. Variation of characteristic periods of seismic response spectra of different models with different input ground motion levels.

Table 13. Fitting results of characteristic periods of models 1–6 at different input levels.

Analytical Model	α	β	R^2
1	0.5015	0.0009	0.9624
2	0.5884	0.0010	0.9937
3	0.7015	0.0009	0.9624
4	0.7515	0.0009	0.9624
5	0.8015	0.0009	0.9624
6	0.8254	0.0012	0.9953

6. Conclusions

Based on the established 12 site models with soft soil layers, the one-dimensional equivalent linearization site seismic response analysis is carried out under different input ground motion levels. The effects on the soft soil layer thickness, buried depth, and input ground motion intensity of the on-site seismic response, as well as the characteristic period of the site acceleration response spectrum, have been discussed. In light of these findings, a method for adjusting the characteristic period of the site acceleration response spectrum with soft soil layers is proposed. The main conclusions obtained are as follows:

1. Under the same input ground motion level, the burial depth of the soft soil layer influences the peak ground acceleration. Specifically, as the burial depth increases, the peak ground acceleration decreases, and the rate of attenuation is faster near the surface. Conversely, as the burial depth increases further, the rate of attenuation becomes slower. Under different input ground motion levels, higher input peak accelerations result in faster attenuation of the peak ground acceleration. In other words, as the input peak acceleration increases, the rate of attenuation becomes more rapid. Similarly, the thickness of the soft soil layer impacts the peak ground acceleration. A thicker soft soil layer leads to a smaller peak ground acceleration, and the attenuation is faster near the shallow surface. The difference becomes more prominent as the input ground motion peak acceleration increases.
2. At the same input ground motion, the dynamic amplification coefficient of the peak surface acceleration decreases with an increase in the burial depth of the soft soil layer. Additionally, the attenuation of the dynamic amplification coefficient is faster near the surface and slower as the burial depth increases. Under different input ground motion

levels, higher input peak acceleration results in a faster attenuation of the dynamic amplification coefficient. In other words, as the input peak acceleration increases, the dynamic amplification coefficient decreases at a faster rate. Likewise, the thickness of different soft soil layers has an effect on peak surface acceleration. A thicker soft soil layer corresponds to a smaller dynamic amplification coefficient, and the attenuation is faster near the shallow surface. The difference becomes more obvious as the input ground motion peak acceleration increases.

3. The characteristic period of the site seismic acceleration response spectrum progressively increases with the increase in burial depth or thickness of the weak soil layer. It also increases with the increase in the input ground motion peak. Subsequently, the influence characteristics of soft soil layer thickness, buried depth, and input ground motion intensity on the characteristic period of the site seismic acceleration response spectrum are analyzed. Furthermore, a method for adjusting the characteristic period of the site seismic acceleration response spectrum with the soft soil layer is proposed.

Author Contributions: Conceptualization, Z.Z. and Z.C.; methodology, Y.L.; software, B.H.; validation, Y.L., B.H. and Y.H.; formal analysis, Y.L.; investigation, Z.B.; resources, C.P.; data curation, Y.L.; writing—original draft preparation, Y.L.; writing—review and editing, Y.L. and Z.C.; visualization, B.H.; supervision, Z.B., Y.H. and C.P.; project administration, Z.Z.; funding acquisition, Z.Z. All authors have read and agreed to the published version of the manuscript.

Funding: This research was funded by the [National Natural Science Foundation of China], grant numbers [U2039208] and [U1839202].

Institutional Review Board Statement: Not applicable.

Informed Consent Statement: Not applicable.

Data Availability Statement: Not applicable.

Conflicts of Interest: The authors declare no conflict of interest.

References

1. Bo, J.S.; Li, Q.; Qi, W.H.; Wang, Y.T.; Zhao, X.L.; Zhang, Y.Y. Research progress and discussion of the site condition effect on ground motion and earthquake damage. *J. Jilin Univ. (Earth Sci. Ed.)* **2021**, *51*, 1295–1305. [CrossRef]
2. Molly, M.G.; Leah, S.; Madeleine, C.L.; James, N.; Seth, S.; Norman, A.; Hough, S.E. A study on the effect of site response on California seismic hazard map assessment. *Front. Earth Sci.* **2022**, *10*, 93134. [CrossRef]
3. Chi, M.J.; Li, X.J.; Chen, X.L.; Ma, S.J. Problems and suggestions on site classification. *Acta Seismol. Sin.* **2021**, *43*, 787–803+817. [CrossRef]
4. Li, W.H.; Zhao, C.G. Effect of groundwater level variation on earthquake ground motions. *Acta Seismol. Sin.* **2015**, *37*, 482–492+532. [CrossRef]
5. Wang, W.; Liu, B.D.; Liu, P.X.; Wang, Z.Y.; Liu, X. Analyses on the effect of the local site conditions on the strong motion based on the array records. *Acta Seismol. Sin.* **2016**, *38*, 307–317. [CrossRef]
6. Li, P.; Bo, J.S.; Xiao, R.J.; Zhang, Y.D. The study of effect by the Valley Site on ground motion. *Technol. Earthq. Disaster Prev.* **2018**, *13*, 331–341. [CrossRef]
7. Zhou, Z.H.; Zhang, Y.M.; Sun, P.S.; Yang, B.P. Study on effect of fault seismic damage. *J. Nat. Disastersi* **2003**, *12*, 20–24. [CrossRef]
8. Alemayehu, A.; Matebie, M.; Kifle, W. Site response and liquefaction hazard analysis of Hawassa town, Main Ethiopian Rift. *Front. Built Environ.* **2022**, *195*, 101421. [CrossRef]
9. Li, X.J.; Li, N.; Rong, M.S.; Dong, Q. Novel Evaluation Method for Site Effect on Earthquake Ground Motion Based on Modified Horizontal to Vertical Spectral Ratio. *Front. Earth Sci.* **2022**, *10*, 93851. [CrossRef]
10. Li, P.; Bo, J.S.; Qi, W.H.; Liu, D.D.; Xiao, R.J. Effects of soil structure on abnormal intensity in Hanyuan old town. *Acta Seismol. Sin.* **2012**, *34*, 851–857+880. [CrossRef]
11. Wang, H.Y. Amplification effects of soil sites on ground motion in the Weihe basin. *Chin. J. Geophys.* **2011**, *54*, 137–150. [CrossRef]
12. Zhou, Z.H.; Chen, L.; Zhou, Y.; Li, X.J.; Su, J.; Dong, Q.; Zhong, K.M.; Li, Y.P. The effect of surface hard cover on the site earthquake response. *J. Basic Sci. Eng.* **2020**, *28*, 321–330. [CrossRef]
13. Zhou, Z.H.; Li, Y.P.; Zhou, Y.; Li, X.J.; Chen, L.; Su, J.; Dong, Q.; Wang, Y.F. The effect of hard interlayer thickness on the site seismic response. *Seismol. Geol.* **2019**, *41*, 1254–1265. [CrossRef]
14. Zhang, H.; Li, K.Q.; You, H.B.; Zhou, Z.H. Influence of buried depth of hard interlayer on ground-motion parameters. *China Earthq. Eng. J.* **2016**, *38*, 935–941. [CrossRef]

15. Su, J.; Zhou, Z.H.; Zhou, Y.; Li, X.J.; Dong, Q.; Wang, Y.F.; Li, Y.P.; Chen, L. The Characteristics of Seismic Response on Hard Interlayer Sites. *Adv. Civ. Eng.* **2020**, *2*, 1425969. [CrossRef]
16. Xia, X.; Wang, Y.B.; Xi, S.Y.; Sun, J.; Xu, H.D.; Huang, Y. Study on Effects of Earthquake Response of Soft Interlayer. *Mod. Technol. Mater. Mech. Intell. Syst.* **2014**, *1049*, 205–208. [CrossRef]
17. Xiao, M.; Cui, J.; Li, Y.D.; Nguyen, V.Q. Nonlinear Seismic Response Based on Different Site Types: Soft Soil and Rock Strata. *Adv. Civ. Eng.* **2022**, *2022*, 5370369. [CrossRef]
18. Xu, J.C.; Jian, W.B.; Shang, Y.Q. Study on the seismic failure mechanism of the thick soft soil foundation. *Chin. J. Rock Mech. Eng.* **2005**, *24*, 313–320. [CrossRef]
19. Yao, E.L.; Li, W.C.; Miao, Y.; Lin, Y.; Yang, Z.W. Study on the Influence of a Soft Soil Interlayer on Spatially Varying Ground Motions. *Appl. Sci.* **2022**, *12*, 1322. [CrossRef]
20. Cao, Z.X. Influence of Properties of soil Layer on site amplification effect for SH waves. *Trans. Shenyang Ligong Univ.* **2006**, *25*, 88–91.
21. Wang, Y.H.; Sun, D.F.; Wei, D.X. Influence of buried depth of soft interlayer on ground motion parameters. *Gansu Sci. Technol.* **2019**, *35*, 65–67+111. [CrossRef]
22. Tian, S.Q. Analysis of influence of soft soil layer on seismic response of site. *China Sci. Technol. Inf.* **2013**, *16*, 39. [CrossRef]
23. Yan, K.M.; Liu, F.C.; Zhu, C.H.; Wang, Z.J.; Zhang, J.J. Dynamic response of slops with intercalated soft layers under seismic excitations. *Chin. J. Rock Mech. Eng.* **2017**, *36*, 2686–2698. [CrossRef]
24. Li, M.J.; Xia, X. Research on the effect of soft clay interlayer thickness on ground seismic response characteristics. *Earthq. Resist. Eng. Retrofit.* **2017**, *39*, 149–153. [CrossRef]
25. Wang, J.; Wang, S.Y.; Pan, Y.J.; Yan, J.X. Influence of soft interlayer at different depth on seismic site response based on ideal site model. *J. Inst. Disaster Prev.* **2022**, *24*, 33–41. [CrossRef]
26. Ministry of Housing and Urban-Rural Construction of the People's Republic of China. *Code for Seismic Design of Buildings (GB 50011-2010)*; China Architecture Building Press: Beijing, China, 2016.
27. Li, B.; Xie, W.C.; Pandey, M.D. Newmark Design Spectra Considering Earthquake Magnitudes and Site Categories. *Earthq. Eng. Eng. Vib.* **2016**, *15*, 519–535. [CrossRef]
28. Gaxiola-Camacho, J.R.; Azizsoltani, H.; Villegas-Mercado, F.J.; Haldar, A. A Novel Reliability Technique for Implementation of Performance-Based Seismic Design of Structures. *Eng. Struct.* **2017**, *142*, 137–147. [CrossRef]
29. Yaghmaei-Sabegh, S.; Makaremi, S. Development of Duration-Dependent Damage-Based Inelastic Response Spectra. *Earthq. Eng. Struct. Dyn.* **2017**, *46*, 771–789. [CrossRef]
30. Hjörtur, T.; Anne, S.K. Simulation of digital earthquake accelerograms using the inverse discrete Fourier transform. *Earthq. Eng. Struct. Dyn.* **2002**, *31*, 2023–2048. [CrossRef]
31. Liu, S.; Pan, C.; Zhou, Z.G. Discussion on the response spectral solution and fitting of spectrum-compatible artificial seismic waves. *Acta Seismol. Sin.* **2018**, *40*, 519–530. [CrossRef]
32. Xu, G.D.; Shi, P.J.; Zhou, X.Y. Artifical ground motion based on target power spectra and envelope. *Earthq. Eng. Eng. Dyn.* **2010**, *30*, 1–9. [CrossRef]
33. Cheng, Q.S.; Gao, G.Y.; He, J.F. Three-dimensional nonlinear analysis of seismic ground response of soft soils in Shanghai. *Rock Soil Mech.* **2011**, *32*, 3461–3467. [CrossRef]
34. Jin, X.; Kong, G.; Ding, H.P. Nonlinear seismic response analysis of horizontal layered site. *Earthq. Eng. Eng. Vib.* **2004**, *3*, 38–43. [CrossRef]
35. Fang, Y.; Li, H.; Li, Y.; Chen, G.X.; Lv, Y.J.; Peng, Y.J. Vs30 Prediction Models Based on Measured Shear-Wave Velocities in Tangshan, China. *Sustainability* **2023**, *15*, 3282. [CrossRef]
36. Gao, W.P.; Gao, M.T.; Chen, X.L. Far-field strong earthquake effect in Tianjin coastal soft site. *Acta Seismol. Sin.* **2012**, *34*, 235–243+283. [CrossRef]
37. Rong, M.S.; Li, X.J.; Lu, T.; Huang, Y.H.; Lv, Y.J. Suggestions on determination of design ground motion parameters for offshore engineering sites with deep soft surface layers. *Acta Seismol. Sin.* **2013**, *35*, 262–271+286. [CrossRef]
38. Zhou, J.; Li, X.J.; Li, Y.Q.; Gang, C.C. Compariative analysis and transformation relations between China and US site classification systems in building seismic code provisions. *Acta Seismol. Sin.* **2021**, *43*, 521–532+534. [CrossRef]
39. Li, X.J.; Peng, Q. Calculation and analysis of earthquake ground motion parameters for different site categories. *Earthq. Eng. Eng.* **2001**, *21*, 29–36. [CrossRef]

sustainability

MDPI

Article

Spatiotemporal Analysis of Earthquake Distribution and Associated Losses in Chinese Mainland from 1949 to 2021

Tongyan Zheng [1], Lei Li [2,3,4], Chong Xu [2,3,*] and Yuandong Huang [2,3,5]

1 China Earthquake Networks Center, Beijing 100045, China
2 National Institute of Natural Hazards, Ministry of Emergency Management of China, Beijing 100085, China
3 Key Laboratory of Compound and Chained Natural Hazards Dynamics, Ministry of Emergency Management of China, Beijing 100085, China
4 School of Engineering and Technology, China University of Geosciences (Beijing), Beijing 100083, China
5 School of Earth Sciences and Resources, China University of Geosciences (Beijing), Beijing 100083, China
* Correspondence: chongxu@ninhm.ac.cn

Abstract: A comprehensive earthquake hazard database is crucial for comprehending the characteristics of earthquake-related losses and establishing accurate loss prediction models. In this study, we compiled the earthquake events that have caused losses since 1949, and established and shared a database of earthquake hazard information for the Chinese mainland from 1949 to 2021. On this basis, we preliminarily analyzed the spatiotemporal distribution characteristics of 608 earthquake events and the associated losses. The results show the following: (1) The number of earthquakes is generally increasing, with an average of annual occurrence rising from three to twelve, and the rise in the economic losses is not significant. The number of earthquakes occurring in the summer is slightly higher than that in the other three seasons. (2) The average depths of earthquakes within the six blocks display a decreasing trend from west to east, with a majority (63.8%) of earthquakes occurring at depths ranging from 5 to 16 km. (3) Although the number of earthquakes in the east is lower than that in the west, earthquakes in the east are more likely to cause casualties when they have the same epicenter intensity. Southwest China is located in the Circum-Pacific seismic zone where earthquake hazards are highly frequent. The results can provide fundamental data for developing earthquake-related loss prediction models.

Keywords: earthquake; hazard information database; casualties; spatiotemporal distribution characteristics

Citation: Zheng, T.; Li, L.; Xu, C.; Huang, Y. Spatiotemporal Analysis of Earthquake Distribution and Associated Losses in Chinese Mainland from 1949 to 2021. *Sustainability* **2023**, *15*, 8646. https://doi.org/10.3390/su15118646

Academic Editor: George D. Bathrellos

Received: 10 February 2023
Revised: 21 May 2023
Accepted: 23 May 2023
Published: 26 May 2023

1. Introduction

Earthquakes are a natural geological phenomenon that can cause fatal destruction [1]. Due to the unpredictable suddenness and enormous energy released, earthquakes often result in a significant loss of people's lives and property, causing devastating economic consequences [2,3]. According to statistics, more than one million earthquakes occur globally each year, equivalent to one earthquake every thirty seconds [4]. Regarding the earthquake magnitude, a total of 14,588 earthquakes with a magnitude of 4.0 or greater were recorded worldwide in only the year of 2015 (www.usgs.gov, accessed on 15 August 2022). Earthquakes can trigger not only immediate hazards such as fires, floods, and the diffusion of radioactive materials but also secondary disasters such as tsunamis, landslides, and collapses [5]. Although the topic of earthquake prediction has been proposed for a long time, its progress is slow. However, many studies have shown that the impact of earthquake hazards has a certain regularity [6–10]. For instance, Gutenberg et al. [11] evaluated the seismic activity of different plates of the Earth and discussed the topographic and geological characteristics of seismic zones.

The most important aspect after a destructive earthquake is the casualties [12,13]. In addition, statistics on the economic loss caused by the earthquake can help the government more directly understand the situation. How to accurately predict the casualties and economic losses caused by an earthquake and quickly provide post-earthquake emergency response management? Compiling a reliable database of historical earthquake hazard information is both the foundation and the key [14,15]. Based on the database, a series of forecasting models (empirical or analytical) can be established. It is mainly used in two parts: (1) estimating the number of casualties resulted from the earthquake [16–18] and (2) estimating the economic loss after the earthquake [19]. Samardjieva and Oike [20] approximated the number of casualties due to earthquakes using global seismic data and separated the data from the earthquake hazards in Japan. The data were further combined with a quantitative proportional model of the hazard. Samardjieva and Badal [12] analyzed the human losses after destructive earthquakes worldwide in the 20th century and proposed a quantitative model for the preliminary assessment of human casualties by relating the magnitude to the expected number of victims. Badal et al. [21] also considered the correlation of population density and conducted experiments in major cities in seismically active areas of Spain. Additionally, information on earthquake hazards from the Cambridge Earthquake Impacts Database (CEQID, www.ceqid.org) [22], Turkey [23], and Iran [15] was used to establish empirical models for estimating the number of fatalities. Xia et al. [24,25] proposed an earthquake lethality evaluation model based on building damage. Jaiswal and Wald [19] extended the USGS empirical lethality estimation method for immediate assessment of global earthquake response proposed by Jaiswal et al. [26] and rapidly estimated the economic loss after major earthquakes around the globe. Erdik et al. [27] systematically summarized the new methods and applications for estimating earthquake-related losses in real time over the past decades. Gerstenberger et al. [28] reviewed the research on earthquake hazard models in the past 50 years, and proposed future research directions.

China is located in the southeast of the Eurasian continent. Its eastern margin is affected by the Pacific Rim seismic zone, and the southwest and northwest regions are both located in the Eurasian seismic zone (Figure 1). The unique geographical location has made China an earthquake-prone country, and one of the countries that suffer the most severe losses from earthquake hazards [29,30]. On this basis, in this study, we established a database of earthquake hazard information for the Chinese mainland from 1949 to 2021. The database includes 608 earthquake events. We also analyzed their spatiotemporal distribution characteristics. In addition, earthquakes in neighboring countries have also caused some losses to China, and it should be noted that this kind of earthquakes are not included in the 608 seismic events. For example, the M6.8 earthquake in India on 18 September 2011 caused seven deaths and one hundred and thirty-six injuries in the Tibet Autonomous Region [31], and the M8.1 earthquake in Nepal on 25 April 2015 caused twenty-seven deaths and eight hundred and sixty injuries in the Tibet Autonomous Region [32]. A detailed database of earthquake hazard information is an important basis for understanding the characteristics of earthquake-hazard-related losses, establishing loss prediction models, and subsequently carrying out post-earthquake emergency rescues.

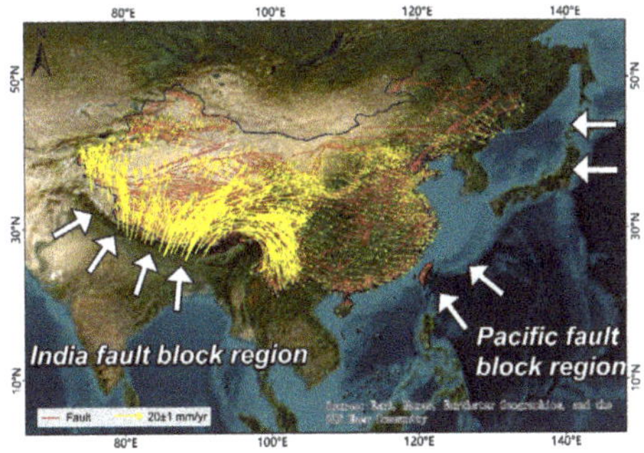

Figure 1. The geographical location of China. White arrows indicate plate forces; yellow arrows indicate regional GPS velocity (time period: 2009–2014) [33].

2. Data Collection and Methodology

The data used in this study come from two sources: (1) the China Earthquake Networks Center (https://www.ceic.ac.cn, accessed on 12 July 2022) that provided information on earthquake magnitude, date of occurrence, macroscopic epicenter, and depth; (2) provincial (autonomous regions and municipality) earthquake agencies that provided the assessment reports of earthquake-hazard-related losses, including both the casualties and economic loss. We also filled in some missing information in other ways [34].

During 1949–2021, earthquakes in the Chinese mainland totally caused 608 hazard events, 386 of which resulted in casualties (371,625 deaths and 1,304,936 injuries), while the remaining 222 earthquakes resulted in economic loss only. The most recent earthquake occurred on 16 September 2021 in Luxian County, Sichuan Province. This event caused three deaths and one hundred and fifty-seven injuries, with an economic loss of 364 million USD. The M7.8 earthquake that occurred in Tangshan, Hebei Province on 28 July 1976 caused the most tragic disaster. Statistically, 242,000 residents were killed, with nearly 170,000 serious injuries and more than 540,000 slight injuries. The economic loss reached nearly 73.45 billion USD. In addition, the M4.8 earthquake that occurred on 22 October 1965 in the south of Dalat Banner in Inner Mongolia caused minimal losses, with no casualties and a relatively low economic loss (14,000 USD).

We used ArcGIS 10.5 to edit the attributes of 608 seismic events, and established the earthquake hazard database for the Chinese mainland from 1949 to 2021. The attributes of the earthquakes in the database include date of occurrence, epicenter latitude and longitude, magnitude, intensity, focal depth, number of casualties, and economic loss. We fit the cumulative magnitude–frequency distribution [35] of 608 earthquakes using the Origin 2018 software, as shown in Figure 2. Using the classification criteria for earthquake events specified in the National Earthquake Emergency Plan, the cumulative magnitude–frequency distribution of earthquakes with general earthquake hazards and above ($M \geq 5$) obeys an exponential distribution. Using statistical methods, we also quantified the characteristics of the spatiotemporal distribution [36] of earthquake events and the associated losses.

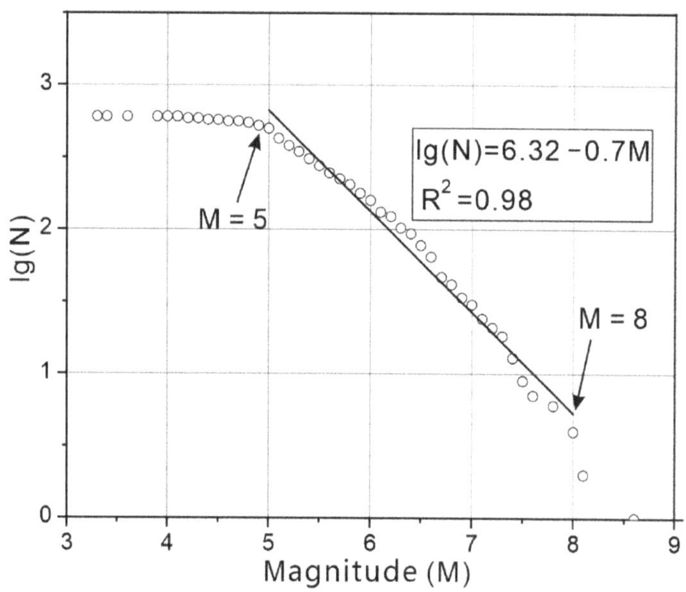

Figure 2. Relationship between earthquake magnitude and frequency. Value lg(N) refers to the logarithm of N. N denotes the cumulative number of earthquakes that resulted in human casualties or economic loss.

3. Results and Analysis

3.1. Characteristics of the Temporal Distribution of Earthquake Hazards and Associated Losses

3.1.1. The Pattern of a Decade

When analyzing the chronological pattern of earthquake events and the associated losses, we merged 1949 into the 1950s and similarly merged 2021 into the previous 10-year time period. In our analysis of economic loss, we converted the annual economic loss based on China's GDP (Gross Domestic Product) data released by the National Bureau of Statistics of China (www.stats.gov.cn, accessed on 15 August 2022), using the economic loss in 2021 as a reference. From the 1950s to the last decade, the number of earthquakes showed a roughly increased trend (Figure 3a). The number of earthquakes has increased from an annual average of three to twelve (in the past 10 years), reaching the peak in the 1990s at an annual average of 12.8. The interdecadal average economic loss showed a jumping trend (Figure 3b), with the 2000s having an average annual economic loss of 33,510.4 million USD, followed by the 1970s with 15,478 million USD. The 1980s and 1990s had the fewest economic losses. The number of casualties is mainly controlled by major catastrophic earthquake events, and there is no obvious trend of an increase or decrease during the study period. For example, the two peaks in Figure 3c,d appeared in the 1970s and early 21st century, corresponding to the Tangshan earthquake in 1976 and the Wenchuan earthquake in 2008. A total of 82 earthquakes occurred in the decade 1970–1980, killing more than 260,000 people and injuring 767,000. The M7.8 earthquake that occurred in the densely populated Tangshan area in 1976 alone killed 242,000 people. In addition, with 110 earthquakes in the decade 2000–2010, the number is significantly higher. However, only about 70,000 people were killed and 390,000 were injured. In the 1980s and 1990s, casualties from earthquake events were minor. Although the economic loss caused by the earthquakes have reached an annual average of 4.21 billion USD in the past decade, the number of casualties has been effectively controlled. Details of the 23 typical earthquakes ($M \geq 7$) that caused fatalities are listed in Table 1. Their geographical locations are shown in Figure 4. There is a clear concentration of developmental features in earthquake events with an $M \geq 6$. Southwest China is a tectonically active earthquake-prone zone. When analyzing the chronological pattern of the number of deaths caused by the earthquakes, we considered the influence of magnitude. As shown in Figure 5, the magnitudes of the earthquakes that caused more than 1000 deaths were all above M6. The most devastating damage was caused by the 1976 Tangshan M7.8 earthquake, which killed 242,000 people and was the only earthquake to kill more than 100,000 people. There are also some unusual earthquakes in Figure 5, such as the 1969 Bohai M7.4 earthquake with a depth of 35 km. This earthquake was located in the Bohai Bay and killed only 10 people. In contrast, the Xiji M5.1 earthquake that occurred on the night of 3 December 1970, with a focal depth of 15 km, resulted in 117 fatalities because the local residents lived in earth kilns with extremely poor seismic resistance. The 1949 Kuche M7.3 earthquake and the 1955 Wuqia M7.0 earthquake both occurred within Xinjiang, which is among the least densely populated areas of the Chinese mainland. The fatalities caused by the two earthquakes were twelve and five, respectively, with a very low number of casualties. It can be seen that these unusual earthquakes are often related to the depth of the epicenter and the distribution of population.

Table 1. Twenty-three typical earthquakes ($M \geq 7$) that caused fatalities.

No.	Date	Epicenter	Magnitude	Intensity	Economic Losses (Million USD)	Fatalities	Injured
1	15 August 1950	Chayu	8.6	XII	51,533.4	3300	260
2	12 May 2008	Wenchuan	8	XI	317,020.7	69,227	375,783
3	5 January 1970	Tonghai	7.8	X+	2985.4	15,621	26,783
4	28 July 1976	Tangshan	7.8	XI	73,447.6	242,000	708,602
5	6 February 1973	Luhuo	7.6	X	16,618.4	2199	2743

Table 1. *Cont.*

No.	Date	Epicenter	Magnitude	Intensity	Economic Losses (Million USD)	Fatalities	Injured
6	18 August 1952	Naqu	7.5	X	12,883.4	54	0
7	14 April 1955	Kangding	7.5	X	6283.9	94	260
8	18 July 1969	Bohai	7.4	IX	593.9	10	353
9	29 May 1976	Longling	7.4	IX	1062.8	98	2442
10	6 November 1988	Lancang	7.4	IX	6676.4	748	7751
11	24 February 1949	Kuche	7.3	IX	-	12	20
12	4 February 1975	Haiyu	7.3	IX+	60,513.3	1328	16,980
13	11 February 1954	Alashan	7.2	X	12,357.0	50	329
14	22 March 1966	Ningjin	7.2	X	12,567.7	8064	38,451
15	16 August 1976	Songfan	7.2	IX	-	41	756
16	11 May 1974	Yanjin	7.1	IX	730.8	1541	1600
17	23 August 1985	Wuqia	7.1	IX	335.4	67	256
18	14 April 2010	Yushu	7.1	IX	7023.5	2698	11,000
19	15 April 1955	Wuqia	7	IX	6283.9	5	13
20	26 April 1990	Gonghe	7	IX	615.0	119	2049
21	3 February 1996	Yulong	7	IX	2878.9	309	17,057
22	20 April 2013	Lushan	7	IX	16,053.7	196	13,019
23	8 August 2017	Jiuzhaigou	7	IX	1483.8	29	543

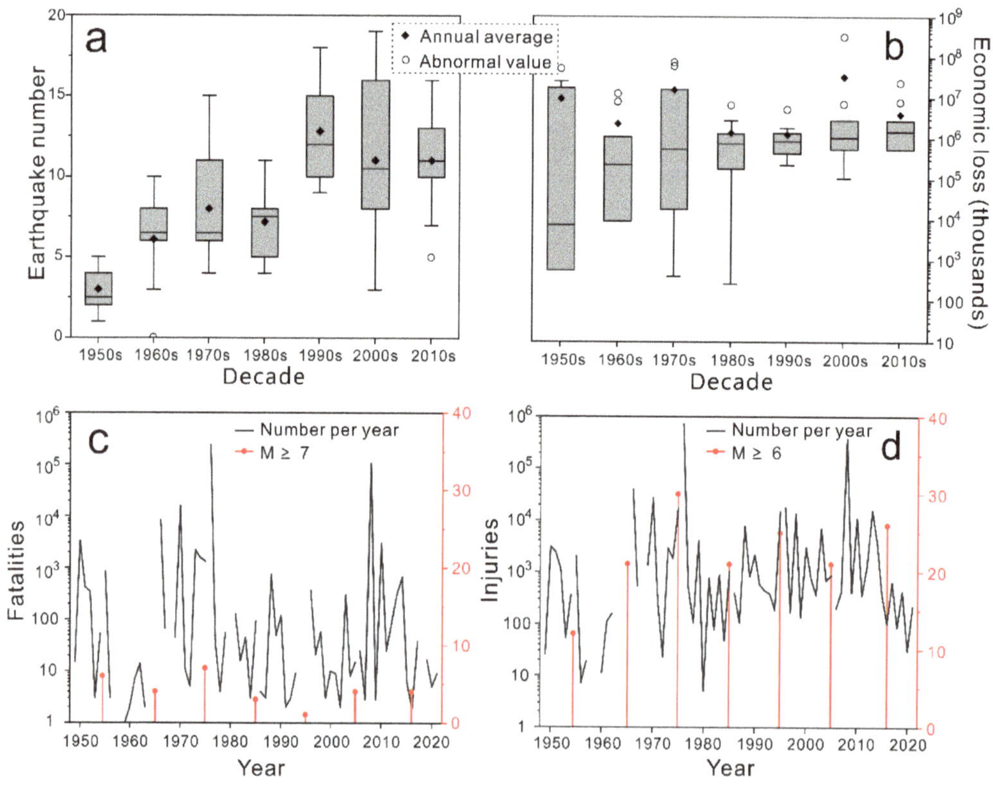

Figure 3. Chronological pattern of earthquake events and statistics of the losses caused: (**a**) number of earthquakes in a decade; (**b**) economic loss in a decade; (**c**) number of fatalities per year (black) and number of the M ≥ 7 earthquakes in a decade (red); (**d**) number of the injured per year (black) and number of the M ≥ 6 earthquakes in a decade (red).

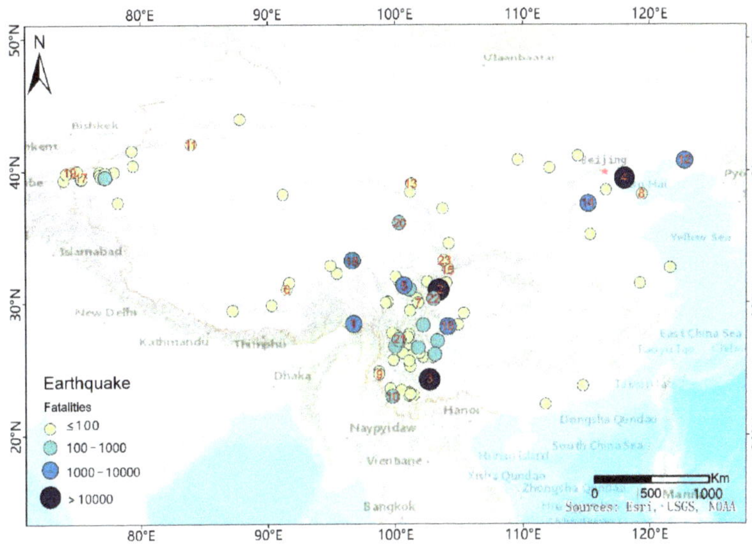

Figure 4. Spatial distribution of seismic events (M $\geq$ 6) that caused fatalities. Numbered in red: locations of 23 typical earthquakes (M $\geq$ 7); unnumbered: M6–7 earthquakes.

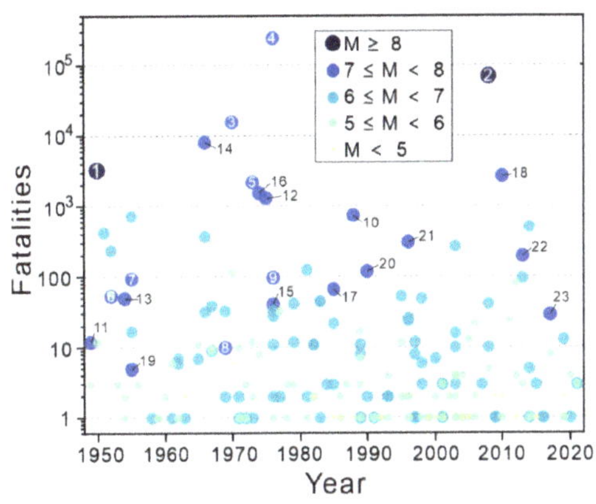

Figure 5. Interannual distribution characteristics of deaths caused by earthquakes with different magnitudes. The number is listed in Table 1.

In conclusion, the number of earthquakes shows a general upward trend, and the relationship between the economic loss and casualties and chronology is not obvious, which is mainly controlled by major catastrophic earthquake events.

3.1.2. Interannual Pattern

A total of 608 earthquake events causing losses occurred from 1949 to 2021, and Figure 6 illustrates the number of earthquake hazards for each year. There is a general upward trend in the number of earthquakes, with a significant decrease in the past two years. The average annual number of earthquakes from 1949 to 1960 is three. The number changes to seven and twelve for 1961–1990 and 1991–2019, respectively. However, there is no obvious stable trend in the annual number of earthquakes; for example, no earthquake event causing casualties occurred in 1968. There was a high incidence of earthquakes around 2000, in which 19 earthquake hazards occurred in 2003. After 2000, the number of earthquake events fluctuated around ten, with less than five in 2002 and 2007.

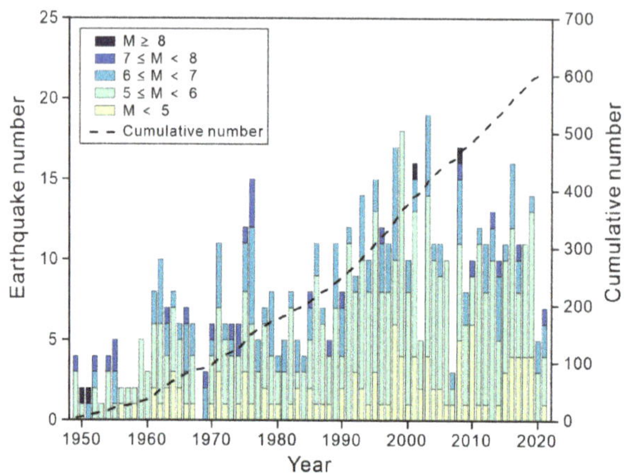

Figure 6. Number of interannual earthquakes.

Overall, the number of earthquakes shows an upward trend, increasing from an annual average of three to twelve. However, there is no clear stable trend.

3.1.3. Seasonal Pattern

The 608 earthquakes were classified according to the number of deaths: ordinary earthquakes (causing less than 20 deaths), moderate earthquake hazards (causing 20–50 deaths), major earthquake hazards (causing 50–300 deaths), particularly major earthquake hazards (causing more than 300 deaths). Table 2 lists the seasonal distribution of each class of earthquake hazards, with six particularly major earthquakes occurring in winters and only two in autumns. Except for particularly major earthquakes, there is not much difference among the four seasons. As can be seen from Figure 7, no matter in which season the moderate or major earthquake hazards occur, Southwest China is always a high-incidence area. Detailed information of earthquake proportion in the Southwest is listed in Table 2. Table 3 lists the earthquake hazards and casualties for the four seasons, with the number of earthquakes occurring slightly higher in the summer than in the other three seasons. The Tangshan M7.8 earthquake on 28 July 1976 killed 242,000 people and injured more than 700,000; the Wenchuan M8 earthquake on 12 May 2008 killed 69,000 people and injured 375,000. These two earthquakes caused the highest number of casualties, which explains why the casualties in the spring and summer are the most serious. As mentioned above, the number of casualties is mainly influenced by major events. Figure 8 shows the number of fatalities caused by earthquakes with different magnitudes in each month. The nine earthquake events that led to more than 1000 fatalities are all distributed between January and August. Earthquake events with less than 100 fatalities are more evenly distributed in the four seasons.

Table 2. Seasonal distribution of earthquake hazards.

	Particularly Major	Major	Moderate	* Ordinary
Spring	4	4	5	130
Summer	3	3	7	156
Autumn	2	4	3	142
Winter	6	2	3	134
Total	15	13	18	562

* Ordinary earthquake events include earthquakes that do not cause casualties.

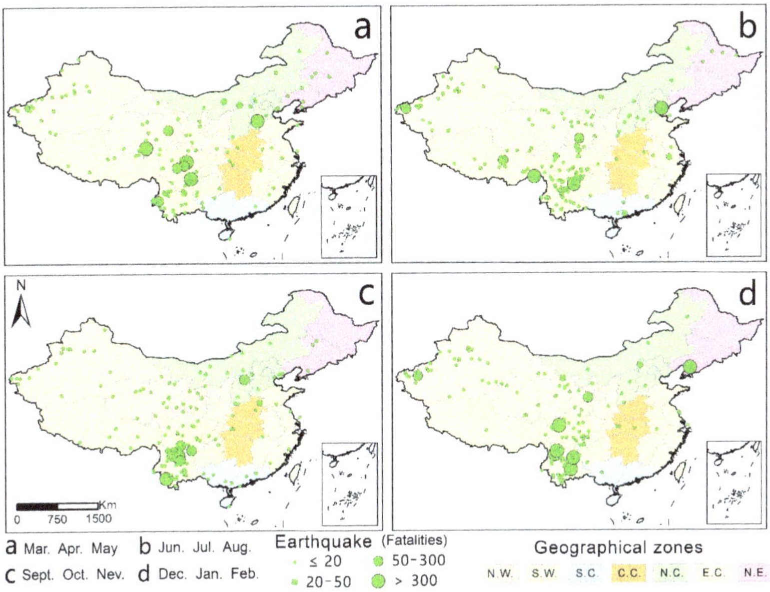

Figure 7. Seasonal distribution of earthquakes. Geographical zones: N.W.—Northwest, S.W.—Southwest, S.C.—South China, C.C.—Central China, N.C.—North China, E.C.—East China, N.E.—northeast. The subfigure in the lower right corner represents South China Sea.

Table 3. Earthquake hazards and casualties in four seasons.

	Number of Earthquakes Causing Casualties	Fatalities
Spring	90	82,283
Summer	110	246,470
Autumn	91	2144
Winter	95	20,998

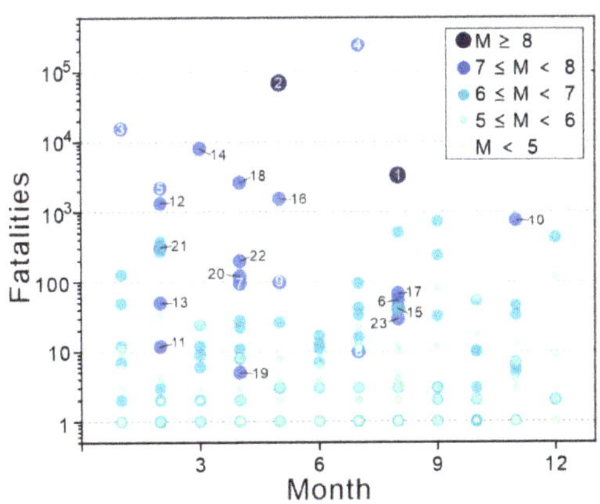

Figure 8. Inter-monthly distribution characteristics of the number of deaths caused by earthquakes with different magnitudes. The number is listed in Table 1.

In conclusion, the number of earthquakes during the studied time period is slightly higher in the summer than in the other three seasons. Specifically, the number of casualties is significantly higher in the summer, which is mainly influenced by major events.

3.2. Spatial Distribution Characteristics of Earthquake Hazards

3.2.1. Spatial Distribution of Earthquake Depths

The depth of the earthquake is the vertical distance from the epicenter to the ground, which is closely related to the tectonic zoning. Under the same conditions, the shallower the depth of the earthquake, the greater its destructive capacity. In this study, only 503 of the 608 earthquake events were recorded with accurate depths. The Chinese mainland is divided into six subregions according to the main tectonic block boundaries [37]. The spatial distribution of earthquake depths is shown in Figure 9. The earthquake depths in the Westfield fault block region are distributed in the range of 4–64 km, with an average depth of 19 km. The earthquake depths in the Tibetan fault block region are distributed in the range of 5–48 km, with an average depth of 17.4 km. Similarly, the earthquake depths of the Dian–Mian fault block region, South China fault block region, North China fault block region, Northeast Asia fault block region are 4–34, 4–48, 5–33, 5–30 km, with an average depth of 15.6, 13.2, 14.7, and 14.4 km, respectively. The earthquake depths between the blocks are distributed in the range of 5–50 km, with an average depth of 18 km. The M6.0 earthquake in Gashi, Xinjiang that occurred on 2 August 1998 is the deepest earthquake (64 km) on record, causing three injuries and an economic loss of 7.65 million USD. Among the 503 earthquake events, there are 321 (63.8%) events with an earthquake depth within 5–16 km. The detailed statistics of earthquake depths are shown in Figure 10. Finally, we analyzed the effect of the earthquake depth on the number of fatalities. As shown in Figure 11, all earthquakes causing more than 100 deaths have a focal depth of less than 25 km, mostly concentrated in the range of 10–20 km. When the focal depth exceeds 30 km, the number of fatalities is no larger than 10. Only two of the earthquake events with focal depths greater than 40 km resulted in fatalities: the Urumqi M6.6 earthquake in 1965 and the Atushi M5.7 earthquake in 1971. Notably, the 1976 M7.8 Tangshan earthquake was smaller in magnitude than the 2008 M8.0 Wenchuan earthquake, but it caused the largest number of fatalities. The epicenter of Tangshan earthquake is located in the downtown area of Tangshan City, with high population density, resulting in destructive housing damage and serious casualties. By contrast, the epicenter of the Wenchuan earthquake is located in a mountainous area, with a much smaller population density, and the regional secondary hazards are mainly landslides. Beichuan county, which has the highest concentration of casualties associated with the Wenchuan earthquake, is more than 100 km from the epicenter.

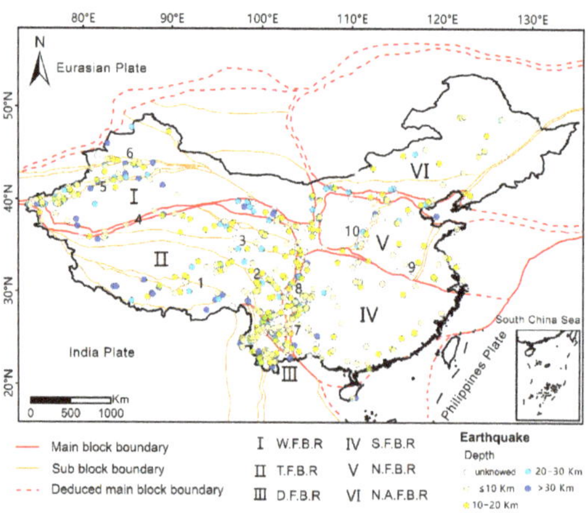

Figure 9. Spatial distribution of earthquake depths. The block information is referenced to Zhang et al. [37]. W.F.B.R.—Westfield fault block region (F.B.R.—fault block region), T.—Tibetan, D.—Dian–Mian, S.—South China, N.—North China, N.A.—Northeast Asia. Faults: 1—Lijia, 2—Yushu–Xianshui river zone, 3—East Kunlun, 4—Alkin, 5—Nantian mountain, 6—Beitian mountain, 7—Anning–Zemu river zone, 8—Longmen mountain, 9—Tango, 10—Shanxi.

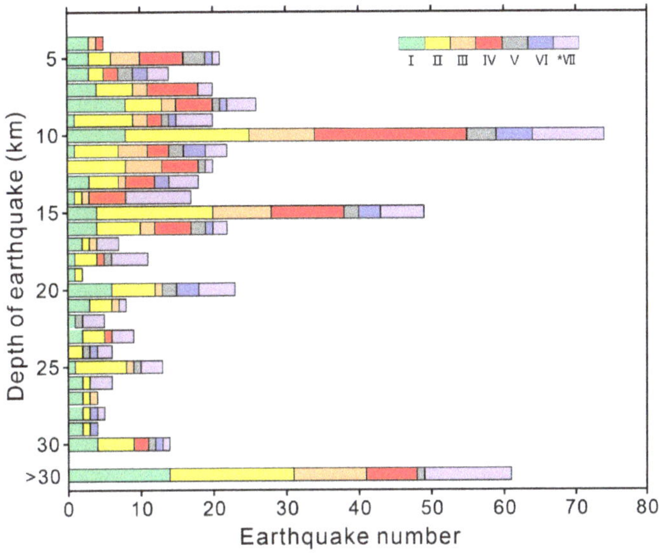

Figure 10. Depths of earthquake. I-VI denote six blocks, see Figure 6, *VII denotes between the blocks.

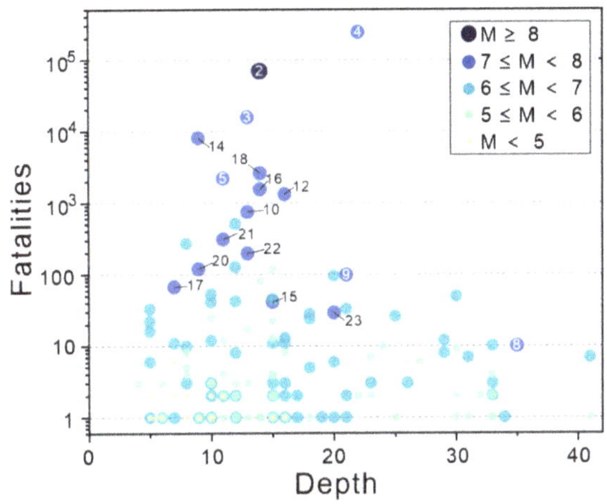

Figure 11. Focal depth distribution characteristics of the number of deaths caused by earthquakes with different magnitudes. The number is listed in Table 1.

In conclusion, the average focal depths of the six plates in the Chinese mainland from west to east show a decreasing trend, and the focal depths of the earthquakes that caused more than 100 deaths are less than 25 km.

3.2.2. Spatial Distribution of Epicentral Intensity

Among the 608 earthquake events, 558 events show clear records of epicentral intensity. The M8.6 earthquake in Chayu, Tibet in 1950 is the only earthquake with an intensity of XII in the extreme seismic zone. The event caused 3300 deaths, 260 serious injuries, and an economic loss of 51,533.4 million USD. There are four earthquakes with an intensity of up to XI. Among them, the 1976 Tangshan earthquake and the 2008 Wenchuan earthquake caused more than 244,000 and 69,000 deaths, respectively, while the M8.0 earthquake in Nagqu, Tibet in 1951 and the M8.1 earthquake in Ruoqiang, Xinjiang in 2001 did not cause any fatalities. The epicenters of both earthquakes are in no-man's land, and the distances to the nearest populated distribution sites are 74 and 400 km, respectively. Referring to Li et al. [36], the Chinese mainland was divided into two parts, East and West, using the "Hu Huanyong line" as the boundary. As shown in Figure 12, there are 40 earthquakes with an epicenter intensity greater than IX. Among them, only thirteen are distributed in the East. Eighty-seven earthquakes possess an epicenter intensity of VIII, and only twenty-seven of them are distributed in the East. Five of the twenty-seven events resulted in more than twenty deaths (including one with fifty-eight deaths), accounting for 18.5%. In contrast, among the sixty earthquake events that occurred in the West, only seven events resulted in more than twenty deaths, accounting for 11.7%. Additionally, Figure 12 shows that earthquakes with an epicentral intensity greater than VIII occur frequently in Sichuan and Yunnan Provinces, especially in Yunnan Province. Among the recorded earthquake events, the lowest intensity is V. Three of them also caused fatalities: the 1970 M4.8 earthquake in Qinglong, Guizhou, the 1986 M4.7 earthquake in Xiaxian, Shanxi, and the 1992 M4.6 earthquake in Sheyang, Jiangsu. In addition to analyzing the spatial distribution of earthquake intensity, the relationship between the intensity and the number of fatalities in combination with different magnitudes is displayed in Figure 13. Among the earthquake events that led to fatalities, those with intensities greater than X resulted in at least 50 deaths. Although the earthquakes with an intensity of V also caused fatalities, the death numbers are all smaller than three. Most of the earthquake events that caused more than 100 deaths have an intensity $\geq$ IX. The two exceptions are the M5.1 Xiji earthquake in 1970 and the M6.9 Daofu earthquake in 1981. In detail, the Daofu earthquake caused nearly 3000 Tibetan-style houses to collapse, killing 126 people. The number of fatalities is controlled by multiple factors, and in terms of intensity alone, the number of fatalities shows a roughly increasing trend as the intensity increases. Some earthquakes in Figure 13 do not satisfy this trend, including the 1952 Naqu earthquake, the 1954 Alashan earthquake, and the 1955 Kangding earthquake. These events are all characterized by an intensity of X, resulting in less than 100 fatalities. The 1950 Chayu earthquake killed only 3300 people, although it reached an intensity of XII. After analyzing the historical data, the seismic macro-damage and intensity distribution of the above events were in the sparsely populated areas.

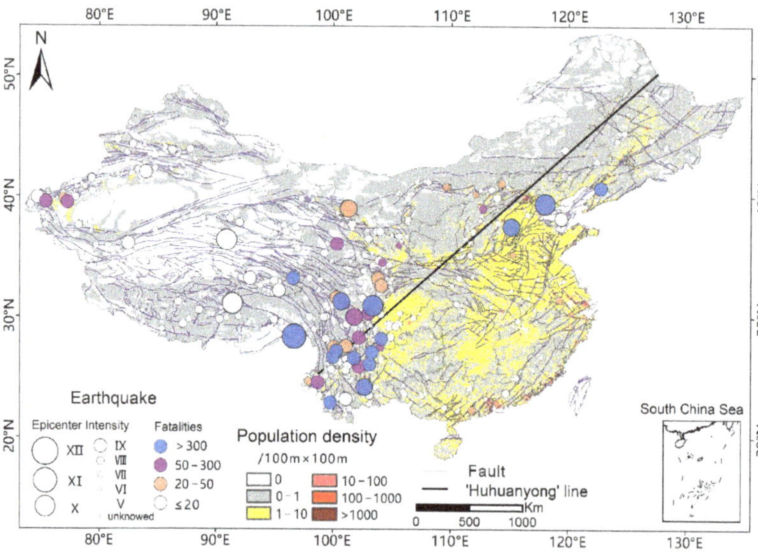

Figure 12. Spatial distribution of epicentral intensity. "Hu Huanyong line" [36]—a population division line separating China into two parts, East and West.

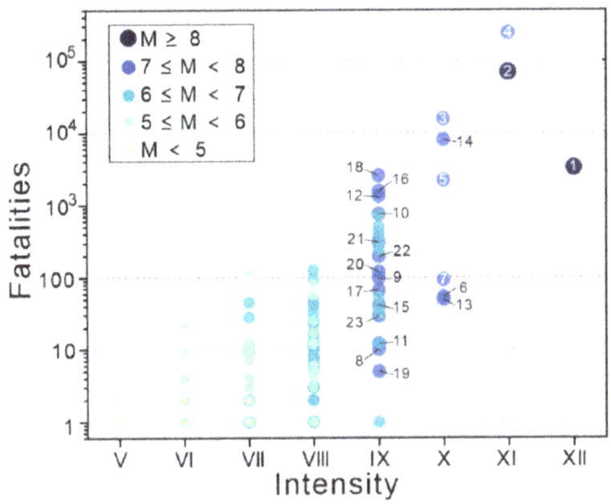

Figure 13. Intensity distribution of the number of deaths caused by earthquakes with different magnitudes. The number is listed in Table 1.

Overall, in terms of intensity alone, the number of fatalities increases with the increasing intensity. The proportion of deaths caused by the earthquakes is higher in the East (bounded by the "Hu Huanyong line") due to the high regional population density.

4. Discussion

Based on the earthquake hazard data from the China Earthquake Networks Center and the provincial (autonomous region and municipality) earthquake agencies, we established a database of earthquake hazard information for the Chinese mainland from 1949 to 2021, and preliminarily analyzed the spatiotemporal distribution characteristics of 608 earthquake events and the associated losses. In this study, we only used basic statistical analysis methods, which have certain limitations, and this will be a direction to be addressed in future studies. The number of earthquakes is basically on the rise, which is related to the continuous improvement of statistical tools. No obvious trend of an increase or decrease in the number of casualties was observed during the study period, which was mainly associated with particularly major events. Although the economic loss caused by the earthquakes has reached an annual average of 4.21 billion USD in the past 10 years, the number of casualties has been effectively controlled. As the economy grows, people begin to invest more in life security. Coupled with the increasing attention to safety by governments, earthquake hazards are gradually changing from a threat to personal safety to a threat to property [36]. Different from Lin et al.'s [38] study that investigated the seasonal distribution characteristics of fatal landslides, the season when earthquakes occur and the influencing factors of losses are not obvious. However, no matter in which season the moderate or major earthquake hazards occur, the Southwest China is always a high-incidence area. The reason is that the western part of the Sichuan–Yunnan block is located at the junction zone of the Indian Ocean plate, Eurasian plate, and Pacific plate. Such a region belongs to the Pacific Rim seismic zone, which is an earthquake-prone area.

The average earthquake depths of the six blocks in the Chinese mainland show a decreasing trend from west to east, which is consistent with Shao et al.'s data [39]. In their study, the Chinese mainland was divided into two subregions (using 105°E as the boundary), eastern and western regions, with the depth of earthquakes in the latter being slightly greater than that in the former. It is also believed that when the earthquake magnitudes are comparable, the damage in the epicenter region will be heavier in the east and lighter in the west. Casualties caused by earthquake hazards are significantly influenced by the population density. Taking the population division line, the "Hu Huanyong line", as the boundary, the number of earthquakes in the East is much lower than that in the West, but under the same epicenter intensity, earthquakes are more likely to cause casualties in the East because the population density is much higher there than that in the West. Chen et al. [40] used indicators such as seismic intensity and population density to present a simplified methodology to assess earthquake risk, which provided effective, if rough, earthquake damage information for initial response and/or risk management planning in the Chinese mainland. This, of course, is also related to aspects such as awareness of earthquake hazard prevention and the earthquake resistance of buildings. However, there are studies with different claims, such as the anti-lethal level proposed by Xia et al. [14], where the average anti-lethal level was higher in the east of China than in the west. Coseismic landslides are also a major cause of a large number of casualties [41,42]. Further analyses need to be completed with the support of more data, which is also the goal of our future study.

The establishment and preliminary analysis of the earthquake hazard information database are the basis for understanding the characteristics of earthquake-related losses and establishing a series of casualty prediction models (e.g., loss prediction models based on factors such as the location of the earthquake occurrence, magnitude, and focal depth). The construction of a population-loss evaluation model will allow for the prediction of population casualties in future earthquake scenarios. The database of earthquake hazard information in this study contains 608 earthquakes, and our data are richer in both the involved years and the number of earthquake events, relative to the studies by Chen et al. [43] and Wu et al. [18]. There are other contributing factors that deserve further studies [44]. Rapid assessment of earthquake casualties can provide a scientific basis and data reference for post-earthquake emergency rescue. However, there are some limitations in this study: the casualties caused by the earthquakes vary with the moment of the earthquake [44]. During the construction of the earthquake hazard information database, specific moments of earthquake occurrence were not included. In addition, the influencing factors of earthquake casualties [2] and the specific geographical location of casualties were not discussed. Hazard analysis of major earthquake events and prediction of earthquake damage and casualties will be the focus of future work.

5. Conclusions

In this study, we collected 608 earthquake events causing losses in the Chinese mainland during the period 1949–2021. Among them, 222 earthquakes caused only economic damage, while the remaining 386 events resulted in more than 370,000 deaths and 1.3 million injuries. We used the ArcGIS 10.5 software to edit the attributes of 608 events and established a database of earthquake hazard information. We further analyzed the spatiotemporal distribution of earthquake hazards and the associated losses. The conclusions of this study are as follows:

(1) From the 1950s to the last decade, the number of earthquakes roughly shows an increasing trend, with the number of earthquakes increasing from an annual average of three to twelve, but there is no clear stable trend. The number of earthquakes during the studied time period is slightly higher in the summer than in the other three seasons. The number of casualties is significantly higher in the summer than in the other three seasons, which is mainly influenced by major events. No matter in which season the moderate or major earthquake hazards occur, Southwest China is always a high-incidence area.

(2) The average earthquake depths of the six blocks from west to east show a decreasing trend. The largest focal depth was found in the Westfield fault block region. The average focal depths of the three blocks in South China, North China, and Northeast Asia are relatively small. In the statistical earthquake events, the focal depth distribution in the range of 5–16 km accounts for a high proportion (63.8%). The focal depths of the earthquakes that caused more than 100 deaths are all within 25 km.

(3) In terms of epicenter intensity alone, the number of deaths caused by the earthquakes shows a basically increasing trend with the increasing intensity. Affected by the population density, the proportion of deaths caused by the earthquakes is higher in the East (bounded by the "Hu Huanyong line").

Author Contributions: C.X. and T.Z. proposed the research concept and provided basic data. L.L. designed the framework and wrote the manuscript. Y.H. provided basic data. All authors have read and agreed to the published version of the manuscript.

Funding: This study was supported by the National Institute of Natural Hazards, Ministry of Emergency Management of China (ZDJ2021-12).

Institutional Review Board Statement: Not applicable.

Informed Consent Statement: Not applicable.

Data Availability Statement: The earthquake hazard information database (in shp. format) used in this study are freely available at this link: https://zenodo.org/record/7800432.

Conflicts of Interest: The authors declare no conflict of interest.

References

1. Shadmaan, S.; Islam, A.I. Estimation of earthquake vulnerability by using analytical hierarchy process. *Nat. Hazards Res.* **2021**, *1*, 153–160. [CrossRef]
2. Shen, Z.; Jackson, D.; Kagan, Y. Implications of geodetic strain rate for future earthquakes in southern California. *Seismol. Res. Lett.* **2007**, *78*, 116–120. [CrossRef]
3. Wang, W.; Chen, H.; Jiang, X.; Ma, L.; Xu, Y.; Qu, M. Lessons from the casualties caused by the Changning M 6.0 earthquake in China. *Nat. Hazards Res.* **2021**, *1*, 81–87. [CrossRef]
4. Dong, L.; Shan, J. A comprehensive review of earthquake-induced building damage detection with remote sensing techniques. *ISPRS J. Photogramm. Remote Sens.* **2013**, *84*, 85–99. [CrossRef]
5. Ren, J.; Zhang, Z.; Gai, H.; Kang, W. Typical Riedel shear structures of the coseismic surface rupture zone produced by the 2021 Mw 7.3 Maduo earthquake, Qinghai, China, and the implications for seismic hazards in the block interior. *Nat. Hazards Res.* **2021**, *1*, 145–152. [CrossRef]
6. Kanamori, H. The energy release in great earthquakes. *J. Geophys. Res.* **1977**, *82*, 2981–2987. [CrossRef]
7. Farolfi, G.; Keir, D.; Corti, G.; Casagli, N. Spatial forecasting of seismicity provided from earth observation by space satellite technology. *Sci. Rep.* **2020**, *10*, 9696. [CrossRef]
8. Nakamura, M.; Kinjo, A. Activated seismicity by strain rate change in the Yaeyama region, south Ryukyu. *Earth Planets Space* **2018**, *70*, 154. [CrossRef]
9. Zeng, Y.; Petersen, M.; Shen, Z. Earthquake potential in California-Nevada implied by correlation of strain rate and seismicity. *Geophys. Res. Lett.* **2018**, *45*, 1778–1784. [CrossRef]
10. Riguzzi, F.; Crespi, M.; Devoti, R.; Doglioni, C.; Pietrantonio, G. Geodetic strain rate and earthquake size: New clues for seismic hazard studies. *Phys. Earth Planet. Inter.* **2012**, *206*, 67–75. [CrossRef]
11. Gutenberg, G.; Richter, C.F. Seismicity of the earth and associated phenomena, Howard Tatel. *J. Geophys. Res.* **1950**, *55*, 97.
12. Samardjieva, E.; Badal, J. Estimation of the expected number of casualties caused by strong earthquakes. *Bull. Seismol. Soc. Am.* **2002**, *92*, 2310–2322. [CrossRef]
13. Nie, G.; An, J.; Deng, Y. Advances in earthquake emergency disaster service. *Seismol. Geol.* **2012**, *34*, 782–791. (In Chinese)
14. Xia, C.; Nie, G.; Li, H.; Fan, X.; Qi, W. A composite database of casualty-inducing earthquakes in mainland China. *Nat. Hazards* **2023**, *116*, 3321–3351. [CrossRef]
15. Firuzi, E.; Amini Hosseini, K.; Ansari, A.; Izadkhah, Y.O.; Rashidabadi, M.; Hosseini, M. An empirical model for fatality estimation of earthquakes in Iran. *Nat. Hazards* **2020**, *103*, 231–250. [CrossRef]
16. Oike, K.; Hori, T. History of earthquakes and seismic disasters in east Asia. *Science* **1998**, *68*, 409–415.
17. Koshimura, S.; Katada, T.; Mofjeld, H.O.; Kawata, Y. A method for estimating casualties due to the tsunami inundation flow. *Nat. Hazards* **2006**, *39*, 265–274. [CrossRef]
18. Wu, X.; Gu, J.; Wu, H. A modified exponential model for reported casualties during earthquakes. *Acta Seism. Sin* **2009**, *31*, 457–463.
19. Jaiswal, K.; Wald, D.J. Estimating economic losses from earthquakes using an empirical approach. *Earthq. Spectra* **2013**, *29*, 309–324. [CrossRef]
20. Samardjieva, E.; Oike, K. Modelling the number of casualties from earthquakes. *J. Nat. Disaster Sci.* **1992**, *14*, 17–28.
21. Badal, J.; Vázquez-Prada, M.; González, Á. Preliminary quantitative assessment of earthquake casualties and damages. *Nat. Hazards* **2005**, *34*, 353–374. [CrossRef]
22. So, E.; Spence, R. Estimating shaking-induced casualties and building damage for global earthquake events: A proposed modelling approach. *Bull. Earthq. Eng.* **2013**, *11*, 347–363. [CrossRef]
23. Turkan, S.; Özel, G. Modeling destructive earthquake casualties based on a comparative study for Turkey. *Nat. Hazards* **2014**, *72*, 1093–1110. [CrossRef]
24. Xia, C.; Nie, G.; Fan, X.; Zhou, J.; Li, H.; Pang, X. Research on the rapid assessment of earthquake casualties based on the anti-lethal levels of buildings. *Geomat. Nat. Hazards Risk* **2020**, *11*, 377–398. [CrossRef]

25. Xia, C.; Nie, G.; Fan, X.; Zhou, J.; Li, H.; Zhou, J.; Zeng, X. A new model for the quantitative assessment of earthquake casualties based on the correction of anti-lethal level. *Nat. Hazards* **2022**, *110*, 1199–1226.
26. Jaiswal, K.; Wald, D.J.; Hearne, M. *Estimating Casualties for Large Earthquakes Worldwide Using an Empirical Approach*; USGS: Columbus, OH, USA, 2009.
27. Erdik, M.; Şeşetyan, K.; Demircioğlu, M.; Hancılar, U.; Zülfikar, C. Rapid earthquake loss assessment after damaging earthquakes. *Soil Dyn. Earthq. Eng.* **2011**, *31*, 247–266. [CrossRef]
28. Gerstenberger, M.C.; Marzocchi, W.; Allen, T.; Pagani, M.; Adams, J.; Danciu, L.; Field, E.H.; Fujiwara, H.; Luco, N.; Ma, K.-F.; et al. Probabilistic seismic hazard analysis at Regional and National Scales: State of the art and challenges. *Rev. Geophys.* **2020**, *58*, e2019RG000653. [CrossRef]
29. Nie, G.; Gao, J.; Ma, Z.; Gao, Q.; Su, G. On the risk of earthquake disaster in China in the coming 10~15 years. *J. Nat. Disasters* **2002**, *11*, 68–73. (In Chinese)
30. Wei, Y. Earthquake Safety Estimation Study for the Urban Rail Transportation Engineering Field. Master's Thesis, Central South University, Changsha, China, 2009. (In Chinese).
31. Zheng, T.; Zheng, Y. Review of earthquake damage losses in mainland China in 2011. *J. Nat. Disasters* **2012**, *21*, 88–97. (In Chinese)
32. Chen, T.; Zheng, T. Review of earthquake damage losses in mainland China in 2015. *J. Catastrophol.* **2016**, *31*, 133–137. (In Chinese)
33. Zhao, B.; Huang, Y.; Zhang, C.; Wang, W.; Tan, K.; Du, R. Crustal deformation on the Chinese mainland during 1998–2014 based on GPS data. *Geod. Geodyn.* **2015**, *6*, 7–15. [CrossRef]
34. Huang, X.; Jin, H. An earthquake casualty prediction model based on modified partial Gaussian curve. *Nat. Hazards* **2018**, *94*, 999–1021. [CrossRef]
35. Lindenfeld, M.; Rümpker, G.; Batte, A.; Schumann, A. Seismicity at the Rwenzori Mountains, East African Rift: Earthquake distribution, magnitudes and source mechanisms. *Solid Earth Discuss.* **2012**, *4*, 565–598.
36. Li, W.; Chen, W.; Zhou, Z.; Gao, N.; Chen, J. Analysis of temporal-spatial distribution of life losses caused by earthquake hazards in Chinese mainland. *J. Catastrophol.* **2019**, *34*, 222–228. (In Chinese)
37. Li, W.; Chen, W.; Zhou, Z.; Gao, N.; Sun, Y. Assessing the applicability of life vulnerability models for earthquake disasters in typical regions of China. *J. Beijing Norm. Univ.* **2019**, *55*, 284–290. (In Chinese)
38. Lin, Q.; Wang, Y. Spatial and temporal analysis of a fatal landslide inventory in China from 1950 to 2016. *Landslides* **2018**, *15*, 2357–2372. [CrossRef]
39. Shao, D.; Sun, Z.; Tian, Q.; Zhang, W. Statistical analysis of historical earthquake intensity in Chinese mainland. *Seismol. Geomagn. Obs.* **2018**, *39*, 135–140. (In Chinese)
40. Chen, Q.; Mi, H.; Huang, J. A simplified approach to earthquake risk in mainland China. *Pure Appl. Geophys.* **2005**, *162*, 1255–1269. [CrossRef]
41. Cui, S.; Wu, H.; Pei, X.; Yang, Q.; Huang, R.; Guo, B. Characterizing the spatial distribution, frequency, geomorphological and geological controls on landslides triggered by the 1933 Mw 7.3 Diexi Earthquake, Sichuan, China. *Geomorphology* **2022**, *403*, 108177. [CrossRef]
42. Xiao, Z.; Xu, C.; Huang, Y.; He, X.; Shao, X.; Chen, Z.; Xie, C.; Li, T.; Xu, X. Analysis of spatial distribution of landslides triggered by the Ms 6.8 Luding earthquake in China on September 5, 2022. *Geoenvironmental Disasters* **2023**, *10*, 3. [CrossRef]
43. Shapira, S.; Aharonson-Daniel, L.; Shohet, I.M.; Peek-Asa, C.; Bar-Dayan, Y. Integrating epidemiological and engineering approaches in the assessment of human casualties in earthquakes. *Nat. Hazards* **2015**, *78*, 1447–1462. [CrossRef]
44. Nan, Y.; Liu, K.; Gao, B.; Zhang, X.; Li, Y. Discussion on the grade scales of natural disasters by the losses in Wenchuan earthquake. *J. Catastrophol.* **2021**, *36*, 42–47. (In Chinese)

sustainability

Brief Report

Theoretical and Experimental Study of Rotational Behaviour of Friction Pendulum Bearings

Tianbo Peng [1,2], Yuxin Liu [2] and Thierno Seydou Ka [2,*]

1 State Key Laboratory of Disaster Reduction in Civil Engineering, Tongji University, Shanghai 200092, China; ptb@tongji.edu.cn
2 College of Civil Engineering, Tongji University, Shanghai 200092, China
* Correspondence: tsk98@tongji.edu.cn

Abstract: Friction Pendulum Bearing (FPB for short) is a kind of widely used device to improve seismic capacities of building and bridge structures. Despite the considerable progress made in developing theoretical models to understand the mechanical behaviour of FPB, these models mostly focus on the horizontal movement behaviour and rarely on the rotational behaviour of FPB. However, rotational displacements of FPBs indeed will occur along with horizontal displacements in earthquakes and may affect the structural seismic performance. Motivated from these findings, a theoretical model of FPB that describes the rotational behaviour is established based on the moment balance theory in this paper. A set of rotational tests are carried out to validate the theoretical model, and comparisons of experimental and theoretical results show that they are in good agreement. The theoretical model developed in this paper will help understand the structural seismic performance more accurately in the case of rotations of FPBs.

Keywords: Friction Pendulum Bearing; theoretical model; rotation angle; rotational behaviour; moment balance theory; seismic isolation

Citation: Peng, T.; Liu, Y.; Ka, T.S. Theoretical and Experimental Study of Rotational Behaviour of Friction Pendulum Bearings. *Sustainability* **2023**, *15*, 7327. https://doi.org/10.3390/su15097327

Academic Editors: Humberto Varum, Chong Xu, Su Chen and Shuang Li

Received: 18 January 2023
Revised: 17 March 2023
Accepted: 13 April 2023
Published: 28 April 2023

1. Introduction

Earthquakes are one of the most devastating natural disasters faced by engineering structures. The end of an earthquake is generally synonymous with the enormous destruction of buildings, bridges, roads, etc. Those structural collapses are the main cause of mortality and economical loss after earthquakes [1,2]. The fatal consequences caused by earthquakes have aroused worldwide attention, especially in the civil engineering field. Nowadays, seismic protection is a crucial aspect to consider in structural design. During the last decades, many seismic protection techniques have been proposed and developed.

Among all these techniques, base isolation techniques such as Lead Rubber Bearings (LRBs), high damping seismic isolation rubber bearings, and Friction Pendulum Bearings (FPBs) have earned significant recognition as efficient seismic isolators for structures [3–5].

Friction Pendulum Bearing (FPB) are reputed to be very efficient at reducing the risk of damage and collapse for structures [6–12].

As presented in Figure 1, the FPB consists of two plates with concave surfaces and a slider to separate the two plates. The upper and lower concave surfaces have respective curvature radii R_1 and R_2 and friction coefficients of μ_1 and μ_2. The function of an FPB is to elongate the structural natural period beyond the predominant period of ground motions.

The first concept of the FPB was presented by Jules Touallion in 1870, conceptualising a rolling ball bearing in his patent [13]. Since then, many prototypes of friction pendulum bearings have been invented, considering different curvature radii, curvature variation of concave surfaces, different friction coefficients of concave surfaces, different thicknesses of the slider, etc. Based on the force balance theory, many scholars have described the operation principle and force-displacement relationship of FPBs. Good fidelity with experimental results has been observed for the proposed theoretical models [14–16].

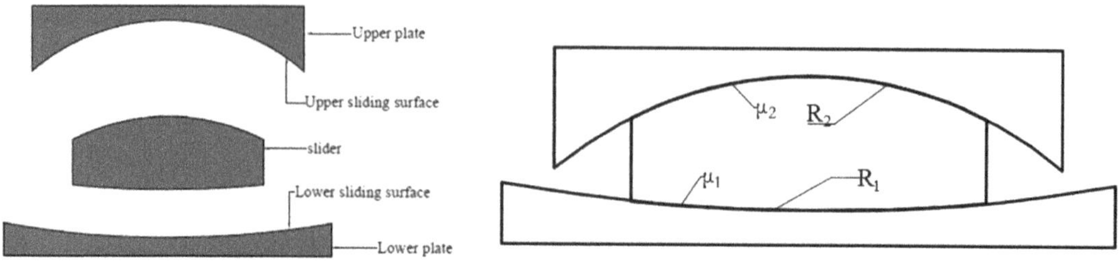

Figure 1. Sketch of an FPB cross-section.

The main advantages of FPBs are the simplicity and efficacy of their principle, good re-centring capacity, and high energy dissipation capacity for both bridge and building structures [17–22].

The common aspect of aforementioned studies is that upper and lower plates of FPBs are assumed to remain horizontal, and the effect of a possible rotation of the two plates in earthquakes is neglected. This assumption may be acceptable if the upper and lower plates can only move in the horizontal direction, such as in the case when the base isolation technique is used for building structures. However, if FPBs are adopted in bridges with flexible piers for seismic isolation purpose, rotations of FPBs are inevitable.

Up until now, a few studies have concerned the issue of rotations of FPBs. Based on the kinematic constitutive model of the triple friction pendulum (TFP) bearing including rotations of lower and upper plates, Becker et al. investigated the effects of constant support rotation and flexible support on the TFP bearing seismic behaviour [23,24]. It is found that, when TFP bearings are subjected to constant rotations, the hysteresis loop is shifted to the positive shear direction, and the displacement capacity of the bearing decreases with the increase in the support rotation. Mosqueda et al. [25] made similar observations by including constant rotation of the plates to the restoring force formula. To minimize the effects of rotation on hysterical performance of FPBs, it is recommended to install the bearing plates to parts of the structure where rotations are less probable to occur.

However, only the influence of a constant rotation on the seismic performance is considered in these references. In this paper, the rotational behaviour of FPBs is described in a distinct way from the previous studies by considering variable rotation angle and moment. Formulas for restoring force and moment are derived based on the moment balance theory [26].

In addition, an experimental study on the rotational behaviour of FPB is conducted. The study includes four different specimens with different friction coefficient combinations of sliding surfaces under three different vertical loads. Experimental results are presented and compared to the theoretical predictions, and some conclusions are drawn.

2. Theoretical Study of the Rotational Behaviour of FPB

The initial state and rotational state of a typical FPB are presented in Figure 2.

It is assumed that the upper plate is rotated at an angle α, and the centre of rotation C_0 is located on the symmetric line of the upper plate; h_c represents the vertical distance between C_0 and the top of the upper plate.

From the free body diagram shown in Figure 3, forces considered in the diagram are:

1. The vertical load W;
2. Rotation moment M_b acting on the upper plate
3. The horizontal restoring force F;
4. The friction stresses acting on the lower and upper sliding surfaces, respectively τ_1 and τ_2;
5. The normal stresses acting on the lower and upper sliding surfaces, respectively n_1 and n_2.

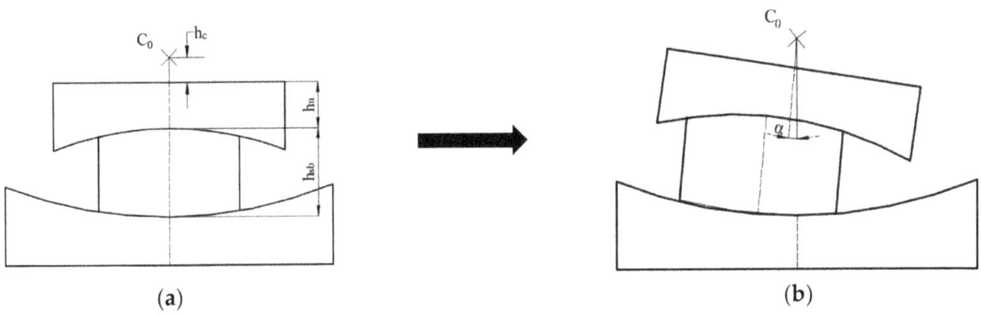

Figure 2. The initial state and rotational state of a typical FPB: (**a**) initial state; (**b**) rotational state.

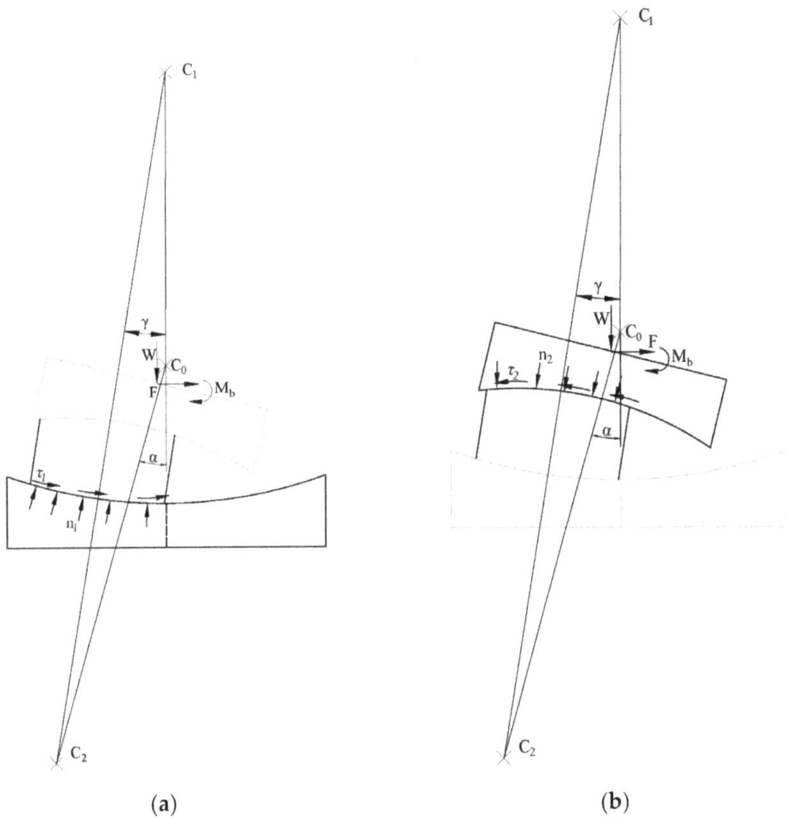

Figure 3. Free body diagram of FPB in rotation: (**a**) lower sliding surface free body diagram; (**b**) upper sliding surface free body diagram.

Setting the sum of the moment about points C_1 and C_2 to zero, the moment equilibrium equations could be written as

$$\sum M(C1) = -Wh_c \sin\alpha - F(|C_0C_1| + h_c \cos\alpha) - R_1 \int |\vec{\tau_1}|dA + d_1 \int |\vec{n_1}|dA + M_b = 0 \quad (1)$$

$$\sum M(C_2) = W(R_2 + h_u)\sin\alpha + F(R_2 + h_u)\cos\alpha - R_2 \int |\vec{\tau_2}|dA - d_2 \int |\vec{n_2}|dA + M_b = 0 \quad (2)$$

where h_u represents the minimum thickness of the upper plate as shown in Figure 2a.

Based on the geometry of Figure 3, the following relation can be obtained:

$$\frac{\sin\alpha}{R_e} = \frac{\sin\gamma}{R_2 + h_u + h_c} = \frac{\sin(\alpha-\gamma)}{|C_0C_1|} \tag{3}$$

where $R_e = R_1 + R_2 - h_{sb}$, h_{sb} represents the maximum thickness of the slider as shown in Figure 2a.

Horizontal displacements are far less than the curvature radii of the sliding surfaces, and the angles α and γ are relatively small. Additionally, assuming the distribution of stresses on the concave surfaces is uniform, the resultants of normal forces of the concave surfaces and the radius of the concave surfaces are superposed; thus, the lever arm of the normal forces on the concave surfaces (d_1 and d_2) equals to zero. Therefore,

$$d_1\int |\vec{n_1}|dA = 0 \tag{4}$$

$$d_2\int |\vec{n_2}|dA = 0 \tag{5}$$

Considering the simplifications and assumptions made above, the following relations are obtained:

$$\sum M(C1) = -Wh_c\alpha - F(|C_0C_1| + h_c) - R_1\int |\vec{\tau_1}|dA + M_b = 0 \tag{6}$$

$$\sum M(C_2) = W(R_2 + h_u)\alpha + F(R_2 + h_u) - R_2\int |\vec{\tau_2}|dA + M_b = 0 \tag{7}$$

$$\frac{\alpha}{R_e} = \frac{\gamma}{R_2 + h_u + h_c} = \frac{\alpha-\gamma}{|C_0C_1|} \tag{8}$$

Combining Equations (6)–(8), equations of the moment M_b and the restoring force F can be determined as follows:

$$M_b = \frac{W(R_2 + h_u)(R_2 + h_u + h_c - R_e)}{R_e}\alpha + CST_1 \tag{9}$$

$$F = -\frac{W(R_2 + h_u - h_c)}{R_e}\alpha + \frac{-\mu_1 R_1 W + \mu_2 R_2 W}{R_e} \tag{10}$$

where

$$CST_1 = (\mu_1 R_1 R_2 - \mu_2 R^2 + \mu_2 R_2 R_e + \mu_1 R_1 h_u - \mu_2 R_2 h_u)W/R_e \approx (\mu_1 + \mu_2)R_1 R_2 W/R_e \tag{11}$$

From Equations (9)–(11), the following conclusions can be obtained:

(1) Without any displacement before loading, the horizontal restoring force and moment are linear functions of the rotation angle α. It means that Equations (9) and (10) can be used more conveniently to compute the restoring force and moment of FPB from a zero initial state. It also means that simple connection elements of FEM software can be used to simulate rotational characteristics of FPB bearing.

(2) From Equation (9), the rotational stiffness k_α of the FPB can be obtained by Equation (12). To keep the sign of k_α positive, the condition shown in Equation (13) must be satisfied.

$$k_\alpha = \frac{W(R_2 + h_u)(R_2 + h_u + h_c - R_e)}{R_e} \tag{12}$$

$$R + h_u + h_c - R_e > 0 \Leftrightarrow h_c > R_e - R_2 - h_u \tag{13}$$

(3) From Equation (9), the values of CST_1 and h_u are relatively small compared to R_1 and R_2, and the sign of CST_1 is determined by the three first terms of Equation (11), knowing that R_2 is inferior to R_e. Therefore, the initial value of the moment of FPB is generally positive. However, if k_α is negative and the FPB rotation angle is large enough, the moment will be negative.

(4) Constant terms of F and M_b are only related to bearing parameters and the vertical load; they are independent of the location of the rotation centre. Therefore, the location of the rotation centre has only influence on the rotational stiffness of FPB. To reduce the moment and make the bearing rotate as flexibly as possible, it is suggested to use a smaller friction coefficient and make the difference between the curvature radii of the two concave surfaces as large as possible.

3. Experimental Study

An experimental study was conducted in the State Key Laboratory of Disaster Reduction in Civil Engineering of Tongji University to validate the proposed theoretical model of the rotational behaviour of FPBs.

3.1. Testing Device

The tests are carried out using a testing device composed by a moment applying system and a vertical load applying system as presented in Figure 4. One end of a rigid loading beam is placed between two identical FPB specimens symmetrically disposed with respect to the horizontal plane, and the vertical load is applied by the vertical load applying system; the other end of the loading beam is progressively vertically loaded to create a moment by displacement control mode.

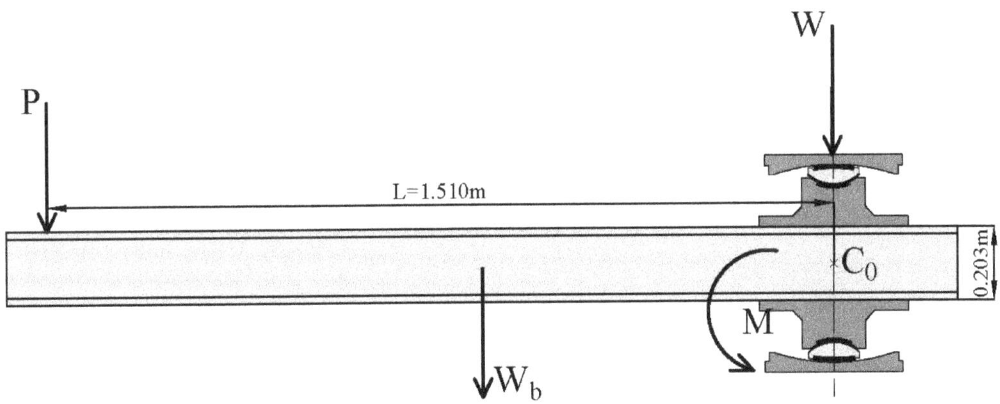

Figure 4. Testing device free body diagram.

3.2. Test Specimen and Loading Procedure

The FPB specimen used in the tests is illustrated in Figure 5. The tests are carried out according to the Chinese specifications on Friction Pendulum isolation bearing for highway bridges (JJT 852-2013) [27]. It consists of four different specimens with different friction coefficient combinations of sliding surfaces under three different vertical loads. Different friction coefficients are obtained by lubrication with silicone grease or non-lubrication. For each friction coefficient combination, three vertical loads of 20 kN, 30 kN, and 40 kN are loaded. The test cases are shown in Table 1. The friction coefficients are determined after each test case.

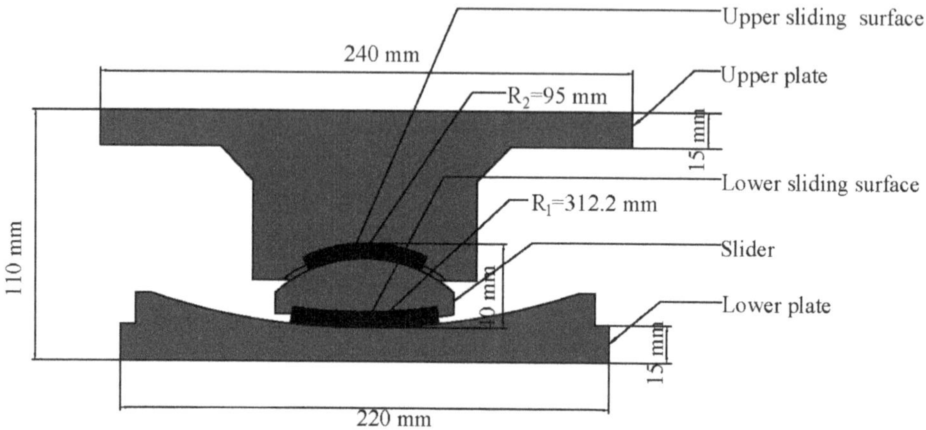

Figure 5. FPB specimen.

Table 1. Test cases.

Configuration	Vertical Load	Friction Coefficient
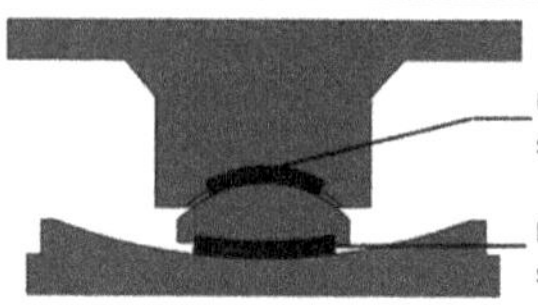	20 kN 30 kN 40 kN	$\mu_1 = \mu_2 = 0.0925$ $\mu_1 = \mu_2 = 0.0827$ $\mu_1 = \mu_2 = 0.0738$
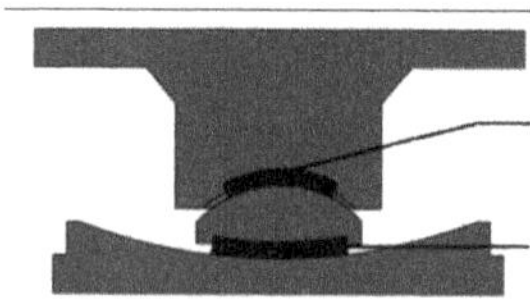	20 kN 30 kN 40 kN	$\mu_1 = 0.0925,\ \mu_2 = 0.0759$ $\mu_1 = 0.0827,\ \mu_2 = 0.0764$ $\mu_1 = 0.0738,\ \mu_2 = 0.0706$
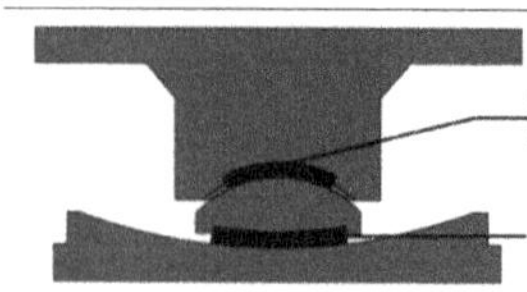	20 kN 30 kN 40 kN	$\mu_1 = 0.0759,\ \mu_2 = 0.0925$ $\mu_1 = 0.0671,\ \mu_2 = 0.0827$ $\mu_1 = 0.0618,\ \mu_2 = 0.0738$
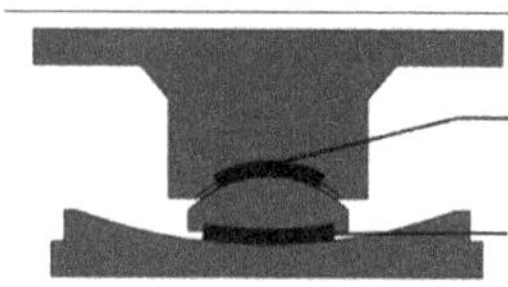	20 kN 30 kN 40 kN	$\mu_1 = \mu_2 = 0.0759$ $\mu_1 = \mu_2 = 0.0671$ $\mu_1 = \mu_2 = 0.0618$

A photograph of the testing device is shown in Figure 6. Moment is applied by controlling the left actuator in displacement control mode at 30 mm/min velocity to rotate the specimens, and the reaction force is measured at each time increment until the vertical displacement of the loading point reaches 150 mm. The moment of a specimen is given by

$$M_\theta = \frac{Pl}{2} + \frac{W_b L}{4} \quad (14)$$

where P is the load applied by the actuator at each increment; L represents the lever arm (L = 1.510 m); and W_b is the weight of the loading beam. For each test case, three identical tests are conducted, and the average value of moments are calculated for analysis.

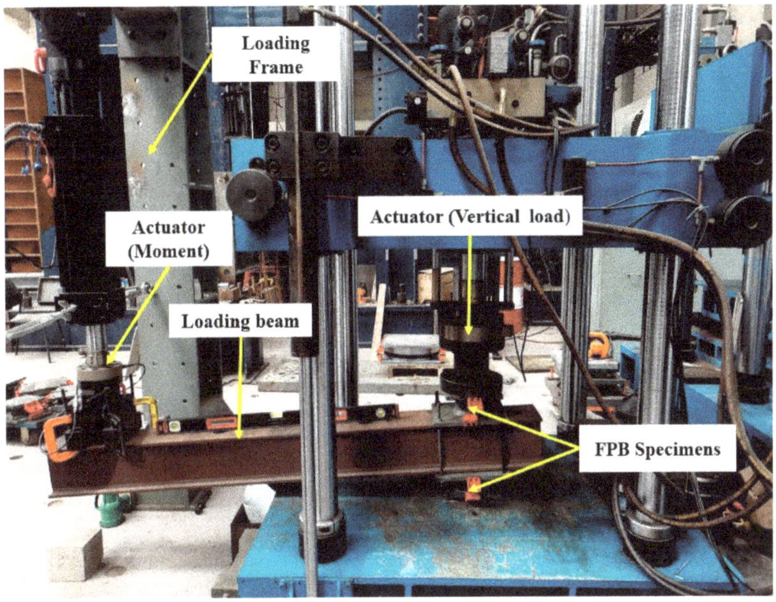

Figure 6. Photograph of the testing device.

The angle of rotation, α, for each vertical displacement of the left actuator, δ, is given by

$$\alpha = \arctan\frac{\delta}{L} \quad (15)$$

3.3. *Comparison of Theoretical and Experimental Results*

The moment–rotation angle curves from both the experimental data and theoretical model for each configuration are shown in Figure 7.

The comparisons of rotational stiffnesses from both the experimental data and theoretical model for each case are shown in Table 2. The theoretical model shows the ability to accurately describe the rotational behaviour of FPB. The predictions of the theoretical model are in very good agreement with the experimental results.

However, some errors are noted. From comparison of theoretical and experimental results, it is noticed that the theoretical model generally slightly underestimates the rotational stiffness of the bearing. The error between the experimental and theoretical results increases with the increase in the rotational angle.

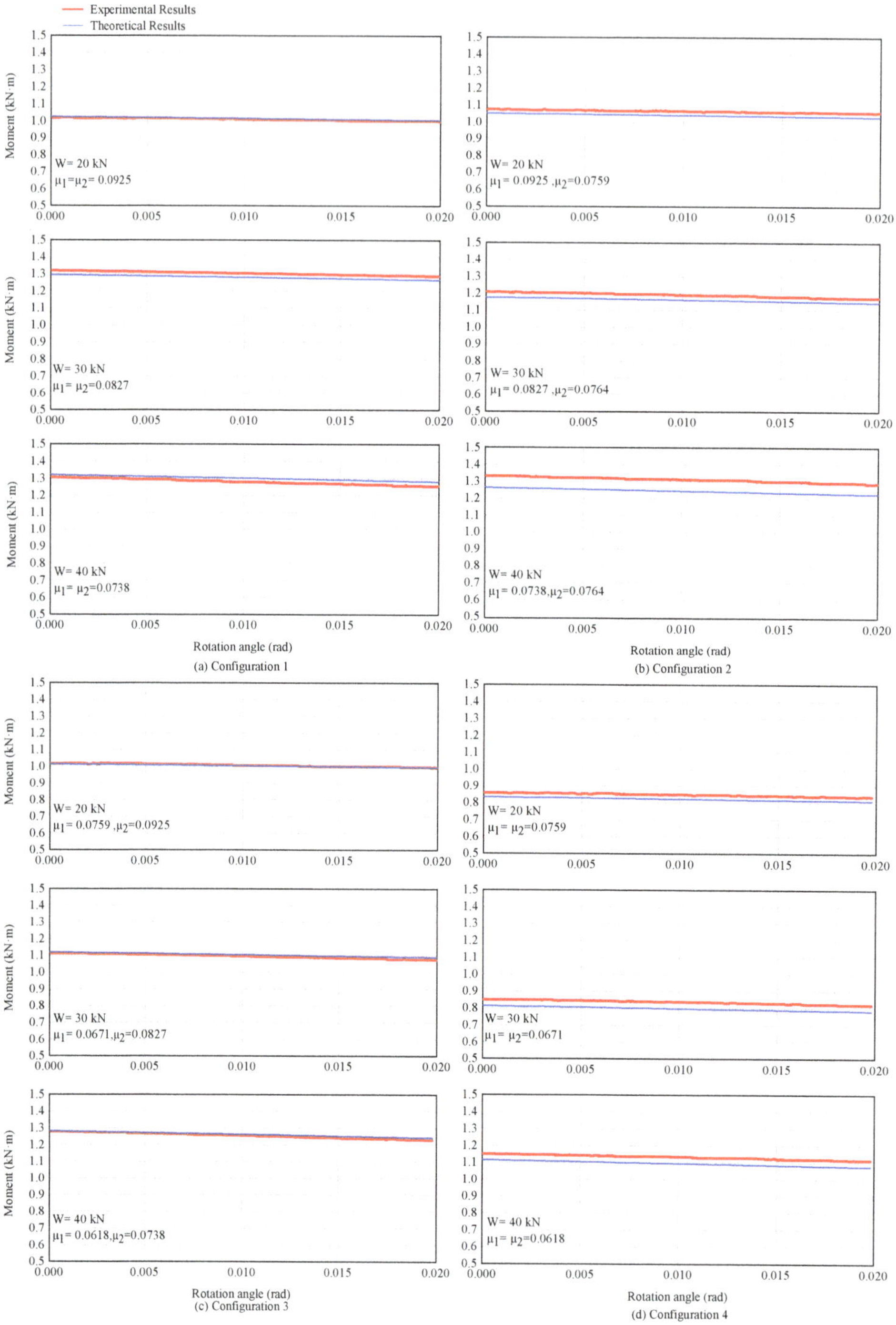

Figure 7. Experimental and theoretical results.

Table 2. Comparisons of Experimental and Theoretical rotational stiffness.

Configuration	Vertical Load (kN)	Experimental Rotational Stiffness (kN·m/rad)	Theoretical Rotational Stiffness (kN·m/rad)	Error
1	20	1.122	1.030	8%
	30	1.612	1.510	6%
	40	2.432	1.986	18%
2	20	0.941	1.030	−9%
	30	1.851	1.510	18%
	40	2.192	1.986	9%
3	20	1.203	1.030	14%
	30	1.872	1.510	19%
	40	2.423	1.986	18%
4	20	1.020	1.030	−1%
	30	1.428	1.510	−6%
	40	1.729	1.986	−15%

4. Conclusions

This paper presents a theoretical model of FPB based on the moment balance theory. The model describes the rotational behaviour of friction pendulum bearings considering the variation of rotation angle. Additionally, extension to include the coupling of rotational and horizontal displacements in the mechanical model, not incorporated in this model, is suggested for future research.

An experimental study of the FPB rotational behaviour is conducted in order to validate the theoretical model. The experimental results show that the proposed theoretical model is accurate enough to describe the rotational behaviour of FPBs. The theoretical approach used in this paper is a powerful tool to describe the rotational behaviour of FPBs and can help to evaluate the structural performance in earthquakes in case of variable rotations of FPBs.

Author Contributions: Conceptualization, T.P.; methodology, T.P. and Y.L.; experimental study, T.P. and T.S.K.; writing—original draft preparation, T.P. and T.S.K.; writing—review and editing, T.P. and T.S.K.; supervision, T.P.; funding acquisition, T.P. All authors have read and agreed to the published version of the manuscript.

Funding: This work was supported partly by the Ministry of Science and Technology of China, Grant No. SLDRCE19-B-21 and the National Natural Science Foundation of China (No. 51278372 and No. 51878489).

Data Availability Statement: Some or all data, models, or code that support the findings of this study are available from the corresponding author upon reasonable request.

Acknowledgments: The authors would acknowledge the financial support the Ministry of Science and Technology of China, Grant No. SLDRCE19-B-21 and the National Natural Science Foundation of China (No. 51278372 and No. 51878489).

Conflicts of Interest: The authors declare no conflict of interest.

References

1. Coburn, A.W.; Spence, R.J.; Pomonis, A. Factors Determining Human Casualty Levels in Earthquakes: Mortality Prediction in Building Collapse. In *Proceedings of the Tenth World Conference on Earthquake Engineering*; Balkema: Rotterdam, The Netherlands, 1992.
2. Cao, X.; Shen, D.; Feng, D.; Wang, C.; Qu, Z.; Wu, G. Seismic Retrofitting of Existing Frame Buildings through Externally Attached Sub-Structures: State of the Art Review and Future Perspectives. *J. Build. Eng.* **2022**, *57*, 104904. [CrossRef]
3. Foti, D. Response of Frames Seismically Protected with Passive Systems in Near-Field Areas. *Int. J. Struct. Eng.* **2014**, *5*, 326. [CrossRef]
4. Foti, D. Rolling Devices for Seismic Isolation of Lightweight Structures and Equipment. Design and Realization of a Prototype. *Struct. Control Health Monit.* **2019**, *26*, e231. [CrossRef]

5. Habieb, A.B.; Valente, M.; Milani, G. Effectiveness of Different Base Isolation Systems for Seismic Protection: Numerical Insights into an Existing Masonry Bell Tower. *Soil Dyn. Earthq. Eng.* **2019**, *125*, 105752. [CrossRef]
6. Kim, Y.-S.; Yun, C.-B. Seismic Response Characteristics of Bridges Using Double Concave Friction Pendulum Bearings with Tri-Linear Behavior. *Eng. Struct.* **2007**, *29*, 3082–3093. [CrossRef]
7. Deringöl, A.H.; Güneyisi, E.M. Effect of Friction Pendulum Bearing Properties on Behaviour of Buildings Subjected to Seismic Loads. *Soil Dyn. Earthq. Eng.* **2019**, *125*, 105746. [CrossRef]
8. Imran, I.; Siringoringo, D.M.; Michael, J. Seismic Performance of Reinforced Concrete Buildings with Double Concave Friction Pendulum Base Isolation System: Case Study of Design by Indonesian Code. In *Proceedings of the Structures*; Elsevier: Amsterdam, The Netherlands, 2021; Volume 34, pp. 462–478.
9. Scheaua, F. Improvement of Structures Seismic Response Based on Pendulum Systems with Double Sliding Surface. In *Proceedings of the IOP Conference Series: Materials Science and Engineering*; IOP Publishing: Bristol, UK, 2020; Volume 916, p. 012102.
10. Ates, S. Investigation of Effectiveness of Double Concave Friction Pendulum Bearings. *Comput. Concr.* **2012**, *9*, 195–213. [CrossRef]
11. Calvi, P.M.; Calvi, G.M. Historical Development of Friction-Based Seismic Isolation Systems. *Soil Dyn. Earthq. Eng.* **2018**, *106*, 14–30. [CrossRef]
12. Avinash, A.R.; Krishnamoorthy, A.; Kamath, K.; Chaithra, M. Sliding Isolation Systems: Historical Review, Modeling Techniques, and the Contemporary Trends. *Buildings* **2022**, *12*, 1997. [CrossRef]
13. Touaillon, J. Improvement in Buildings. Patent No. 99,973, 15 February 1870.
14. Fenz, D.M.; Constantinou, M.C. Spherical Sliding Isolation Bearings with Adaptive Behavior: Experimental Verification. *Earthq. Eng. Struct. Dyn.* **2008**, *37*, 185–205. [CrossRef]
15. Lu, L.; Lee, T.; Yeh, S. Theory and Experimental Study for Sliding Isolators with Variable Curvature. *Earthq. Eng. Struct. Dyn.* **2011**, *40*, 1609–1627. [CrossRef]
16. Malekzadeh, M.; Taghikhani, T. Adaptive Behavior of Double Concave Friction Pendulum Bearing and Its Advantages over Friction Pendulum Systems. *Sci. Iran.* **2010**, *17*, 81–88.
17. Tsai, C.; Chiang, T.; Chen, B. Experimental Evaluation of Piecewise Exact Solution for Predicting Seismic Responses of Spherical Sliding Type Isolated Structures. *Earthq. Eng. Struct. Dyn.* **2005**, *34*, 1027–1046. [CrossRef]
18. Lomiento, G.; Bonessio, N.; Benzoni, G. Friction Model for Sliding Bearings under Seismic Excitation. *J. Earthq. Eng.* **2013**, *17*, 1162–1191. [CrossRef]
19. Fenz, D.M.; Constantinou, M.C. Behaviour of the Double Concave Friction Pendulum Bearing. *Earthq. Eng. Struct. Dyn.* **2006**, *35*, 1403–1424. [CrossRef]
20. Shang, J.; Tan, P.; Zhang, Y.; Han, J.; Mi, P. Seismic Isolation Design of Structure Using Variable Friction Pendulum Bearings. *Soil Dyn. Earthq. Eng.* **2021**, *148*, 106855. [CrossRef]
21. Ansari, S.; Fallah, N.; Tashakori, J. Optimal Reliability Design of Pure Friction Isolators Using Asymptotic Sampling. *Asian J. Civ. Eng.* **2019**, *20*, 911–924. [CrossRef]
22. Castaldo, P.; Palazzo, B.; Ferrentino, T. Seismic reliability-based ductility demand evaluation for inelastic base-isolated structures with friction pendulum devices. *Earthq. Eng. Struct. Dyn.* **2017**, *46*, 1245–1266. [CrossRef]
23. Becker, T.C.; Mahin, S.A. Effect of Support Rotation on Triple Friction Pendulum Bearing Behavior. *Earthq. Eng. Struct. Dyn.* **2013**, *42*, 1731–1748. [CrossRef]
24. Becker, T.C.; Mahin, S.A. Correct Treatment of Rotation of Sliding Surfaces in a Kinematic Model of the Triple Friction Pendulum Bearing. *Earthq. Eng. Struct. Dyn.* **2013**, *42*, 311–317. [CrossRef]
25. Mosqueda, G.; Whittaker, A.S.; Fenves, G.L.; Mahin, S.A. *Experimental and Analytical Studies of the Friction Pendulum System for the Seismic Protection of Simple Bridges*; UCB/EERC 2004/01; Earthquake Engineering Research Center, University of California: Berkeley, CA, USA, 2004.
26. Peng, T.; Yan, B.; Li, F. The Hysteresis Model of the Friction Pendulum Bearing Based on the Moment Balance Theory. *Ain Shams Eng. J.* **2022**, *13*, 101707. [CrossRef]
27. *JT-T852-2013*; Friction Pendulum Isolation Bearing for Highway Bridges. Ministry of Transportation of the People's Republic of China: Beijing, China, 2013. (In Chinese)

sustainability

MDPI

Article

Effects of the 2011 Mw 9.0 Tohoku-Oki Earthquake on the Locking Characteristics and Seismic Risk of the Yishu Fault Zone in China

Cunpeng Du [1], Haitao Yin [2,*], Shengwen Yu [1], Le Yang [2] and Yuan Jia [2]

[1] College of Geodesy and Geomatics, Shandong University of Science and Technology, Qingdao 266510, China
[2] Shandong Earthquake Agency, Jinan 250014, China
* Correspondence: yinhaitao121@163.com

Abstract: To ascertain the 2011 Tohoku-Oki Mw 9.0 earthquake's impact on the stability of the Yishu fault zone, this study inverts the fault locking degree and slip rate defect distribution of the Yishu fault zone using GPS horizontal velocity field data covering the period from 1999 to 2009 and from April 2011 to 2019. This is based on the block negative dislocation model. Combined with the *b* value and strain field characteristics, the properties of deformation of the Yishu fault zone before and after the earthquake are comprehensively analyzed. The results show that before the 2011 Tohoku-Oki Mw 9.0 earthquake, the locking degree and depth of the northern segment of the Yishu fault zone were higher, while the locking depth of the southern segment of the fault was shallower. The 2011 Tohoku-Oki earthquake produced different coseismic effects on the southern and northern sections of the Yishu fault zone. The extension on the southern section and compression on the northern section caused the strain release in the southern sections of the Yishu fault zone following the earthquake. After it, the regional locking degree of the southern section of the Yishu fault zone was relieved. However, the locking degree of the northern segment of the fault zone was still high and the depth was deep, at about 26 km. In addition, the northern section of the Yishu fault zone was a section with an abnormally low *b* value and small earthquakes in the northern segment are sparse at present. The coseismic compression of the northern section caused by the 2011 Tohoku-Oki earthquake is conducive to its locking and easily accumulates stress, so it is necessary to pay attention to its seismic risk.

Keywords: 2011 Tohoku-Oki earthquake; Yishu fault zone; deep locking degree; GPS velocity field; strain rates; seismic risk

Citation: Du, C.; Yin, H.; Yu, S.; Yang, L.; Jia, Y. Effects of the 2011 Mw 9.0 Tohoku-Oki Earthquake on the Locking Characteristics and Seismic Risk of the Yishu Fault Zone in China. *Sustainability* **2023**, *15*, 4321. https://doi.org/10.3390/su15054321

Academic Editors: Chong Xu, Su Chen and Shuang Li

Received: 4 January 2023
Revised: 24 February 2023
Accepted: 27 February 2023
Published: 28 February 2023

1. Introduction

The subduction of the Pacific plate affected the Tohoku-Oki Mw9.0 earthquake in Japan that occurred on 11 March 2011 [1–3]. Previous studies have shown that the 2011 Tohoku-Oki earthquake had an essential impact on the seismicity in the areas on both sides of the Yishu Fault Zone, which caused the coseismic horizontal displacement at a millimeter-to-centimeter level in North China and Northeast China [4–6]. The Yishu fault zone, the central segment of the Tanlu Fault Zone, is in the Shandong Peninsula, eastern China [7]. The Tanlu fault zone is characterized by linearly steep-dipping faults traversing the region in an SSW–NNE direction with a clear physiognomy and a synchronous dextral displacement of short gullies [8]. The Tanlu fault zone is the most important active fault zone and strong seismic tectonic zone in the eastern part of the Chinese mainland [9,10]. Historically, there was the Anqiu Ms 7.0 earthquake in 70 BC, the 1668 Tancheng Ms 8.5 earthquake, the 1888 Bohai Bay Ms 7.5 earthquake, the 1969 Bohai Bay Ms 7.4 earthquake, and the 1975 Haicheng Ms 7.3 earthquake.

The Shandong area is located at the southeastern margin of the Sino–Korean Block and the eastern end of the central orogenic area in the primary tectonic area of the Chinese continent. It has complex geological structures and a long evolutionary history. The modern tectonic movement is at the intersection of multiple active blocks and the dynamic environment is affected by the interaction of the Pacific, North American, and Eurasian Plates [9,11–13]. The strong earthquake activity in Tohoku-Oki, which is in the subduction zone of the Pacific Plate, has a significant effect on the earthquake activity in North China [4,14]. The Yishu fault zone and the Tohoku-Oki Trench belong to the same geological tectonic system and are also affected by the subduction of the Pacific Plate [15]. After the 2011 Tohoku-Oki earthquake, the seismicity of the Yishu fault zone and its surrounding areas increased significantly, which was related to the regional coseismic deformation and post-earthquake deformation characteristics [16]. Therefore, the impact of the 2011 Tohoku-Oki earthquake on the Yishu fault zone deserves attention.

The Global Positioning System (GPS) can provide comprehensive velocity field information and can be used to study large-scale tectonic deformation [17]. Fan et al. (2022) [18] investigated the adjustment of the in situ stress field of the Beijing Plain after the 2011 Tohoku-Oki earthquake based on in situ stress monitoring data. It shows that the stress accumulation level of faults in the Beijing Plain area increased in a short time after the earthquake and then gradually decreased. There have been many studies on the crustal deformation of the Yishu fault zone using GPS which have shown that the fault zone is characterized by right-lateral slip and compression [17,19–25]. Li et al. (2016) [26] used the GPS velocity field data from 2009 to 2014 as a constraint to invert the locking degree of the central and southern segment of the Tanlu fault zone and obtained the locking depth of the fault from north to south, first deep then shallow, along the strike. However, the existing studies have not investigated the degree of change of the deep locking in the Yishu fault zone before and after the 2011 Tohoku-Oki earthquake. An earthquake happens because of the continuous accumulation of strain on the active fault under the action of regional tectonic stress and the sudden instability and fractures after reaching the limit state [2,27,28]. How the strain characteristics change before and after the earthquake characterizes them. The slip inversions have successfully modelled the transient deformations [29,30], coseismic slips [31,32], and post-seismic slip [32] in different tectonic regions. Prioritizing earthquake risk is determined using the Turkish rapid assessment method, which shows that the study of micro-zoning is very important in minimizing earthquake damage [33]. In this paper, the horizontal strain field of the Yishu fault zone is analyzed based on GPS data and the locking depth and degree of the Yishu fault zone before and after the 2011 Tohoku-Oki earthquake are inverted using the TDEFNODE negative dislocation inversion model [30,34]. Based on seismic activity, the future seismic risk of the Yishu fault zone is discussed.

2. Overview of the Study Area

The Yishu fault zone is in the Shandong section of the Tanlu fault zone (Figure 1). It has the best exposure and strongest neotectonic activity in the Tanlu fault zone. The Yishu fault zone forms a compound graben structure of "two grabens sandwiched by a barrier" in the Yi River and Shu River Valleys of Luzhong. It starts from Laizhou Bay of Bohai Sea in the north and extends to the Xinyi Area of Jiangsu Province in the south. It consists of the Anqiu–Juxian fault (F5), the Changyi–Dadian fault (F1), the Baifenzi–Fulaishan fault (F2), the Yishui–Tangtou fault (F3), and the Tangwu–Gegou fault (F4), and includes five nearly parallel faults with a length of 330 km. The width of the northern section of the fault zone is about 55 km, which is reduced to about 25 km in the south, forming the structural boundary between the Luxi block and the Jiaoliao block. Among them, the Anqiu–Juxian fault is the most important Late Quaternary active fault in the Yishu fault zone, characterized by dextral strike-slip and thrust activity, and is a strong active zone of neotectonic movement, which controls the occurrence of earthquakes [35]. The north and south sections of the Anqiu–Juxian fault have different earthquake characteristics and can be divided into two permanent north–south earthquake rupture sections [36].

The Yishu fault zone is a deep fault zone. Since the Quaternary, the neotectonic movement in the fault zone has been strong, and the seismicity is characterized by large magnitude earthquakes and long active periods. The Anqiu earthquake with Ms 7.0 occurred in 70 BC and the Tancheng earthquake with Ms 8.5 occurred in 1668. The seismicity of the Yishu fault zone is segmented [37,38], which is mainly concentrated in the south section and less in the north section.

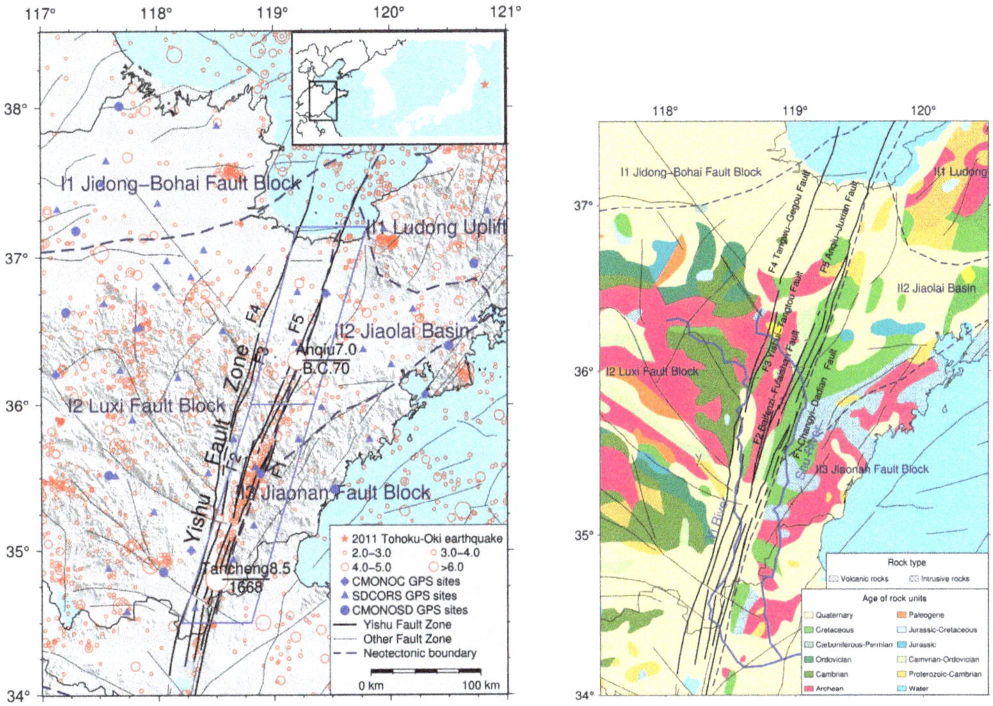

(a) (b)

Figure 1. Tectonic setting and earthquake activity distribution in the study area. (**a**) Tectonics and historical earthquakes in the study area. Inset map on the upper right shows the study area. The blue box divides the Yishu fault zone into north and south segments. The size of the red circles is scaled to the magnitude of the earthquakes (from 2011 to 2019, ML > 2.0). The blue solid circle, triangle and diamond are the selected GPS sites (I1 is the Jidong–Bohai fault block, I2 is the Luxi fault block, II1 is the Ludong uplift, II2 is the Jiaolai basin, II3 is the Jiaonan fault block, II1, II2, and II3 jointly constitute the Ludong fault block). (**b**) Geological setting of Yishu fault zone and its vicinity [39] (F1 is the Changyi–Dadian fault, F2 is the Baifenzi–Fulaishan fault, F3 is the Yishui–Tangtou fault, F4 is the Tangwu–Gegou fault, and F5 is the Anqiu–Juxian fault).

3. Data and Research Method

3.1. Data Processing

In this paper, the GPS observation site data of the Crustal Movement Observation Network of China (CMONOC), the Crustal Movement Observation Network of Shandong (CMONOSD), and the Shandong Continuously Operating Reference Station System (SDCORS) (Figure 1) were used. Additionally, GAMIT/GLOBK software was used to solve the data and obtain GPS site coordinates and time series data (Figure 2).

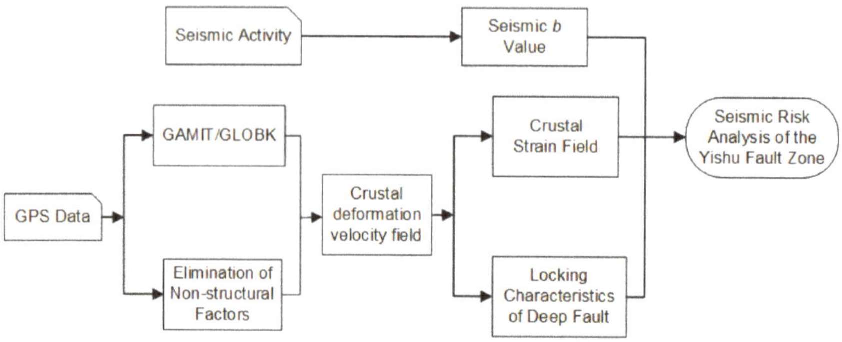

Figure 2. Schematic diagram of the research technical route.

When solving the GPS data, the elevation mask angle of the satellite was set to 15° and the data sampling rate is 30 s. The global atmospheric pressure model and temperature models GPT (Global Pressure Temperature) and GMF (Global Mapping Function) were used as the prior tropospheric dry delay model and convection projection function. The FES2004 ocean tide loading model was used to correct the ocean tide to weaken the influence of ocean tide loading on the station. The IGS (International Global Navigation Satellite System Service) tracking stations around the Shandong area were selected and the one-day relaxation constraint solutions of all the stations were obtained by combining the one-day solutions of the stations in the study area with the existing relaxation solutions of the IGS stations by using the GLOBK software. The data were processed using GLOBK software with the continuous Kalman filter to obtain the coordinate time series and velocity field results for each station under the ITRF2014 (International Terrestrial Reference Frame 2014) framework. The GPS velocity field to the velocity field under the Eurasian framework was converted and the GPS stations whose direction and size deviated were removed from the regional motion background (Figures 3 and 4).

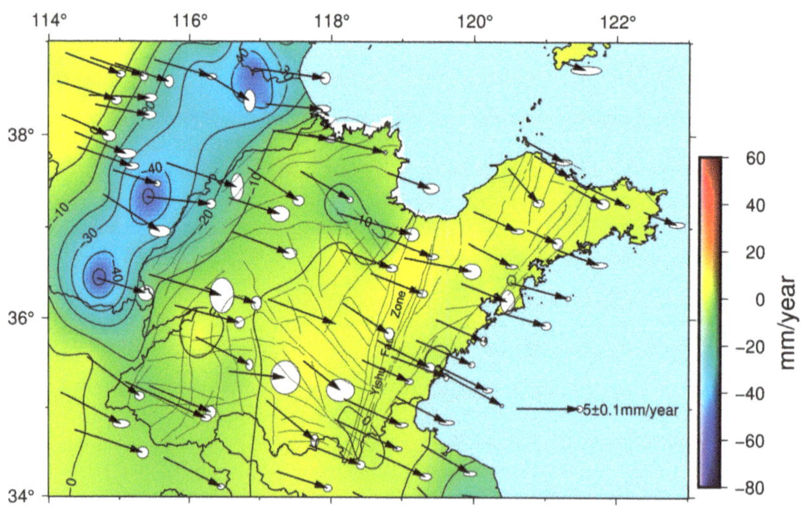

Figure 3. Velocity field of the study area relative to the Eurasian Plate before the 2011 Tohoku-Oki earthquake (1999 to 2009). The background color indicates the vertical deformation velocity field. The confidence interval of the error ellipse is 95%. Velocity field data were obtained from the Shandong Earthquake Agency.

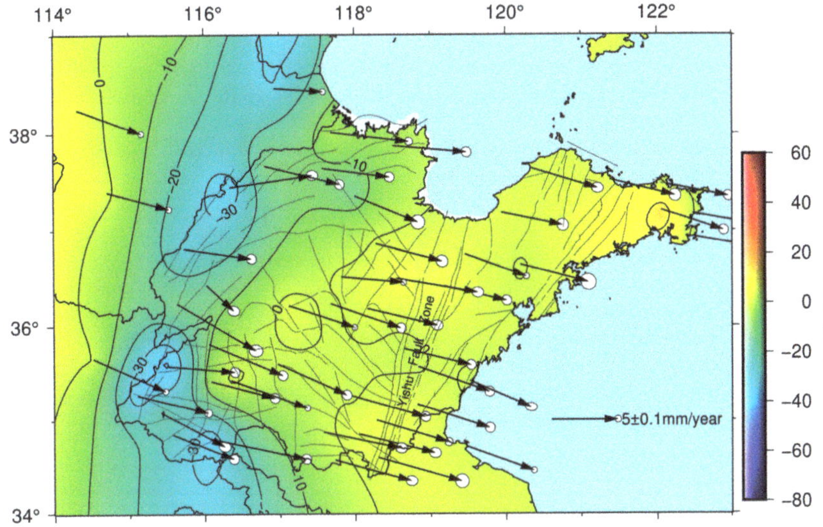

Figure 4. Velocity field of the study relative to the Eurasian Plate after the 2011 Tohoku-Oki earthquake (April 2011 to 2019). The background color indicates the vertical deformation velocity field. Velocity field data were obtained from the Shandong Earthquake Agency.

Since the study area is far from the epicenter of the 2011 Tohoku-Oki earthquake (more than 2200 km), nonlinear effects such as creep-slip during the short period can be neglected. Considering the accuracy of GPS observation, a continuous GPS observation sequence of 10 days each before and after the 2011 Tohoku-Oki earthquake was used to estimate the coseismic displacement and its error using the least squares method [6]. The coseismic displacements calculated are shown in Figure 5.

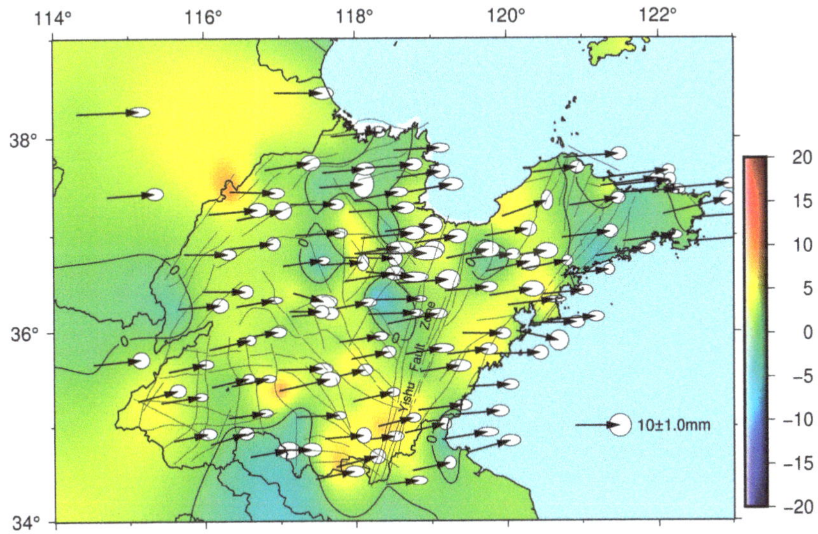

Figure 5. Coseismic displacement of study area caused by 2011 Tohoku-Oki earthquake. The background color indicates vertical coseismic displacement.

The crustal strain fields from 1999 to 2010 and from April 2011 to 2019 were calculated using the GPS horizontal velocity field and the deformation characteristics of the Yishu fault zone before and after the 2011 Tohoku-Oki earthquake were analyzed. The 3D *b* value distribution was calculated using the seismic activity catalogue from 1970 to 2019. Finally, the seismic hazard of the Yishu fault zone was studied by determining the degree and extent of the Yishu fault zone's locking using inversion. The research technical route is shown in Figure 2.

3.2. Calculation of Crustal Strain Field

The crustal strain field analysis is independent of the spatial datum analyzing the variation of deformation within the region and can more comprehensively express the intensity of crustal deformation. To overcome the influence of earth curvature or projection deformation, a more accurate calculation method in a spherical coordinate system is used to calculate the strain (rate) field [40]. In practical applications, only the horizontal strain field is considered due to the poor reliability of the GPS vertical velocity component, and the formula is as follows [40]:

$$\begin{aligned} u_\varphi &= -\overline{\omega}_\theta r_0 - \overline{\omega}_\varphi \cos\theta_0 \Delta\varphi + \overline{e}_{\varphi\varphi} r_0 \Delta\varphi + \overline{e}_{\theta\varphi} r_0 \Delta\theta + \overline{\omega}_r r_0 \Delta\theta \\ u_\theta &= -\overline{\omega}_\theta r_0 \cos\theta_0 \Delta\varphi + \overline{\omega}_\varphi r_0 + \overline{e}_{\theta\varphi} r_0 \sin\theta_0 \Delta\varphi + \overline{e}_{\theta\theta} r_0 \Delta\theta - \overline{\omega}_r r_0 \sin\theta_0 \Delta\varphi \end{aligned} \quad (1)$$

where φ, θ, and r are the longitude angle, the colatitude angle, and the earth radius in the earth coordinate system, respectively. φ_0, θ_0 represents the plane coordinates of the regional geometric center point and r_0 represents the earth radius of the regional geometric center point. When φ, θ, and r are used as subscripts, they represent the longitude direction (east direction), the latitude direction (north direction), and the radial direction (vertical direction), respectively. u represents the displacement of each observation point. $\Delta\varphi$, $\Delta\theta$, and Δr represent the three-direction distance from the observation point to the center point of the observation network, respectively, $\overline{e}_{\varphi\varphi}$, $\overline{e}_{\theta\theta}$, and $\overline{e}_{\theta\varphi}$ represent the strain components, and $\overline{\omega}_\varphi$, $\overline{\omega}_\theta$, and $\overline{\omega}_r$ represent the Euler rotation vectors, respectively. Therefore, as long as the horizontal displacements of more than three GPS observation points are known, the above six unknown parameters ($\overline{e}_{\varphi\varphi}$, $\overline{e}_{\theta\theta}$, $\overline{e}_{\theta\varphi}$, $\overline{\omega}_\varphi$, $\overline{\omega}_\theta$, and $\overline{\omega}_r$) can be solved using Equation (1) and the least square method.

3.3. Calculation of Seismic b Value

The earthquake *b* value is derived from the Gutenberg–Richter magnitude–frequency relationship ($\text{lgN} = a - bM$) [41]. Starting from the laboratory discovery of the relationship between rock fracture stress and the *b* value, many observations have shown that there is a negative correlation between *b* value and differential stress [27,28]. Because the seismic *b* value is closely related to seismic activity and the crustal stress state, it is often used as a crustal "stress gauge" [42].

The earthquake catalogue's completeness is particularly significant for the calculation of the *b* value. The *b* value will be low due to the lack of small earthquake data caused by the weak monitoring ability of the network. The Mc (Magnitude of completeness) can be used to evaluate the seismic measuring ability of the station. The Mc is regarded as a critical magnitude, which represents the minimum complete magnitude that can be monitored by the network. When calculating the *b* value, only events greater than or equal to the Mc value are selected. In this paper, the Mc values were calculated for earthquakes with ML > 2.0 from 1970 to 2019 using the ZMAP program [43]. The maximum curvature method was used to calculate the Mc in the study area and the magnitude corresponding to the maximum of the first derivative of the magnitude frequency curve function was regarded as the Mc.

The maximum likelihood estimation method is a method of estimating model parameters using observation data generated with high probabilities. Based on the maximum likelihood method and according to the exponential distribution relationship between earthquake frequency and magnitude [41],

$$n(M) = e^{(a-bM)} \tag{2}$$

$$b = \frac{\text{lge}}{M - M_0}. \tag{3}$$

In Equations (2) and (3), M is the starting magnitude, M_0 is the average magnitude, lge =0.4343, and when n is the total number of earthquakes, the standard deviation of the 95% confidence level [41] is

$$\sigma = 1.96\frac{b}{\sqrt{n-1}}. \tag{4}$$

3.4. Fault Slip Inversion Model

The occurrence of a large earthquake is closely related to the seismogenic fault's locking degree. As per the classical dislocation theory, the fault's earthquake preparation is directly related to the fault's locking depth and the urgency and risk of seismicity are higher when the fault's earthquake preparation is close to the late stage [44,45]. The TDEFNODE negative dislocation inversion model can use the simulated annealing method and grid search method to estimate the block rotation, internal strain, and fault locking degree under the constraints of GPS velocity field and seismic data. The depth along the longitudinal strike of the fault is specified by the node [30,34]. The program assumes that the movement of points inside the plate is the sum of surface elastic deformation caused by block rotation and the sliding deficit at the block boundary due to fault locking. The inversion result has a certain reliability and good stability [30,34].

$$\overline{V}_{sf} = \overline{V}_{br} + \overline{V}_{is} + \overline{V}_{fs} \tag{5}$$

where $\overline{V}_{sf}$ is the measured surface velocity, $\overline{V}_{br}$ is the velocity caused by block rotation, $\overline{V}_{is}$ is the velocity caused by the internal strain of the block, and $\overline{V}_{fs}$ is the velocity caused by fault-locked negative dislocations. The fault locking coefficient value is represented by Phi [30,34].

$$\text{Phi}(\Sigma) = \Sigma^{-1}\int_{\Sigma}\left[\frac{1 - V_C(s)}{V(s)}\right] \tag{6}$$

where V is the long-term fault slip rate and V_C is the short-term slip rate; Σ is the defined grid area on the fault, and, using Σ to be larger than the characteristic wavelength of the V_C variation, we obtain a continuous approximation of the distribution. When Phi = 0, it indicates that the fault is in a complete creep state and there is no locking. When Phi = 1, it indicates that the fault is completely locked. Generally, the range of values 0–1 is used on the fault to represent the locking state of the fault. When Phi is between 0 and 1, it indicates that the fault is not completely locked and there is a certain creep.

4. Experimental Results

4.1. Strain Field near the Yishu Fault Zone

Considering the impact of the 2011 Tohoku-Oki earthquake, the strain field and the maximum shear strain field in the study area were obtained based on the GPS horizontal velocity field from 1999 to 2009 and from April 2011 to 2019, as shown in Figures 6 and 7. The regional strain of the Yishu fault zone is relatively weak, the overall upper north section is compressive, the southern section is tensile, the differential movement of the crust is weak, and the magnitude of the strain rate is small. The northeastward pushing of the western Indian Ocean Plate and the westward pushing of the eastern North American Plate and the Pacific Plate control the distribution characteristics of the national strain field [16]. These cause the strain rate field in the Chinese mainland to show an obvious feature of strength in the west and weakness in the east. Comparing Figure 6a with Figure 7a, the northern segment of the Yishu fault zone is evidently still in a compressive state, while the southern segment is in an extensional state, but it is slightly smaller than it was before the 2011 Tohoku-Oki earthquake.

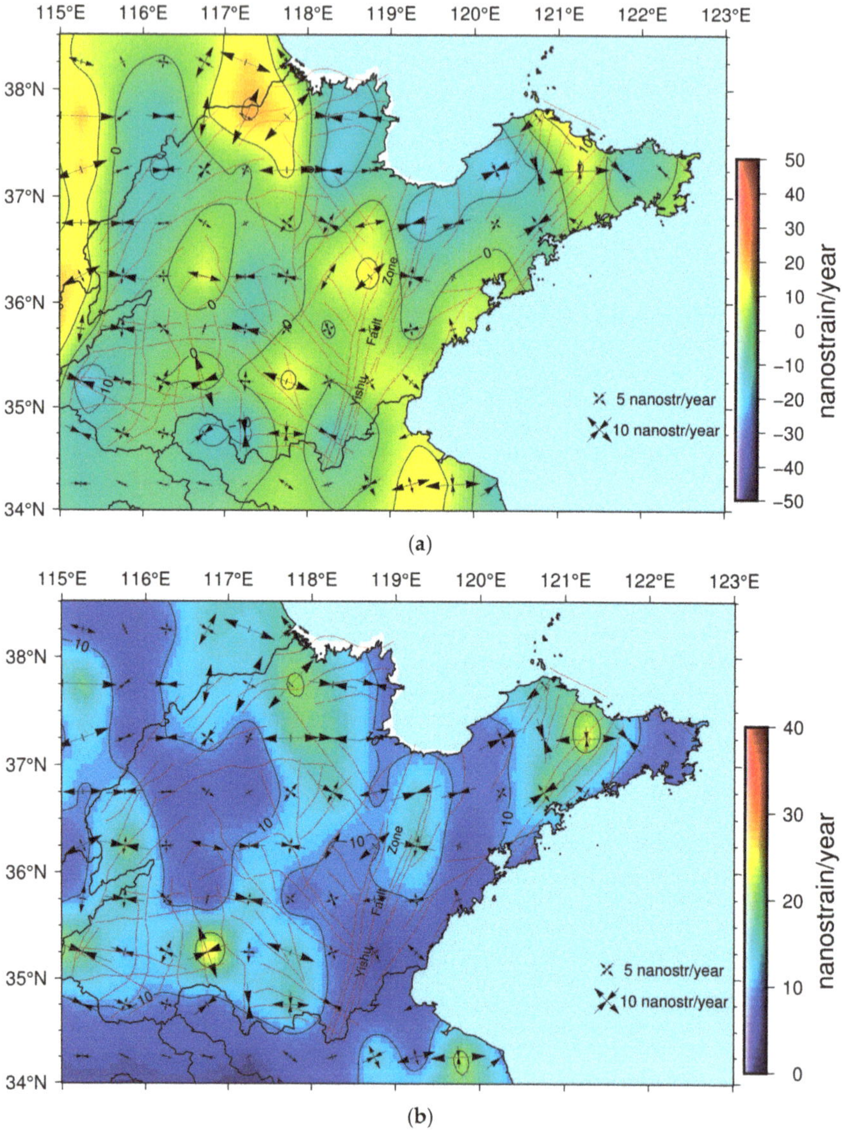

Figure 6. Strain rates in the study area from 1999 to 2009. (**a**) Principal strain rate vectors and dilatation rates. The crossed arrows are the principal strain rate vectors, and the background color indicates the dilation strain rates. Positive dilatation rates show extension, while negative show compression. (**b**) Maximum shear strain rates.

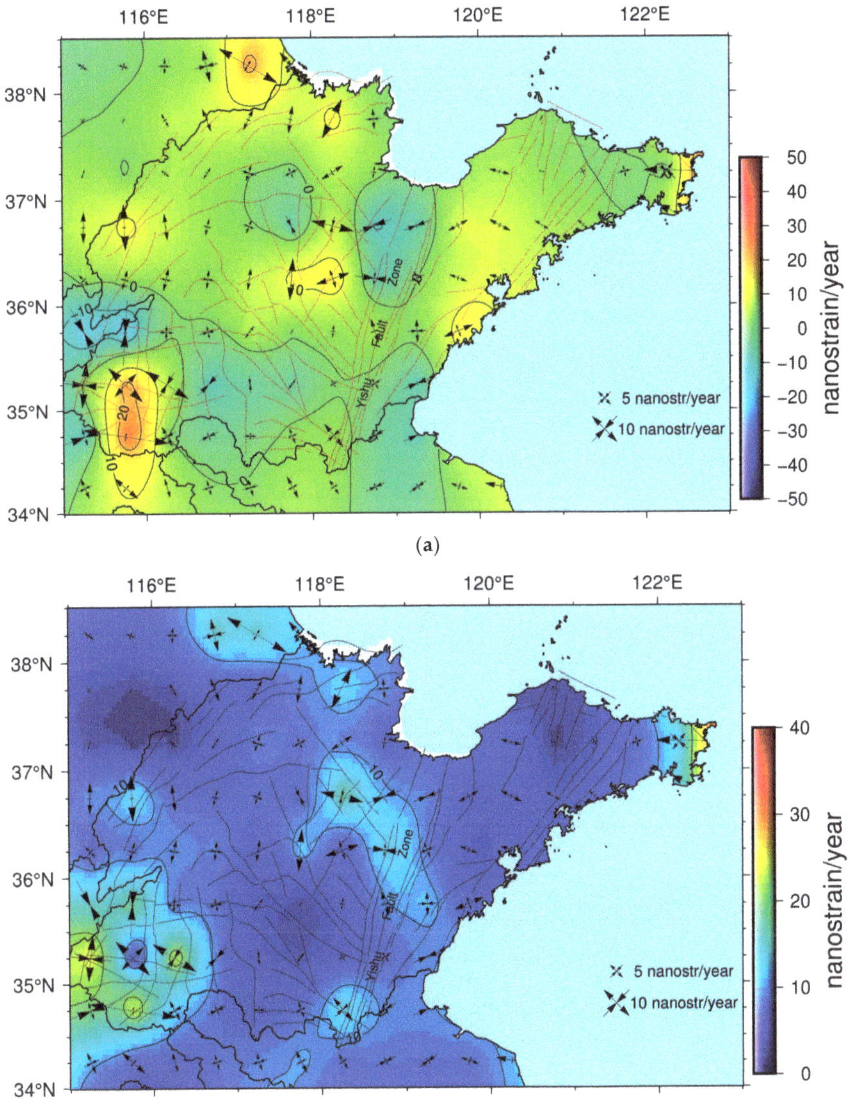

Figure 7. Strain rates in the study area from April 2011 to 2019. (**a**) Principal strain rate vectors and dilatation rates. (**b**) Maximum shear strain rates.

In rock experiments, the differential stress is expressed by the difference between the maximum principal stress and the minimum principal stress. Correspondingly, when the vertical crustal deformation is not considered, the plane maximum shear strain can be used to approximately reflect the regional differential stress state. According to the comparison between Figures 6b and 7b, the maximum shear strain in the Yishu fault zone did not significantly change before and after the 2011 Tohoku-Oki earthquake but decreased on both sides of the fault zone.

Considering the GPS observation accuracy, the continuous GPS observation sequences 10 days before and 10 days after the 2011 Tohoku-Oki earthquake were selected; the coseismic displacement was estimated by linear fitting the time series before and after the earthquake and the coseismic strain field and maximum shear strain field of the earthquake are obtained, as shown in Figure 8. The coseismic strain field (Figure 8a) shows that the eastward tension caused by the 2011 Tohoku-Oki earthquake causes the entire study area to be in a tensile strain state. The maximum shear strain field (Figure 8b) shows that the coseismic deformation field of the 2011 Tohoku-Oki earthquake produced obvious shear strain in the Ludong uplift (II1) area and Luxi fault block (I2, Luxi uplift), a certain shear strain in the southern section of Yishu fault zone, and gentle shear strain in the northern section.

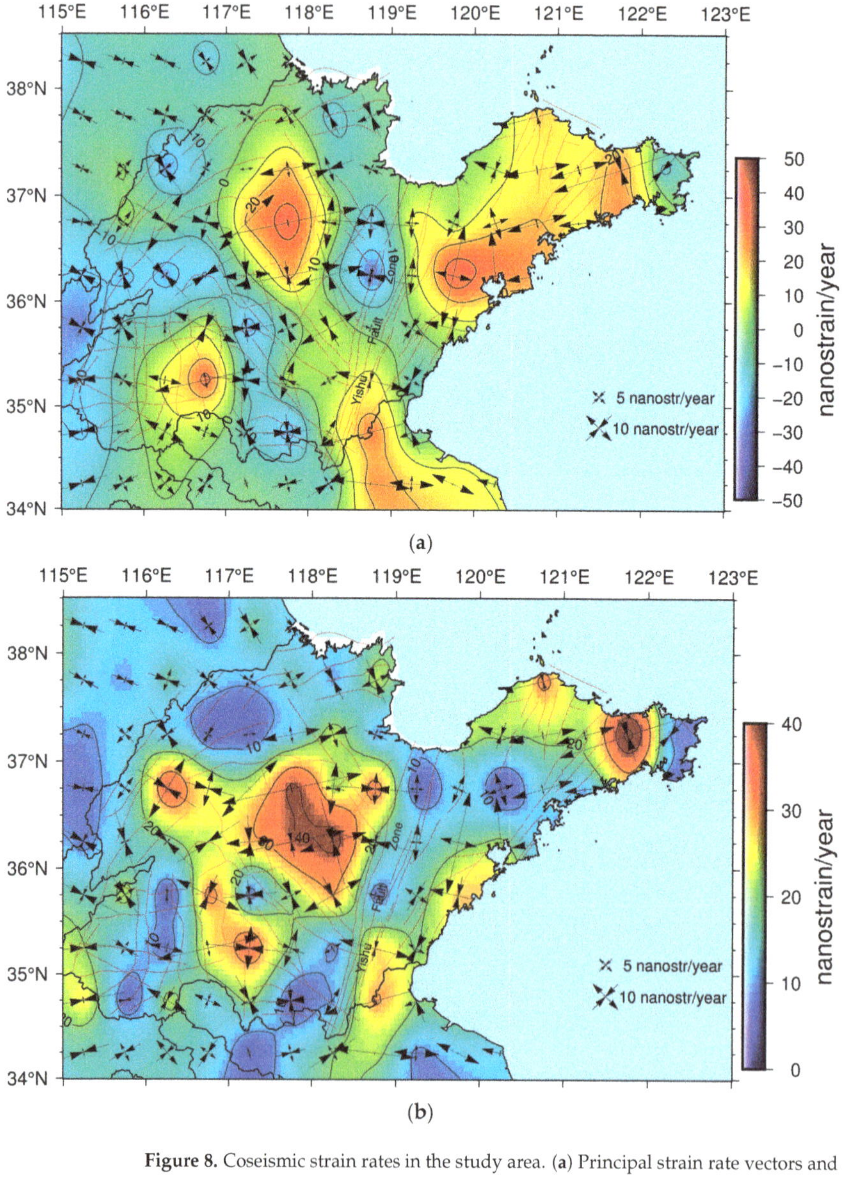

Figure 8. Coseismic strain rates in the study area. (**a**) Principal strain rate vectors and dilatation rates. (**b**) Maximum shear strain rates.

4.2. Three-Dimensional Seismic b Value

In this paper, the maximum likelihood method was used for scanning the three-dimensional *b* value along the Yishu fault zone to analyze the current stress level of the fault zone. Since the maximum likelihood method has high requirements for the earthquake catalogue's integrity and there is a lack of small earthquake data, the *b* value will be low; hence, the Mc is usually used to evaluate the seismic measurement capacity of the station [43]. As shown in Figure 9, the magnitude of completeness, Mc = 2.0, can be obtained. Therefore, it is more reliable to select the seismic data with $M_L > 2.0$ to calculate the *b* value in this paper.

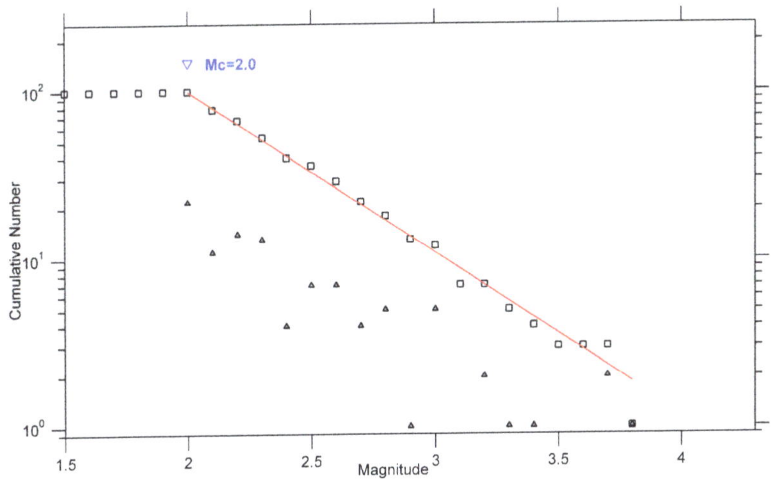

Figure 9. Magnitude of completeness (Mc).

The study area is divided into a grid with a spacing of 0.1° × 0.1°, taking the grid node as the center, the earthquakes larger than Mc in the circular area with a radius of 20 km are selected, and calculate the b value of each grid point by the least square method. The minimum number of earthquakes is set to be no less than 15. Since there are fewer earthquakes in some areas, the radius is then increased to a maximum of no more than 40 km, and the *b* value is null when the number of earthquakes is still less than 15. When calculating the *b* value of depth, the maximum radius of grid nodes is 5 km, and calculate the distribution of *b* values at different depths.

Using the catalogue data of $M_L > 2.0$ earthquakes from 1970 to 2019 derived from the Shandong Earthquake Agency, we calculated the *b* value of the Yishu fault zone and obtained the *b* value at depths of 0 km (Figure 10b) and 20 km (Figure 10c). The *b* value of the cross section of the fault zone is calculated, as shown in Figure 10. It can be seen from Figure 10 that there is an area with a low *b* value in the northern segment of the Yishu fault zone, at about 0.6–0.8, reflecting that the stress level of this segment is relatively high today, that this segment is the unruptured segment of the 1668 Tancheng Ms 8.5 earthquake, and that similar seismic gap characteristics are observable. Some studies or articles [35,46] speculate that the region is currently locked, as it has accumulated more strain energy, so its seismic risk is worthy of attention.

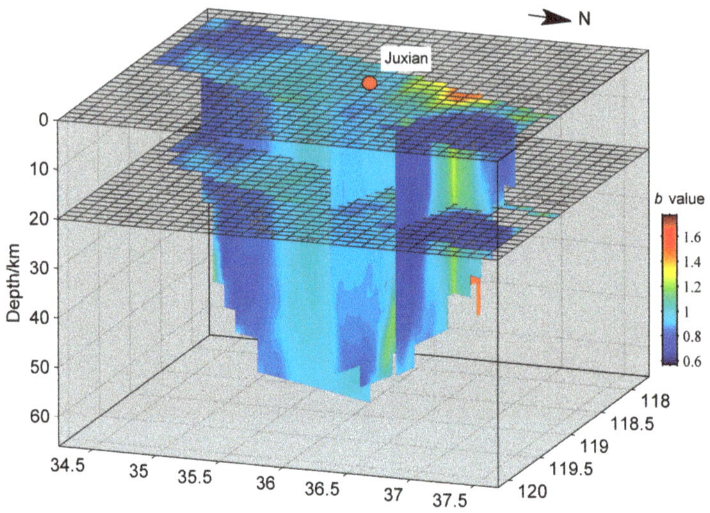

(**a**)

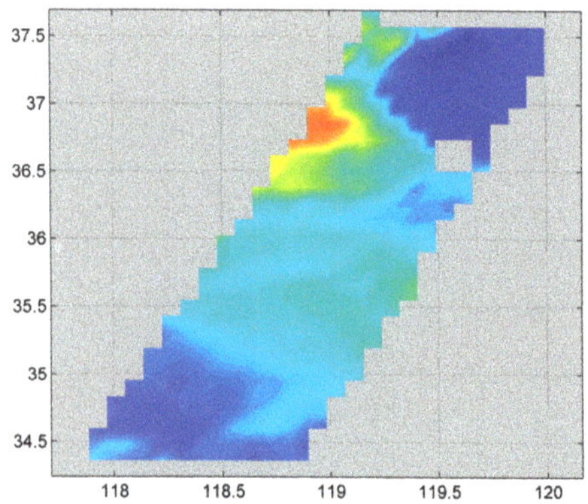

(**b**)

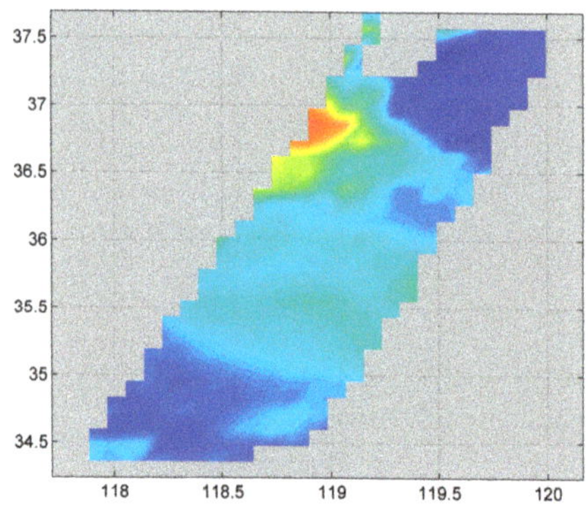

(**c**)

Figure 10. Three-dimensional *b* value spatiotemporal scan (from 1970 to 2019). (**a**) The 3D *b* value distribution. (**b**) The *b* value plane scanning in the study area. (**c**) The *b* value scanning at the depth of 20 km in study area.

4.3. Locking Characteristics of Deep Fault

The occurrence of major earthquakes is closely related to seismogenic faults' locking degrees. Based on the GPS velocity field results, the distribution of locking depths or locking degrees on the fault planes can be obtained using inversion or other means.

Considering that the Anqiu–Juxian fault is the most active in the Yishu fault zone, the setting of the fault model mainly refers to the section parameters of this fault. Wang et al. [47] completed the zonal mapping of the 1:50,000 active faults of the Anqiu–Juxian fault and determined the geometric structure, spatial distribution characteristics, and active nature of the fault. The overall strike of the Anqiu–Juxian fault is NNE 10°–20°, the inclination is NW, and the inclination angle is 70°–80°. Many studies or articles have used geophysical methods such as magnetotelluric sounding and artificial seismic sounding to detect that the crustal thickness of the Yishu fault zone is about 31–36 km [48,49]; the three-dimensional *b* value also indicates that there are low-value areas within 40 km of the deep fault depth. Therefore, we established the fracture zone model based on the above research results, as shown in Figure 11. The Luxi fault block is set as the hanging wall, the Ludong block is the footwall, the dip angle is set to 70°, and a total of nine nodes are arranged along the fault strike, which is distributed between the sections of the Anqiu–Juxian fault. The model depth is set to 40 km and a total of nine isobaths are set at 0.1 km, 5 km, 10 km, 15 km, 20 km, 25 km, 30 km, 35 km, and 40 km, respectively, along with the fault inclination.

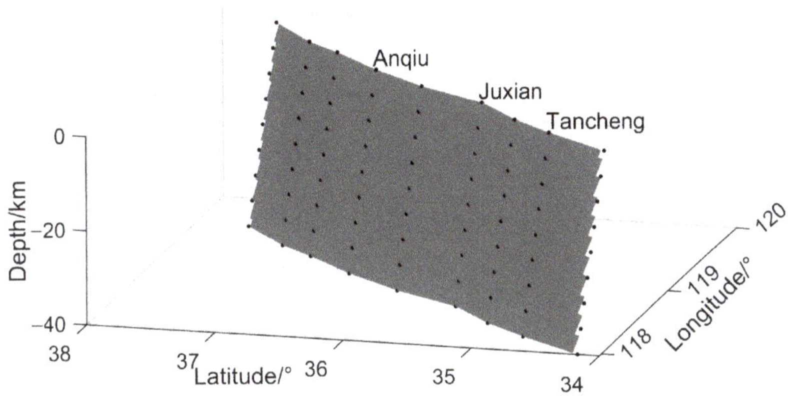

Figure 11. Sketch of simplified geometry for the Yishu fault zone. Block dots are fault nodes.

4.3.1. Locking Degree of Deep Fault

According to the above fault model, the spatial variation distribution of the locking degree of the Yishu fault zone before and after the 2011 Tohoku-Oki earthquake was obtained. Figure 12 demonstrates that, before the 2011 Tohoku-Oki earthquake, the north of Juxian in the Yishu fault zone's locking was high degree, deep, and had a maximum locking depth of about 26 km. Additionally, Figure 12 indicates that there was a locking area in the south of Juxian, but the locking depth was shallow, about 6 km, and the deep area was in creep state. The results in Figure 13 show that the locking area changed after the 2011 Tohoku-Oki earthquake. The northern section of the Yishu fault zone still has a high locking degree and a deep locking depth, with a maximum locking depth of about 26 km. The south section's conditions may be due to the tensile effect of the 2011 Tohoku-Oki earthquake on the area (Figure 8) or because the accumulated strain energy is released frequently by small earthquakes that alleviate the locking degree in the area.

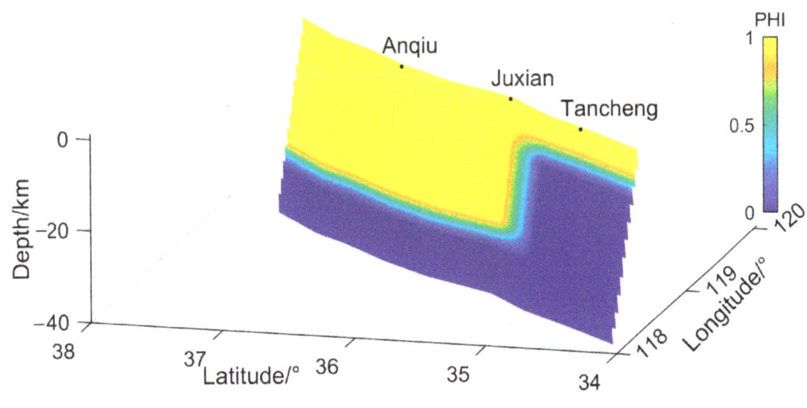

Figure 12. Locking degree of Yishu fault zone before the 2011 Tohoku-Oki earthquake (from 1999 to 2009).

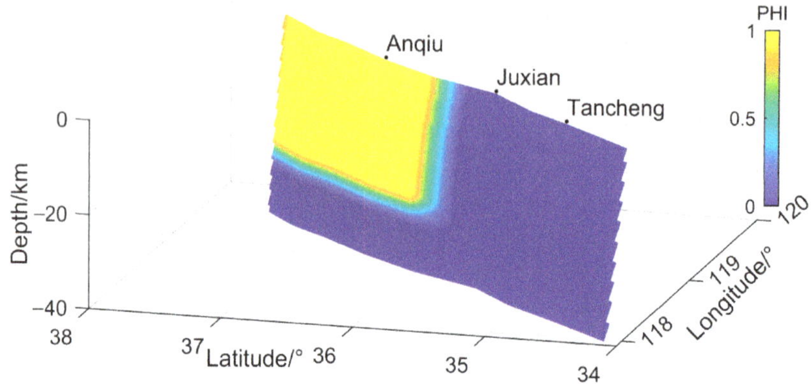

Figure 13. Locking degree of Yishu fault zone after the 2011 Tohoku-Oki earthquake (from April 2011 to 2019).

4.3.2. Fault Slip Rate Deficit Distribution

The relative motion between the faults is determined by the Euler poles of the blocks on both sides of the fault and the fault sliding loss rate is the product of the locking degree and the relative motion on the fault between the two blocks [29–31,34]. The sliding loss caused by fault locking accumulates as strain energy at and near the fault plane [31]. Figure 14 shows the slip rate deficit distribution in the Yishu fault zone from 1999 to 2009. The north of Juxian has a dextral compression slip rate deficit from the surface to the depth of 26 km, which is about 0.4~0.8 mm/a, and 26–40 km is basically a creep state without the slip rate deficit. The locking depth of the southern section of the fault zone is small, the deficit of sliding rate is large, about 1.1 mm/a, and the fault depth is 5–40 km, which is basically a creep state. The results in Figure 15 show that the locking range of the Yishu fault zone decreased after the 2011 Tohoku-Oki earthquake and the south section basically entered a creep state. The sliding rate deficit is mainly concentrated in the north section of the fault and the rate deficit reduced to about 0.4–0.6 mm/a.

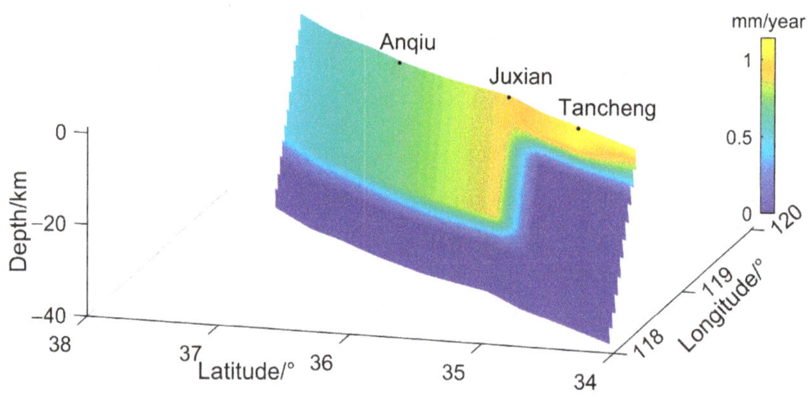

Figure 14. The slip rate deficit of the Yishu fault zone before the 2011 Tohoku-Oki earthquake (from 1999 to 2009).

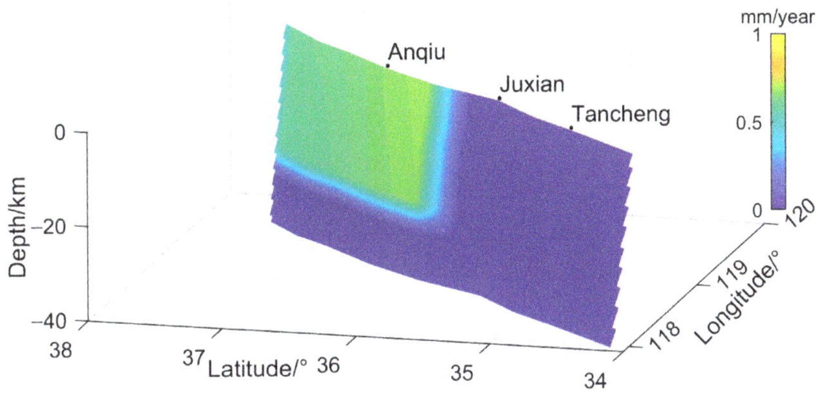

Figure 15. The slip rate deficit of the Yishu fault zone after the 2011 Tohoku-Oki earthquake (from April 2011 to 2019).

4.3.3. Uncertainty Analysis of Inversion Results

By comparing the residuals of the model fitting (comparing the total GPS rate values at each point obtained from the model inversion with the rate values obtained from the direct fitting of the time series), the quality of the fitting of the inversion results can be checked to some extent [50]. Figure 16 compares the residual fitting velocity of the model. The residuals before and after the 2011 Tohoku-Oki earthquake are small, both less than 0.5 mm/year, and the direction of the residuals is random, indicating that the inversion results are good.

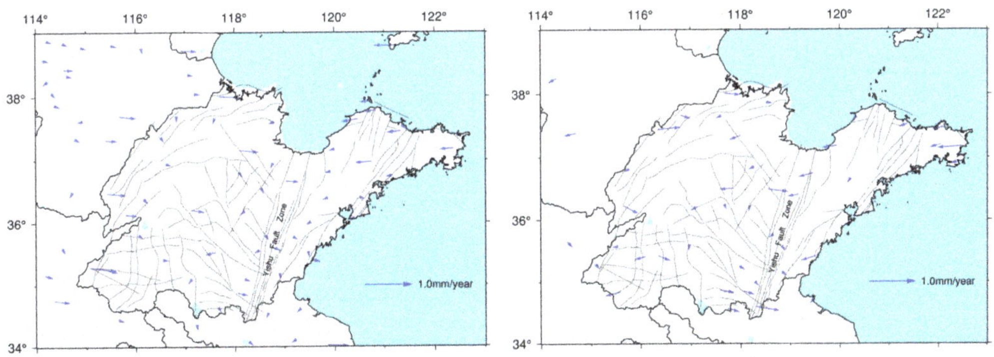

(**a**) (**b**)

Figure 16. Comparing the residual fitting velocity of the model. (**a**) Distribution of residual velocity before the 2011 Tohoku-Oki earthquake (from 1999 to 2009). (**b**) Distribution of residual velocity after the 2011 Tohoku-Oki earthquake (from April 2011 to 2019).

Since the model assumes that the internal deformation of the block is uniform, the inhomogeneity of the spatial distribution of deformation, crustal medium, wave velocity structure, and so on are not considered, which is a source of uncertainty in the inversion results [2]. The distribution and number of near-field GPS observation stations of faults have a certain influence on the inversion results [50]. Whether the block is rigid or not has a larger influence on the inversion results. Different reference frames have relatively little influence on the inversion results and different stable blocks as reference frames have almost no influence on the inversion results [26]. In conclusion, we have obtained good results for the inversion of the locking degree in the Yishu Fault Zone, but this is still affected by uncertainties.

5. Discussion

The 2011 Tohoku-Oki earthquake caused significant coseismic and post-earthquake effects on eastern mainland China [21,51], which influenced the crustal deformation and seismic risk of the Yishu fault zone [5,52]. The results for the fault locking degrees obtained from two phases of GPS horizontal velocity field inversion show that the Anqiu–Juxian section of the Yishu fault zone has a high locking degree and deep locking depth, which is due to the compression sliding deficit. After being affected by the 2011 Tohoku-Oki earthquake, the locking state of the southern section of the Yishu fault zone has eased, the compression activity has weakened, and the sliding rate deficit has decreased. However, the coseismic compression of the northern section caused by the 2011 Tohoku-Oki earthquake is conducive to its locking, so the northern section of the Yishu Fault Zone is still in a strong locking state. Due to the complexity of the crustal structure, the block negative dislocation model cannot fully simulate the crustal deformation in this area. At the same time, the number and distribution of GPS observation stations near the fault will also have a certain impact on the results. Therefore, there is some uncertainty in the inversion of the fault locking degree. Compared with the results of Li et al. (2016) [26] and Tao et al. (2022) [25], the main difference is observable in southern Tancheng. The former uses GPS data from 2009 to 2014, without considering the impact of the 2011 Tohoku-Oki earthquake on the Yishu fault zone. The difference from the latter is small; we believe that this is due to the near-field GPS data integration we use and that this is caused by the integration of near-field data. The results of each study are different from the fracture zone segmentation nodes used in this paper, so there will be some differences in the results.

The 2011 Tohoku-Oki earthquake caused by the subduction of the Pacific Plate has a direct impact on the dynamic environment of the region [53,54]. The state of the horizontal strain field after the 2011 Tohoku-Oki earthquake has weakened. Combined with the seismic strain field (Figure 8), the eastward tension generated by the 2011 Tohoku-Oki earthquake causes the entire study area to be in a tensile strain state. Since the Yishu fault zone is characterized by strike-slip movement, the 2011 Tohoku-Oki earthquake has a strain energy release and delayed seismic latent effect on it. However, due to the complex structural characteristics in the study area, especially the structural differences between the neotectonic units [24], the drag from below the crust may produce uneven tension on the upper crust and compressive strain in local areas. Such drag and strain have different coseismic effects on the south and north segments of the fault zone, including tensile effects on the south segment and a compressive effect on the north segment. The tensioning effect on the southern section is more significant, given the activity magnitude of the southern section is larger than for the northern section [16]. The Coulomb stress changes induced by the 2011 Tohoku-Oki earthquake on the northern segment of the Yishu fault zone are positive with a value of 0.23 KPa while the changes are smaller on the southern segment with a value of 0.08 KPa. The 2011 Tohoku-Oki earthquake helped to accumulate stress on the northern segment of the Yishu fault zone [55].

According to the earthquake distribution statistics in the study area, 285 earthquakes with $M_L > 2.0$ occurred in the southern section of the Yishu fault zone and 75 earthquakes occurred in the northern section of the Yishu fault zone from 1970 to 2019, accounting for only one-quarter of the southern section of the Yishu fault zone. The small earthquake activity in the northern section presents an abnormally sparse state; a seismic gap [56] formed in the northern segment of the fault zone that easily accumulates stress. Additionally, there is a low *b* value area in the north section, indicating that the stress level is relatively high. To study the current seismicity of the Yishu fault zone, the earthquakes from 2011 to 2019 were selected from the earthquake catalogue and the seismic distribution profile was drawn (Figure 17). Small earthquakes occurred within the depth of 30 km. At present, there are still fewer small earthquakes in the north section of the Yishu fault zone and the strain energy continues to accumulate, which is in good agreement with the fault locking range. The accumulated strain energy in the south section of the fault zone can be released in time through medium and small earthquakes, which is conducive to delaying the risk of strong earthquakes in the surrounding areas.

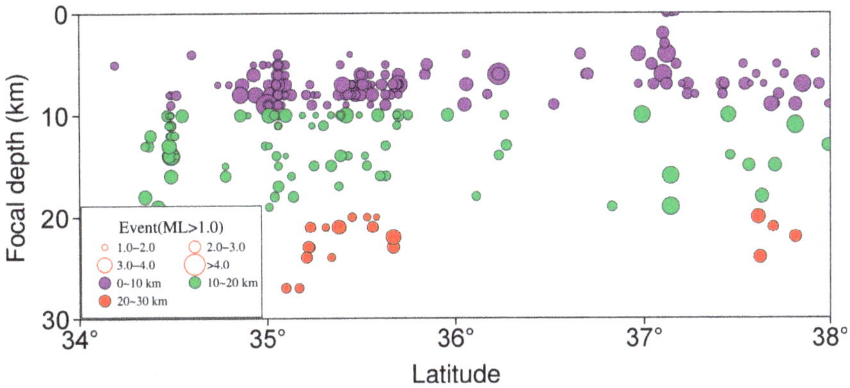

Figure 17. Depth distribution of the earthquakes (from 2011 to 2019; ML > 1.0) occurred in the location shown with blue box in Figure 1. Different colors indicate different depths of earthquakes. The size of the circle indicates the magnitude of the earthquake.

The northern and southern sections of the Yishu fault zone have different seismic risks and the seismic risk of the northern section is still relatively high. An earthquake of Ms 7.0 occurred in Anqiu in 70 BC, but no ruptured segment corresponding to the earthquake has been found [35]. In 1668, an earthquake of Ms 8.5 occurred in Tancheng in the southern segment and the northern segment was the unruptured segment of the earthquake. Several studies have obtained the segmental characteristics of the upper- and middle-crustal mediums in the southcentral section of the Tanlu fault zone using the double-difference seismic tomography method and found that the velocity and Poisson's ratio of the crustal medium in the section north of Tancheng in the Yishu fault zone show relatively high values [57–59]. According to the recurrence period of 3500a [60], the last occurrence time accounted for 60% [6], meaning the earthquake risk is relatively high.

6. Conclusions

By using the GPS-derived horizontal velocity field and historical earthquake distribution in the Shandong Province in the period from 1999 to 2009 and April 2011 to 2019, the crustal strain field was calculated and the deformation difference characteristics in the Yishu fault zone were analyzed. The 3D *b* values were also calculated using the historical seismic distribution to present a preliminary determination of the fault locking depth. Finally, based on the block negative dislocation model, the locking degree and sliding rate deficit of the Yishu fault zone before and after the 2011 Tohoku-Oki earthquake were inverted. The results show the following.

Before the 2011 Tohoku-Oki earthquake, there was a high locking degree area in the south section of the Yishu fault zone, with a shallow depth of about 6 km, while the locking depth in the north section of the Yishu fault zone was about 26 km. The 2011 Tohoku-Oki earthquake alleviated the westward subduction and compression of the Pacific Plate. The eastward tension caused the entire study area to be a tensile strain state. Combined with the negative dislocation inversion results, the southern section's locking in the Yishu fault zone changed to a creep state after the 2011 Tohoku-Oki earthquake and the locking range was relatively reduced. However, the locking degree in the northern section of the Yishu fault zone is still high and deep. Moreover, the *b* value in the north section of the Yishu fault zone is low and the small earthquake activity is sparse. This is the unbroken section of the Tancheng Ms 8.5 earthquake from 1668, which easily accumulates stress, so its seismic risk is worthy of attention.

Author Contributions: Conceptualization, C.D. and H.Y.; methodology, C.D.; validation, S.Y. and L.Y.; formal analysis, C.D. and Y.J.; investigation, H.Y. and L.Y.; data curation, L.Y. and Y.J.; writing—original draft preparation, C.D.; writing—review and editing, C.D., H.Y. and S.Y.; funding acquisition, H.Y. and Y.J. All authors have read and agreed to the published version of the manuscript.

Funding: The research was funded by the National Natural Science Foundation of China, grant number 41974105 and the Spark Program of Earthquake Technology of China Earthquake Administration, grant number XH21019Y.

Data Availability Statement: The data used during the study are available from the first author by request.

Acknowledgments: We acknowledge the Shandong Earthquake Agency for providing earthquake observation data and GPS data. Some figures are generated using the Generic Mapping Tools (GMT) software [61].

Conflicts of Interest: The authors declare no conflict of interest.

Abbreviations

GPS	Global Positioning System
CMONOC	Crustal Movement Observation Network of China
CMONOSD	Crustal Movement Observation Network of Shandong
B.C./BC	Before Christ
SDCORS	Shandong Continuously Operating Reference Station System
ITRF 2014	International Terrestrial Reference Frame 2014
IGS	International Global Navigation Satellite System Service
Mc	Magnitude of completeness
GPT	Global Pressure Temperature
GMF	Global Mapping Function
3D	Three-dimensional

References

1. Kato, A.; Obara, K.; Igarashi, T.; Tsuruoka, H.; Nakagawa, S.; Hirata, N. Propagation of slow slip leading up to the 2011 Mw 9.0 Tohoku-Oki earthquake. *Science* **2012**, *335*, 705–708. [CrossRef] [PubMed]
2. Zhang, P.; Deng, Q.; Zhang, Z.; Li, H. Active fault, earthquake hazards and associated geodynamic processes in continental China. *Sci. Sin. Terrae* **2013**, *43*, 1607–1620. [CrossRef]
3. Shao, Z.; Wang, P.; Li, H. Review on Researches Associated with the 2011 Mw 9.0 Tohoku-Oki Earthquake. *Earthquake* **2016**, *36*, 1–21.
4. Zheng, J.; Jiang, H. Correlation analysis and causality test between Ludong-Huanghai Block and South Japan. *Acta Seismol. Sin.* **2007**, *29*, 358–368. [CrossRef]
5. Yin, H.; Gan, W.; Huang, B.; Xiao, G.; Li, J.; Zhu, C. Study on the effects of Japan M9.0 huge earthquake on the crustal movement of Shandong area. *Chin. J. Geophys.* **2013**, *56*, 1497–1505. [CrossRef]
6. Zhu, C.; Gan, W.; Li, J.; Jia, Y.; Wang, Q.; Yin, H. Relative motion between the two blocks on either side of the Yishu fault zone after the 2011 Japan Mw 9.0 earthquake and its effect on seismic activity. *Chin. J. Geophys.* **2018**, *61*, 988–999. [CrossRef]
7. Lin, A.; Rao, G.; Yan, B. Dynamic characteristics of the Yishu Fault Zone, central segment of the Tan-Lu Fault Zone, Shandong Province, China: Inferred from the distribution patterns of drainages. *Earth Sci. Front.* **2013**, *20*, 125–136.
8. Lin, A.; Miyata, T.; Wan, T. Tectonic characteristics of the central segment of the Tancheng–Lujiang fault zone, Shandong Peninsula, eastern China. *Tectonophysics* **1998**, *293*, 85–104. [CrossRef]
9. Zhang, P.; Deng, Q.; Zhang, G.; Ma, J.; Gan, W.; Min, W.; Mao, F.; Wang, Q. Active tectonic blocks and strong earthquakes in the continent of China. *Sci. Sin. Terrae* **2003**, *33*, 12–20. [CrossRef]
10. Zhang, G.; Ma, H.; Wang, H.; Wang, X. Boundaries between active tectonic blocks and strong earthquakes in the China mainland. *Chin. J. Geophys.* **2005**, *48*, 602–610. [CrossRef]
11. Deng, Q.; Zhang, P.; Ran, Y.; Yang, X.; Min, W.; Chu, Q. Basic characteristics of active tectonics in China. *Sci. Sin. Terrae* **2002**, *32*, 1020–1030. [CrossRef]

12. Zhang, J.; Hao, T.; Dong, S.; Chen, X.; Cui, J.; Yang, X.; Liu, C.; Li, T.; Xu, Y.; Huang, S.; et al. The structural and tectonic relationships of the major fault systems of the Tan-Lu fault zone, with a focus on the segments within the North China region. *J. Asian Earth Sci.* **2015**, *110*, 85–100. [CrossRef]
13. Miao, Q.; Lei, J.; He, J.; Du, M.; Yin, W.; Zhang, Z.; Shi, Y. Pn velocity and anisotropy beneath the Yishu fault zone and surrounding areas. *Chin. J. Geophys.* **2021**, *64*, 2324–2335. [CrossRef]
14. Ma, H.; Zhang, G.; Liu, J.; Li, L.; Chen, H. Correlation between strong earthquake activity and active crustal-block in China mainland and its adjacent regions. *Earth Sci. Front.* **2003**, *10*, 74–80.
15. Li, Y.; Zhang, J.; Li, Z.; Guo, L.; Zhang, Z. The underthrust of Pacific Plate to Eurasian and its effect on Chinese mainland. *Acta Geod. Cartogr. Sin.* **2006**, *35*, 99–105. [CrossRef]
16. Zhu, C.; Gan, W.; Jia, Y.; Yin, H.; Xiao, G.; Li, J.; Liang, S.; Zhang, H. Coseismic effects of the Tohoku Mw 9.0 earthquake in 2011 on the crustal movement of Yishu fault zone and its bilateral areas. *Chin. J. Geophys.* **2020**, *63*, 3698–3711. [CrossRef]
17. Wang, M.; Shen, Z. Present-day crustal deformation of continental China derived from GPS and its tectonic implications. *J. Geophys.Res. Solid Earth* **2020**, *125*, e2019JB018774. [CrossRef]
18. Fan, Y.; Feng, C.; Zhang, P.; Qi, B.; Meng, J.; Tan, C. Impact of Tohoku-Oki 3.11 M9.0 Earthquake on the Fault Slip Potential of the Active Quaternary Faults in Beijing City: New Insights from In Situ Stress Monitoring Data. *Sensors* **2022**, *22*, 4888. [CrossRef]
19. Wang, Q.; Zhang, P.; Freymueller, J.T.; Bilham, R.; Larson, K.M.; Lai, X.; You, X.; Niu, Z.; Wu, J.; Li, Y.; et al. Present-day crustal deformation in China constrained by global positioning system measurements. *Science* **2001**, *294*, 574–577. [CrossRef]
20. Gan, W.; Zhang, P.; Shen, Z.; Niu, Z.; Wang, M.; Wan, Y.; Zhou, D.; Cheng, J. Present-day crustal motion within the Tibetan Plateau inferred from GPS measurements. *J. Geophys.Res. Solid Earth* **2007**, *112*, B08416. [CrossRef]
21. Wang, M.; Li, Q.; Wang, F.; Zhang, R.; Wang, Y.; Shi, H.; Zhang, P.; Shen, Z. Far-field coseismic displacements associated with the 2011 Tohoku-oki earthquake in Japan observed by Global Positioning System. *Chin. Sci. Bull.* **2011**, *56*, 2419–2424. [CrossRef]
22. Zhou, X.; Sun, W.; Zhao, B.; Fu, G.; Dong, J. Geodetic observations detecting coseismic displacements and gravity changes caused by the Mw 9.0 Tohoku-Oki earthquake. *J. Geophys. Res. Solid Earth* **2012**, *117*, 81–88. [CrossRef]
23. Zhang, Y.; Zheng, W.; Wang, Y.; Zhang, D.; Tian, Y.; Wang, M.; Zhang, Z.; Zhang, P. Contemporary deformation of the North China Plain from Global Positioning System data. *Geophys. Res. Lett.* **2018**, *45*, 1851–1859. [CrossRef]
24. Li, L.; Li, Y.; Zhang, F.; Chen, C.; Yin, H.; Jia, Y. Fault blocking characteristics and seismic hazard analysis in the middle and southern segments of the Tanlu Fault Zone. *Acta Geol. Sin.* **2020**, *9*, 467–479. [CrossRef]
25. Tao, T.; Chen, H.; Li, S.; Qu, X.; Zhu, Y. Interseismic Fault Coupling and Slip Rate Deficit on the Central and Southern Segments of the Tanlu Fault Zone Based on Anhui CORS Measurements. *Remote Sens.* **2022**, *14*, 1093. [CrossRef]
26. Li, Y.; Shan, X.; Song, X.; Jiang, Y.; Gan, W.; Qu, C.; Wang, Z. Fault locking and slip rate deficit on the middle and southern segment of the Tancheng-Lujia fault inverted from GPS data. *Chin. J. Geophys.* **2016**, *59*, 4022–4034. [CrossRef]
27. Scholz, C.H. *The Mechanics of Earthquakes and Faulting*; Cambridge University Press: New York, NY, USA, 1990.
28. Scholz, C.H. Earthquakes and friction laws. *Nature* **1998**, *391*, 36–42. [CrossRef]
29. Chamoli, A.; Lowry, A.R.; Jeppson, T.N. Implications of transient deformation in the northern Basin and Range, western United States. *J. Geophys. Res. Solid Earth* **2014**, *119*, 4393–4413. [CrossRef]
30. McCaffrey, R. Time-dependent inversion of three-component continuous GPS for steady and transient sources in northern Cascadia. *Geophys. Res. Lett.* **2009**, *36*, 2497–2502. [CrossRef]
31. Lay, T. A review of the rupture characteristics of the 2011 Tohoku-oki Mw 9.1 earthquake. *Tectonophysics* **2018**, *733*, 4–36. [CrossRef]
32. Ozawa, S.; Nishimura, T.; Suito, H.; Kobayashi, T.; Tobita, M.; Imakiire, T. Coseismic and postseismic slip of the 2011 magnitude-9 Tohoku-Oki earthquake. *Nature* **2011**, *475*, 373–376. [CrossRef]
33. Aydin, B.; Ercan, I.; Ehsan, H. A case study for determination of seismic risk priorities in Van (Eastern Turkey). *Earthq. Struct.* **2021**, *20*, 445–455. [CrossRef]
34. McCaffrey, R. Crustal block rotations and plate coupling. *Plate Bound. Zones* **2002**, *30*, 101–122. [CrossRef]
35. Wang, Z.; Wang, D.; Xu, H.; Ge, F.; Yang, C.; Li, J. Geometric features and latest activities of the north segment of the Anqiu–Juxian fault. *Seismol. Geol.* **2015**, *37*, 176–191. [CrossRef]
36. Ji, H.; Li, A.; Zhang, S.; Zhang, J.; Liu, Q. Geometric Distribution and Earthquake Rupture Characteristics of the orthern Anqiu–Juxian Fault in the Tan–Lu Fault Zone, Eastern China. *Front. Earth Sci.* **2022**, *10*, 7666222. [CrossRef]
37. Zhang, P.; Wang, L.; Zhong, K.; Ding, Z. Research on the Segmentation of Tancheng-Lujiang Fault Zone. *Geol. Rev.* **2007**, *53*, 586–591.
38. Liu, X.; Jiang, Z.; Wu, Y. Motion and Deformation State of the Tancheng-Lujiang Fault Zone Derived from GPS Data. *Earthquake* **2012**, *32*, 1–10.
39. Steinshouer, D.; Qiang, J.; McCabe, P.; Ryder, R. *Maps Showing Geology, Oil and Gas Fields, and Geologic Provinces of the Asia Pacific Region: U.S. Geological Survey*; Open-File Report 97-470-F; U.S. Geological Survey: Reston, VA, USA, 1999; p. 16. [CrossRef]
40. Savage, J.C.; Gan, W.; Svarc, J.L. Strain accumulation and rotation in the Eastern California Shear Zone. *J. Geophys. Res.* **2001**, *106*, 21995. [CrossRef]
41. Gutenberg, B.; Richter, C.F. Seismicity of the earth. *Geol. Soc. Am. Spec. Pap.* **1941**, *34*, 1–126. [CrossRef]
42. Liu, Z.; Feng, J.; Zhang, S. The progress on *b*-value of earthquake. *Plateau Earthq. Res.* **2019**, *31*, 9–13.
43. Wiemer, S. A software package to analyze seismicity: ZMAP. *Seismol. Res. Lett.* **2001**, *72*, 373–382. [CrossRef]

44. Savage, J.C.; Burford, R.O. Geodetic determination of relative plate motion in central California. *J. Geophys. Res.* **1973**, *78*, 832–845. [CrossRef]
45. Meade, B.J.; Hager, B.H. Block models of crustal motion in southern California constrained by GPS measurements. *J. Geophys. Res. Atmos.* **2005**, *110*, 353. [CrossRef]
46. Yang, X.; Song, F.; Zhang, L.; He, H.; Li, C.; Wang, Z. A recent paleoearthquake on Qingfengling seismic fault of Tanlu fault zone. *Acta Seismol. Sin.* **2006**, *19*, 225–230. [CrossRef]
47. Wang, Z.; Wang, D.; Xu, H.; Ge, F.; Yang, C.; Wang, H.; Wang, J. *Description of Anqiu–Juxian Fault Active Fault Distribution Map (1:50000)*; Seismological Press: Beijing, China, 2017.
48. Xiao, Q.; Zhao, G.; Wang, J.; Zhan, Y.; Chen, X.; Tang, J.; Cai, J.; Wan, Z.; Wang, L.; Ma, W.; et al. Deep electrical structure of the Sulu orogen and neighboring areas. *Sci. Sin. Terrae* **2009**, *52*, 420–430. [CrossRef]
49. Liu, B.; Feng, S.; Ji, J.; Shi, J.; Tan, Y.; Li, Y. Fine lithosphere structure beneath the middle-southern segment of the Tan-Lu fault zone. *Chin. J. Geophys.* **2015**, *58*, 1610–1621. [CrossRef]
50. Zhao, J.; Zhan, W.; Ren, J.; Jiang, Z.; Gu, T.; Liu, J.; Niu, F.; Yuan, Z. GPS time series inversion of the healing process of the middle segment of the Longmenshan fault after the 2008 Wenchuan earthquake. *Acta Geod. Cartogr. Sin.* **2021**, *50*, 37–51. [CrossRef]
51. Liu, T.; Fu, G.; Zhou, X.; Su, X. Mechanism of post-seismic deformations following the 2011 Tohoku-Oki MW9.0 earthquake and general structure of lithosphere around the source. *Chin. J. Geophys.* **2017**, *60*, 3406–3417. [CrossRef]
52. Sun, Y.; Huang, Y.; Liu, Z.; Zheng, J.; Jiang, H.; Li, T.; Yang, H.; Wang, J. Dynamic variation characteristics of tectonic stress field in the middle-southern segment of Tan-Lu fault zone before and after the 2011 Tohoku-Oki MW9.0 earthquake. *Chin. J. Geophys.* **2022**, *65*, 2124–2136. [CrossRef]
53. Yang, S.; Nie, Z.; Jia, Z.; Peng, M. Far-field coseismic surface displacement caused by the Mw9.0 Tohoku earthquake. *Geomat. Inf. Sci. Wuhan Univ.* **2011**, *36*, 1336–1339. [CrossRef]
54. Hao, M.; Zhuang, W. The impact of great 2011 Tohoku-Oki earthquake on crustal deformation in eastern China. *J. Geod. Geodyn.* **2020**, *40*, 555–558.
55. Feng, C.; Zhang, P.; Tan, C.; Qi, B.; Wang, L. Co-seismic and post-seismic effects of the Tohoku-Oki Mw 9.0 earthquake in North and Northeast China. *J. Seismol.* **2016**, *20*, 333–359. [CrossRef]
56. Mogi, K. Two kinds of seismic gaps. Pure and Applied Geophysics. *Pure Appl. Geophys.* **1979**, *117*, 1172–1186. [CrossRef]
57. Gu, Q.; Ding, Z.; Kang, Q.; Zhao, Q. Pn wave velocity and anisotropy in the middle-southern segment of the Tan-Lu fault zone and adjacent region. *Chin. J. Geophys.* **2016**, *59*, 504–515. [CrossRef]
58. Lei, J.; Zhao, D.; Xu, X.; Du, M.; Lu, M. P-wave upper-mantle tomography of the Tanlu fault zone in eastern China. *Phys. Earth Planet. Inter.* **2020**, *299*, 106402. [CrossRef]
59. He, Y.; Fan, X.; Zhao, Q.; Huo, Z.; Yang, C.; Zheng, L.; Qian, H.; Zheng, T. Segmentation of crustal structure beneath the middle-south segment of Tan-Lu Fault Zone. *Chin. J. Geophys.* **2021**, *64*, 3164–3178. [CrossRef]
60. Chao, H.; Li, J.; Cui, Z. Tectonic conditions of strong earthquakes with $M \geq 6$ in Shandong province and its adjacent seas. *J. Seismol. Res.* **1995**, *18*, 188–196.
61. Wessel, P.; Luis, J.F.; Uieda, L.; Scharroo, R.; Wobbe, F.; Smith, W.H.F.; Tian, D. The Generic Mapping Tools version 6. Geochemistry. *Geophys. Geosyst.* **2019**, *20*, 5556–5564. [CrossRef]

sustainability

Article

V_{s30} Prediction Models Based on Measured Shear-Wave Velocities in Tangshan, China

Yi Fang [1,2], Hao Li [1], Yu Li [2], Guoxing Chen [1,*], Yuejun Lv [2] and Yanju Peng [2]

1 Institute of Geotechnical Engineering, Nanjing Tech University, Nanjing 211816, China
2 National Institute of Natural Hazards, Ministry of Emergency Management of China, Beijing 100085, China
* Correspondence: gxc6307@163.com

Abstract: V_{s30} (equivalent shear-wave velocity of soil layers within a depth of 30 m underground) is widely used in the field of seismic engineering; however, due to the limitation of funds, time, measuring devices, and other factors, the depth for testing shear-wave velocity in an engineering site rarely reaches 30 m underground. Therefore, it is necessary to predict Vs30 effectively. We analyzed the existing models using 343 boreholes with depths greater than 30 m in Tangshan, China. It shows that the topographic slope method is not suitable for predicting V_{s30} in Tangshan. The Boore (2011) model overestimates, while Boore (2004) underestimates V_{s30} in Tangshan, while Junju Xie's (2016) model has ideal prediction results. We propose three new models in this paper, including the bottom constant velocity (BCV) model, linear model, and conditional independent model. We find that the BCV model has limited prediction ability, and the linear model is more suitable when $z \leq 18$ m, while the conditional independent model shows good performance under conditions where $z > 18$ m. We propose that the model can be accurately and effectively applied in Tangshan and other regions with low shear-wave velocity.

Keywords: shear-wave velocity; Tangshan; topographic slope; prediction model; log-likelihood

Citation: Fang, Y.; Li, H.; Li, Y.; Chen, G.; Lv, Y.; Peng, Y. V_{s30} Prediction Models Based on Measured Shear-Wave Velocities in Tangshan, China. *Sustainability* **2023**, *15*, 3282. https://doi.org/10.3390/su15043282

Academic Editor: Claudia Casapulla

Received: 14 December 2022
Revised: 3 February 2023
Accepted: 3 February 2023
Published: 10 February 2023

1. Introduction

Earthquakes have brought a series of engineering challenges for the design and service of constructions and buildings [1,2]. Soil, as a wave dispersion medium and bearing layer for structures, has dynamic characteristics that have important implications for the seismic design and structural analysis of construction materials under ground motion [3–5]. Shear-wave velocity, V_s, is one of the most basic parameters for characterizing the dynamic behaviors of soils. V_{s30} has been widely applied in the field of seismic engineering for inground vibration prediction [6,7], site classification [8], site effect evaluation [9], seismic performance assessment, or designing structures on a regional scale [10]; however, due to the limitation of funds, time, and measuring devices, the depth for testing shear-wave velocity generally cannot reach 30 m and sometimes there is no shear-wave velocity available. At present, V_{s20} (equivalent shear-wave velocity of soil layers within a depth of 20 m underground) is still used to describe the characteristics of site soil in China. Although some studies have suggested that seismic motion can predict V_s without drilling, due to the scarcity of seismic records and the limited effect of this method itself, this method is not widely used. Therefore, it is very important to predict V_{s30} rapidly, accurately, and effectively.

When the project site lacks the measured data of shear-wave velocity, other indicators that can reflect the geological characteristics of a site are usually used to predict V_{s30}, such as surface geology [11–13], topographic classification [14,15], topographic slope [16], category of geotechnical engineering works [17,18], mixed geological features [19], and Quaternary System isopach [20]. These methods of using other indicators to characterize V_{s30} can quickly and widely predict V_{s30}, but the accuracy is low, and the methods are significantly affected by the accuracy of original geological features. While, for an engineering site with shear-wave velocity but the depths are less than 30 m underground, the shear-wave velocities (V_{sz}, $z < 30$ m) at known depths can be used to predict V_{s30} more accurately, such as the BCV model (a constant recursive model) [21] and a linear model [22–24]. Although such models using the known data of shear-wave velocity to predict V_{s30} show a high accuracy, there is certain regional applicability in the established V_{s30} models due to the limitations of the basic data, especially in areas with low shear-wave velocity, there is still no effective prediction method.

In this paper, the measured data of shear-wave velocities from boreholes in Tangshan were used for V_{s30} prediction. Based on the data of shear-wave velocities of 343 boreholes with depths more than 30 m in the city, the topographic slope was first used as an alternative indicator for rapid prediction of V_{s30}. Then, three new V_{s30} prediction models were established based on V_{sz}, and the goodness of fit of the new models was ranked by using the log-likelihood (LLH) method. The prediction model of V_{s30} proposed in this paper can be accurately and effectively applied to areas with low shear-wave velocity, such as Tangshan.

2. Introduction of V_{s30} Prediction Models and Sorting Method of the Goodness of Fit

2.1. Prediction Using Topographic Slope Method

Geomorphologic fluctuations indicate topography and lithology. For example, the steep peaks suggest the rock-dominated geological property of the site, while the nearly flat basins indicate soil-dominated geological properties. Hard sites tend to have steep slopes, while thicker sedimentary basins generally show shallower slopes, indicating there is a certain correlation between topographic slope and V_{s30}. Based on this, Wald et al. [16] proposed a V_{s30} prediction model based on the topographic slope in 2007. This study analyzed the corresponding relationship between V_{s30} and NHRP different site categories, including a tectonically stable zone and a tectonically active zone, respectively. The United States Geological Survey constructed an empirical model of the relationship between global V_{s30} and topographic slope using this model.

2.2. V_{s30} Prediction Models Based on V_{sz}

When the depth for testing shear-wave velocity does not reach 30 m, V_{sz} can be generally used to predict V_{s30}. At present, there are two categories of commonly used models for predicting V_{s30} based on V_{sz}: the constant-velocity extrapolation model and the gradient-velocity extrapolation model.

For the constant-velocity extrapolation model, the main assumption of the BCV model is that V_s is constant from z to 30 m [21] as given by:

$$V_{s30} = \frac{30}{\Delta t_z + \dfrac{30 - z}{V_{sz}}} \tag{1}$$

where z, Δt_z, and $V_{s(z)}$ represent the depth of the hole bottom, the travel time of shear waves, and the instantaneous shear-wave velocity at the depth (z m) of the bottom, respectively. For the convenience of description in the following sections, the model is referred to as the BCV model.

The assumption of the constant-velocity extrapolation model is contrary to common sense that shear-wave velocities increase with the burial depth of the soil layers, so the model generally underestimates V_{s30}. Boore found that there is a strong linear correlation between log V_{s30} and log V_{sz} in 2004 and established a corresponding relationship between V_{s30} and V_{sz} by using Formula (2) based on data collected from 135 boreholes in California [22].

$$\log V_{s30} = a_0 + a_1 \log(V_{sz}) \tag{2}$$

where a_0 and a_1 are regression coefficients. This study is known as the Boore (2004) model.

Furthermore, based on Kik-net borehole data from Japan, Boore et al. [23] established the following new linear relationship between V_{s30} and V_{sz}.

$$\log V_{s30} = b_0 + b_1 \log(V_{sz}) + b_2(\log V_{sz})^2 \quad (3)$$

where b_0 and b_1 are regression coefficients. The regression coefficient of 5 m $\leq z \leq$ 29 m is finally determined. The model is referred to as the Boore (2011) model.

Dai et al. [25] built a conditional independent prediction model of V_{s30} based on a Markov process. This model supposes that shear-wave velocity from the bottom of boreholes to 30 m underground is unrelated to that from zero to z m underground. Therefore, this model was the first to establish the corresponding relationship (Formula (4)) between V_{sz} and $V_{s(z,30)}$ (equivalent shear-wave velocity from z to 30 m underground), and then V_{s30} was predicted using Formula (5):

$$\log V_{s(z,30)} = c_0 + c_1 \log(V_{sz}) \quad (4)$$

$$V_{s30} = \frac{30}{\Delta t_z + \frac{30 - z}{V_{s(z,30)}}} \quad (5)$$

where c_0 and c_1 are regression coefficients. The model is called the conditional independent model.

2.3. Sorting Method for Goodness of Fit of the Prediction Models Based on the LLH Method

At present, R^2 (the adjusted coefficient of determination) and similar parameters are used to sort the goodness of fit of prediction models; however, Weimin He et al. [26] found that there is a misjudgment that arises when using the adjusted coefficient of determination to determine the regression effects. Based on this, we adopted a more reliable LLH method [27] to sort goodness of fit data about the prediction models.

The LLH method was proposed based on information theory, allowing the selection of the optimal model based on data [27]. The LLH value of the prediction model is given by:

$$LLH(g, x) = -\frac{1}{N}\sum_{i=1}^{N} \log_2[g(x_i)] \quad (6)$$

where $g(x)$ and N indicate the probability density function (PDF) of the model with samples from x_1 to x_N and sample size, respectively. The smaller the LLH value of the model, the better the prediction.

Based on the measured data of shear-wave velocities in Iran, Shafiee et al. [28] analyzed and compared hypothesis testing and the likelihood method of optimal fitting of standard residuals. Different sorting results of V_{s30} prediction models using the LLH method show that LLH methods can accurately sort the goodness of fit of V_{s30} prediction models.

3. Engineering Geology beneath Tangshan City

Tangshan City is located at the border of the North China Plain and the southern slope of the Yanshan Mountains, with eroded low mountains and hills in the northeast that extend to the Tangshan uplift zone in the southwest and belongs to a zone with shallow bedrock. The other part presents flat alluvial plains of the Luanhe River system, sporadically distributed relic mountains of bedrock and striking to the southeast on the whole. Due to the combined effects of rivers and transgression, the coastal region is low and flat, where sand dunes and many shell ridges remain (Figure 1).

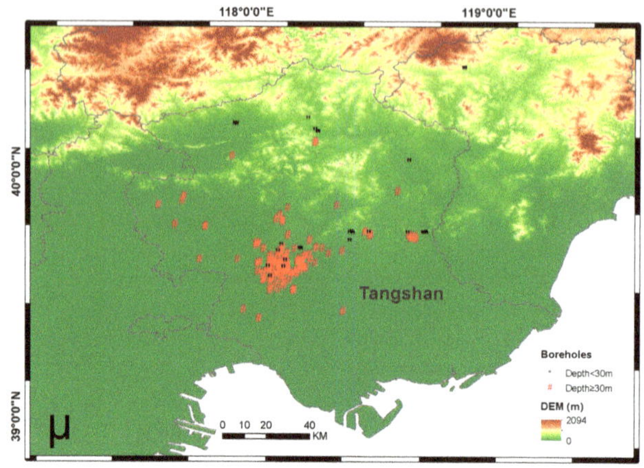

Figure 1. Digital elevation and borehole distribution in Tangshan City.

The Quaternary strata in this region are extremely well developed, accounting for about four-fifths of the total area. An area of low mountains and hills is found in the northeast, and the central region is dominated by a mixture of slope deposits and diluvial deposits from the late Pleistocene, showing a small thickness therein. Late Pleistocene alluvial and diluvial deposits are developed in the piedmont belt, and the southern region is dominated by alluvial and diluvial strata in the Holocene epoch, interspersed with relatively thick alluvial deposits of modern rivers. The underlying bedrock is generally high in the north and low in the south on the whole, and the thickness of the Quaternary strata gradually increases from north to south, while sediment particles gradually decrease in prevalence from north to south. The surface is covered by Holocene strata.

4. Analysis of the Existing Shear-Wave Velocity Data in Tangshan City

The shear-wave velocities data of 418 boreholes in Tangshan City (Figure 1) were mainly obtained from seismic safety evaluation and seismic micro-zoning in the region. Most of the data were obtained using the excitation method in a single borehole using an XG-I instrument, and the test depth of shear-wave velocity varied from 6 m to 110 m (Figure 2a). The 343 boreholes with depths greater than 30 m accounted for about 82% and formed the basis for the dataset used in this research. The V_{s30} of borehole data used in this study ranges from 169 to 325 m s^{-1} (Figure 2b), and V_{s30} > 300 m s^{-1} was found in four boreholes.

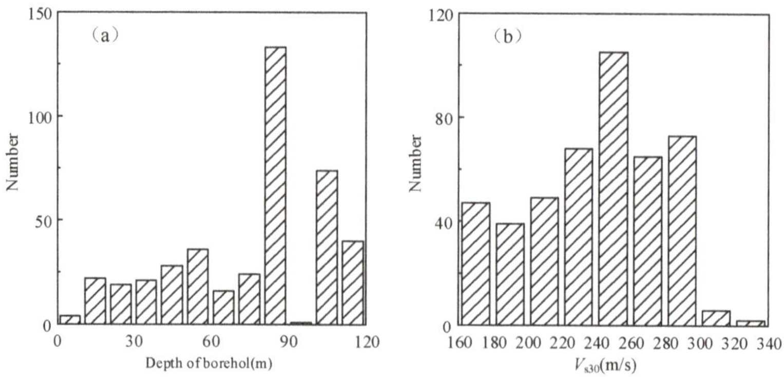

Figure 2. Statistics of the depth of boreholes and V_{s30} values in Tangshan City: (**a**) statistics of boreholes depth; (**b**) Statistics of V_{s30} values.

5. Vs_{30} Prediction Based on Topographic Slope

Wald et al. [16] studied the correlations between two topographic parameters (surface slope and elevation) with V_{s30}. The results demonstrate that the effects of using surface slope were better than that of using surface slope and elevation. Therefore, here, we directly used topographic slope as a proxy to predict V_{s30}. As the topographic slope cannot be directly obtained, it needs to be converted from digital elevation model (DEM) data, so the accuracy of DEM data can directly affect the results of the topographic slope. In addition, considering the current DEM data sources in Tangshan City, we separately downloaded DEM data at 30 m and 90 m over this region from the Geospatial Data Cloud website. Moreover, using the "Slope" function in the "Spatial Analyst" module of ArcGIS, geographic information system software, DEM data were converted into topographic slope through the fitting surface method. After obtaining the topographic slope in the city, they were allocated to 343 boreholes deeper than 30 m.

Figure 3a,b show site categories of 343 boreholes deeper than 30 m and a box plot of topographic slopes with 30 m and 90 m in Tangshan City. Using these kinds of figures, Wald et al. [16] analyzed the correlation between site category and topographic slope in California. In the figure, shear-wave velocities correspond to the boundary value of V_{s30} in NEHRP site categories, and the site categories are subdivided. V_{s30} ranges of site categories E, D1, D2, and D3 are <180, 180 to 240, 240 to 300, and >300 m s^{-1}, respectively. It can be seen from the figure that for the topographic slope with different resolutions, the scopes of site categories are different. However, as the site soil gradually hardens, the corresponding topographic slope shows a slightly upward trend. Figure 4 shows the relationships between V_{s30} of the 343 boreholes and topographic slope with resolutions of 30 m and 90 m. The plot relates to V_{s30}, and the topographic slope was fitted by the least squares method. Although this curve gradually rises with increasing topographic slope, data points in the figure are very discrete, showing no statistically significant corresponding relationship. Moreover, the ascent stage of the curve appears when the slope exceeds 0.02, and a sharp rise appears when the slope exceeds 0.05, but there are few data points; therefore, this trend is not reliable.

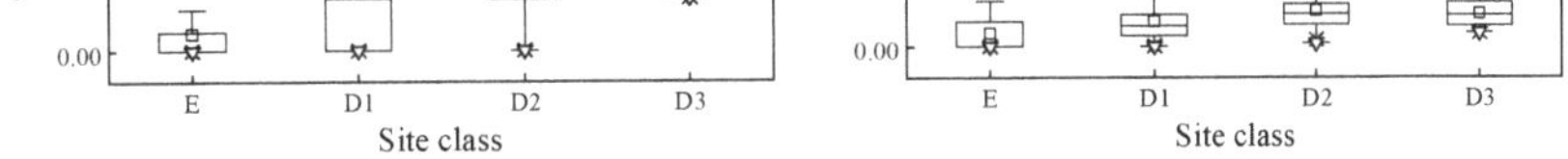

Figure 3. Correlation between site category and topographic slope: (**a**) Correlation between site category and topographic slope in 30 m; (**b**) Correlation between site category and topographic slope in 90 m.

Figure 5 compares the research results in this study with those obtained by Wald et al. [16]. In the figure, polygons represent the scope of V_{s30} in the tectonically unstable zone determined by Wald and Allen corresponding to the topographic slope. It can be seen from the figure that the topographic slope corresponding to V_{s30} in Tangshan City is systematically lower than the results obtained by Wald and Allen in some regions of the United States. This is similar to the research results obtained by Wald in Salt Lake City in the United States in 2007.

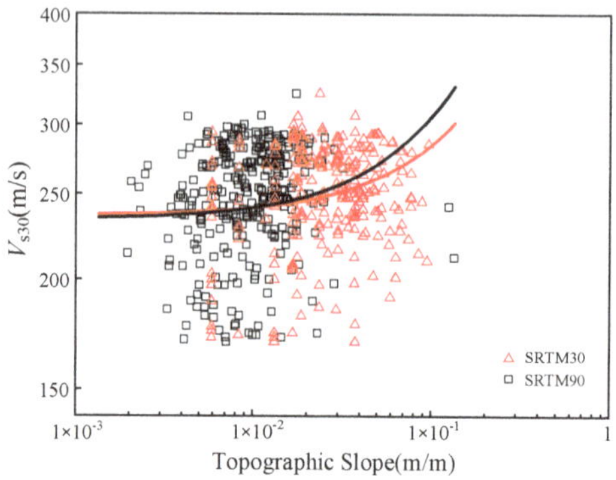

Figure 4. Correlation between topographic slope and V_{s30}.

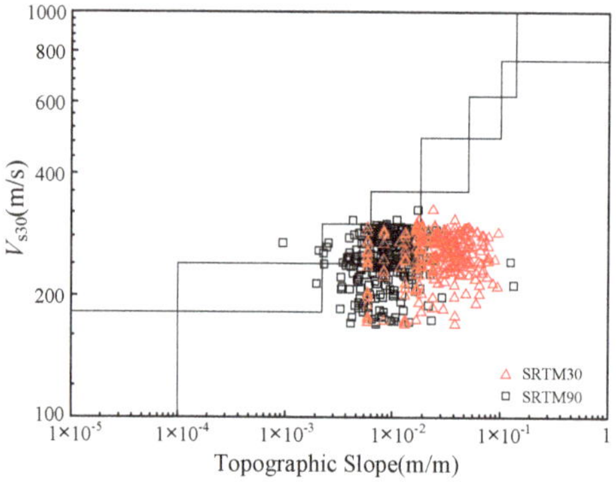

Figure 5. Comparison of conclusions obtained by Wald et al. with V_{s30} and topographic slope in Tangshan.

In conclusion, it can be found that there is a weak correlation between V_{s30} and topographic slopes with multiple resolutions in Tangshan City; that is, the topographic slope is not suitable for predicting V_{s30} in this region. The main reasons are as follows: (1) the prediction capacity of the topographic slope for V_{s30} is weaker than other geological features, which has been demonstrated in many studies; (2) Tangshan City is mainly located in the North China plain with a flat topography, where the topographic gradient changes are less obvious than that in hilly and mountainous areas; (3) the slope calculated by DEM in urban areas is usually significantly affected by the canopy effects; (4) most of the boreholes collected in this project are located in plain areas, while there are only a few boreholes in hilly and mountainous areas.

6. V_{s30} Prediction Models Based on V_{sz}

6.1. Establishment of V_{s30} Prediction Models

Considering that V_{s30} prediction based on V_{sz} has not been carried out in Tangshan, this paper first analyzed the correlation between V_{sz} and V_{s30} in this region. Figure 6 shows the relationship between V_{sz} (z = 10, 15, 20, and 25 m) and V_{s30} in the city. In the figure, the circle indicates the measured shear-wave velocities, while the dotted line represents $V_{s30} = V_{sz}$. Moreover, r is Pearson's correlation coefficient between them. As displayed in the figure, even when $z = 10$ m, r is still greater than 0.9, indicating that it has a very strong correlation between V_{sz} and V_{s30}. Furthermore, r increases with z, indicating that the correlation between V_{sz} and V_{s30} increases with z increasing.

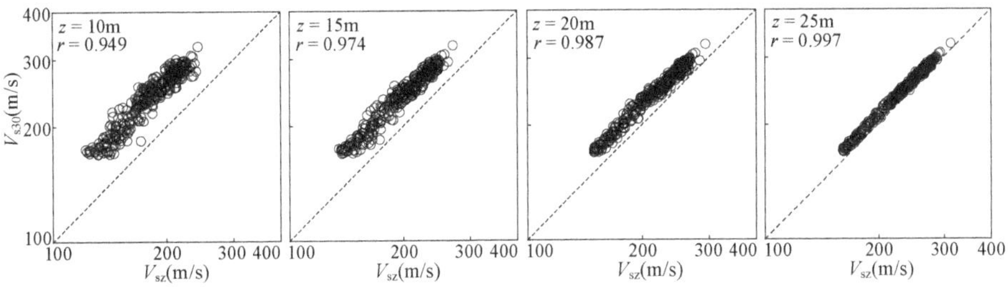

Figure 6. Correlation between V_{s30} and V_{sz}.

The above analysis shows that V_{sz} is strongly correlated with V_{s30} in the Tangshan area, so three new V_{s30} prediction models were established based on the data of shear-wave velocities measured in 343 boreholes across the city and compared with existing prediction models. Finally, the new models in this paper were prioritized.

(1) BCV model

The simplest model of predicting V_{s30} is to assume bottom constant velocity; therefore, we built the BCV prediction model for V_{s30} in Tangshan. Figure 7 demonstrates the comparison between $V_{s30est.}$ (predicted value of V_{s30} using the BCV model) and $V_{s30obv.}$ (measured value of V_{s30}) when z is 10, 15, 20, and 25 m. The dotted line represents $V_{s30est.} = V_{s30obv}$. As is shown in the figure, when z = 10 m, $V_{s30est.}$ is lower than $V_{s30obv.}$ With the constant increase in z, $V_{s30est.}$ is gradually close to $V_{s30obv.}$, indicating that the predicted results using the BCV model are gradually close to the measured values. This is because the BCV model assumes that $V_{s(z,30)}$ (equivalent shear-wave velocity from the bottom of boreholes to 30 m underground) is equal to $V_{s(z)}$; when z is small, $V_{s(z,30)}$ is also correspondingly small, resulting in small $V_{s30est.}$ With the gradual increase in z, $V_{s(z)}$ rises correspondingly; therefore, the predictive capacity of the model is enhanced. In addition, when the shear-wave velocity in the shallow parts of the boreholes is low or changes rapidly with z, the BCV model underestimates V_{s30} more obviously. As most borehole data used in this study are from regions with thick overburden in the North China Plain where shear-wave velocities obtained from boreholes in shallow strata are generally low, the phenomenon whereby the BCV model underestimates V_{s30} is significant when z is small.

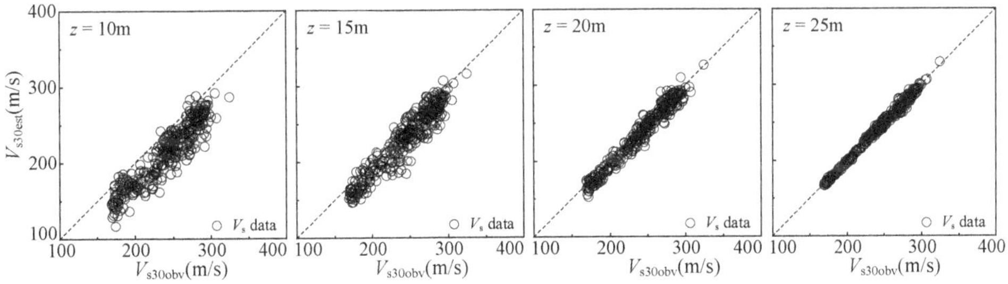

Figure 7. Comparisons between $V_{s30est.}$ obtained using the BCV model and the measured $V_{s30est.}$ at different depths.

(2) Linear prediction model

Generally, in a soil profile, the shear-wave velocity linearly increases with depth. Based on the data of shear-wave velocities in 343 boreholes deeper than 30 m in Tangshan, we built new V_{s30} linear prediction models using Formula (2). Table 1 presents the regression parameters of the prediction models when 10 m $\leq z \leq$ 29 m. In the models, a and b are regression coefficients; r represents Pearson's correlation coefficient between $V_{s30est.}$ and $V_{s30obv.}$ at different depths, and σ_{resi} is the standard deviation of the fitted residuals. It can be observed from Table 1 that when the depth gradually increases from 10 m to 29 m, r rises from 0.952 to 1, while σ_{resi} decreases from 0.022 to 0.01, indicating that the linear model shows good predictive capacity when z is small and the predicted results gradually approach the measured values with increasing z.

Table 1. Regression parameters of the V_{s30} linear prediction model (Formula (2)).

Depth (m)	a_0	a_1	r	σ_{resi}	Depth (m)	a_0	a_1	r	σ_{resi}
10	0.288	0.925	0.952	0.022	20	0.207	0.932	0.988	0.011
11	0.281	0.925	0.959	0.021	21	0.185	0.939	0.991	0.010
12	0.272	0.926	0.964	0.019	22	0.157	0.949	0.993	0.009
13	0.262	0.926	0.968	0.018	23	0.134	0.957	0.994	0.008
14	0.232	0.926	0.968	0.018	24	0.114	0.963	0.996	0.007
15	0.263	0.921	0.975	0.016	25	0.097	0.969	0.997	0.005
16	0.255	0.921	0.978	0.015	26	0.080	0.969	0.997	0.005
17	0.247	0.922	0.980	0.014	27	0.059	0.981	0.999	0.003
18	0.236	0.924	0.983	0.013	28	0.039	0.987	0.999	0.002
19	0.224	0.927	0.986	0.012	29	0.020	0.993	1.000	0.001

Figure 8 displays the comparisons of predicted results of V_{s30} using the Boore (2004) model [22], Boore (2011) model [23], Junju Xie (2016) model [20], and the new linear model built in this study when z = 10, 15, 20, and 25 m. As shown in Figure 8, at any depth, the Boore (2011) model overestimates V_{s30} in Tangshan. By comparing different V_{s30} prediction models based on shear-wave velocities measured in Greece, Stewart et al. [29] obtained similar conclusions because the measured V_{s30} in basic shear-wave velocity data from the boreholes used by Boore (2011) is generally large. However, the Boore (2004) model underestimates V_{s30} in Tangshan to some extent. These results demonstrate that shear-wave velocities in the region change faster with depth compared to those in California while changing slower than those in Japan. In addition, the boreholes selected in this study and by the Junju Xie (2016) model are located in the North China Plain, and the measured values of V_{s30} are relatively close, so the prediction results are also similar.

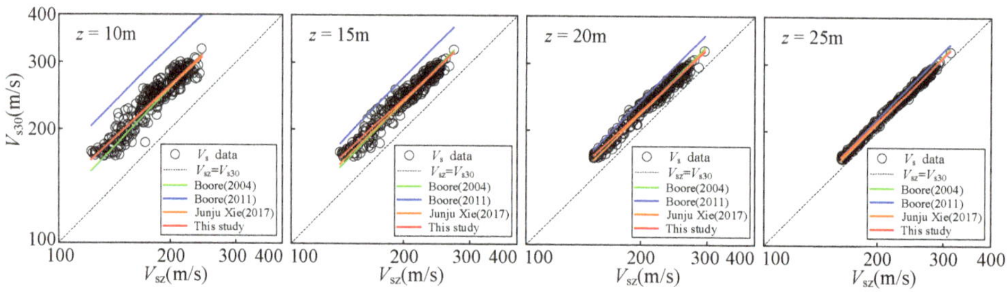

Figure 8. Comparisons of multiple prediction models at different depths.

(3) Conditional independent model

This model supposes that shear-wave velocity from the bottom of boreholes to 30 m underground is unrelated to that from zero to z m underground. In this paper, the corresponding relationship between $V_{s(z)}$ and $V_{s(z,30)}$ at 10 m $\leq z \leq$ 29 m is first analyzed using Formula (4). Table 2 shows the regression coefficients of $V_{s(z)}$ and $V_{s(z,30)}$. Figure 9 presents the relationship between $V_{s(z)}$ and $V_{s(z,30)}$ when z is 10, 15, 20, and 25 m, and the dotted line represents $V_{s(z,30)} = V_{s(z)}$. As is shown in Figure 9, (1) the convergence of data points of the measured shear-wave velocities at any depth is basically consistent, and (2) observing the fitting parameters of the prediction models, it can be found that the standard deviation of residuals of the prediction models at different depths is approximately 0.04, indicating that the influence of z on the linear relationship between $V_{s(z)}$ and $V_{s(z,30)}$ is limited, and

(3) some measured shear-wave velocities fall below the dotted line, that is, $V_{s(z)} > V_{s(z,30)}$, which is more prominent when $z > 10$ m. In general, shear-wave velocities of soil layers gradually will increase with the increase in burial depth, so $V_{s(z)}$ will be less than $V_{s(z,30)}$. However, due to the existence of soft soil, the shear-wave velocity of many boreholes in Tangshan decreases with burial depth. Figure 10 shows the schematic diagrams of shear-wave velocities as they change with depths shallower than 30 m in the No. 76 and No. 112 boreholes. It can be seen that shear-wave velocities decrease monotonically when the depths of the two boreholes increase from 25 m to 30 m, resulting in $V_{s(25)}$ exceeding $V_{s(25,30)}$.

Table 2. Regression parameters of the linear prediction model of $V_{s(z,30)}$.

Depth (m)	c_0	c_1	r	σ_{resi}	Depth (m)	c_0	c_1	r	σ_{resi}
10	0.938	0.643	0.830	0.043	20	0.919	0.642	0.804	0.042
11	0.947	0.637	0.838	0.042	21	0.798	0.690	0.818	0.041
12	0.937	0.641	0.827	0.043	22	0.751	0.708	0.812	0.042
13	0.959	0.631	0.813	0.044	23	0.772	0.699	0.821	0.042
14	1.134	0.558	0.794	0.045	24	0.843	0.670	0.815	0.041
15	1.124	0.561	0.777	0.046	25	0.827	0.677	0.820	0.040
16	1.106	0.569	0.766	0.047	26	0.767	0.701	0.803	0.042
17	1.090	0.575	0.762	0.047	27	0.698	0.728	0.815	0.040
18	0.951	0.631	0.800	0.043	28	0.466	0.818	0.839	0.039
19	0.927	0.639	0.801	0.043	29	0.267	0.896	0.886	0.036

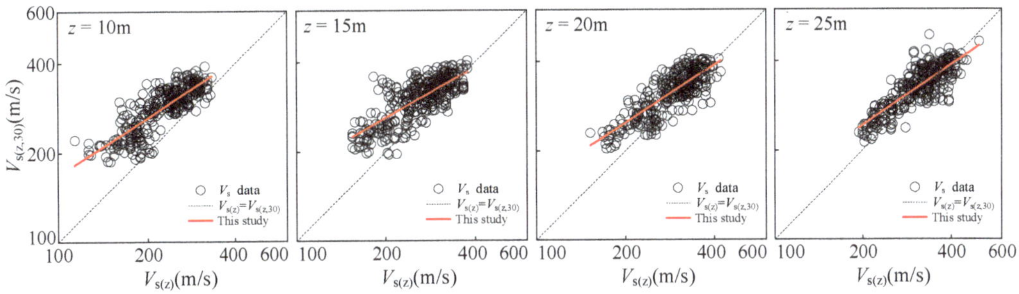

Figure 9. The relationship between $V_{s(z)}$ and $V_{s(z,30)}$ at different depths.

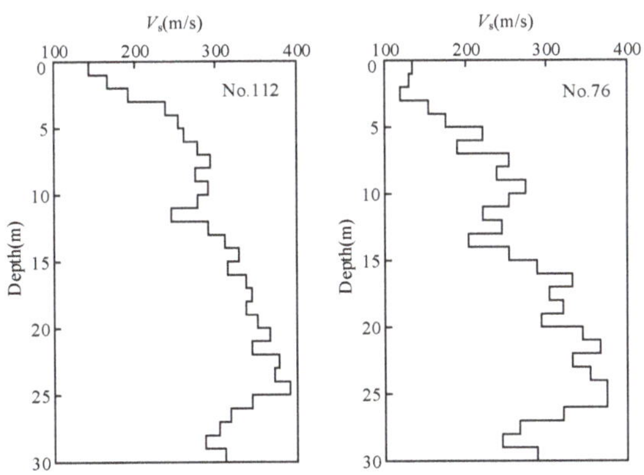

Figure 10. Changes in shear-wave velocities at depths of less than 30 m in some boreholes in Tangshan.

After obtaining the corresponding relationship between $V_{s(z)}$ and $V_{s(z,30)}$ at different depths, this paper predicted V_{s30} based on Formula (5). Figure 11 displays the comparison between $V_{s30eat.}$ from the new conditional independent model and $V_{s30obv.}$ when z values are 10, 15, 20, and 25 m. It can be seen from the figure that data points are uniformly distributed on both sides of the straight line, denoting $V_{s30est.} = V_{s30obv.}$ when $z = 10$ m. With the increase in z, the data points become more convergent, indicating that the prediction result of V_{s30} is better.

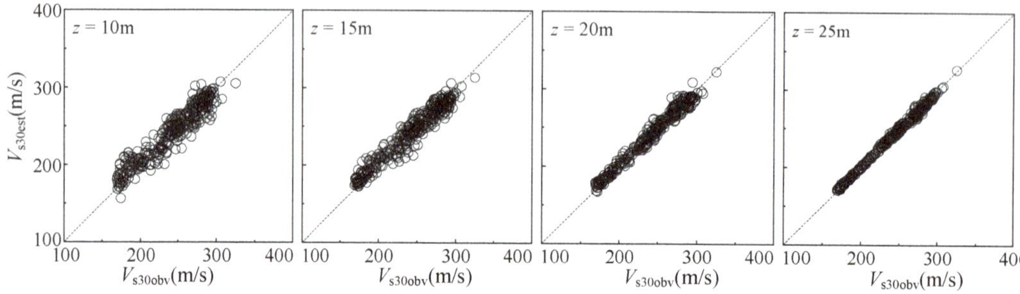

Figure 11. Comparisons between $V_{s30est.}$ based on the conditional independent model and $V_{s30obv.}$ at different depths.

6.2. Sorting of Goodness of Fit of the New Vs30 Prediction Models

As described above, three new V_{s30} prediction models, the BCV model, the linear model, and the conditional independent model, are established in this study, based on the data of shear-wave velocities of 343 boreholes deeper than 30 m in Tangshan.

Then, the LLH values were calculated at different depths using the three new V_{s30} prediction models, i.e., the BCV model, the linear model, and the conditional independent model (CIP model) (Figure 12).

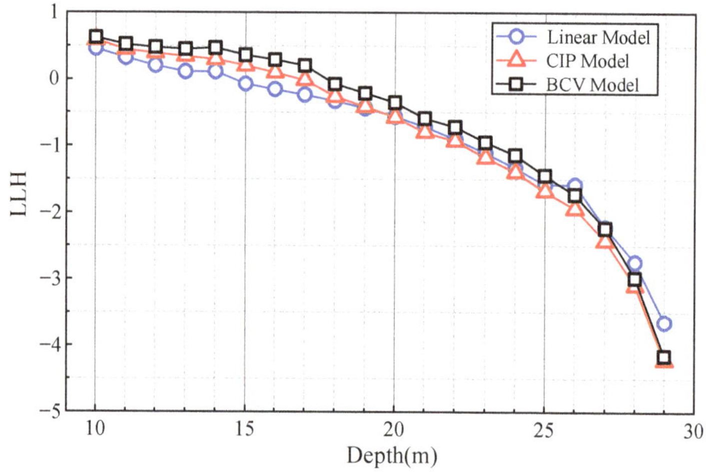

Figure 12. Comparisons of LLH of the three V_{s30} prediction models (the BCV model, the linear model, and the conditional independent model).

(1) It can be observed from the figure that when the depth z of the boreholes changes from 10 m to 25 m, the LLH value of the BCV model is greater than that of the other models. When z is not less than 26 m, the LLH value of the BCV model gradually decreases; that is, when $z \geq 26$ m, the predicted value of the BCV model approaches the measured value. This is because the BCV model assumes that shear-wave velocities from the bottom of the boreholes to 30 m underground are equal to that at the bottom. As z approaches 30 m, the better the prediction effect of this model is.

(2) The LLH value of the linear model is smaller than that of the other two models when $z < 19$ m, but it gradually increases and even becomes the largest when $z > 19$ m. This is mainly because the model can only perform simple linear estimation when predicting shear-wave velocities from z to 30 m underground, which cannot reflect the complex correspondence between the shear-wave velocity and the burial depth, especially in Tangshan where the thick Quaternary layer makes shear-wave velocities decrease with increasing burial depth.

(3) When $10 \text{ m} \leq z \leq 18$ m, the LLH value of the conditional independent model is greater than that of the linear model, but when $z > 18$ m, its LLH value is the minimum of the three models. It suggests that Formula (4) based on a Markov process can better predict $V_{s(z,30)}$ when z is large. In other words, when z is large, the difference between $V_{s(z)}$ and V_{sz} may be large, and the correlation between them is stronger.

In conclusion, it is suggested to use the linear model and the conditional independent model separately as V_{s30} prediction models in Tangshan when z is not larger than 18 m and exceeds 18 m.

7. Conclusions

When V_s measurements are not available in an area, proxy-based relationships can be used. By collecting the data from 343 boreholes with a depth greater than 30 m in Tangshan, this paper predicted V_{s30} first using a topographic slope with resolutions of 30 m and 90 m as an alternative indicator. Then, the three new V_{s30} prediction models were established based on the measured V_{sz} in the region. Finally, the goodness of fit of the new models was sorted. The main conclusions are as follows:

(1) Topographic slope is not suitable for the prediction of V_{s30} in Tangshan, mainly because of the limitations of the topographic slope method itself. Additionally, the variation of the topographic slopes in this region is smaller than that in hilly and mountainous areas. The slope calculated by DEM in urban regions is generally significantly affected by canopy effects. In addition, the relatively low V_{s30} of the borehole data collected in this study is one of the reasons.

(2) The Boore (2011) model established based on Japanese Kik-net data overestimates V_{s30} in Tangshan, while the Boore (2004) model based on data from California underestimates V_{s30} to some extent. This indicates that the gradient change of shear-wave velocities with depth in Tangshan is smaller than that in Japan and larger than that in California. In addition, the boreholes selected in this study and by the Junju Xie (2016) model are located in the North China Plain, and the measured values of V_{s30} are relatively close, so the prediction results are also similar. The prediction results between this study and the Junju Xie (2016) model are also similar because the boreholes selected are both located in the North China Plain. It also shows that the V_{s30} prediction models are regionally applicable.

(3) Three new prediction models (the BCV model, the linear model, and the conditional independent model) established in this study based on shear-wave velocity data from Tangshan all have a certain prediction ability. The BCV model has limited prediction ability at all depths. The linear model is suitable when $z \leq 18$ m, while the conditional independent model shows good predictive capacity when $z > 18$ m. Therefore, it is suggested to use the linear model and the conditional independent model separately as V_{s30} prediction models. A possible limitation of this study is that our models should be used with care because of the relatively small range of V_s values used.

Author Contributions: Conceptualization, Y.F. and H.L.; methodology, Y.F.; software, Y.F.; validation, Y.L. (Yu Li), Y.L. (Yuejun Lv) and Y.P.; investigation, Y.F.; resources, Y.F.; data curation, H.L.; writing—original draft preparation, Y.F.; writing—review and editing, Y.F. and G.C.; visualization, Y.L. (Yu Li); supervision, G.C.; project administration, Y.F.; funding acquisition, Y.P. All authors have read and agreed to the published version of the manuscript.

Funding: This research was funded by the National Key R&D Program of China (2022YFC3003503).

Institutional Review Board Statement: Not applicable.

Informed Consent Statement: Not applicable.

Data Availability Statement: The data used during the study are available from the first author by request.

Conflicts of Interest: The authors declare no conflict of interest.

References

1. Luo, L.; Qin, J.G.; Zhao, D.Z.; Wu, Z.W. Seismic Behavior of Extended End-Plate Connections Subjected to Cyclic Loading on the Top-Side of the Column. *Materials* **2020**, *13*, 3724. [CrossRef] [PubMed]
2. Wu, Q.; Wang, Z.F.; Qin, Y.; Yang, W.B. Intelligent model for dynamic shear modulus and damping ratio of undisturbed marine clay based on Back-Propagation neural network. *J. Mar. Sci. Eng.* **2023**, *11*, 249. [CrossRef]
3. Shi, K.; Zhang, M.Y.; Zhang, T.; Li, P.F.; Zhu, J.P.; Li, L. Seismic Performance of Steel Fiber Reinforced High–Strength Concrete Beam-Column Joints. *Materials* **2021**, *14*, 3235. [CrossRef] [PubMed]
4. Ding, W.; Jia, S.Z. Numerical Investigation of the Seismic Performance of Steel Frames with Energy-Dissipating Composite Walls. *Materials* **2022**, *15*, 828. [CrossRef]
5. Yin, Z.Y.; Sun, H.F.; Jing, L.P.; Dong, R. Geotechnical Seismic Isolation System Based on Rubber-Sand Mixtures for Rural Residence Buildings: Shaking Table Test. *Materials* **2022**, *15*, 7724. [CrossRef]
6. Abrahamson, N.A.; Silva, W.J. Summary of the Abrahamson & Silva NGA ground-motion relations. *Earthq. Spectra* **2008**, *24*, 67–97.
7. Zafarani, H.; Hassani, B.; Ansari, A. Estimation of earthquake parameters in the Alborz seismic zone, Iran using generalized inversion method. *Soil Dynam. Earthq. Eng.* **2012**, *42*, 197–218. [CrossRef]
8. Building Seismic Safety Council (BSSC). *NEHRP Recommended Seismic Provisions for New Buildings and Other Structures*; Building Seismic Safety Council, Federal Emergency Management Agency: Washington, DC, USA, 2015.
9. Silva, W.; Darragh, R.; Gregor, N.; Martin, G.; Abrahamson, N.; Kircher, C. *Reassessment of Site Coefficients and Near-Fault Factors for Building Code Provisions, Program Element II: 98-HQ-GR-1010*; Technical Report; Pacific Engineering and Analysis: El Cerrito, CA, USA, 1998.
10. Xu, J.-G.; Feng, D.-C.; Mangalathu, S.; Jeon, J.-S. Data-driven rapid damage evaluation for life-cycle seismic assessment of regional reinforced concrete bridges. *Earthq. Eng. Struct. D* **2022**, *51*, 2730–2751. [CrossRef]
11. Park, S.; Elrick, S. Prediction of shear wave velocities in southern California using surface geology. *Bull. Seismol. Soc. Am.* **1998**, *88*, 677–685. [CrossRef]
12. Wills, C.J.; Clahan, K.B. Developing a map of geologically defined site-condition categories for California. *Bull. Seismol. Soc. Am.* **2006**, *96*, 1483–1501. [CrossRef]
13. Wills, C.J.; Gutierrez, C.I.; Perez, F.G.; Branum, D.M. A next generation V_{s30} map for California based on geology and topography. *Bull. Seismol. Soc. Am.* **2015**, *105*, 3083–3091. [CrossRef]
14. Yong, A. Comparison of measured and proxy-based V_{s30} values in California. *Earthq. Spectra* **2015**, *32*, 171–192. [CrossRef]
15. Yong, A.; Hough, S.E.; Iwahashi, J.; Braverman, A. A terrain-based site-conditions map of California with implications for the contiguous United States. *Bull. Seismol. Soc. Am.* **2012**, *102*, 114–128. [CrossRef]
16. Wald, D.J.; Allen, T.I. Topographic slope as a proxy for seismic site conditions and amplification. *Bull. Seismol. Soc. Am.* **2007**, *97*, 1379–1395. [CrossRef]
17. Chiou, B.; Darragh, R.; Dregor, N.; Silva, W. NGA project strong-motion database. *Earthq. Spectra* **2008**, *24*, 23–44. [CrossRef]
18. Seyhan, E.; Stewart, J.P.; Ancheta, T.D.; Darragh, R.B.; Graves, R.W. NGA-West2 site database. *Earthq. Spectra* **2014**, *30*, 1007–1024. [CrossRef]
19. Thompson, E.M.; Wald, D.J.; Worden, C.B. A V_{s30} map for California with geologic and topographic constraints. *Bull. Seismol. Soc. Am.* **2014**, *104*, 2313–2321. [CrossRef]
20. Xie, J.J.; Zimmaro, P.; Li, X.J.; Wen, Z.P.; Song, Y.S.H. V_{S30} Empirical Prediction Relationships Based on a New Soil-Profile Database for the Beijing Plain Area, China. *Bull. Seismol. Soc. Am.* **2016**, *97*, 2843–2854. [CrossRef]
21. Kuo, C.H.; Wen, K.L.; Hsieh, H.H.; Chang, T.M.; Lin, C.M.; Chen, C.T. Evaluating empirical regression equations for VS and estimatingV_{S30} in northeastern Taiwan. *Soil Dynam. Earthq. Eng.* **2011**, *31*, 431–439. [CrossRef]
22. Boore, D.M. Estimating V_{S30} (or NEHRP site classes) from shallow velocity models (depths < 30 m). *Bull. Seismol. Soc. Am.* **2004**, *94*, 591–597.
23. Boore, D.M.; Thompson, E.M.; Cade, H. Regional correlations of V_{S30} and velocities averaged over depths less than and greater than30 meters. *Bull. Seismol. Soc. Am.* **2011**, *101*, 3046–3059. [CrossRef]
24. Wang, H.Y.; Wang, S.Y. A New Method for Estimating V_{S30} from a Shallow Shear-wave velocity Profile (Depth < 30 m). *Bull. Seismol. Soc. Am.* **2015**, *105*, 1359–1370.
25. Dai, Z.; Li, X.; Hou, C. A shear-wave velocity model for V_{S30} estimation based on a conditional independence property. *Bull. Seismol. Soc. Am.* **2013**, *103*, 3354–3361. [CrossRef]

26. He, W.M.; Liu, M.J.; Yang, J. Statistical analysis and application of the relationship between shear-wave velocity and buried depth of soil layers. *Seismol. Geol.* **2016**, *38*, 937–949. (In Chinese)
27. Burnham, K.P.; Anderson, D.R. *Model Selection and Multimodel Inference: A Practical Information-Theoretic Approach*, 2nd ed.; Springer: New York, NY, USA, 2002.
28. Shafiee, A.H.; Hamid, Z.; Mojtaba, J. Model Selection for Correlating V_{S30} with Average Shear-Wave Velocities at Lower Depths Based on the Iranian Data. *Bull. Seismol. Soc. Am.* **2016**, *106*, 289–299. [CrossRef]
29. Stewart, J.P.; Klimis, N.; Savvaidis, A.; Theodoulidis, N.; Zargli, E.; Athanasopoulos, G.; Pelekis, P.; Mylonakis, G.; Marga, B. Compilation of a Local VS Profile Database and Its Application for Inference of V_{S30} from Geologic-and Terrain-Based Proxies. *Bull. Seismol. Soc. Am.* **2014**, *104*, 2827–2841. [CrossRef]

sustainability

Article

Preliminary Analysis of Coseismic Landslides Induced by the 1 June 2022 Ms 6.1 Lushan Earthquake, China

Xiaoyi Shao [1,2,3], Chong Xu [1,2,3,*] and Siyuan Ma [4]

1 National Institute of Natural Hazards, Ministry of Emergency Management of China, Beijing 100085, China
2 Key Laboratory of Compound and Chained Natural Hazards Dynamics, Ministry of Emergency Management of China, Beijing 100085, China
3 Key Laboratory of Landslide Risk Early-Warning and Control, Ministry of Emergency Management of China, Chengdu 610059, China
4 Institute of Geology, China Earthquake Administration, Beijing 100029, China
* Correspondence: chongxu@ninhm.ac.cn or xc11111111@126.com

Abstract: At 17:00 (UTC+8) on 1 June 2022, an Ms 6.1 reverse earthquake struck Lushan County, Ya'an City, Sichuan Province. This earthquake event had a focal depth of 10 km and the epicenter was located at 30.37° N and 102.94° E. The purpose of this study is to document a comprehensive coseismic landslide inventory for this event and analyze the distribution pattern and factors controlling the landslides. After careful visual interpretations, this quake event was determined to have in total triggered about 2352 landslides in an area of 3900 km^2, including both shallow disrupted landslides and collapses, for which the spatial distribution was statistically related to regional topography, geology, and seismicity. Notably, a vast majority of the landslides were located on the NW plate of the seismogenic fault, and were distributed in the area with a seismic intensity of VII. In addition, coseismic landslides were more likely to appear in areas with high altitude, relief, and large slope. The landslide area density (LAD) increased with an increase in the above factors and is explained by an exponential relationship, indicating that the occurrence of coseismic landslides in this area was more easily affected by topographic factors than seismic factors. Most small-scale landslides were clustered in the ridge area, which shows the seismic amplification effects of mountain slopes. Due to the impact of seismic wave propagation direction, hillslopes facing northeast-east (NE-E) were more prone to collapse than southwest-facing ones. Based on the distribution pattern of the landslides, we suggest that the seismogenic fault of this event was NW dipping. These findings indicate that it is effective to identify the dipping of seismogenic faults using the spatial distribution pattern of coseismic landslides.

Keywords: coseismic landslide; Lushan earthquake; landslide inventory; spatial analysis; influencing factors

Citation: Shao, X.; Xu, C.; Ma, S. Preliminary Analysis of Coseismic Landslides Induced by the 1 June 2022 Ms 6.1 Lushan Earthquake, China. *Sustainability* **2022**, *14*, 16554. https://doi.org/10.3390/su142416554

Academic Editor: Antonio Miguel Martínez-Graña

Received: 18 October 2022
Accepted: 1 December 2022
Published: 9 December 2022

1. Introduction

Strong earthquakes often trigger massive geological disasters in mountainous areas, causing grievous casualties and economic losses [1–5]. Coseismic landslides are considered one of the most destructive seismic secondary disasters, often characterized by large quantity and scale, wide distribution, and long duration of post-quake effects [5–9]. In recent years, earthquakes have happened frequently in the mainland of China, especially in the eastern Tibet Plateau, such as the 2008 Ms 8.0 Wenchuan earthquake, the 2013 Ms 7.0 Lushan earthquake, the 2017 Ms 7.0 Jiuzhaigou earthquake, etc. Among them, the 2008 Wenchuan earthquake was the most catastrophic and triggered nearly 200,000 landslides in an area of more than 44,000 km^2 [10,11]; these landslides directly caused about 20,000 fatalities, accounting for about 30% of the total deaths triggered by the earthquake [4,12,13]. The 2013 Lushan earthquake in the same active fault zone also induced a large number of geological disasters, severely limiting local socioeconomic development [6,14]. The 2017 Jiuzhaigou earthquake triggered about 5000 coseismic landslides, causing serious damage to Jiuzhaigou tourist landscapes like Panda Sea, as well as a significant influence on the local natural and ecological environment [7,15,16].

A coseismic landslide inventory is the foundation for conducting multiple types of relevant analyses, including landslide occurrence, distribution pattern, and the corresponding influencing factors. These types of useful landslide information can provide enormous support for enhancing seismic disaster mitigation and prevention, as well as understanding the disaster chain and regional landscape evolution [9,17–20]. At the same time, a complete and detailed landslide mapping is also an important basis for landslide susceptibility and hazard assessment [21,22].

At 17:00 CST (UTC+8) on 1 June 2022, an Ms 6.1 earthquake occurred in Lushan County (30.37° N, 102.94° E), Ya'an, Sichuan area, with a focal depth of 17 km. The seismogenic fault was a branch fault of the Shuangshi–Dachuan fault, with a thrust focal mechanism. The earthquake had a maximum intensity of VIII and was 9 km away from the location of the Lushan earthquake on April 20, 2013. In this event, at least four people were killed, 42 were injured (31 in Baoxing County and 11 in Lushan County) and a total of 14,427 people were severely affected. The earthquake triggered a larg number of landslides, which attracted extensive attention. Since it is difficult to obtain clear satellite remote sensing images within 72 h after an earthquake, Fan et al. [23] employed a near real-time assessment model to forecast occurrence probability of the earthquake-induced landslides. The results showed that the highly landslide-prone areas were mainly distributed within 15 km of the epicenter and the seismogenic fault. For different earthquake events, due to the multiple influences of topography, earthquake magnitude, properties of seismogenic faults, and lithology, the main controlling factors of landslides are quite different [24,25]. Thus, details of coseismic landslides should be documented for this earthquake for analyzing distribution patterns and influencing factors of coseismic landslides.

To identify the landslide characteristics induced by the Lushan earthquake, a detailed and complete database of coseismic landslides is required. Thus, the purpose of this study is to establish a landslide inventory for this earthquake event. High resolution (3 m) remote-sensing images were used to interpret the seismic landslides based on visual interpretation, in order to obtain a detailed coseismic landslide panorama. Particularly, we analyzed the distribution pattern and its relationship with geology, topography, and seismicity. This study can provide sufficient background information for post-disaster reconstruction and help deepen the understanding of the occurrence laws of coseismic landslides. Moreover, this study is conducive to in-depth analyses of the seismic disaster chain and evolution, as well as the evolution of rivers and landforms in the earthquake area.

2. Tectonic Setting of the Study Area

The collision between the Indian and Eurasian plates causes the uplift of the Tibetan plateau; as a result, a sequence of blocks in this highland move eastward [26]. The Longmenshan fault zone is essential for understanding the Tibetan Plateau's uplifting mechanism because the eastern extrusion of the Tibetan Plateau, which controls its formation, is closely related to the formation of the Longmenshan Mountains. Attributed to the obstruction of the Sichuan Basin, the Longmenshan thrust area is developed at the border between the Tibetan Plateau and Sichuan Basin [27], where the accumulated strain is released during an earthquake occurrence. The 2008 Wenchuan and 2013 Lushan earthquakes both occurred on this NE-trending fault zone. The epicenter of the 2013 Lushan seismic event was approximately 85 km southwest of the 2008 Wenchuan earthquake (Figure 1). The Lushan earthquake occurred in the southwestern segment of the Longmenshan thrust zone, whereas the Wenchuan earthquake was located in the center of the belt. Previous studies [28,29] suggested that the coseismic coulomb stress loading of both of the earthquakes increased the likelihood of future earthquakes.

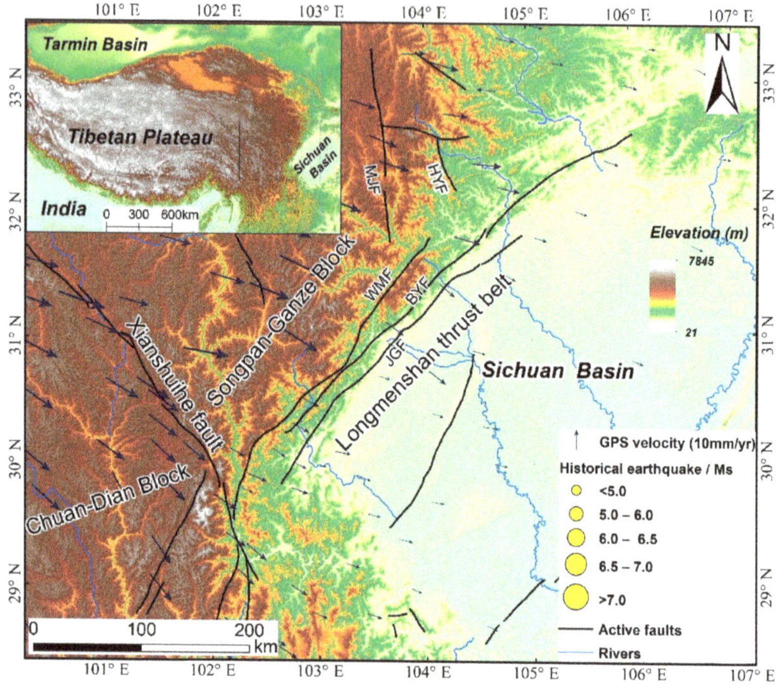

Figure 1. Map showing the active faults, historical earthquakes, and topography of the eastern margin of Tibetan Plateau. JGF = Jiangyou–Guanxian fault, WMF = Wenchuan–Maoxian fault, BYF = Beichuan–Yingxiu fault, HYF = Huya fault, MJF = Minjiang fault The active fault lines are derived from [30]; the GPS velocity during 1998–2014 is obtained from [31].

On 1 June 2022, a strong earthquake hit Lushan County in Ya'an, Sichuan Province, China. The earthquake occurred at around 17:00 p.m. local time. According to the information provided by US Geological Survey (USGS), this event had a moment magnitude of (Mw) 5.8 and a focal depth of 10 km. The China Earthquake Networks Center (CENC) also reported that the earthquake had a surface-wave magnitude (Ms) of 6.1 and a focal depth of 17 km. A smaller magnitude 4.5 earthquake occurred three minutes later. The main shock was felt over an estimated radial area of 3900 km^2 and was suggested to have an intensity of at least VI. Notably, the Taiping, Shuangshi, Lingguan, and Muping regions had the highest intensity, reaching VIII.

Based on the influence range of seismic intensity from CENC, an elliptical region with seismic intensity greater than VI was selected as the study area, which is 3900 km^2 and spans from 30°00′ N to 30°40′ N of latitude and from 102°35′ to 103°25′ E of longitude (Figure 2). Notably, the northwest region of the study area is characterized by high mountains, while the southeast is almost lowland. The elevation ranges from 500 to 3650 m, with an average elevation of 1280 m. The study area has a subtropical monsoon climate with an average precipitation of 1300 mm, and the highest rainfall is concentrated between June and September. The strata range from Quaternary to Precambrian, with lithology dominated by mudstone, sandstone, limestone, and shale. Simultaneously, abundant granite bodies are formed in this area. Additionally, the active structure and climate have resulted in well-developed rock joints and fissures, as well as strong weathering [32].

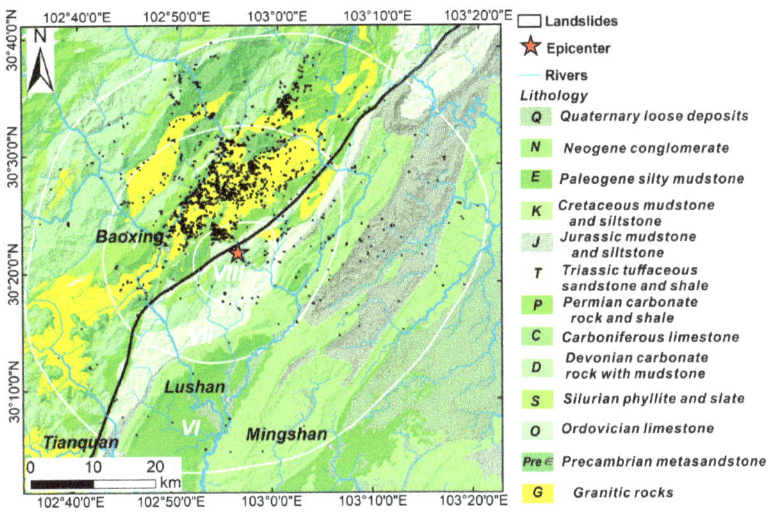

Figure 2. The lithology distribution of Lushan area obtained from 1:200,000 published geological maps from China Geological Survey.

3. Data and Methods

3.1. Data

The high-resolution satellite images both after and before the earthquake allowed us to conduct a complete visual interpretation of the coseismic landslides. On this basis, we successfully completed the landslide inventory covering the entire affected area based on high-resolution optical satellite images (Planet) with 3 m resolution acquired on 6 July 2022. To prevent the pre-earthquake landslides from being erroneously identified as coseismic landslides, a series of pre-quake remote-sensing images were derived from Planet images for May, and these images were subsequently spliced and fused using QGIS.

To analyze the impact of terrain, geology, and seismicity on the distribution of coseismic landslides, we collected the relevant data regarding the three aspects. The elevation data were obtained from ALOS PALSAR DEM with 12.5 m resolution and then resampled to a resolution of 5 m using a linear interpolation algorithm (Figure 3a). The slope angle, aspect, and topographic relief were obtained from the elevation data (Figure 3b–d). The distances from the centroid of every grid cell to the seismogenic fault were calculated by ArcGIS software. The distribution of Peak Ground Acceleration (PGA) was downloaded from the USGS, and then the raster format of the PGA was obtained by the Kriging interpolation method. In addition, we estimated the topographic relief from the altitude range within a 2.5 km radius [33]. The stratigraphic data were acquired from the published 1:2,000,000 geological map of the China Geological Survey (http://dcc.cgs.gov.cn/ (accessed on 9 October 2022)). We divided the lithology into 13 categories according to stratigraphic chronology in order of youngest to oldest, which are Quaternary (Q), Neogene (N), Paleogene (E), Cretaceous (K), Jurassic (J), Triassic (T), Permian (P), Carboniferous (C), Devonian (D), Silurian (S), Ordovician (O), Precambrian (Pre∈), and Granitic rocks (G) (Figure 2). All influencing factors were then resampled to have a 5 m resolution.

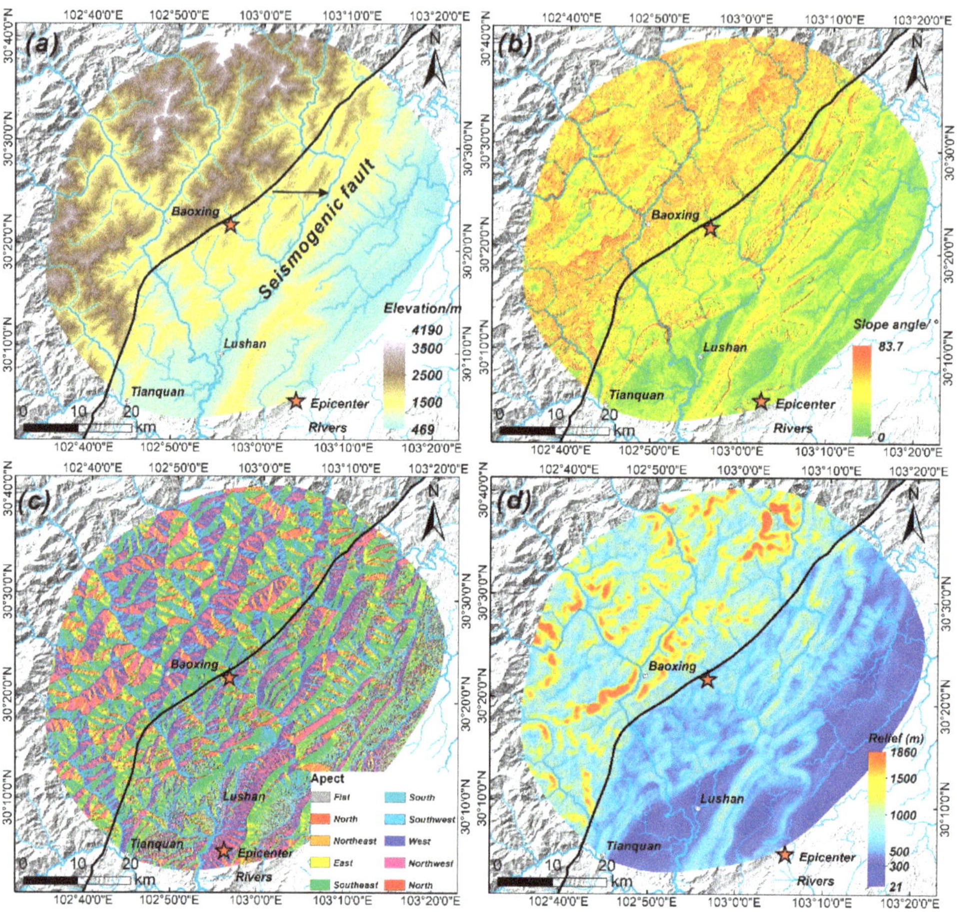

Figure 3. Spatial distribution of topographic factors: (**a**) elevation; (**b**) slope angle; (**c**) slope aspect; (**d**) topographic relief.

3.2. Method

Coseismic landslides affect different portions of slopes unevenly [16,34]. In this study, we calculated the positions of coseismic landslides with respect to the ridge and the valley of the slope. The entire length of the slope, as proposed by Meunier, Hovius, and Haines [34], was used to standardize the measured distances by Equation (1).

$$|d_{st,top}| = \frac{d_{st,top}}{d_{st} + d_{top}}. \quad (1)$$

where the value of ranges from 0 (i.e., a grid situated on a mountain ridge) to 1 (i.e., a river channel). The opposite variation is seen in d_{st}.

The spatial distribution of landslides and landscape area for each influencing factor was described by frequency density, which can also express the occurrence frequency of landslides for different influence factors. The larger the value, the higher the occurrence frequency of landslide or landscape in this interval (Equation (2)).

$$\text{Frequency density} = \frac{\frac{A_{i\_ls}}{A_{i\_total}}}{A_{i\_inter}} * 100\% \quad (2)$$

where i represents landslide type (landslides, landscape area); A_{i_ls} is the i's area of each class; A_{i_total} represents the total area of the i-th influencing factor; and A_{i_inter} represents the i-th classification interval.

We also calculated the landslide area density (LAD) and landslide number density (LND). The LND and LAD represent the density of landslide distribution and the scale of landslide development in the specific area, respectively. The two parameters are defined as the landslide number per square kilometer (number/km^2) and the proportion of the area that has experienced landslides (%), respectively. The corresponding parameters can be calculated by Equations (3) and (4).

$$\text{LAD} = \frac{landslide\ area\ of\ the\ class}{total\ landslides\ area} \quad (3)$$

$$\text{LND} = \frac{Landslide\ number\ of\ the\ class}{total\ landslides\ area} \quad (4)$$

The flow chart of this work is shown in Figure 4.

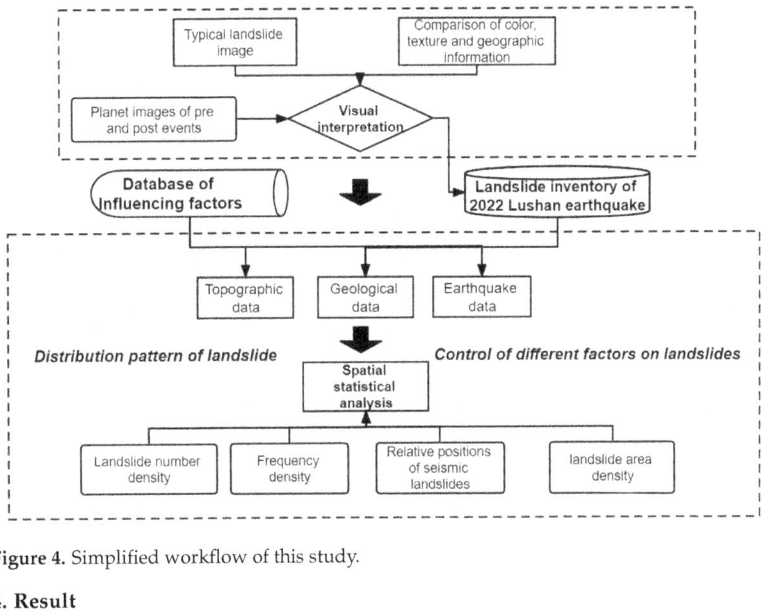

Figure 4. Simplified workflow of this study.

4. Result

4.1. Coseismic Landslide Inventory

The 2022 Lushan earthquake triggered in total 2352 landslides over an area of 3900 km^2 (Figure 5). These landslides included massive shallow disrupted landslides and collapses, and most landslides were concentrated in the area to the northwest of the seismogenic fault. The landslides had an area of 5.51 km^2, and the average area of each landslide was 2300 m^2. The area of these landslides had a concentration in the 600–2500 m^2 range, accounting for half of the total number (Figure 6). Furthermore, 234 landslides were larger than 5000 m^2, making up 12.8% of the total, while 69 landslides were larger than 10,000 m^2.

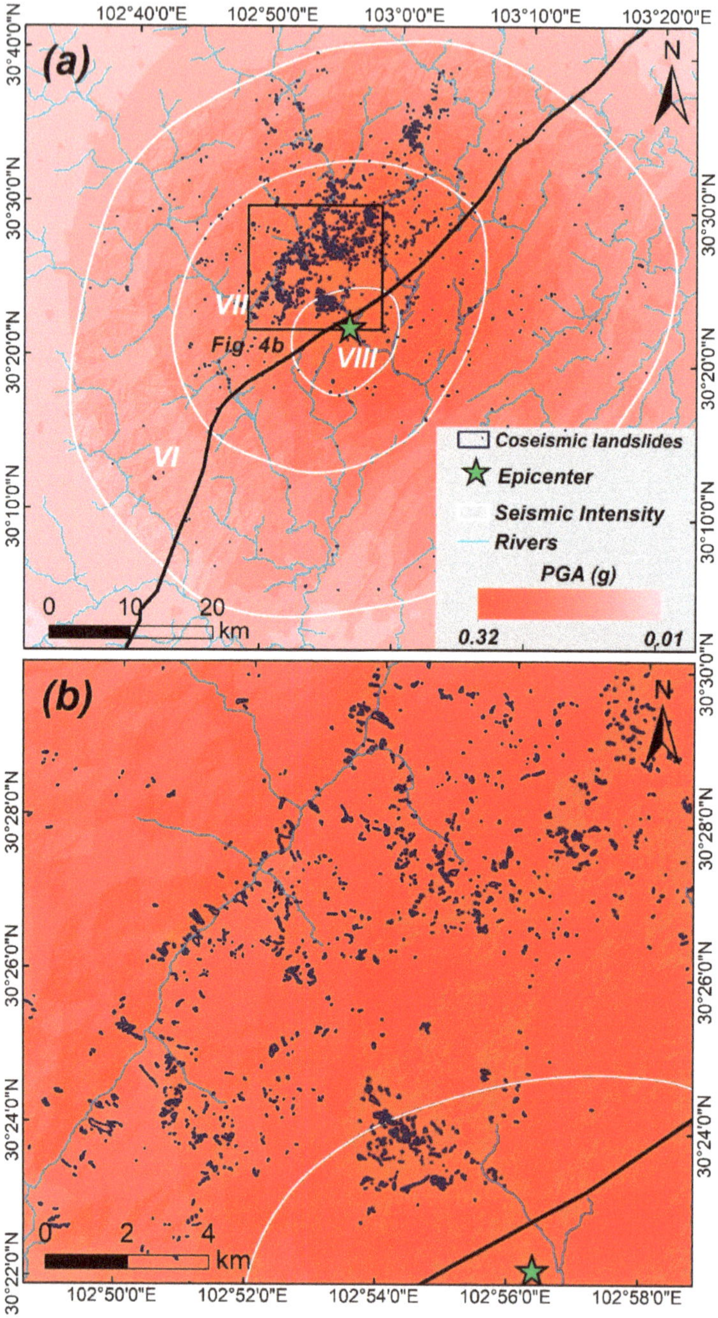

Figure 5. Coseismic landslide inventory for the Ms 6.1 Lushan earthquake; the white line is seismic intensity downloaded from CENC. (**a**) Panorama of landslide distribution. (**b**) Enlarged view of landslides.

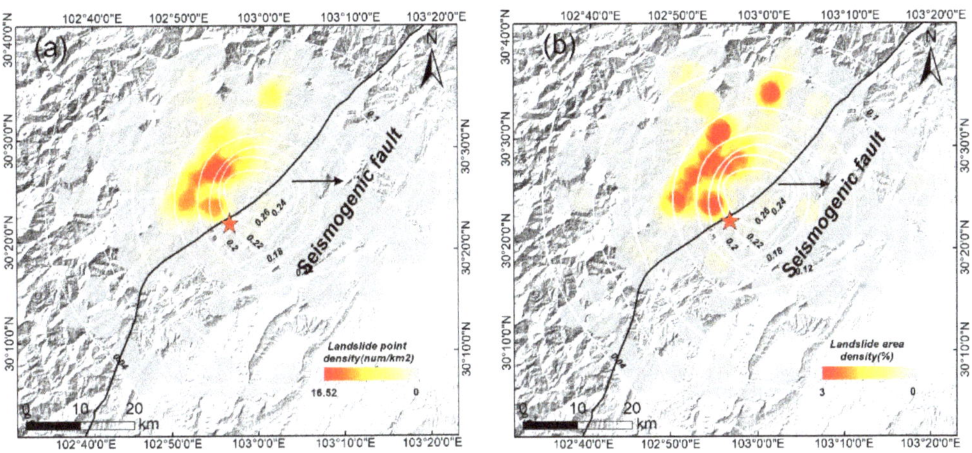

Figure 6. Spatial distribution of LND and LAD of 2022 Lushan earthquake. (**a**) LND; (**b**) LAD. The black solid line represents the seismogenic fault. The red star marks the epicenter.

We calculated the LND and LAD within a 2.5 km moving window by a Gaussian kernel density estimator. The spatial distribution of LND and LAD indicates that the landslides were primarily clustered in the northwest area of Baoxing County (Figure 6). The largest LND reaches 16.52/km^2, and the maximum LAD is 3%, which appears in the PGA range of 0.2–0.22 g.

4.2. Influencing Factor Analysis

In order to investigate the correlation between various factors and landslide occurrence, we computed the frequency distribution of landslide and non-landslide areas, as well as the LAD for different influence factors. Figure 7 shows the frequency density of both landsliding and non-landsliding (landscape) areas for four different influencing factors, and Figure 8 shows the LAD for influencing factors in different intervals. In terms of elevation, the landscape area was clustered between 500~1500 m, while landslides were most prone to occur in the elevation range between 1300 and 2500 m. Overall, the LAD increases with the increase of elevation and is described by an exponential relationship of $y = 0.011e^{0.0015*x}$ (where x is the elevation (m) and y is the LAD (%)). As for slope angle, the majority of the landscape area was affected by slopes with a slope angle of 15° to 30°, while most of the landslides were clustered between 40° and 50°, indicating that landslides were more frequent in areas with larger slope angle. Similarly, we found that with the increase of slope, the LAD also increased and showed a good exponential relationship of $y = 0.0159e^{0.0707*x}$ (where x is the slope gradient (°) and y is the LAD (%)). As for relief, the tendency of frequency density in relief follows the same pattern as the slope angle. Most of the coseismic landslides were situated in a high relief area (800–1200 m), while landscape areas are mainly seen in a low relief area from 500 to 800 m (Figure 7c). The LAD and relief have a positive exponential relationship (Figure 8c). For the distance to rivers, with the increase in distance, the frequency density of landslides and landscape area rapidly drops (Figure 8d), and the LAD and distance to rivers have a positive exponential relationship of $y = 0.0579e^{0.0021*x}$ (where x is the distance to rivers (m) and y is the LAD (%)). These results indicate that the landslides were prone to occur in locations distant from the rivers.

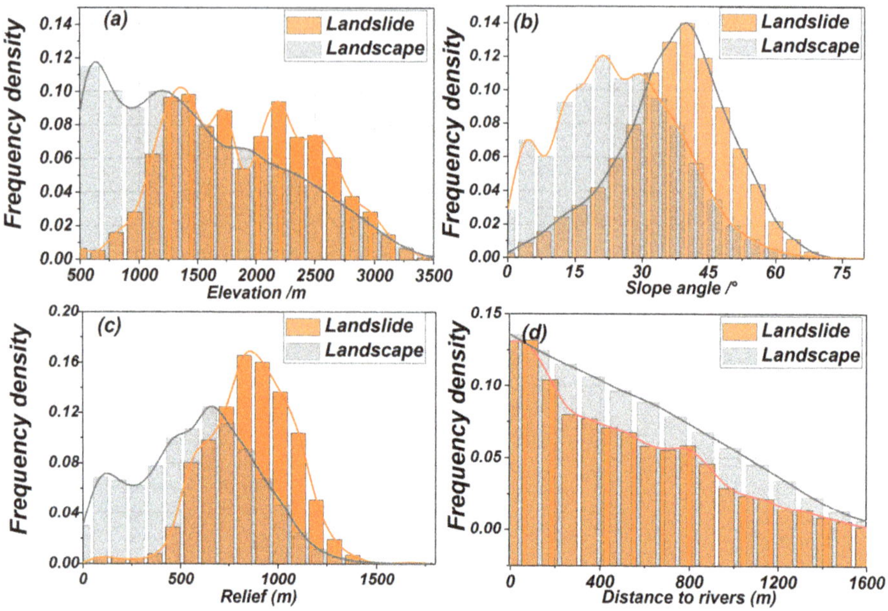

Figure 7. Frequency density estimates of landsliding and non-landsliding areas for different factors; (**a**) elevation; (**b**) slope gradient; (**c**) terrain relief; (**d**) distance to rivers.

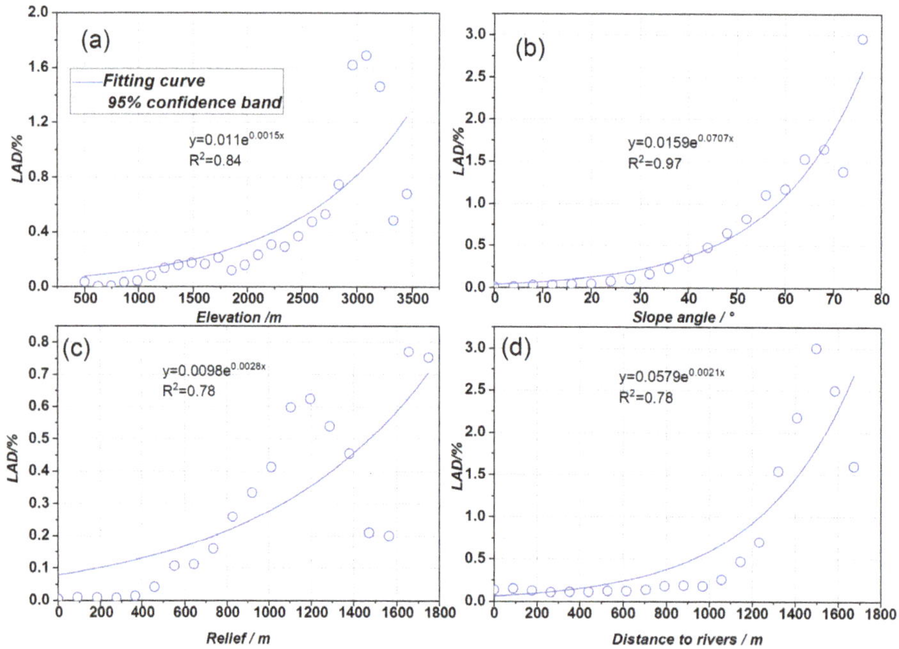

Figure 8. The relationship between the factors and the LAD; (**a**) elevation; (**b**) slope angle; (**c**) relief; (**d**) distance to rivers.

Figure 9 depicts the distribution of areal coverage in landsliding and non-landsliding areas, LAD and average landslide area. Among them, Cretaceous (K) and Granitic (G) rocks are the most widely distributed, accounting for 19.9% and 13.5% of the total area, respectively. In addition, 59.7% of coseismic landslides occurred in Granitic rock (G). The Permian (P) was the second most prevalent lithological unit connected with landslides, accounting for 6% of the total area. The LAD of Granitic rock (G) was the highest, accounting for 0.62%, indicating that slope failures occurred mainly in the Granitic rock area during the earthquake.

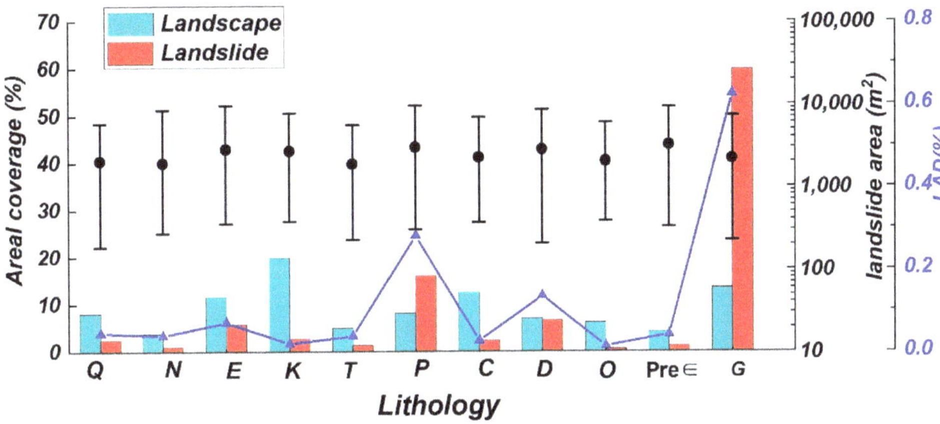

Figure 9. Distribution of areal coverage (%) LAD and mean landslide area for landslide and non-landslide areas, in different lithologic units.

The association between the landslides and the PGA distribution is displayed in Figure 10, in which 28% of coseismic landslides occurred in 0.2~0.24 g. As the PGA increases, the LAD and LND increase first and subsequently drop. The LND and LAD reach their maximums in the PGA range of 0.16–0.2 g, peaking at 3.9/km^2 and 0.86% respectively.

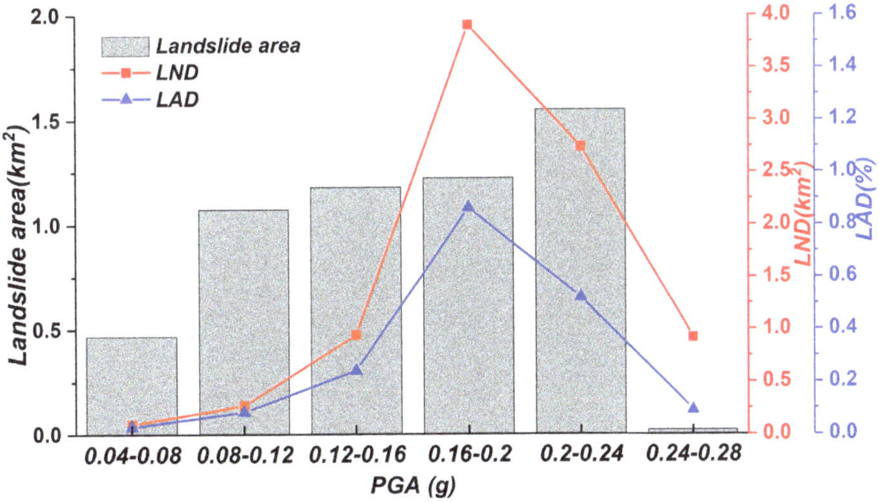

Figure 10. Classifications of PGA and its relationship with LND and LAP (right).

Coseismic landslides are controlled by seismogenic faults in earthquake events [33,35–37]. We constructed 60 bins at 1 km intervals along the strike of the seismogenic fault to analyze the relationships between landslide abundance and distance to the seismogenic fault. As shown in Figure 11, the abundance of landslides to the NW of the seismogenic fault was much higher than that to the SE of the fault. This 31 km wide NW plate recorded the most coseismic landslides with a number of 2162 and an area of 5.03 km^2, which account for 91.9% and 91.2% of the total, respectively. The largest landslide abundance was observed in the region 8 km away from the seismogenic fault, with LND and LAD values of 3/km^2 and 0.79 %, respectively.

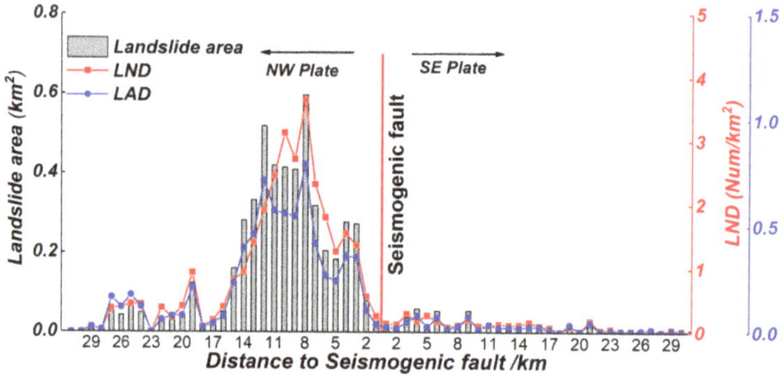

Figure 11. Correlations between landslide abundance proxies (LND and LAD) and distance to the seismogenic fault.

We plotted the landslide density, area, and relative distance to the ridge and river on a fictitious slope profile to visualize the position of each landslide (Figure 12). The results show that nearly 60% of the landslides (as indicated by the red spot at the apex of the simulated slope) occurred in the ridge (relative distance to the top less than 0.3) and were located far from the rivers (relative distance to the stream greater than 0.6), with the remainder occurring in the middle of the slope. Massive small-scale landslides, in particular, were concentrated in the ridge area.

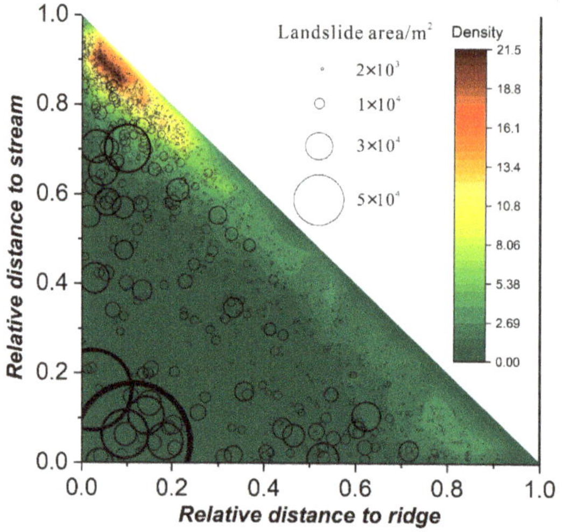

Figure 12. Coseismic landslides' location in relation to the ridge crest and streams. The relative LND is represented by different background colors. Green indicates low density while red represents high density.

Figure 13 depicts the relationship between landslides and slope aspect. Figure 13a characterizes the frequency density distribution of landslide and landscape zones for various aspects. The landscape area was distributed in all aspects evenly, but the frequency densities in the 280–310° (northwest) and 90–130° (southeast) aspects were slightly higher than those in other aspects. For landslide distribution, the frequency densities of landslide areas were remarkably high in the 30–90°and 280–310° aspects. The results for LAD show that the peak LAD (0.25%) was present in the aspects from 30 to 90° (NE to N), illustrating that slopes with aspects in this range were more prone to landslides.

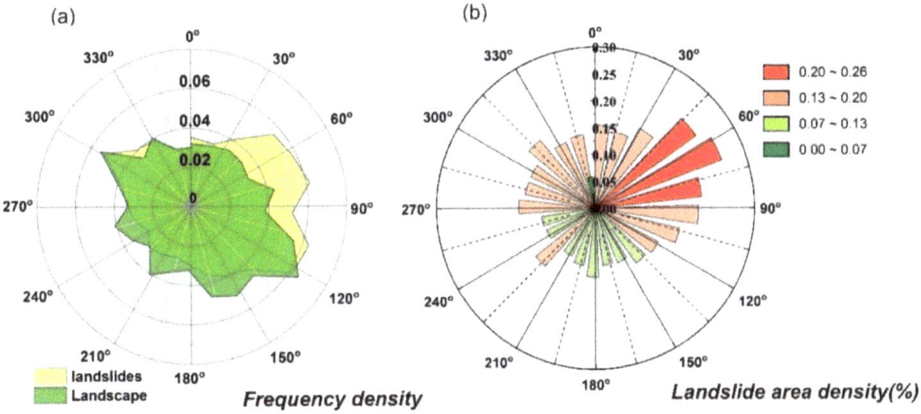

Figure 13. (**a**) The aspect distribution for landslides and non-landsliding areas; (**b**) relation between LAD and aspect.

5. Discussion

The intensity of ground motion is affected not only by the earthquake itself (magnitude, focal depth, etc.) [38], but also by the medium where seismic waves propagate (rock type, fractures, etc.) [39], and the topographic conditions (slope angle, aspect, elevation, etc.) [40]. Based on the interaction between the coseismic landslides and the slope aspect of these events, we can observe that the landslide distribution has a clear directional effect, known as the "back slope effect" [41,42]. In detail, slopes facing the epicenter are less likely to collapse than those parallel to the seismic wave direction. According to the wave propagation direction of the seismic wave of this earthquake event, the southwest (SW) slopes faced the epicenter, while northeast (NE) was a back slope. Figure 13 show that northeast-east (NE-E)-aspect hillslopes were more prone to collapse than the southwest-west (SW-W)-aspect ones, that is, the number of landslides on the side of the back slope where the seismic wave propagates was greater than that on the side of the facing slope. Based on the stress wave theory, Tang et al. [43] proposed that the "back slope effect" was caused by the multiplication of reflected stretching waves generated by compressional waves when they encountered the slope free surface. According to the statistical results of the lithological data, more than 60% of the landslides occurred in Granitic rock (G) due to the reduced rock strength that is caused by the broadly developed rock joint fissures associated with frequent tectonic activity in this area [6,37,44]. In addition, the slope of Granitic rock in the Lushan area is generally steep, making it easy to collapse or slide under the action of ground motions.

Coseismic landslides are significantly controlled by active faults, and their distribution patterns around seismogenic faults with different movement properties are different [6,45]. For example, the landslides triggered by thrust earthquakes are prone to occur in the hanging wall of the seismogenic fault, and the attenuation rate of the landslides in the hanging wall is significantly lower than that in the footwall with the increase in the distance from the fault [10,46]. However, for strike-slip earthquakes, there is little difference in the development of landslides on both sides of the fault, and landslides are often densely distributed along the seismogenic fault [24,35,45]. The landslides caused by the 1999 Chichi earthquake were mostly concentrated in the hanging wall, while there were fewer landslides in the footwall [47]. Other reverse earthquake events, such as the 2002 Denali fault earthquake in Alaska [48] and the 1994 Northridge earthquake in the United States [49], also show that seismic landslides are primarily distributed in the hanging wall, that is, the abundance mutation characteristics of landslides across reverse seismogenic faults. The seismogenic fault of this earthquake event was a branch fault of the Shuangshi–Dachuan fault. The focal mechanism solution and precise location of the aftershocks reveals that the seismogenic fault was an SE-trending thrust fault (https://www.cea-igp.ac.cn/kydt/279024.html (accessed on 9 October 2022)). However, based on the spatial distribution of the coseismic landslides, 90% of the landslides developed in the NW plate of the seismogenic fault, while few landslides developed in the SE plate. Such observations may indicate that the seismogenic fault has an NW dip. Meanwhile, the Granitic rock and the steeper landform in the hanging wall (i.e., the NW plate of the seismic fault) further increased the occurrence possibility of coseismic landslides.

The seismic amplification effects of mountainsides have been known for several decades [50–52], and this phenomenon has been confirmed in many earthquake events [11,36,53]. According to the landslide distribution, 60% of the coseismic landslides were concentrated near ridges and far away from rivers, demonstrating that the topographic site effects on seismic waves predominantly cause landslides on ridges [34,39,54]. This unusual occurrence suggests that seismic wave diffraction and interference in topographic ridges may increase the ground accelerations close to ridge crests [34,40,54]. Meanwhile, the statistical results of LAD also showed that landslide abundance areas were more likely to be concentrated in regions with higher elevation and far away from the rivers, indicating that coseismic landslides triggered by this earthquake were more likely to occur in the upper slopes.

The interpretation criteria of coseismic landslide mapping and the quality of remote-sensing images are important factors to ensure the objectivity and accuracy of the landslide inventory [55,56]. It should be admitted that the visual interpretation based on remote-sensing images in this study has some limitations. First, some small-scale landslides were not included in the landslide inventory due to inadequate resolution. Therefore, the spatial distribution pattern of small-scale landslides may have some deviation compared with the actual results. Second, landslide identification is heavily reliant on human experience, and there is no universal interpretation standard for landslide visual interpretation. Our future research will focus on overcoming the limitations mentioned above. This study primarily identified landslides using satellite images, but it is still difficult to distinguish between different types of landslides with this method. Therefore, we will supplement the landslide type records with detailed field investigation information in the following work, in order to enhance the quality of the landslide inventory.

6. Conclusions

Using 3 m resolution Planet satellite images, a landslide database for the 1 June 2022 Ms 6.1 Lushan earthquake in China was obtained in this work. Based on the spatial analysis of the coseismic landslides caused by this earthquake, the following understandings are preliminarily presented. The results show that the 2022 Lushan earthquake generated around 2352 coseismic landslides over a 3900 km^2 area. The total area of the landslides was approximately 5.51 km^2, with an average area of approximately 2300 m^2 for each individual landslide. The landslide occurrence was greatly affected by topographic factors including elevation, slope angle, and relief. Landslides were clustered in the slopes of 40–50°, elevation range of 1300~2500 m, and relief of 800–1200 m. The LAD increased as the aforementioned three factors increased and was explained by an exponential relationship, indicating that the occurrence possibility of landslide rapidly increased with the increase of these three factors. Notably, 59.7% of co-seismic landslides occurred in Granitic rock (G).

The landslide area density (LAD) of this lithologic unit was also the highest, accounting for 0.62%, indicating that slope failures occurred mainly in Granitic rock during the earthquake. According to the landslide distribution, the source area of landslides usually occurred in a ridge, showing an obvious seismic amplification effect of mountain slopes. In addition, the landslides in the NW plate of the fault were more densely distributed than those in the SE plate, suggesting a hanging wall effect.

The Longmenshan fault zone has experienced multiple moderate and strong earthquakes in the past decades. Further research work is needed to investigate the possible cumulative damage to the slope and the shattered mountain body due to the strong ground motions released by earthquakes. In addition, further analyses of long-term vegetation restoration and landform evolution are also necessary due to the post-quake effect of earthquake-induced landslides. Therefore, in the follow-up studies, the gradually completed landslide inventory of the Longmenshan area can provide abundant data for us to establish a near real-time model for seismic landslide hazard assessments and related landform evolution analyses.

Author Contributions: X.S. proposed the research concept, processed the relevant data and analysis, and wrote the manuscript. C.X. participated in data curation and analysis and revised the manuscript. S.M. designed the framework of this research and participated in data analysis. All authors have read and agreed to the published version of the manuscript.

Funding: This study was supported by the National Natural Science Foundation of China (41941016) and the Lhasa National Geophysical Observation and Research Station (NORSLS20-07).

Data Availability Statement: The coseismic landslide inventory of this earthquake event are available from the corresponding author upon request. Contact address: xc11111111@126.com; xiaoyishao@ninhm.ac.cn.

Conflicts of Interest: The authors declare that they have no known competing financial interests or personal relationships that could have appeared to influence the work reported in this paper.

References

1. Budimir, M.; Atkinson, P.M.; Lewis, H.G. Earthquake-and-landslide events are associated with more fatalities than earthquakes alone. *Nat. Hazards* **2014**, *72*, 895–914. [CrossRef]
2. Cui, P.; Zhu, Y.; Han, Y.; Chen, X.; Zhuang, J. The 12 May Wenchuan earthquake-induced landslide lakes: Distribution and preliminary risk evaluation. *Landslides* **2009**, *6*, 209–223. [CrossRef]
3. Roback, K.; Clark, M.K.; West, A.J.; Zekkos, D.; Li, G.; Gallen, S.F.; Chamlagain, D.; Godt, J.W. The size, distribution, and mobility of landslides caused by the 2015 Mw7.8 Gorkha earthquake, Nepal. *Geomorphology* **2018**, *301*, 121–138. [CrossRef]
4. Huang, R.; Fan, X. The landslide story. *Nat. Geosci.* **2013**, *6*, 325–326. [CrossRef]
5. Koi, T.; Hotta, N.; Ishigaki, I.; Matuzaki, N.; Uchiyama, Y.; Suzuki, M. Prolonged impact of earthquake-induced landslides on sediment yield in a mountain watershed: The Tanzawa region, Japan. *Geomorphology* **2008**, *101*, 692–702. [CrossRef]
6. Rodrguez, C.E.; Bommer, J.J.; Chandler, R.J. Earthquake-induced landslides: 1980–1997. *Soil Dyn. Earthq. Eng.* **1999**, *18*, 325–346. [CrossRef]
7. Hu, X.; Hu, K.; Zhang, X.; Wei, L.; Tang, J. Quantitative assessment of the impact of earthquake-induced geohazards on natural landscapes in Jiuzhaigou Valley. *J. Mt. Sci.* **2019**, *16*, 441–452. [CrossRef]
8. Serey, A.; Piñero-Feliciangeli, L.; Sepúlveda, S.A.; Poblete, F.; Petley, D.N.; Murphy, W. Landslides induced by the 2010 Chile megathrust earthquake: A comprehensive inventory and correlations with geological and seismic factors. *Landslides* **2019**, *16*, 1153–1165. [CrossRef]
9. Dai, L.; Scaringi, G.; Fan, X.; Yunus, A.P.; Liu-Zeng, J.; Xu, Q.; Huang, R. Coseismic Debris Remains in the Orogen Despite a Decade of Enhanced Landsliding. *Geophys. Res. Lett.* **2021**, *48*, e2021GL095850. [CrossRef]
10. Huang, R.Q.; Li, W.L. Analysis of the geo-hazards triggered by the 12 may 2008 wenchuan earthquake, china. *Bull. Eng. Geol. Environ.* **2009**, *68*, 363–371. [CrossRef]
11. Gorum, T.; Fan, X.; van Westen, C.J.; Huang, R.Q.; Xu, Q.; Tang, C.; Wang, G. Distribution pattern of earthquake-induced landslides triggered by the 12 May 2008 Wenchuan earthquake. *Geomorphology* **2011**, *133*, 152–167. [CrossRef]
12. Yin, Y.P.; Wang, F.W.; Sun, P. Landslide hazards triggered by the 2008 Wenchuan earthquake, Sichuan, China. *Landslides* **2009**, *6*, 139–152. [CrossRef]
13. Cui, P.; Chen, X.; Zhu, Y.; Su, F.; Wei, F.; Han, Y.; Liu, H.; Zhuang, J. The Wenchuan Earthquake (May 12, 2008), Sichuan Province, China, and resulting geohazards. *Nat. Hazards* **2009**, *56*, 19–36. [CrossRef]

14. Zhang, J.; Su, F.; Fan, J. Distribution of landslides and collapses induced by 2013 "4.20" Lushan earthquake and hazards assessment: A case study of S210 highway. *J. Mt. Sci.* **2013**, *31*, 616–623.
15. Wu, C.-H.; Cui, P.; Li, Y.-S.; Ayala, I.A.; Huang, C.; Yi, S.-J. Seismogenic fault and topography control on the spatial patterns of landslides triggered by the 2017 Jiuzhaigou earthquake. *J. Mt. Sci.* **2018**, *15*, 793–807. [CrossRef]
16. Fan, X.; Scaringi, G.; Xu, Q.; Zhan, W.; Dai, L.; Li, Y.; Pei, X.; Yang, Q.; Huang, R. Coseismic landslides triggered by the 8th August 2017 Ms 7.0 Jiuzhaigou earthquake (Sichuan, China): Factors controlling their spatial distribution and implications for the seismogenic blind fault identification. *Landslides* **2018**, *15*, 967–983. [CrossRef]
17. Kargel, J.S.; Leonard, G.J.; Shugar, D.H.; Haritashya, U.K.; Bevington, A.; Fielding, E.J.; Fujita, K.; Geertsema, M.; Miles, E.S.; Steiner, J.; et al. Geomorphic and geologic controls of geohazards induced by Nepal's 2015 Gorkha earthquake. *Science* **2016**, *351*, aac8353. [CrossRef]
18. Ling, S.; Sun, C.; Li, X.; Ren, Y.; Xu, J.; Huang, T. Characterizing the distribution pattern and geologic and geomorphic controls on earthquake-triggered landslide occurrence during the 2017 Ms 7.0 Jiuzhaigou earthquake, Sichuan, China. *Landslides* **2020**, *18*, 1275–1291. [CrossRef]
19. Zhao, B.; Li, W.; Wang, Y.; Lu, J.; Li, X. Landslides triggered by the Ms 6.9 Nyingchi earthquake, China (18 November 2017): Analysis of the spatial distribution and occurrence factors. *Landslides* **2019**, *16*, 765–776. [CrossRef]
20. Parker, R.N.; Densmore, A.L.; Rosser, N.J.; de Michele, M.; Li, Y.; Huang, R.; Whadcoat, S.; Petley, D.N. Mass wasting triggered by the 2008 Wenchuan earthquake is greater than orogenic growth. *Nat. Geosci.* **2011**, *4*, 449–452. [CrossRef]
21. Azarafza, M.; Azarafza, M.; Akgün, H.; Atkinson, P.M.; Derakhshani, R. Deep learning-based landslide susceptibility mapping. *Sci. Rep.* **2021**, *11*, 24112. [CrossRef]
22. Nanehkaran, Y.; Mao, Y.; Azarafza, M.; Kockar, M.; Zhu, H.-H. Fuzzy-based multiple decision method for landslide susceptibility and hazard assessment: A case study of Tabriz, Iran. *Geomech. Eng.* **2021**, *24*, 407–418. [CrossRef]
23. Fan, X.; Fang, C.; Dai, L.; Wang, X.; Luo, Y.; Wei, T.; Wang, Y. Near real time prediction of spatial distribution probability of earthquake-induced landslides-Take the Lushan Earthquake on June 1, 2022 as an example. *J. Eng. Geol.* **2022**, *30*, 729–739. [CrossRef]
24. Shao, X.; Ma, S.; Xu, C. Distribution and characteristics of shallow landslides triggered by the 2018 Mw 7.5 Palu earthquake, Indonesia. *Landslides* **2022**, 1–19. [CrossRef]
25. Zhao, B.; Su, L.; Wang, Y.; Ji, F.; Li, W.; Tang, C. Insights into the mobility characteristics of seismic earthflows related to the Palu and Eastern Iburi earthquakes. *Geomorphology* **2021**, *391*, 107886. [CrossRef]
26. Tapponnier, P.; Zhiqin, X.; Roger, F.; Meyer, B.; Arnaud, N.; Wittlinger, G.; Jingsui, Y. Oblique stepwise rise and growth of the Tibet plateau. *Science* **2001**, *294*, 1671–1677.
27. Xu, X.W.; Wen, X.Z.; Yu, G.H.; Chen, G.H.; Klinger, Y.; Hubbard, J.; Shaw, J. Coseismic reverse- and oblique-slip surface faulting generated by the 2008 Mw 7.9 Wenchuan earthquake, China. *Geology* **2009**, *37*, 515–518.
28. Liu, M.; Luo, G.; Wang, H. The 2013 Lushan Earthquake in China Tests Hazard Assessments. *Seismol. Res. Lett.* **2014**, *85*, 40–43. [CrossRef]
29. Liu, C.; Zhu, B.; Yang, X.; Shi, Y. Crustal rheology control on earthquake activity across the eastern margin of the Tibetan Plateau: Insights from numerical modelling. *J. Asian Earth Sci.* **2015**, *100*, 20–30. [CrossRef]
30. Deng, Q.D. *Chinese Active Tectonic Map*; Seismological Press: Beijing, China, 2007.
31. Zhao, B.; Huang, Y.; Zhang, C.; Wang, W.; Tan, K.; Du, R. Crustal deformation on the Chinese mainland during 1998–2014 based on GPS data. *Geod. Geodyn.* **2015**, *6*, 7–15. [CrossRef]
32. Gallen, S.F.; Clark, M.K.; Godt, J.W. Coseismic landslides reveal near-surface rock strength in a high-relief, tectonically active setting. *Geology* **2015**, *43*, 11–14.
33. Gorum, T.; van Westen, C.J.; Korup, O.; van der Meijde, M.; Fan, X.; van der Meer, F.D. Complex rupture mechanism and topography control symmetry of mass-wasting pattern, 2010 Haiti earthquake. *Geomorphology* **2013**, *184*, 127–138. [CrossRef]
34. Meunier, P.; Hovius, N.; Haines, J.A. Topographic site effects and the location of earthquake induced landslides. *Earth Planet. Sci. Lett.* **2008**, *275*, 221–232. [CrossRef]
35. Xu, C.; Xu, X. Statistical analysis of landslides caused by the Mw 6.9 Yushu, China, earthquake of 14 April 2010. *Nat. Hazards* **2014**, *72*, 871–893.
36. Zhuang, J.; Peng, J.; Xu, C.; Li, Z.; Densmore, A.; Milledge, D.; Iqbal, J.; Cui, Y. Distribution and characteristics of loess landslides triggered by the 1920 Haiyuan Earthquake, Northwest of China. *Geomorphology* **2018**, *314*, 1–12. [CrossRef]
37. Dai, F.C.; Xu, C.; Yao, X.; Xu, L.; Tu, X.B.; Gong, Q.M. Spatial distribution of landslides triggered by the 2008 Ms 8.0 Wenchuan earthquake, China. *J. Asian Earth Sci.* **2011**, *40*, 883–895. [CrossRef]
38. García, D.; Mah, R.T.; Johnson, K.L.; Worden, C.; Hearne, M.; Marano, K.D.; So, E.; Lin, K.-W.; Wald, D.J. ShakeMap Atlas 2.0: An improved suite of recent historical earthquake ShakeMaps for global hazard analyses and loss model calibration. In Proceedings of the World Conference on Earthquake Engineering, Lisbon, Portugal, 24–28 September 2012.
39. Gischig, V.S.; Eberhardt, E.; Moore, J.R.; Hungr, O. On the seismic response of deep-seated rock slope instabilities—Insights from numerical modeling. *Eng. Geol.* **2015**, *193*, 1–18. [CrossRef]
40. Hailemikael, S.; Lenti, L.; Martino, S.; Paciello, A.; Rossi, D.; Mugnozza, G.S. Ground-motion amplification at the Colle di Roio ridge, central Italy: A combined effect of stratigraphy and topography. *Geophys. J. Int.* **2016**, *206*, 1–18. [CrossRef]

41. Sato, H.P.; Hasegawa, H.; Fujiwara, S.; Tobita, M.; Koarai, M.; Une, H.; Iwahashi, J. Interpretation of landslide distribution triggered by the 2005 Northern Pakistan earthquake using SPOT 5 imagery. *Landslides* **2007**, *4*, 113–122. [CrossRef]
42. Xu, Q.; Li, W. Study on the Direction Effects of Landslides Triggered by Wenchuan Earthquake. *J. Sichuan Univ.* **2010**, *42*, 7–14.
43. Tang, C.; Zuo, Y.; Qin, S.; Yang, S.; Wang, D.; Li, L.; Xu, T. Spalling and slinging pattern of shallow slope and dynamics explanation in the 2008 Wenchuan earthquake. In Proceedings of 10th Conference on Rock Mechanics and Engineering, Weihai, China, 28–30 July 2008; pp. 258–262.
44. Ma, S.Y.; Xu, C. Assessment of co-seismic landslide hazard using the Newmark model and statistical analyses: A case study of the 2013 Lushan, China, Mw6.6 earthquake. *Nat. Hazards* **2019**, *96*, 389–412. [CrossRef]
45. Gorum, T.; Korup, O.; van Westen, C.J.; van der Meijde, M.; Xu, C.; van der Meer, F.D. Why so few? Landslides triggered by the 2002 Denali earthquake, Alaska. *Quat. Sci. Rev.* **2014**, *95*, 80–94. [CrossRef]
46. Xu, C.; Xu, X.; Yao, X.; Dai, F. Three (nearly) complete inventories of landslides triggered by the May 12, 2008 Wenchuan Mw 7.9 earthquake of China and their spatial distribution statistical analysis. *Landslides* **2014**, *11*, 441–461.
47. Wang, W.; Wu, H.; Nakamura, H.; Wu, S.; Ouyang, S.; Yu, M. Mass movements caused by recent tectonic activity: The 1999 Chi-chi earthquake in central Taiwan. *Island Arc* **2003**, *12*, 325–334. [CrossRef]
48. Jibson, R.W.; Harp, E.L.; Schulz, W.; Keefer, D.K. Landslides Triggered by the 2002 Denali Fault, Alaska, Earthquake and the Inferred Nature of the Strong Shaking. *Earthq. Spectra* **2004**, *20*, 669–691. [CrossRef]
49. Harp, E.L.; Jibson, R.W. Inventory of Landslides Triggered by the 1994 Northridge, California Earthquake. *Bull. Seismol. Soc. Am.* **1996**, *86*, S319–S332.
50. Leonov, N.N. The Khait, 1949 earthquake and geological conditions of its origin. *Proc. Acad. Sci. USSR Geophys. Ser.* **1960**, *3*, 409–424.
51. Ashford, S.; Sitar, N.; Lysmer, J.; Deng, N. Topographic Effects on the Seismic Response of Steep Slopes. *Bull. Seismol. Soc. Am.* **1997**, *87*, 701–709.
52. Davis, L.; West, L. Observed effects of topography on ground motion. *Bull. Seismol. Soc. Am.* **1973**, *63*, 283–298. [CrossRef]
53. Lee, S.; Chan, Y.; Komatitsch, D.; Huang, B.; Tromp, J. Effects of Realistic Surface Topography on Seismic Ground Motion in the Yangminshan Region of Taiwan Based Upon the Spectral-Element Method and LiDAR DTM. *Bull. Seismol. Soc. Am.* **2009**, *99*, 681–693. [CrossRef]
54. Luo, Y.; Fan, X.; Huang, R.; Wang, Y.; Yunus, A.P.; Havenith, H.B. Topographic and near-surface stratigraphic amplification of the seismic response of a mountain slope revealed by field monitoring and numerical simulations. *Eng. Geol.* **2020**, *271*, 105607. [CrossRef]
55. Guzzetti, F.; Mondini, A.C.; Cardinali, M.; Fiorucci, F.; Santangelo, M.; Chang, K.-T. Landslide inventory maps: New tools for an old problem. *Earth-Sci. Rev.* **2012**, *112*, 42–66. [CrossRef]
56. Xu, C. Preparation of earthquake-triggered landslide inventory maps using remote sensing and GIS technologies: Principles and case studies. *Geosci. Front.* **2015**, *6*, 825–836. [CrossRef]

sustainability

Article

Soil Dynamic Constitutive Considering Post-Liquefaction Deformation and Reversible Pore-Water Pressure

Qing Dong [1], Zhenghua Zhou [2], Xiaojun Li [1,*], Bing Hao [2] and Liguo Jin [3]

1 Key Laboratory of Urban Security and Disaster Engineering of China Ministry of Education, Beijing University of Technology, Beijing 100020, China
2 College of Transportation Science & Engineering, Nanjing Tech University, Nanjing 210000, China
3 Institute of Geophysics, China Earthquake Administration, Beijing 100081, China
* Correspondence: lixiaojun@bjut.edu.cn

Abstract: In the seismic response analysis of liquefiable sites, the existing soil dynamic constitutive model is challenging to simulate saturated sand's post-liquefaction deformation, and the current pore-water pressure buildup model cannot reflect the decrease in the actual pore-water pressure under unloading stress. We aim at these problems to propose a feasible and straightforward time-domain post-liquefaction deformation constitutive model through experimental analysis and theoretical research, consisting of reversible pore-water pressure. According to the dynamic triaxial test data, the regularities of large deformation stress and strain behavior of the saturated sand after liquefaction are obtained, and the corresponding loading and unloading criteria are summarized. Combined with the effective stress constitutive model proposed by the author, a soil dynamic constitutive that can describe saturated sand's post-liquefaction deformation path is obtained. According to the test results, the model can simulate the deformation of saturated sand during the whole liquefaction process. The self-developed program Soilresp1D realized the dynamic response analysis of the liquefiable site, and the results were compared with the experimental results. It shows that the model based on the effective stress-modified logarithmic dynamic skeleton and post-liquefaction deformation constitutive can be directly applied to the dynamic response analysis of the liquefiable site.

Keywords: post-liquefaction deformation; soil dynamic constitutive model; effective stress; reversible pore-water pressure; loading-unloading rules

Citation: Dong, Q.; Zhou, Z.; Li, X.; Hao, B.; Jin, L. Soil Dynamic Constitutive Considering Post-Liquefaction Deformation and Reversible Pore-Water Pressure. *Sustainability* **2022**, *14*, 16512. https://doi.org/10.3390/su142416512

Academic Editor: Jeroen Meersmans

Received: 10 November 2022
Accepted: 6 December 2022
Published: 9 December 2022

1. Introduction

Some important marine and coastal infrastructures, such as undersea tunnels, cross-sea bridges, offshore oil platforms, artificial reclamation islands, and infrastructure on coral reefs, are located on or within the liquefiable soil layer [1–4]. Many liquefaction damage phenomena of deep sand (10–26 m) have occurred in recent large earthquakes [4–7]. Several factors affect the occurrence of liquefaction, such as the load amplitude, soil type, initial shear stress, shear strain amplitude, age, and hydraulic conditions [8,9]. When determining the seismic fortification parameters of such sites, a soil dynamic constitutive model considering sand liquefaction should be established [10–12]. Large deformation caused by liquefaction mainly causes the disasters induced by the liquefaction of saturated sand layers [12–15]. Therefore, selecting a reasonable dynamic constitutive model suitable for saturated sand is essential when studying liquefaction [16–19]. Some nonlinear or elastic-plastic constitutive models used for liquefaction analysis have been proposed successively [7,20–25]. However, these models can only simulate the dynamic response of a small deformation before the initial liquefaction and are not suitable for simulating the dynamic response of a large deformation after the initial liquefaction. The soil's stress–strain characteristics and boundary value conditions will change when a severe liquefaction and large deformation occur under earthquake action [7,26–28]. Elgamal and Yang developed a plasticity-based constitutive model with emphasis on simulating the cyclic mobility response mechanism and associated pattern of shear strain accumulation; this constitutive model is incorporated into a two-phase (solid-fluid), fully coupled finite element code (PDMY02) [22,29,30]. The model incorporates shear-induced contractive, perfectly plastic, and dilative response phases implemented through an appropriate non-associative flow rule. At the same time, Dafalias and Manzari proposed a stress-ratio controlled, critical state compatible sand plasticity model (SANISAND) in the triaxial and generalized stress space [31]. They illustrated the model's simulative ability by a comparison with the data over an extensive range of pressures and densities.

Following the basic framework of SANISAND, Boulanger and Ziotopoulou presented the sand plasticity model PM4Sand (the model follows the basic framework of the stress-ratio controlled, critical state compatible, and bounding surface plasticity model for sand presented by Dafalias and Manzari (2004) [32]) for geotechnical earthquake engineering applications, which provides reasonable approximations of the desired behaviors and is relatively easy to calibrate. Asgaria applied the model to the numerical simulation of the liquefaction site [33]. Shamoto and Zhang constructed an elastoplastic constitutive model suitable for the dynamic reaction calculation of the sand liquefaction with a large deformation. Still, it relied on sizeable finite element calculation software and had a poor general applicability [7,26,34]. Shong proposes a completely explicit finite element method for solving dynamic u-p equations of fluid-saturated porous media [35,36]. These studies provide a theoretical basis for establishing a unified constitutive model for a post-liquefaction deformation. However, these models are almost realized by empirically reducing the modulus or introducing mathematical techniques lacking the objective physical mechanisms. The description of a time-domain constitutive relationship changes before and after the liquefaction is not perfect.

There are few studies on the time-domain post-liquefaction deformation constitutive for the liquefiable site seismic response analysis. It is necessary to construct a model that can be directly applied to the liquefiable site and describe the development process of pore-water pressure and the post-liquefaction stress–strain relationship. Additionally, in light of that, the main objectives of this study are as follows: (1) summarizing the fundamental law of a large deformation and post-liquefaction stress–strain relationship basic formula; (2) deciphering the loading and unloading criteria for the large liquefaction deformation of saturated sand; (3) combining with the primary stress–strain relationship formula and the loading and unloading criterion, establishing a large deformation constitutive model of saturated sand liquefaction with a simple expression and accessible parameters, and calibrating the parameter; and (4) embedding the large deformation constitutive model into the site seismic response analysis self-compiled program Soilresp1D (based on the boundary conditions, initial conditions, motion equation, and LDSCM, a 1D sites time-domain nonlinear seismic response analysis program Soilresp1D by Microsoft Visual c++ 6.0 platform was developed. The program can easily and quickly calculate the seismic response at any site depth [37].) and developing a set of numerical analysis methods and calculation programs for the large deformation analysis of the liquefiable site.

2. Model Formulations

2.1. Demonstration of Test Results

New methods and protocols should be described in detail, while well-established methods can be briefly described and appropriately cited. In the cyclic shear liquefaction test, when the effective stress of saturated sand is zero for the first time, the soil is liquefied, and the state of the soil is an initial liquefaction, which is called before the liquefaction and after the liquefaction [19,38]. More studies have been conducted on the soil liquefaction tests and post-liquefaction character [39–43]. Based on the pore-water pressure model, which can simulate the reversible pore water pressure proposed by the author, combined with a 1D time-domain dynamic nonlinear constitutive model based on the logarithmic dynamic skeleton, the effective stress constitutive model of saturated sand under a strong ground motion before liquefaction is established [37,44]. However, this model can only simulate the dynamic reaction of a small deformation before sand liquefaction but cannot affect the large deformation after the sand liquefaction.

In this paper, we studied the stress–strain curves of saturated sand before and after liquefaction under the condition of a thecyclic triaxial test. The dynamic triaxial control system of DYNTTS-60KN was used to apply a vibration stress load to saturated coral sand. The changes in the pore water pressure were observed and recorded, or when the shear deformation of the sample reaches a critical value, it is judged that liquefaction occurs. Some cyclic triaxial test was conducted on saturated coral with a relative density of 45%. We performed a stress-controlled test (constant CSR) until reaching initial liquefaction and then switched to a strain-controlled test. Before reaching initial liquefaction, the effective confining pressure was 100 kPa, CSR was 0.325, 0.3, 0.25, 0.22, and the loading frequency was 0.5 Hz, respectively. The results are shown in Figures 1–4, where (a) is the overall stress–strain curve; (b) is the time history of the pore pressure change; (c) is the stress–strain curve before liquefaction; (d) is the stress–strain curve after liquefaction; and (c) and (d) are obtained by the decomposition of (a). According to the test results, the cyclic loads required for sand to reach the initial liquefaction state under various CSRs are 11, 35, 37, and 85 cycles, respectively.

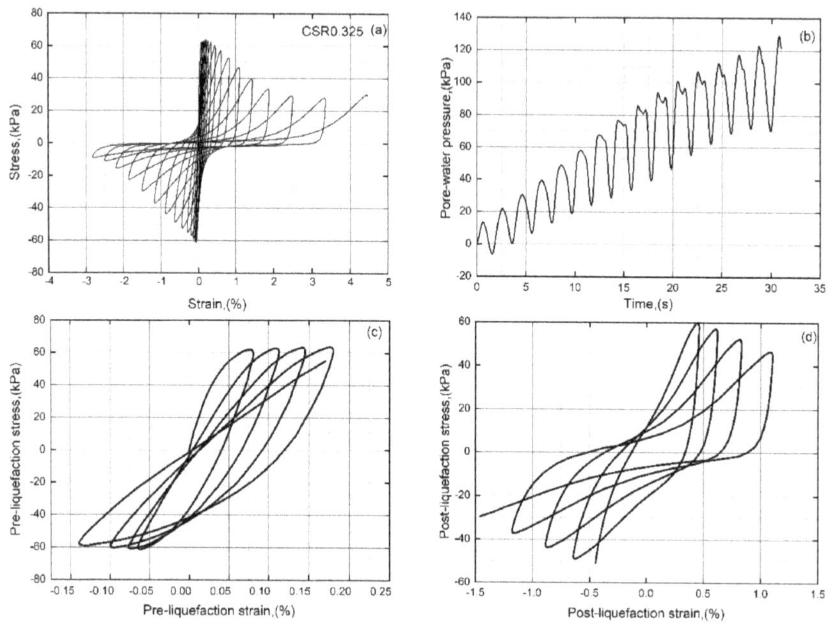

Figure 1. Stress–strain curves of CSR 0.325 saturated coral sand before and after liquefaction. (**a**) overall stress–strain curve (**b**) time-history of pore-water pressure ratio change (**c**) stress–strain curve before liquefaction (**d**) stress–strain curve after liquefaction.

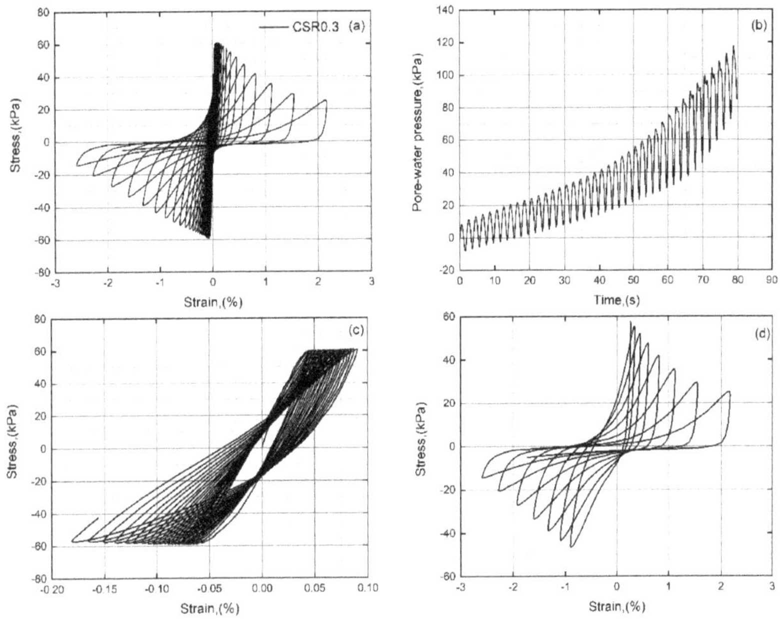

Figure 2. Stress–strain curves of CSR 0.3 saturated coral sand before and after liquefaction. (**a**) overall stress–strain curve (**b**) time-history of pore-water pressure ratio change (**c**) stress–strain curve before liquefaction (**d**) stress–strain curve after liquefaction.

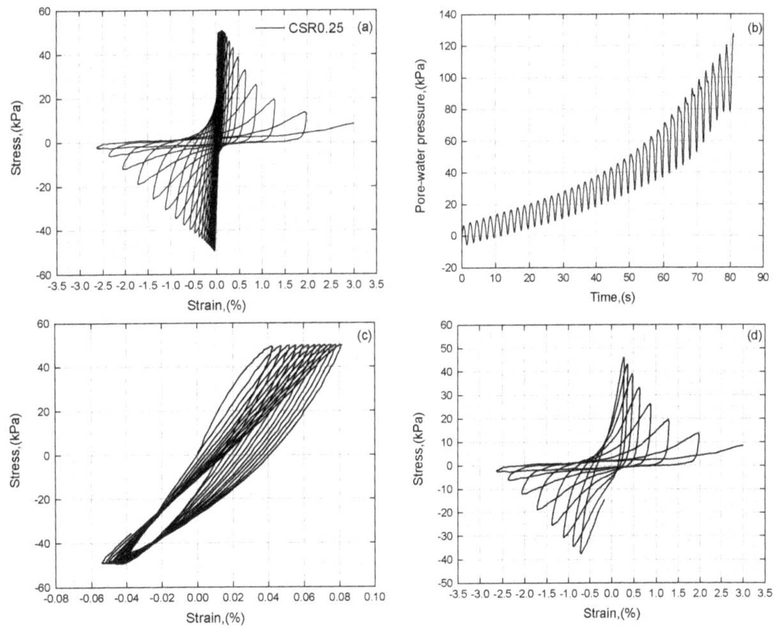

Figure 3. Stress–strain curves of CSR 0.25 saturated coral sand before and after liquefaction. (**a**) overall stress–strain curve (**b**) time-history of pore-water pressure ratio change (**c**) stress–strain curve before liquefaction (**d**) stress–strain curve after liquefaction.

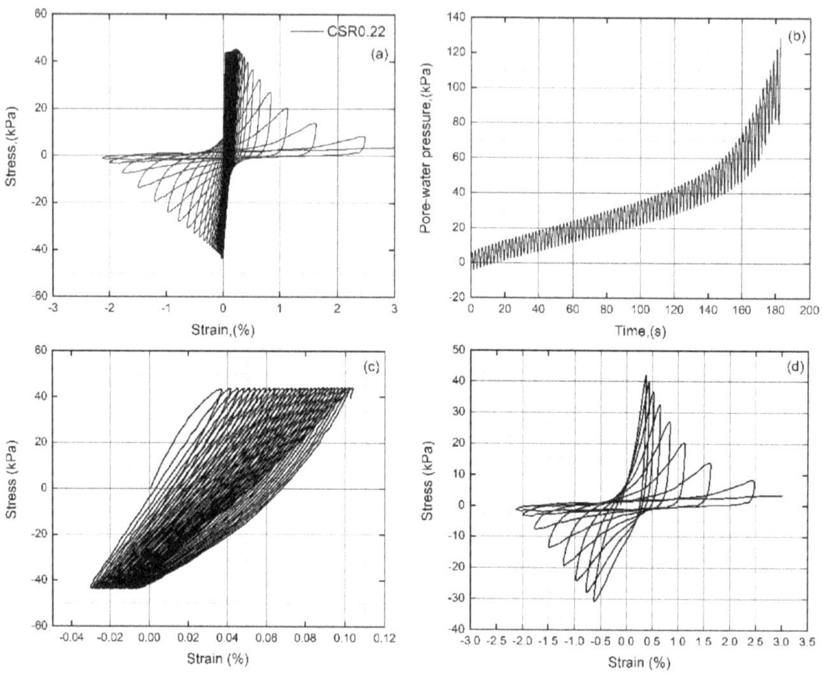

Figure 4. Stress–strain curves of CSR 0.22 saturated coral sand before and after liquefaction. (**a**) overall stress–strain curve (**b**) time-history of pore-water pressure ratio change (**c**) stress–strain curve before liquefaction (**d**) stress–strain curve after liquefaction.

2.2. Analysis of Test Results

As can be seen from the overall stress–strain curve in Figures 1a, 2a, 3a and 4a, with the increase in the number of cycles, each cycle's maximum shear strain amplitude gradually increases [24,40]. In contrast, the shear modulus gradually decreases, and the shear stress amplitude of the soil in the initial liquefaction state gradually decreases. The time history of the pore-water pressure shown in Figures 1b, 2b, 3b and 4b shows that the pore-water pressure in saturated coral sand increases monotonously during a continuous accumulation and fluctuates periodically with the transformation of shear stress. When the pore-water pressure reaches the confining pressure, the soil presents zero effective stress and an initial liquefaction.

Figures 1c,d, 2c,d, 3c,d and 4c,d, respectively, give the stress–strain curves of the soils before and after liquefaction. Because the mechanical properties of saturated sand are inconsistent before and after liquefaction, the shape of the stress–strain curves is also different. By comparing Figures 1c,d, 2c,d, 3c,d and 4c,d, it shows that the skeleton curve of the upper half before liquefaction is convex upward, while that of the upper half after liquefaction is concave downward. Figures 1c, 2c, 3c and 4c shows that the shear modulus and shear strength are related to effective stress due to compressive hardening characteristics. Before the initial liquefaction, the shear modulus decreases gradually as the effective stress decreases.

Figures 1d, 2d, 3d and 4d shows that the soil properties changed after liquefaction, and the soil behaves like a fluid. After liquefaction, the pore-water pressure still fluctuates, and a zero effective stress state appears twice in one cycle. According to the cyclic stress–strain curves after liquefaction, it can be divided into unloading and loading states. After liquefaction, the initial shear modulus decreases significantly with the number of cycles, which is more evident than before liquefaction, and the cyclic shear strain peak of the soil after liquefaction increases significantly. Unlike before the liquefaction, the peak shear stress of each cycle of soil decreases gradually after liquefaction, and the peak shear stress of the soil cannot reach the value of the stress applied. The experimental phenomenon above is typical according to the undrained cyclic triaxial test results of saturated coral sand under different CSR conditions in Figures 1–4.

Based on the above analysis, the variation characteristics of the shear modulus, shear stress, and shear strain of the soil before and after the liquefaction are summarized as follows: (1) before liquefaction, the initial shear modulus of each cycle decreased with the decrease in the effective stress. After liquefaction, it decreased significantly with the increase in the cycles. (2) Before liquefaction, the soil's cyclic shear stress amplitude was constant and did not increase with the increase in the cyclic times. After liquefaction, it decreases significantly with the increase in the cycles. (3) Before liquefaction, the amplitude of the cyclic shear strain increased with the decrease in the effective stress. After liquefaction, it increases significantly with the increase in the cyclic number. (4) The skeleton curve in the upper half of the stress–strain curve is convex before liquefaction and concave after liquefaction, which shows an antisymmetric relationship.

Therefore, the abrupt change in the stress–strain relationship occurs when the effective stress reaches zero states for the first time. The key to simulating the stress–strain response is to judge the soil's liquefaction state and the stress–strain curve's development law. The above is only a limited description of the experimental results, which need to be expressed by a certain functional expression or mathematical model to establish a large deformation constitutive model.

2.3. Determine Loading and Unloading Criteria

The stress–strain curves of saturated sand during the whole liquefaction process can be divided into two stages: small deformation before liquefaction and large deformation after liquefaction. The stress–strain relationship of small deformation before the liquefaction has been described in the literature [37,44]. The effective stress constitutive model can simulate the reversible pore-water pressure, combined with a 1D time-domain dynamic nonlinear constitutive model (LDSCM) based on the logarithmic dynamic skeleton. It shows as in Equations (1) and (2), Equation (1) is the effective stress model; Equation (2) is the pore-water pressure model. The LDSCM LSC-based introduces the concepts of a modified dynamic skeleton curve and the damping ratio degradation coefficient, determined by the test data G/Gmax-γ and ζ-γ of soil (a, b, a_0, b_0), which can fully consider the damping effect in the soil's dynamics problems. The pore-water pressure model shown in Equation (2), containing reversible pore-water pressure, can reflect the increase in the pore-water pressure under stress loading and its decrease under stress unloading, and ($c_{1,0}$, $c_{1,a}$, $c_{1,b}$, $A_{4,0}$) are main material constants of the model, which are related to the density of the soil. The derivation process of Equations (1) and (2) is rather complicated. Equation (1) can be seen from Equations (1)–(12) in [37], and Equation (2) can be seen from Formulas (1)–(4) in [44]. Since this part is not the focus of this article, it will not be repeated here.

$$\left\{\begin{array}{l} \tau(\gamma)=\begin{cases} K(\gamma_0)\cdot\left[\pm 2ln\left(1+\frac{b'}{a'}\left|\frac{\gamma-\gamma_c}{2}\right|\right)/b'-\frac{\pm\tau_m-\tau_c}{\pm\gamma_m-\gamma_c}\right. & \\ \left.\cdot(\gamma-\gamma_c)\right]+\frac{\pm\tau_m-\tau_c}{\pm\gamma_m-\gamma_c}\cdot(\gamma-\gamma_c)+\tau_c & |\gamma|\leq\gamma_m \\ \frac{\pm ln(1\pm b\gamma/a)}{b} & |\gamma|\geq\gamma_m \end{cases} \\ K(\gamma_0)=\frac{\pi b\gamma_0^2}{2(a_0+b_0\gamma_0)[(2a+b\gamma_0)-2b\gamma_0/ln(1+b\gamma_0/a)]} \\ G_{max,N}=\frac{1}{a}=G_{max,0}(1-U_N)=G_{max,0}\left(1-\frac{u_{re,N}+u_{ir,N}}{\overline{\sigma}_0}\right) \\ \tau_{ult,N}=\frac{1}{b}=\tau_{ult,0}(1-U_N)=\tau_{ult,0}\left(1-\frac{u_{re,N}+u_{ir,N}}{\overline{\sigma}_0}\right) \end{array}\right. \tag{1}$$

$$\begin{cases} u_{ir,0} & = 0 \\ u_N & = u_{re,N} + u_{ir,N} \\ & = \Delta u_{ir,N}\left(\frac{\tau}{\tau_{N,max}} - 1\right) + u_{ir,N-1} + \Delta u_{ir,N} \\ & = \Delta u_{ir,N}\frac{\tau}{\tau_{N,max}} + u_{ir,N-1}, N = 1,2,3,\ldots \\ \Delta u_{ir,N} = & \frac{c_{1,0}}{\sqrt{N_{ep}}}\left(\frac{\tau_N}{\overline{\sigma}_{N-1}}\right)^{A_{4,0}} \cdot \left[1 - c_{1,a}(K_c - 1)^{c_{1,b}}\right](\overline{\sigma}_0 - u_{ir,N-1}) \\ N_{ep} & = \sum_{i=1}^{N}\left[\frac{\tau_i}{\tau_N}\right]^{\alpha} \end{cases} \quad (2)$$

where $u_{re,N}$ is the reversible pore-water pressure in the Nth stress cycle; $\Delta u_{ir,N}$ is the irreversible pore-water pressure increment in the Nth stress cycle; $u_{ir,N-1}$ is the irreversible pore-water pressure in the $N-1$st stress cycle; $\overline{\sigma}_0$ is the initial effective confining pressure; $c_{1,0}$, $c_{1,a}$, $c_{1,b}$, $A_{4,0}$ are the material constants, which are related to the density of the soil; K_c is the consolidation ratio; N_{ep} is the number of equivalent actions; τ_i is the shear stress amplitude in the Ith stress cycle ($1 \leq i \leq N$); α is the material constant; $G_{max,N}$, $\tau_{ult,N}$ are the maximum shear modulus and ultimate shear stress after the $N-1$th stress cycle; $|\gamma|$ is the absolute value of the shear strain; and τ_c and γ_c are the values of the shear stress and the shear strain, respectively, at the last reversal. τ_m, γ_m are the values of the shear stress and the shear strain at the biggest reversal, which are positive; $K(\gamma_0)$ is the damping degradation coefficient; a, b are related to the maximum shear modulus $G_{max} = 1/a$ and the ultimate shear stress $\tau_{ult} = 1/b$ can be obtained by fitting the $G/G_0 - \gamma$ experimental data; a', b' are the parameters derived from a, b and τ_m, γ_m; and a_0, b_0 can be obtained by fitting the $\lambda - \gamma$ experimental data. See the literature for the values of all the parameters [37,44].

After liquefaction, the shear modulus of the saturated sand decreases sharply, and the shear stress cannot reach the applied stress amplitude, resulting in a large shear deformation. In this summary, the regularities of the large deformation of saturated sand after the liquefaction are summarized. Their basic function expressions are determined through decomposition analysis of the large deformation stress–strain curves obtained from the liquefaction tests.

2.3.1. The Basic Function of Stress–Strain Relationship after Liquefaction

The soil constitutive model determination advance determines the skeleton curve and the basic functional relationship. Figure 5 shows the large deformation stress–strain of saturated sand under different cycles (N) of action measured by the test. As shown in Figure 6, the stress–strain curve (A-B-C-D-A) of each cycle can be decomposed into three processes: forward loading, unloading, and reloading. It can be disassembled into four stages: forward loading AB, first unloading stage BC, second unloading stage CD, and then loading DA. The following is the determination method of the basic functional relationship of the four stages.

- Initial loading curve AB

First, to determine the basic function of the skeleton curve with a large deformation, the relation of the AB segment should be determined.

It is not difficult to see from Figure 7a that section AB after the liquefaction is symmetric with the curve before the liquefaction ($\tau = \gamma/(a + b\gamma)$) about the line $\tau = \gamma$. Then, γ and τ can be interchanged in the stress–strain relation $\tau = \gamma/(a + b\gamma)$ of the hyperbolic model before the liquefaction, and the basic function of the liquefaction is shown in Equation (3).

$$\gamma = \frac{\tau}{a + b\tau} \quad (3)$$

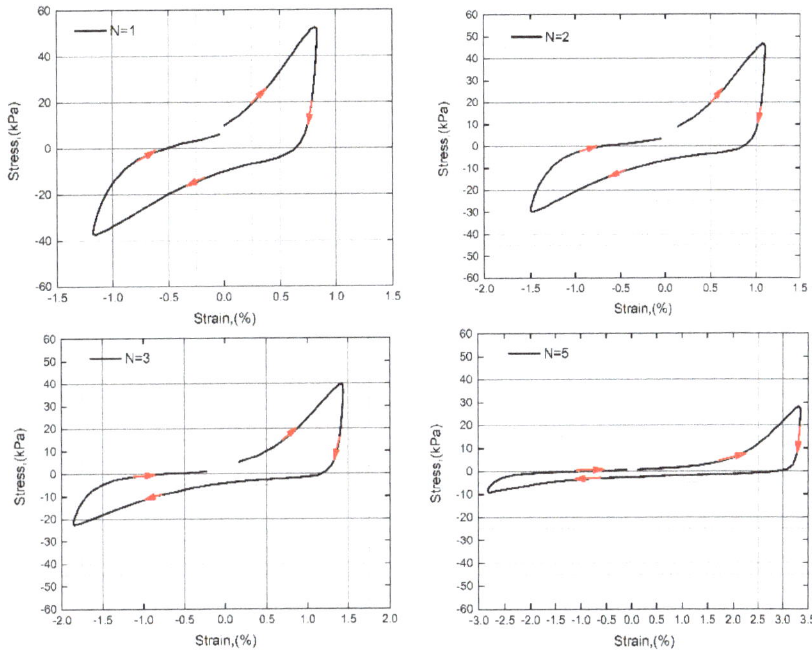

Figure 5. Stress–strain relationship of saturated sand with large deformation under different cycles, N is the number of the cycle.

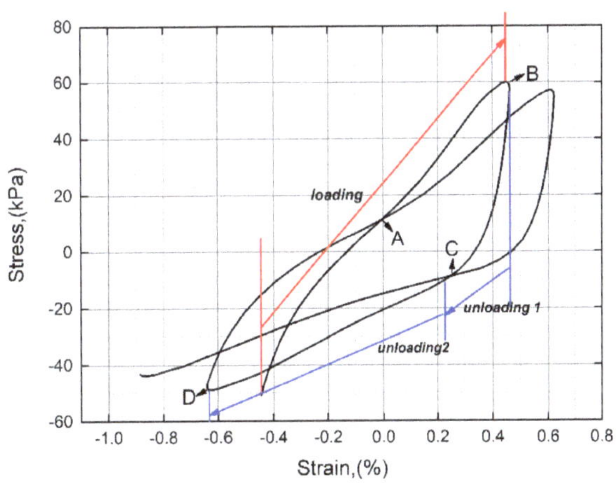

Figure 6. Stress–strain disassembly diagram for the first two cycles after liquefaction.

It can be written as:

$$\tau = \frac{\gamma}{1/a - \gamma b/a} = \frac{\gamma}{a_1 - b_1\gamma} \tag{4}$$

An inverse hyperbolic function (IHF) can express the initial skeleton curve. The parameter $a_1 = 1/a$ is the initial tangential modulus, as shown in Figure 7c. a_1 gradually decreases with the increase in the cycles. The difference between the skeleton curves of large deformation and small deformation also lies in the asymptotes. Before liquefaction, there is a transverse asymptotic line $\tau = 1/b$, which means the ultimate shear strength of the soil. According to Equation (4), after liquefaction, there is a longitudinal asymptote

$\gamma = a_1/b_1$, whose physical meaning is the ultimate shear strain of saturated sand after liquefaction. With the increase in the number of cycles and the complexity of the soil deformation, the initial skeleton curves of each cycle generally cannot reach the origin. The general function of the loading curve is:

$$\tau = \frac{\gamma - \gamma_d}{a - b(\gamma - \gamma_d)} - \tau_d \tag{5}$$

where the coordinate of A is (γ_d, τ_d). As shown in Figure 7c, γ_d is approximately zero, and τ_d decreases linearly with the accumulation of the cycles. Figure 7b has an excellent fitting result through Equation (5), which can be used as the initial skeleton curve of the large deformation.

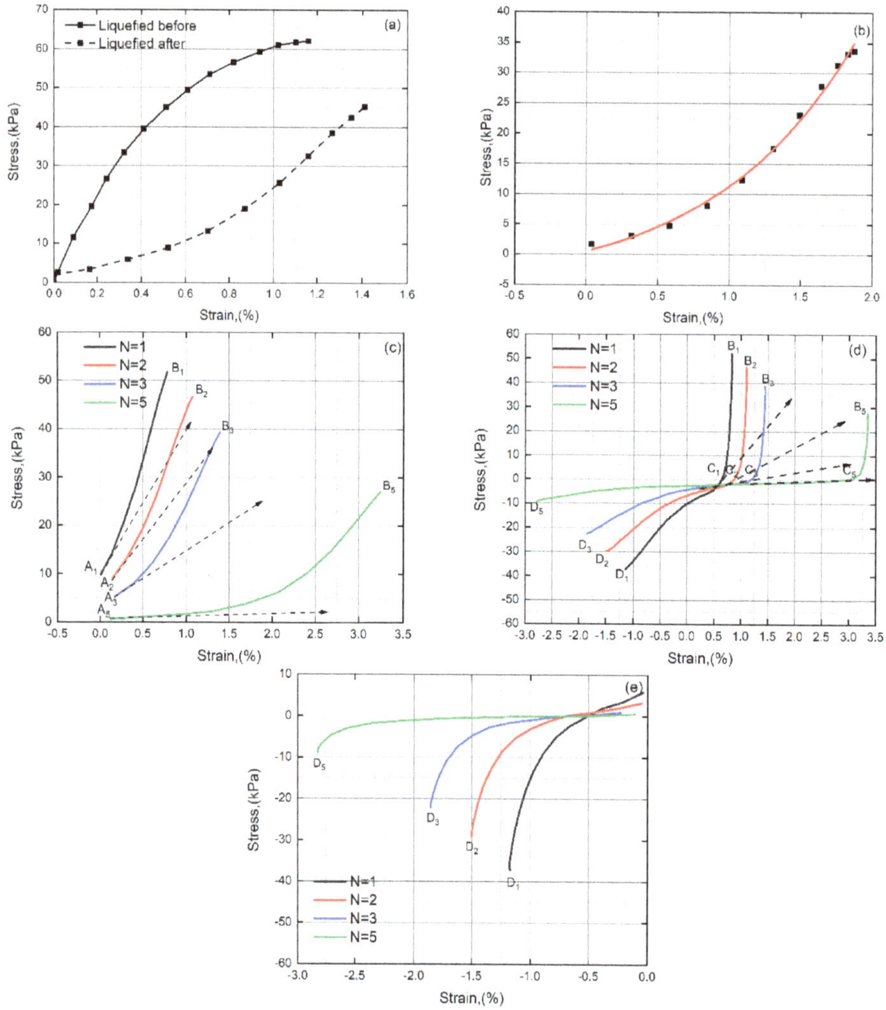

Figure 7. Analysis of experimental results of each section of curves: (**a**) comparison of trunk curves AB before and after liquefaction (**b**) test curve AB fitting results (**c**) comparison of AB trunk curves at the initial loading stage under different cycles (**d**) comparison of unloading curves BCD under different cycles (**e**) comparison of reloading curves DA under different cycles.

- Unloading curve BCD

As shown in Figure 7d, the unloading curve BCD is divided into two sections: in unloading BC, the pore-water pressure decreases, the soil stress decreases sharply, and the stress–strain relationship presents a nonlinear state; when in unloading CD, the pore-water pressure continues to fall, and the stress–strain relationship decreases linearly. Since the shape of BC is familiar to AB, it can also be expressed by the IHF of the AB above:

$$\tau = \frac{\gamma - \gamma_{d1}}{a_1 - b_2(\gamma - \gamma_{d1})} - \tau_{d1} \tag{6}$$

where the coordinate of B is (γ_c, τ_c) and the coordinate of C is (γ_{d1}, τ_{d1}). Point B has a large slope, while point C is the turning point of the two unloading stages and has a small slope.

Because CD is tangent to BC at C, CD can be expressed as follows:

$$\tau = (\gamma - \gamma_{d1})/a_1 - \tau_{d1} \tag{7}$$

- Reloading curve DA

As shown in Figure 7e, the stress at the initial loading point D (γ_c', τ_c') gradually decreases with the accumulation of the number of cycles, and the strain value increases substantially. The stress–strain curve of the DA section can be expressed as:

$$\tau = \frac{\gamma - \gamma_d}{a_2 - b_3(\gamma - \gamma_d)} - \tau_d \tag{8}$$

where (γ_d, τ_d) is the stress–strain of the initial point A of the next cycle. As the number of cycles increases, A's strain can be regarded as a constant value and the stress decreases linearly. When the curve reaches the initial point of the next cycle, the reloading stage ends, the stress cycle ends, and the next cycle begins.

To sum up, BC and AB are similar in shape, DA and AB are symmetric about zero, CD is a straight line, and all the curves are connected by the control points (A, B, C, D). The two loading stages are similar in shape and can be expressed in a functional form. The unloading stage presents different shapes, divided into two unloading stages: (1) the unloading stress–strain curve has the maximum tangent modulus in the initial unloading curve, and the tangent modulus of the curve decreases gradually with the decrease in the strain. (2) In the second stage of the unloading curve, the stress–strain curve decreases linearly.

2.3.2. Loading and Unloading Criteria

The liquefaction of pore-water pressure in the stress cycle has the corresponding fluctuation. The effective stress in the zero value fluctuates up and down. In this paper, when the effective stress reaches zero, the irreversible pore-water pressure does not rise, and the reversible pore-water pressure fluctuates up and down with the change in the cyclic load. The relationship between the reversible pore-water pressure and cyclic stress is similar to before liquefaction. That is, the modulus and strength of the soil after liquefaction are not modified by the pore-water pressure. Based on the above analysis, the following loading and unloading criteria are determined:

The stress–strain relationship satisfies IHF in the first cyclic forward loading process AB after the liquefaction. The curve of AB and the soil before the liquefaction is tangent at point A.

(1) The BC segment satisfies IHF. The slope of point C is equal to the slope of point A. The stress–strain value of point C is related to point B. The relation can be obtained by fitting the test data.
(2) CD section stress–strain relationship is linear. The line is tangent to curve BC at point C.

(3) DA satisfies IHF and goes through the initial point A of the next cycle. This curve is tangent to the initial loading curve of the next cycle.
(4) At the beginning of the new cycle, the tangent modulus at point A is related to soil properties and the number of cycles. The curve is symmetric with the DA curve at about point A.

In the above cyclic loading and unloading process, the tangent modulus a_1 of point A and point C should be obtained by fitting the test data after liquefaction, and the other parameters (b_1, b_2, b_3) can be determined by the inflection point in the stress–strain history.

The large post-liquefaction deformation stress–strain relationship of saturated sand is related to the soil properties and the number of stress cycles after liquefaction. The lower the stress amplitude of the soil, the lower the reduction rate of the tangent modulus of the stress–strain curve is, and the more intensive the peak points B and D of each cycle are.

2.4. The Constitutive Equation

2.4.1. Calibration Parameter

To get the relationship between a_1 and the cyclic action times N, according to the results of the cyclic triaxial test of the saturated coral with a relative density of 45%, the four groups of CSRs were 0.325, 0.3, 0.25, and 0.22. As shown in Figure 8, the variation in a_1 with N after the liquefaction is drawn. The four points in each group of cycles, respectively, represent the test results of four groups of different CSRs. The eight points in the same color are the fitting points of the eight cycles after the sand liquefaction under the action of the same CSR. The abscissa is the number of cycles, and the ordinate is the ratio $a/a_{1,n}$ of the soil pre-liquefaction parameter a to the tangent modulus of each cycle $a_{1,n}$.

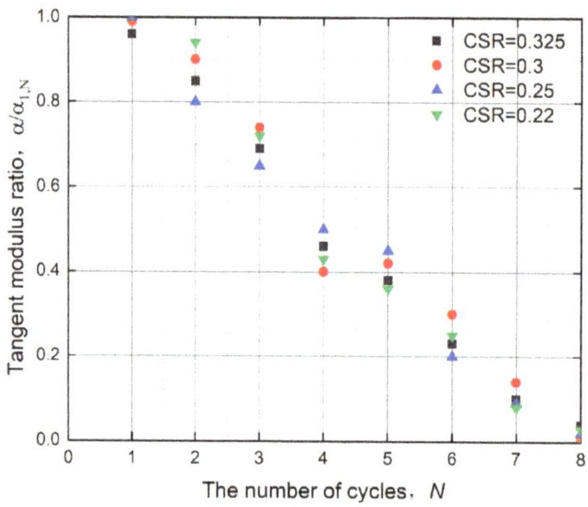

Figure 8. Variation in soil liquefaction parameters with cycles under different CSRs.

The above figure shows that the tangential modulus $a_{1,n}$ of the soil after liquefaction has a consistent relationship with the number of cycles, which does not change with the change in the CSR and is a linear reduction relationship with a reduction rate of about 0.21. The modulus tends to zero at each CSR at the eighth cycle after the liquefaction.

Figure 9a shows the strain ratio at point C and point B in each cycle under different CSRs. Four points in each cycle number represent the test results of the four groups of different CSRs. The points in the same color are the fitting points of eight cycles after liquefaction. The abscissa is the number of cycles, and the ordinate $\frac{\gamma_{d1}}{\gamma_c}$ is the strain ratio at point C and point B. Figure 9b shows the stress ratio τ_{d1}/τ_c at point C and point B in different CSRs.

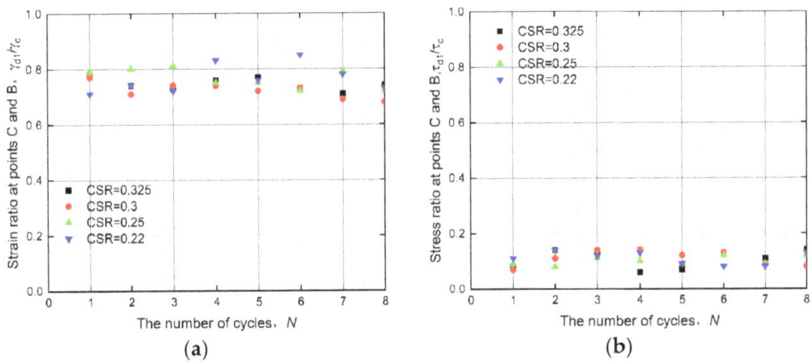

Figure 9. Relationship between point C and point B in each cycle under different CSRs: (**a**) the strain ratio of at point C and point B (**b**) the stress ratio of at point C and point B.

The stress–strain ratio at points C and B of each cycle can be approximately regarded as a constant value, and the relationship does not change with the change in the CSR. The tangent modulus of the curve at point C is consistent with the initial tangent modulus of the cycle.

At this point, we have completed the calibration of the parameters of the large deformation constitutive model of saturated sand. According to the results, the concrete calculation formula of the large deformation constitutive model can be determined by combining the loading and unloading criterion and the primary function of the stress–strain relationship.

2.4.2. Constitutive Relation

After saturated sand is subjected to cyclic loading, the pore-water pressure increment can be obtained by introducing the model. Then, the pore-water pressure increment can be accumulated to obtain the effective stress. The stress–strain relationship at this stage is shown in Equation (1). When the effective stress is zero, the soil layer is considered liquefied, and the stress–strain relationship after liquefaction follows the five large deformation loading and unloading criteria. The stress–strain constitutive relationship of saturated sand after liquefaction is obtained by combining the basic function formula of the stress–strain relationship, as shown in Equation (9):

$$\left.\begin{array}{c} \tau = \frac{\gamma-\gamma_d}{a-b(\gamma-\gamma_d)} - \tau_d, n=1, \Delta\gamma \geq 0, \gamma \geq \gamma_d \\ \tau = \frac{\gamma-\gamma_{d1}}{a_1-b_2(\gamma-\gamma_{d1})} - \tau_{d1}, \Delta\gamma < 0, \gamma \geq \gamma_{d1} \\ \tau = \frac{(\gamma-\gamma_{d1})}{a_1} - \tau_{d1},\ \Delta\gamma < 0, \gamma < \gamma_{d1} \\ \tau = \frac{\gamma-\gamma_d}{a_1-b_3(\gamma-\gamma_d)} - \tau_d, \Delta\gamma \geq 0, \gamma < \gamma_d \\ \tau = \frac{\gamma-\gamma_d}{a_1-b_1(\gamma-\gamma_d)} - \tau_d, n > 1, \Delta\gamma \geq 0, \gamma \geq \gamma_d \end{array}\right\} \quad (9)$$

where:

$$\left.\begin{array}{c} b_2 = \frac{a_1}{\gamma_c-\gamma_{d1}} - \frac{1}{\tau_c-\tau_{d1}} \\ b_3 = \frac{a_1}{\gamma_d-\gamma'_c} + \frac{1}{(\tau'_c-\tau_d)} \\ b_1 = \frac{a_1}{\gamma_d-\gamma'_c} - \frac{1}{2\tau_d-\tau'_c} \end{array}\right\} \quad (10)$$

where n is the number of cycles after liquefaction, the tangential modulus $a_{1,n}$ has a consistent relationship with n, and there is a linear reduction relationship with a reduction rate of about 0.21. $a_{1,n}$ tends to zero at each CSR at the eighth cycle after liquefaction. $A(\gamma_d, \tau_d)$, $B(\gamma_c, \tau_c)$, $C(\gamma_{d1}, \tau_{d1})$, $D(\gamma'_c, \tau'_c)$ and the stress–strain ratio at C and B of each cycle can be approximately regarded as a constant value, and the relationship does not change with the change in the CSR. The tangent modulus of the curve at point C is consistent with $a_{1,n}$. The parameters (b_1, b_2, b_3) can be determined by the inflection point in the stress–strain history.

Equations (1) and (9) constitute the saturated sand's nonlinear dynamic constitutive equations before and after the liquefaction. Their constitutive parameters are few, which can be obtained from undrained cyclic shear test data.

3. Calculation Method

3.1. Time-Domain Seismic Response Analysis

Computational methods to analyze the liquefaction problems solve the equations of motion. The stresses and displacements of the solid particle framework satisfy the equations of the motion and conservation of the mass. The two classic approaches used to discretize a domain in space for a numerical solution are finite differences and finite elements [45–47]. The time-domain finite differences method is used to analyze site nonlinear seismic response [48], but its use is limited due to the soil's lack of post-liquefaction stress–strain relationship. Given this, we develop a program, Soilresp1D, and the large deformation constitutive is embedded into the program. It can calculate the 1D liquefiable site seismic response and analyze the dynamic reaction of saturated sand after liquefaction. In Soilresp1D, finite element and difference methods are applied, and the boundary condition is a multi-transmitting formula [49,50]. The initial conditions are:

$$\left.\begin{array}{r}\tau(0,\ z)=0\\ \gamma(0,z)=0\\ v(0,z)=0\\ \tau(\mathrm{t},0)=0\end{array}\right\} \tag{11}$$

In the formula, $v(t,z)$ is the velocity; $\tau(t,z)$ and $\gamma(t,z)$ are the shear stress and shear strain; and $\rho(z)$ is the density. This paper uses the staggered grid method to solve the initial-boundary value problem (Liao, 2002). Based on the free boundary condition Equation (11), the central difference discrete form of the motion balance equation is as follows:

$$\left.\begin{array}{c}v_1^{p+1}=v_1^p+\frac{\Delta t}{m_1}\tau_1^p\\ v_n^{p+1}=v_n^p+\frac{\Delta t}{m_n}\left(\tau_n^p-\tau_{n-1}^p\right),n=2,3,\ldots,N\\ m_n=\frac{1}{2}(\rho_n h_n+\rho_{n-1}h_{n-1}),n=2,3,\ldots,N\\ m_1=\frac{1}{2}\rho_1 h_1\end{array}\right\} \tag{12}$$

The central difference discrete form of the constraint conditions is as follows:

$$\gamma_n^{p+1}=\frac{\Delta t}{h_n}\left(v_{n+1}^{p+1}-v_n^{p+1}\right)+\gamma_n^p,n=1,\ldots,N \tag{13}$$

The relationship between the shear stress and shear strain proposed in this paper can be expressed as:

$$\tau_n^{p+1}=\tau\left(\gamma_n^{p+1}\right) \tag{14}$$

Equations (12)~(14) give an explicit recursive calculation method for the node velocity of each layer and the stress–strain relationship between the layers.

The recursive formula of a boundary node is as follows:

$$v_{N+1}^{p+1}=v_N^p-v_{I,N}^p+v_{I,N+1}^{p+1} \tag{15}$$

where v_{N+1}^{p+1} is the velocity of the artificial boundary node at time $p+1$, v_N^p is the velocity of the node N at time p, $v_{I,N}^p$ is the velocity of the incident wave at time p at the node N, and $v_{I,N+1}^{p+1}$ is the velocity of the incident wave at time $p+1$ at the artificial boundary.

The above time-domain analysis method is a second-order calculation precision, which is conditionally stable. The stability condition is:

$$\Delta t \leq \min\left(\frac{h_n}{c_n}\right), n = 1, \ldots, N \tag{16}$$

where c_n is the shear wave velocity of the medium in the NTH layer, and h_n is the thickness of the layer, determined by the following equation:

$$h_n \leq \left(\frac{1}{6} \sim \frac{1}{10}\right) T_{min} c_n \tag{17}$$

where T_{min} is the shortest period of input fluctuation with an engineering signification.

According to the given input ground motion, the seismic response of the soil layer at any depth can be calculated by Soilresp1D.

3.2. Algorithm Implementation Process

In the case of the known soil parameters, combined with the above large deformation constitutive relation of saturated sand, the time-domain nonlinear dynamic response analysis method can be used to gradually calculate the saturated sand layer's dynamic response at any time at any section. The detailed operation process of the program can be summarized as follows:

- Read site information. For non-liquefiable clay and other soils, the constitutive parameters were a, b. So, the saturated sand and the liquefaction of the former constitutive parameters a, b, are the liquefied constitutive parameters after $a/a_{1,n}$, γ_{d1}/γ_c, and τ_{d1}/τ_c.
- According to the site model, the stratification of the site, the calculated thickness of each layer, and the calculated time step are determined. Record the number of layers and time nodes.
- According to the staggered grid method, the interlayer node velocity and strain are calculated by displaying the recursive formula.
- According to the stress–strain change trend, judge whether it is the stress–strain inflection point or peak point. If so, record the stress–strain value and calculate the state parameters required in the transportation process.
- They are determining the category of the soil layer. If it is non-liquefiable soil, according to Equation (1), the stress is calculated according to Equation (1). If the saturated sand is not liquefied, the pore-water pressure is calculated according to Equation (2), and the relevant parameters are modified. The stress is calculated according to the effective stress constitutive Equation (1). In liquefied saturated sand, the stress is calculated through the large deformation constitutive Equation (9).

Determine whether this calculation step is the last time node of the topsoil layer. If not, continue to calculate the amount of motion state of the upper soil layer at the next moment. If it is the last time node of the topsoil layer, the calculation ends.

4. Verification

To illustrate our method's reasonability, Soilresp1D is used to analyze the nonlinear seismic response of the experimental model site. By comparing the numerical simulation results and experimental results, the verification of ours constitutive in this paper is demonstrated.

The overlayer of the site consists of clay, saturated sand, and underlying bedrock 32 m thick; the site parameters are given in Table 1, and the nonlinear characteristic parameters are listed in Table 2. In the numerical simulation, the distance Δz is 1 m, and the time step is 0.0025 s. 7–9 m from the top to the bottom is the saturated pine fine sand. The input sine wavelength is 48 s and the period is 1 s, with a peak acceleration of 70 cm/s^2. The undrained cyclic triaxial test of the saturated Nanjingpine fine sand is carried out to compare with the test results. In the test, the effective confining pressure is 100 kPa, the CSR is 0.13, and the loading frequency is 1 Hz.

Table 1. Site calculation model.

Soil	Thickness (m)	V_s (m/s)	Density (t/m^3)
Clay	6	120–142	1.95
Pine fine sand	3	142–153	1.49
Clay	21	153–231	1.95
Bedrock	2	511	2.65

Table 2. Nonlinear characteristic parameters of the site.

Soil	a_1an	γ_d/γ_c	τ_d/τ_c	b/a	$a^1(10^{-3})$	b^1	$c_{1,0}$	$c_{1,a}$	$c_{1,b}$	$A_{4,0}$	C_3	B_3
pine fine sand	0.34	0.75	0.11	1934	2.24	5.20	33.71	0.38	0.56	2.61	0.49	1.0
clay				1160	0.82	5.90						

The stress of the sand at 7 m is consistent with that of the test sand. Now, the simulation results of this soil layer are compared with the test results, as shown in Figure 10. The calculated pore-water pressure is consistent with the measured, which can reflect how the irreversible pore-water pressure gradually increases with the accumulation of cycles and reflect the fluctuation of the reversible pore-water pressure under the action of each cycle load. The sand layer at 7 m began to liquefy at 16.5 s, while the sand at 8 m and 9 m did not liquefy, and the pore-water pressure rose slowly after the liquefaction at 7 m. Due to the small confining pressure of the upper saturated sand layer, it is also liquefied. After the liquefaction of the upper layer, the pore-water pressure of the lower saturated sand layer tends to be gentle, which is consistent with the actual observation phenomenon.

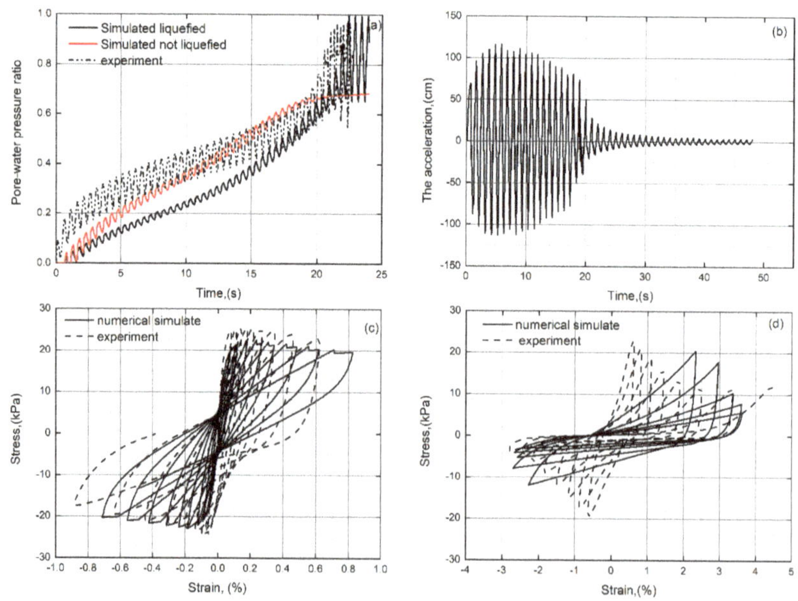

Figure 10. Comparison between site dynamic response and test results under sinusoidal incident: (**a**) time-history of measured and calculated pore-water pressure ratio; (**b**) ground acceleration time-history; (**c**) stress–strain of saturated fine sand before liquefaction; (**d**) stress–strain of saturated fine sand after liquefaction.

Figure 10b: the time-series peak value of ground acceleration is 124 cm/s^2, and the amplification coefficient is 1.77. The ground acceleration attenuates significantly after the liquefaction. Figure 10c: the stress–strain results before the liquefaction show that the numerical simulation results are similar to the test results. The shear modulus decreases with the increase in the number of cycles. Figure 10d shows the large-deformation stress–strain hysteresis loop after the liquefaction, and its variation trend is similar to the test results.

The numerical simulation results show that the amplitude of the ground acceleration reaction (PGA) before the liquefaction is significantly larger than that after the liquefaction. The liquefaction of saturated sand has a noticeable damping effect on the dynamic response of the site, which is consistent with some in situ observations.

5. Conclusions

This paper studied the stress–strain variation, loading, and unloading criterion, and the primary function formula of the saturated sand after the liquefaction, according to the dynamic triaxial test of saturated sand, and proposed a systematic post-liquefaction deformation constitutive model. This paper also set up a time-domain analysis method for simulating the dynamic response of saturated sand before and after liquefaction, combined with the staggered grid method, the multi-transmitting boundary condition, and the effective stress constitutive model before the liquefaction. A 1D time-domain site dynamic response analysis program, Soilresp1D, was developed using the method. According to the experimental results, the main findings are as follows:

(1) The stress–strain curve after the liquefaction can be divided into four loading and unloading stages. IHF can describe the two loading and the first unloading stages. The second unloading stage is linear.
(2) The tangent modulus of the unloading curve connection point is equal to that of the initial loading point. The two unloading curves are tangent at the connection point. The reloading curve is symmetric with the initial loading curve of the next cycle about the initial loading point.
(3) The tangent modulus of the initial loading point decreases linearly with the increase in the number of cycles. The test results show that the initial loading curve of the soil is close to the level after eight times of cyclic loading.

The numerical simulation results show that the liquefaction has a damping effect on the dynamic response of the site, which verifies the model's reliability. This model can be used to analyze the post-liquefaction deformation seismic response in liquefiable sites. It is worth noting that the post-liquefaction deformation constitutive has been verified by the dynamic response under symmetric cyclic loads, which can provide a basis for the study of large deformation constitutive suitable for asymmetric ground motion loads.

Author Contributions: Study concept and design: Q.D., Z.Z. and L.J. Analysis and interpretation of data: Q.D. Drafting of the manuscript: Q.D. Critical revision of the manuscript for important intellectual content: Z.Z. and X.L. Statistical analysis: B.H. Obtained funding: X.L. Study supervision: X.L. All authors have read and agreed to the published version of the manuscript.

Funding: This work was supported by the Major Program of the National Natural Science Foundation of China (52192671) the National Natural Science Foundation of China (U1839202).

Data Availability Statement: The data used to support the findings of this study are available from the corresponding author upon request.

Conflicts of Interest: The authors declare that there is no conflict of interest regarding the publication of this paper.

References

1. Dobry, R.; Powell, D.J. Liquefaction potential of saturated sand-the stiffness method. In Proceedings of the Seventh World Conference on Earthquake Engineering, Istanbul, Turkey, 8–13 September 1980; Volume 3, pp. 25–32.
2. Shahir, H.; Mohammadi, H.B.; Ghassemi, A. Employing a variable permeability model in numerical simulation of saturated sand behavior under earthquake loading. *Comput. Geotech.* **2014**, *55*, 211–223. [CrossRef]
3. Ross, W.B.; Katerina, Z. A constitutive model for clays and plastic silts in plane-strain earthquake engineering applications. *Soil Dyn. Earthq. Eng.* **2019**, *127*, 1–15.
4. Erlei, Y.; Suyang, W.; Bin, R.; Yu, M.; Luhua, Z. Numerical study on site response considering ground motion spatial variation. *Soil Dyn. Earthq. Eng.* **2019**, *127*, 361–366.
5. Ishibashi, I.; Sherif, M.A.; Tsuchiya, C. Pore-Pressure Rise Mechanism and Soil Liquefaction. *J. Jpn. Soc. Soil Mech. Found. Eng.* **1977**, *17*, 17–27. [CrossRef] [PubMed]
6. Ishibashi, M.; Tsuchiya, C. Pore-pressure prediction during earthquake loadings. *Soils Found.* **1978**, *18*, 19–30.
7. Wang, R.; Zhang, J.M.; Wang, G. A unified plasticity model for large post-liquefaction shear deformation of sand. *Comput. Geotech.* **2014**, *59*, 54–66. [CrossRef]
8. Kavazanjian, E.; Andrade, J.E.; Arulmoli, K.A.; Atwater, B.F.; Wang, Y. *State of the Art and Practice in the Assessment of Earthquake-Induced Soil Liquefaction and Its Consequences*; National Academies of Sciences, Engineering, and Medicine: Washington, DC, USA, 2016.
9. Alba, P.D.; Seed, H.B.; Chan, C.K. Sand Liquefaction in Large-Scale Simple Shear Tests. *J. Geotech. Eng. Div.* **1976**, *102*, 909–927. [CrossRef]
10. Chen, G.; Jin, D.; Zhu, J.; Shi, J.; Li, X. Nonlinear analysis on seismic site response of fuzhou basin, China. *Bull. Seismol. Soc. Am.* **2015**, *105*, 928–949. [CrossRef]
11. Chen, G.; Zhao, D.; Chen, W.; Juang, C.H. Excess pore-water pressure generation in cyclic undrained testing. *J. Geotech. Geoenviron.* **2019**, *145*, 04019022. [CrossRef]
12. Chen, G.; Ruan, B.; Zhao, K.; Chen, W.; Zhuang, H.; Du, X.; Khoshnevisan, S.; Juang, C.H. Nonlinear response characteristics of undersea shield tunnel subjected to strong earthquake motions. *J. Earthq. Eng.* **2020**, *24*, 351–380. [CrossRef]
13. Olgun, C.; Martin, I.; LaVielle, T. *Award Number 07HQGR0043*; Liquefaction Susceptibility of Calcareous Sediments along the Coastal Plains of Puerto Rico. Virginia Polytechnic Institute and State University Blacksburg: Blacksburg, VA, USA, 2009.
14. Finn, W.D.L. State-of-the-art of geotechnical earthquake engineering practice. *Soil Dyn. Earthq. Eng.* **2000**, *20*, 1–15. [CrossRef]
15. Ruan, B.; Zhao, K.; Wang, S.; Chen, G.; Hai-Yun, W. Numerical modeling of seismic site effects in a shallow estuarine bay (Suai Bay, Shantou, China). *Eng. Geol.* **2019**, *260*, 105233. [CrossRef]
16. Sturm, H. Numerical investigation of the stabilisation behaviour of shallow foundations under alternate loading. *Acta Geotech.* **2009**, *4*, 283–292. [CrossRef]
17. Sfriso, A.; Weber, G. Formulation and validation of a constitutive model for sands in monotonic shear. *Acta Geotech.* **2010**, *5*, 257–272. [CrossRef]
18. Mele, L.; Tian, J.T.; Lirer, S.; Flora, A.; Koseki, J. Liquefaction resistance of unsaturated sands: Experimental evidence and theoretical interpretation. *Geotechnique* **2019**, *69*, 541–553. [CrossRef]
19. Cubrinovski, M.; Rhodes, A.; Ntritsos, N.; Van Ballegooy, S. System response of liquefiable deposits. *Soil Dyn. Earthq. Eng.* **2019**, *124*, 212–229. [CrossRef]
20. Wu, J.; Kammerer, A.M.; Riemer, M.F. Laboratory study of liquefaction triggering criteria. In Proceedings of the 13th World Conference on Earthquake Engineering, Vancouver, BC, Canada, 1–6 August 2004.
21. Sawicki, A.; Morland, L.W. Pore pressure generation in a saturated sand layer subjected to a cyclic horizontal acceleration at its base. *J. Mech. Phys. Solids* **1985**, *33*, 545–558. [CrossRef]
22. Elgamal, A.; Yang, Z.H.; Parra, E. Computational modeling of cyclic mobility and post-liquefaction site response. *Soil Dyn. Earthq. Eng.* **2002**, *22*, 259–271. [CrossRef]
23. Yuan, X.; Rui, S.; Chen, L.; Tang, F. A method for detecting site liquefaction by seismic records. *Soil Dyn. Earthq. Eng.* **2010**, *30*, 270–279. [CrossRef]
24. Sun, R.; Li, X.F.; Chen, L.W. Effects of increase in pore water pressure on dynamic parameters of hyperbolic model describing stress-strain relation of liquefiable soil. *J. Vib. Shock.* **2018**, *37*, 1–6.
25. Chen, G.; Wang, Y.; Zhao, D.; Zhao, K.; Yang, J. A new effective stress method for nonlinear site response analyses. *Earthq. Eng. Struct. Dyn.* **2021**, *50*, 1595–1611. [CrossRef]
26. Wang, R.; Fu, P.; Zhang, J.M. Finite element model for piles in liquefiable ground. *Comput. Geotech.* **2016**, *72*, 1–14. [CrossRef]
27. Lu, Y.; Zhu, W.X.; Ye, G.L.; Zhang, F. A unified constitutive model for cemented/non-cemented soils under monotonic and cyclic loading. *Acta Geotech.* **2021**, *17*, 1–21. [CrossRef]
28. Zhi, H.; Liang, W.; Yannis, F.D.; Chih, H.; Kang, S. Bounding surface hypoplasticity model for sand. *J. Eng. Mech.* **1990**, *116*, 983–1001.
29. Yang, Z.; Elgamal, A. Computational Model for Cyclic Mobility and Associated Shear Deformation. *J. Geotech. Geoenviron. Eng.* **2003**, *129*, 1119–1127. [CrossRef]
30. Elgamal, A.; Yang, Z.; Parra, E.; Ragheb, A. Modeling of cyclic mobility in saturated cohesionless soils. *Int. J. Plast.* **2003**, *19*, 883–905. [CrossRef]

31. Dafalias, Y.F.; Manzari, M.T. Simple Plasticity Sand Model Accounting for Fabric Change Effects. *J. Eng. Mech.* **2004**, *130*, 622–634. [CrossRef]
32. Boulanger, R.W.; Ziotopoulou, K. *PM4Sand Version 3: A Sand Plasticity Model for Earthquake Engineering Applications*; University of California: Los Angeles, CA, USA, 2015.
33. Asgaria, A.; Oliaei, M.; Bagheri, M. Numerical simulation of improvement of a liquefiable soil layer using stone column and pile-pinning techniques. *Soil Dyn. Earthq. Eng.* **2013**, *51*, 77–96. [CrossRef]
34. Shamoto, Y.; Zhang, J.M.; Tokimatsu, K. New charts for predicting large residual post-liquefaction ground deformation. *Soil Dyn. Earthq. Eng.* **1998**, *17*, 427–438. [CrossRef]
35. Xu, C.; Song, J.; Du, X.; Zhong, Z. A completely explicit finite element method for solving dynamic up equations of fluid-saturated porous media. *Soil Dyn. Earthq. Eng.* **2017**, *97*, 364–376. [CrossRef]
36. Song, J.; Gao, Y.; Feng, T. Influence of interactions between topographic and soil layer amplification on seismic response of sliding mass and slope displacement. *Soil Dyn. Earthq. Eng.* **2020**, *129*, 1–17. [CrossRef]
37. Dong, Q.; Zhou, Z.; Su, J.; Li, X. Constitutive model for effective stress based on logarithmic skeleton curve considering reversible pore pressure. *Chin. J. Geotech. Eng.* **2020**, *42*, 2322–2329.
38. Chu, F.; Shao, S.J.; Chen, C.L. Experimental study of dynamic deformation and dynamic strength properties of saturated silty sand. *Chin. J. Rock Mech. Eng.* **2014**, *33* (Suppl. S1), 3299–3305.
39. Rahman, M.; Seed, H.; Booker, J. Pore pressure development under offshore gravity structures. *J. Geotech. Eng. Div.* **1977**, *103*, 1419–1436. [CrossRef]
40. Sun, R.; Zhao, Q.; Yuan, X. Comparison between acceleration response spectra on liquefaction and non-liquefaction sites. *Rock Soil Mech.* **2014**, *35* (Suppl. S1), 299–305.
41. Bin, R.; Zhao, D.; Chen, G.; Zhu, X. A New Effective Stress Method Using the Integration Algorithm of the Modified Davidenkov Constitutive Model and the Byrne Pore Pressure Increment Model, and Validation and Calibration. *J. Basic Sci. Eng.* **2017**, *25*, 956–966.
42. Basu, D.; Montgomery, J.; Stuedlein, A.W. Observations and challenges in simulating post-liquefaction settlements from centrifuge and shake table tests. *Soil Dyn. Earthq. Eng.* **2022**, *153*, 107089. [CrossRef]
43. Ishihara, K.; Yasuda, S.; Nagase, H. Soil characteristics and ground damage. *Soils Found. Tokyo* **1996**, *36*, 109–118. [CrossRef]
44. Dong, Q.; Su, J.; Zhou, Z.; Li, X. Time-domain constitutive model based on logarithmic skeleton curve and its application. *Chin. J. Geotech. Eng.* **2020**, *42*, 1491–1498.
45. Ziotopoulou, K.; Boulanger, R.W.; Kramer, S.L. Site Response Analysis of Liquefying Sites. *GeoCongress* **2012**, *2012*, 1799–1808.
46. Zienkiewicz, O.C.; Chan, A.; Pastor, M.; Schrefler, B.A.; Shiomi, T. *Computational Geomechanics with Special Reference to Earthquake Engineering*; Wiley: Hoboken, NJ, USA, 1999; p. 80.
47. Moin, P. Fundamentals of engineering numerical analysis. *Can. J. Civ. Eng.* **2002**, *29*, 344–345. [CrossRef]
48. Lu, T.; Zhou, Z.; Huo, J. 1D nonlinear seismic response analysis of soil layers in time domain. *Rock Soil Mech.* **2008**, *8*, 2170–2176.
49. Liao, Z.P.; Yang, G. Multi-directional transmitting boundaries for steady-state SH waves. *Earthq. Eng. Struct. Dyn.* **2010**, *24*, 361–371. [CrossRef]
50. Liao, Z. *Introduction to Wave Motion Theories in Engineering*; Science Press: Beijing, China, 2002; pp. 62–63.

sustainability

Article

A Structure Economic Loss Optimization Method with the Uncertainty of Ground Motion Amplitude for Chinese Masonry Building

Jinpeng Zhao [1], Xiaojun Li [2] and Chen Liu [3,*]

[1] Faculty of Urban Construction, Beijing University of Technology, Beijing 100124, China
[2] Key Laboratory of Urban Security and Disaster Engineering of China Ministry of Education, Beijing University of Technology, Beijing 100124, China
[3] China Re Catastrophe Risk Management Company Ltd., Chongqing 400025, China
* Correspondence: liuchen@chinarecrm.com.cn

Abstract: In the catastrophe insurance industry, it is impractical for a catastrophe model to simulate millions of sites' environments in a short time. Hence, the attenuation relation is often adopted to simulate the ground motion on account of calculation speed, and both ground motion expectations and uncertainties must be calculated. Due to the vulnerability curves of our model being based on simulations with a large number of deterministic ground motions, it is necessary but not efficient for loss assessment to analyze all possible ground motion amplitudes and their corresponding loss rates. This paper develops a simplified method to rapidly simulate loss expectations and uncertainties. In this research, Chinese masonry buildings are the focus. The result shows that the modified method gives accurate loss results quickly.

Keywords: ground motion intensity uncertainty; masonry building; economic vulnerability curves; seismic risk management; earthquake catastrophe model

Citation: Zhao, J.; Li, X.; Liu, C. A Structure Economic Loss Optimization Method with the Uncertainty of Ground Motion Amplitude for Chinese Masonry Building. *Sustainability* **2022**, *14*, 13860. https://doi.org/10.3390/su142113860

Academic Editor: Claudia Casapulla

Received: 30 September 2022
Accepted: 21 October 2022
Published: 25 October 2022

1. Introduction

Since the beginning of the 21st century, the frequency of natural disasters has been on the rise, and with the development of urbanization and industrialization, economic losses and casualties caused by natural disasters have become more and more serious. According to the report issued by the United Nations International Disaster Reduction Agency [1], from 1998 to 2017, climate and geological disasters caused 1.3 million deaths, more than 4.4 billion injuries, and even homelessness, resulting in direct economic losses of up to USD 298 billion, which is 251% higher than those caused by natural disasters in the past 20 years (1978–1997). The seismic intensity of the Wenchuan Earthquake in 2008 reached XI degree, and the building collapse caused serious casualties and infrastructure damage that could not provide normal service functions, which led to the shutdown of enterprises and huge economic losses. From the perspective of industry, the impact of the Great East Japan Earthquake in 2011 was more significant. After the earthquake, the production capacity of the affected areas decreased by 30~50% directly, the recovery time of production capacity was slow, and even 4 years after the earthquake, the industrial production capacity of some affected areas still had not recovered to the pre-earthquake level [2]. The risk management of natural disasters has been paid more and more attention to, and how to scientifically estimate the losses caused by natural disasters has become one of the urgent issues to be solved.

Loss assessment in catastrophe modelling has some differences from that in theoretical research. The loss needs to be analyzed and calculated by using the earthquake risk module and earthquake vulnerability module in combination with the object value [3]. As insurance and reinsurance claims are based on events, for a single object, the output of the earthquake risk module is the estimated result of the ground motion parameters of each simulated event. The earthquake vulnerability is the damage rate of corresponding ground motion parameters, and these curves composed of the economic loss ratio and ground motion parameters are called vulnerability curves in catastrophe modelling [4]. These are different from the structural vulnerability curves in earthquake engineering but are related to structural vulnerability. In seismic engineering, the structural vulnerability curve is usually the statistical analysis of the correlation between the seismic damage investigation data of the structure under a large number of ground motions or the performance data of the structure after elastic–plastic analysis and the ground motion parameters [5]. In catastrophe modelling, the structural vulnerability in earthquake engineering is often referred to as fragility. Vulnerability in the insurance industry represents the weighted sum of the damage ratio and the probability of all different damage levels of a single structure under different earthquake incentives [3].

There are many uncertainties in earthquake risk management, which need to be considered as much as possible when estimating the loss model. Abrahamson [6] and FEMA-P58 [7] indicated that uncertainty may exist in the characteristics of earthquake hazards as well as in the properties of geometric shape and material. Cutfield and Ma [8] studied three fragility curves under different cases, which included two curves without uncertainty and one curve with demand uncertainty, and recommended considering the demand uncertainty when dealing with the post-earthquake data. In addition, De Risi et al. [9] indicated that, when processing fragility curves, the uncertainty of input data should be adequately considered; otherwise, the uncertainty of the output result may be amplified. International scholars have carried out some research on the uncertainty of earthquake ground motion and structural response. In terms of the influence of uncertainties of different factors, Kwon and Elnashai [10], Soleimani [11], and Pan et al. [12] have different thoughts. Kwon and Elnashai [10] indicated that the influence of the randomness of materials on the fragility curve is much less than that of the randomness of ground motion properties for reinforced concrete buildings. Soleimani [11] stated that, compared to the uncertainty of ground motion and materials, the geometric characteristics were the most sensitive for the high box beam concrete bridge. In contrast, Pan et al. [12] found that, in the elastic model, the influence of the uncertainty of the structural geometry on the structural earthquake response is much less than the uncertainty of the material properties, such as damping ratio and yield strength for the transmission tower. The reason for the aforementioned result may be caused by their different structure types. Hence, this study pays more attention to the uncertainty of ground motion parameters. Yamin et al. [13] carried out non-linear time-history analysis based on a 3D structure model and used Monte Carlo simulation to simulate the time and cost of building restoration, ground motion properties, structural response, and building destruction, and finally obtained the fragility of reinforced concrete frame structures. Yazgan [14] provided an empirical method to build a fragility model of the reinforced concrete structure that considers the uncertainty caused by the lack of peak ground motion data. Ansari et al. [15] studied the impact of base types on the fragility of high-rise reinforced concrete buildings, and the seismic uncertainty was considered by adopting a certain number of near-field and far-field earthquakes. Choudhury and Kaushik [16] studied the fragility curve of the steel–concrete frame structure with and without brickwork infill wall, and the open steel–concrete frame structure on the ground floor by analyzing three uncertainties, including the performance difference of non-linear materials, the structural geometric characteristics, and the load situation. Kim et al. [17] analyzed the impact of construction quality defects on the seismic fragility of reinforced concrete frame structures and stated that the seismic fragility of the selected reinforced concrete frame is sensitive to the strength of concrete and the volumetric ratio of transverse reinforcement. Saloustros et al. [18], using the finite element method and the Monte Carlo method, investigated the influence of uncertainty on materials of masonry structure components and present the fragility curve of Santa Maria del Mar church. However, European historical masonry buildings, such as churches and cathedrals, are different from Chinese masonry buildings, which are built with clay bricks instead of stone. In China, many masonry buildings in big cities, such as Beijing and Shanghai, were built from the 1970s to the 1990s to solve the residential problem during urbanization. However, the seismic capacity of a multistory masonry building is poorer than the reinforcement concrete (RC) structure, which means the loss of a masonry building will be more than for RC under a damaging earthquake. Hence, as a more vulnerable structure, the Chinese masonry buildings are selected as calculation examples. Jiang et al. [19] studied four unreinforced masonry structures with different floors, established the finite element model by OpenSees, and discussed the influence of uncertainty of ground motion and structural parameters on fragility by the IDA method and the one-variable quadratic matrix method. It was pointed out that, with the increase of failure degree, the greater the influence of uncertainty on fragility, and, compared with the seismic uncertainty, the structural parameter uncertainty has a more obvious influence on a structure with a lower number of stories, and the structural damping ratio is the most sensitive among all the influence factors.

Although the aforementioned research considered the uncertainties of ground motion and structural response, the method of amplitude adjustment is generally selected to process different known seismic excitations. The amplitude, spectrum characteristics, and duration of each set of seismic excitations are all bound together and fixed. The amplitude adjustment of ground motion means that only the uncertainty of spectrum characteristics and duration is considered in the fragility and vulnerability curve. Potential future earthquakes will not exactly match the fixed combination of amplitude by ground motion prediction equations (GMPEs), spectrum characteristics, and duration of ground motions for historical earthquakes. Most seismic risk assessments are qualification models presented by earthquake intensity parameters and damage probability, and each step of the risk model has uncertainties that should be considered [20]. In addition, Hwang et al. [21] indicated that all the uncertainties need to be considered in the ideal structure model with possible seismic ground motion when seismic vulnerability assessment was processed. Hence, if GMPEs are used to describe the ground motion intensity and estimate the loss, the uncertainties of the calculated amplitude should be considered. For the insurance industry, simplifying the model and saving calculation time is very important, and it is necessary to use a parametric empirical statistical model, such as GMPEs, to get a loss result as soon as possible. All possible ground motion amplitude should be theoretically simulated in this method. However, it is inefficient to take every sample for each site to calculate the damage ratio. Therefore, the uncertainty of ground motion and vulnerability is coupled to generate a modified vulnerability curve, and the results are expressed by the combined mean and variance. The negative effect of this method is that the combined distribution is different from the sampling distribution, but the mean and variance are consistent, and the calculation speed is greatly improved.

The current study aims to overcome the aforementioned issues, such as the lack of consideration of uncertainties and computational efficiency. In this research, PGA (Peak Ground-motion Acceleration) amplitude distribution is applied to create a modified vulnerability curve to simplify loss ratio expectation and its uncertainty to make rapid calculations. The method proposed in this study can not only calculate massive objects' economic loss and their vulnerability curves in a short time, but also deepen the understanding of how the interaction of uncertainties affects the damage results of objects. It provides a novel idea and a risk-based with the consideration of uncertainties solutions for practical application in the catastrophe insurance industry.

2. Methodology

The attenuation relation model is a parameterized empirical statistical model, which is built by a large amount of data of different magnitudes and distances in a certain area. It has a general rule of seismic response and statistical significance. Specifically, in the insurance industry, the business contract usually calls for loss reports as soon as possible during the renewal season. If using numerical simulation to calculate the ground motion, it usually consumes lots of time. In contrast, attenuation is an efficient and time-saving way to acquire ground motion information. Therefore, this study uses the attenuation relation model to calculate the ground motion parameters.

Due to the seismic source, wave propagation, and site effect, the ground motion produced by the same earthquake in different sites or caused by different earthquakes in the same site will be accompanied by the uncertainty of ground motion. In this study, the samples of PGA distribution are adopted instead of the mean value to analyze the influence of loss assessment results. Abrahamson [22,23] put forward that the influence of the uncertainty of the ground motion acceleration spectrum conforms to a lognormal distribution. In addition, Rossetto and Elnashai [24], Lagomarsino and Giovinazzi [25], Bradley and Dhakal [26], Jayaram and Baker [27], and Lallemant et al. [28] stated that a lognormal cumulative distribution function (CDF) is a frequently used method to model seismic structure fragility curves, and it has a significant representation in earthquake risk analysis. Furthermore, according to the model from the *Seismic ground motion parameters zonation map of China (GB18306-2015)* [29], as shown in Formula (1), the ground motion parameters are also in line with a lognormal distribution:

$$lgY = A + BM + Clg\left(R + De^{EM}\right) \quad (1)$$

where Y is the ground motion parameter, such as peak ground motion acceleration (PGA) or peak ground motion velocity (PGV), M is the magnitude, R is the epicentral distance, and A, B, C, D, and E are regression coefficients.

2.1. Data

The premise of obtaining the vulnerability curve is to have the fragility curve of an object first. The steps of obtaining the fragility curve by the IDA (Incremental Dynamic Analysis) method are usually to build the structural model and calculate the dynamic equation, then use the corresponding calculation and analysis methods and input ground motion parameters; next, calculate the structural response, and finally, according to the relationship between the seismic response and the structural limit under different strength levels, determine the probability of different failure states of the structure under different ground motion parameters [30]. Gattulli et al. [31] indicated that masonry structure is extremely variable in mechanical characteristics due to its features, such as anisotropy, inhomogeneity, and nonlinearity. Hence, the uncertainty of seismic ground motion intensity may cause a greater variation in the loss result for masonry buildings.

The dataset is obtained from the fragility curves of 3–6 floor masonry buildings with different occupancies [32]. In this paper, based on the above dataset, the fragility curve is established by the IDA method with the interlayer elongation coefficient as the index of failure [32]. According to the five elements of floors, design intensity, design code, located area, and building occupancy, a total of 84 fragility curves are formed, and the detailed classification is shown in Table 1. Based on the 84 fragility curves, the vulnerability curves are established, then the uncertainty of seismic ground motion intensity is considered, and the calculation method of loss with uncertainty is established to obtain the modified vulnerability curve. The structures and their curves, listed in Table 2, will be shown as examples in this study. In addition, the Chinese seismic design code has a strict limitation on the height and floors of masonry buildings. The building located zone is applied to distinguish the outside wall thickness, and 240 mm, 370 mm, and 490 mm stand for zone III, II, and I, respectively. Figure 1 shows the typical masonry structure in China, where the wall thickness is usually 240 mm and 370 mm, respectively, in Sichuan and Beijing. Figure 1a is a two-floor masonry structure built following 01-code with 7-degree seismic design, which can represent structure 8 in Table 2. The building in Figure 1b represents structure 26 in this study, which is built following the 01-code.

Table 1. The classification of the classic Chinese masonry building.

Building Occupancy	Residential						Commercial			
Design Code	1989-Code			2001-Code			1989-Code		2001-Code	
Floors	1–3	4	4–6	1–3	4	4–6	1–2	3–5	1–2	3–5
Design Intensity	6, 7, 8	9	6, 7, 8	6, 7, 8	9	6, 7, 8	6, 7, 8, 9	6, 7, 8	6, 7, 8, 9	6, 7, 8
Outside Wall Thickness (mm)	240, 370, 490			240, 370, 490			240, 370, 490		240, 370, 490	

Table 2. The information on structure and curve examples.

Structure and Curve	Building Occupancy	Building Age	Floors	Design Intensity	Outside Wall Thickness
0	Residence	89-code	1–3	6	240
8	Residence	01-code	1–3	7	240
9	Residence	01-code	1–3	8	240
10	Residence	01-code	4–6	6	240
16	Residence	89-code	1–3	8	370
19	Residence	89-code	4–6	8	370
24	Residence	01-code	4–6	6	370
26	Residence	01-code	4–6	8	370
27	Residence	89-code	1–3	6	240
51	Commerce	01-code	1–3	8	240
52	Commerce	01-code	5	6	240
54	Commerce	01-code	5	8	240
66	Commerce	01-code	5	6	370
83	Commerce	01-code	4	9	490

(a) (b)

Figure 1. The typical masonry building in China: (**a**) the self-built masonry house in Sichuan, China; (**b**) the residential masonry building in Beijing, China.

The calculation process is shown in Figure 2. First, each PGA in the original vulnerability curve data is sampled to obtain PGA samples. Next, the PGA samples are transformed into damage ratio samples by the interpolation method. Next, the damage ratio samples are sampled again to get the second damage ratio sample. Finally, the means of second damage ratio samples are calculated to obtain the final damage ratio corresponding to each original PGA.

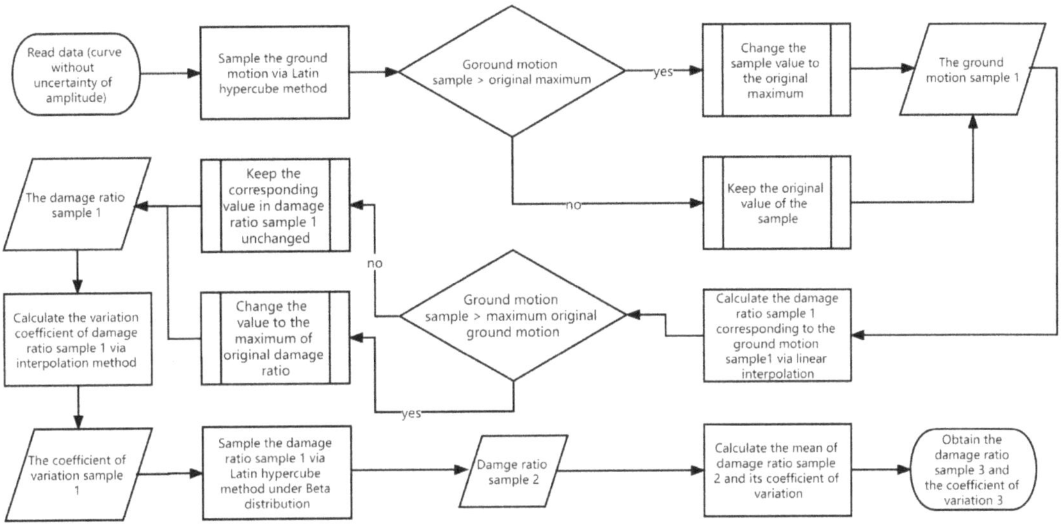

Figure 2. Basic calculation flow of the algorithm.

2.2. Sampling of Ground Motion Parameters

The optimization of ground motion uncertainty originates from quantifying the regular and repeatable parts of random uncertainty and quantifying them in the final risk results. The uncertainty of ground motion is generally reflected by the standard deviation in the attenuation relationship of ground motion. If the standard deviation obtained via ergodic assumption is divided carefully and the identifiable part is quantified, the uncertainty can be reduced, and the calculated result is closer to the real value [33]. In this study, the standard deviation is acquired by the attenuation relation model. However, because the coefficients of the attenuation relation model are diverse, they depend on the geological information of the different areas. Hence, for demonstration, a basic attenuation relation model is adopted in this study. According to the PGA attenuation coefficient and attenuation relationship model (Formula (1)) given in the publicity and implementation textbook of Seismic ground motion parameters zonation map of China [34], the original PGA is sampled. The ground motion parameter PGA conforms to the lognormal distribution, and its standard deviation is a constant that equals 0.236. There are 197 PGA points in the original data, which are arranged from 0.4 m/s^2 to 20.0 m/s^2 in steps of 0.1 m/s^2. Because the Monte Carlo method consumes much computational power, and this algorithm needs to sample twice, the consumption of computational power will increase exponentially. Latin hypercube sampling is adopted to keep the function monotonically increasing, as shown in Figure 3. It is a masonry structure of a residential building with three stories and six degrees of design intensity in site class III with 89-code. Ghotbi and Taciroglu [35] indicated that the Latin hypercube method has a better effect than the Monte Carlo method under the same number of samplings via studying the damage analysis of a normal 4-story non-ductile reinforced concrete structure when selecting seismic ground motion sets.

In addition, the bias of different Latin hypercube sampling times was tested, and the results are shown in Table 3. Compared with 100 sampling times, the bias of 500 sampling times is much smaller.

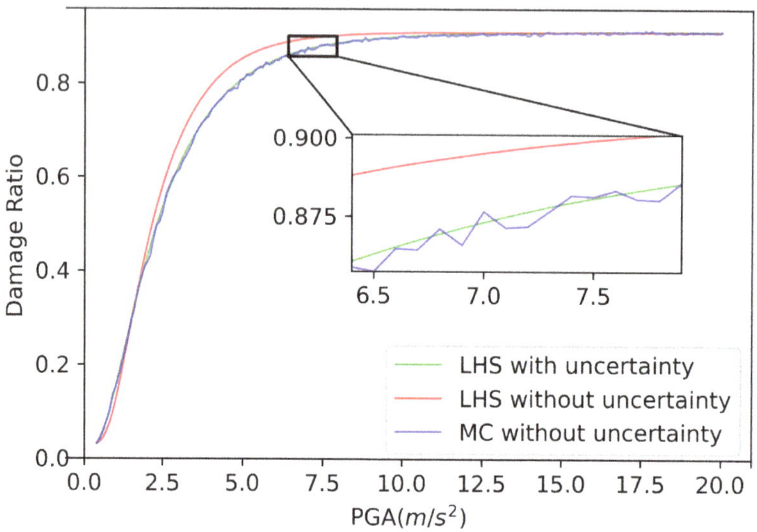

Figure 3. Comparing the curves using LHS (Latin Hypercube Sampling) and MC (Monte Carlo).

Table 3. The bias of different sampling times.

Sampling Times	Bias
100	0.5%
200	0.07~0.08%
500	0.02~0.03%

Therefore, the Latin hypercube sampling method is adopted in this algorithm, which can not only reduce the computational power consumption but also satisfy the empirical judgment. In this study, each PGA point is sampled 20 times with truncation in the corresponding lognormal distribution, and the specific truncation method will be introduced in the following. The 3940 PGA samples obtained from the sampling are named ground motion sample 1. It is substituted into the linear relationship between the original PGA and the original damage ratio, and the corresponding 3940 damage ratios are obtained by using the interpolation method, which is named damage ratio sample 1. In the damage ratio sample 1, if the corresponding PGA is larger than 20.0 m/s^2 (maximum value of the original PGA), the damage ratio takes the value corresponding to 20.0 m/s^2.

Currently, there are 3940 damage ratio samples, but only 197 corresponding coefficients of variation. Therefore, before sampling the damage ratio, the coefficient of variation should be expanded to 3940. Assuming that the corresponding coefficient of variation follows the linear relationship composed of the initial 197 damage ratios and the coefficient of variation, the corresponding 3940 coefficients of variation can be obtained via the linear interpolation method, namely, the coefficient of variation sample 1. In the process of linear interpolation, the sample value of the damage ratio is less than the original value. In this paper, the original minimum damage ratio is connected with the zero point to interpolate the sample value less than the minimum original minimum damage ratio. It is worth noting that the known condition is the coefficient of variation, so the corresponding standard deviation needs to be calculated via Formula (2), and then linear interpolation must be performed. The results of interpolating Cv first and then calculating σ are different from calculating σ and then interpolating.

$$\sigma = Cv \times \mu \tag{2}$$

where σ is the standard deviation, Cv is the coefficient of variation, and μ is the mean value.

According to the attenuation relationship given by the Seismic ground motion parameters zonation map of China (Standardization Administration 2015) [30], the ground motion conforms to the lognormal distribution with a standard deviation equal to 0.236. When the translation between seismic intensity and ground motion parameters is processed, the seismic intensity usually corresponds to a PGA value. In fact, a seismic intensity should correspond to a range of PGA with distribution instead of its mean value. The sample range needs to be truncated because the ground motions cannot tend to infinity or 0. In this study, hence, the correlation between PGA and seismic intensity of site class II in GB 18306-2015 Seismic ground motion parameters zonation map of China [30] was used to determine the truncated range within the corresponding seismic intensity range. Table 4 shows the details of truncation.

Table 4. The comparison table of PGA and seismic intensity for site class II.

PGA in the Site Class II (m/s^2)	$0 \le \alpha_{maxII} < 0.392$	$0.392 \le \alpha_{maxII} < 0.882$	$0.882 \le \alpha_{maxII} < 1.862$	$1.862 \le \alpha_{maxII} < 3.724$	$3.724 \le \alpha_{maxII} < 7.35$	$7.35 \le \alpha_{maxII} < 14.7$	$\alpha_{maxII} \ge 14.7$
Seismic intensity	V and below	VI	VII	VIII	IX	X	XI and above

In this algorithm, for each ground motion parameter P, the algorithm finds seismic intensity level IP using Table 4 and uses [IP − 1, IP + 1] as the initial range. The left endpoint of the new range is the left endpoint of the minimum value of the three seismic intensities, which is recorded as the temporary left endpoint. The right endpoint of the new range is the left endpoint of the maximum value of the three seismic intensities, which is recorded as the temporary right endpoint. For example, the initial range of 6-level seismic intensity is the range of the entirety of the 5-level, 6-level, and 7-level. Next, according to the original ground motion parameter data, its quantile is obtained in its corresponding seismic intensity range (an intensity range). The final truncated range of the original ground motion parameter data is calculated by using the temporary right endpoint, temporary left endpoint, and quantile. The calculation process is shown in Formulas (3)–(5), which are equivalent to the range of two seismic intensities.

$$Q_i = (PGA_i - PGA_{i,j})/(PGA_{i,j+1} - PGA_{i,j}) \tag{3}$$

$$Lower_i = PGA_{i,j-1} + Q_i(PGA_{i,j} - PGA_{i,j-1}) \tag{4}$$

$$Upper_i = PGA_{i,j+1} + Q_i(PGA_{i,j+2} - PGA_{i,j+1}) \tag{5}$$

where Q_i is the quantile of PGA_i in its corresponding seismic intensity range, PGA_i is the ith element in the original PGA,$PGA_{i,j}$ is PGA_i corresponding to the left end of the seismic intensity range, $PGA_{i,j+1}$ is PGA_i corresponding to the right end of the seismic intensity range, $Lower_i$ is the lower limit of the final truncation range, $Upper_i$ is the upper limit of the final truncation range, $PGA_{i,j-1}$ is the temporary left endpoint, and $PGA_{i,j+2}$ is the temporary right endpoint.

When PGA_i is in the range corresponding to seismic intensity V and below, or XI and above, the truncation range will be different from Formulas (3)–(5). When PGA_i is in the range of seismic intensity V and below, due to it is the minimum seismic intensity range, it is unable to obtain a smaller level, so its truncation lower limit is fixed equal to 0, and the truncation upper limit still follows the algorithm of Formula (5). When PGA_i is in the range of seismic intensity XI and above because it is the largest seismic intensity range, it is unable to obtain a higher level, so its upper truncation limit is fixed at 20 (the maximum PGA of the calculation example of this algorithm is 20 m/s^2), and the lower truncation limit is still calculated according to Formula (4).

2.3. Damage Ratio Sampling

Most of the previous studies used lognormal distribution to describe the distribution characteristics of building damage ratio [22–28,36], and this would lead to more than one sample, but this is not logical, because the damage rate cannot exceed 100% of the replacement value. Some scholars pointed out that if the cost of cleaning up the ruins is superimposed, it will exceed the replacement cost. However, if this factor is not considered in the statistical process, then it is not logical to consider it at this time. If truncated sampling is used, the distribution characteristics will lead to a change in mean and standard deviation. Therefore, the β distribution is used to describe the distribution of structural loss. The mean value and standard deviation of the distribution are kept unchanged and, as known conditions, the coefficients α and β of β distribution are calculated by substituting Formulas (6) and (7).

$$\alpha = \frac{\mu^2\sigma^4 - \mu\sigma^4 + \mu^2(1-\mu)^2}{(1-\mu)\sigma^4} \tag{6}$$

$$\beta = \frac{\mu\sigma^4 - \sigma^4 + \mu(1-\mu)^2}{\sigma^2} \tag{7}$$

where μ is the mean value of the structural damage ratio and σ is the standard deviation of the structural damage ratio.

If the α and β corresponding to 3940 damage ratios are known, the second Latin hypercube sampling can be carried out, and the sampling frequency is the same as 20. After sampling, the sample was expanded to 78,800, equivalent to 400 damage ratio samples derived from each original PGA point, which results in damage ratio sample 2.

The 197 final damage ratios are obtained via calculating the mean of corresponding damage ratio sample 2, which is, namely, damage ratio sample 3 with the uncertainty of ground motion intensity. Finally, according to damage ratio sample 2, the coefficient of variation of damage ratio sample 3 considering the coupling effects of various uncertainties is calculated, namely, the coefficient of variation sample 3.

3. Result and Discussion

Vulnerability Curve Analysis

To more intuitively understand the twice sampling process and the distribution of PGA and damage ratio samples, the vulnerability curves obtained from each group of Latin hypercube samples are displayed in three-dimensional (3D) form. Figure 4 shows the 3D vulnerability diagrams of structure 0 and structure 54. Structure 0 is a 3-story masonry structure of a residential building with 6-degree design intensity and site zone III according to 89-code, and structure 54 is a 6-story masonry structure of an office building with 8-degree design intensity and site zone III according to 01-code. The 89-code and 01-code mean the Code for seismic design of buildings in China, which are published in 1989 (GB50011-1989) and 2001 (GB50011-2001), respectively, and the key difference between the above standards is shown in Table 5. In this study, the wall thickness of the masonry structure is 240 mm in site zone III, 370 mm in site zone II, and 490 mm in site zone I. The x-axis of the 3D vulnerability diagram is the original PGA in m/s^2, the y-axis is ground motion sample 1 from the first sampling in m/s^2, and the z-axis is the damage ratio. Each original PGA on the x-axis has 20 PGA samples on the y-axis, and 3940 points on the X-Y plane have 20 damage ratio samples on the z-axis. In the 3D coordinate system, there are 78,800 (197 $\times$ 20 $\times$ 20) points, representing 197 original PGA points sampled separately 20 times, and then the corresponding damage ratio (damage ratio sample 1) of 3940 PGA samples (ground motion sample 1) was sampled, respectively, with 20 sampling frequency, and 78,800 damage ratios obtained from the final sampling are distributed in three-dimensional space.

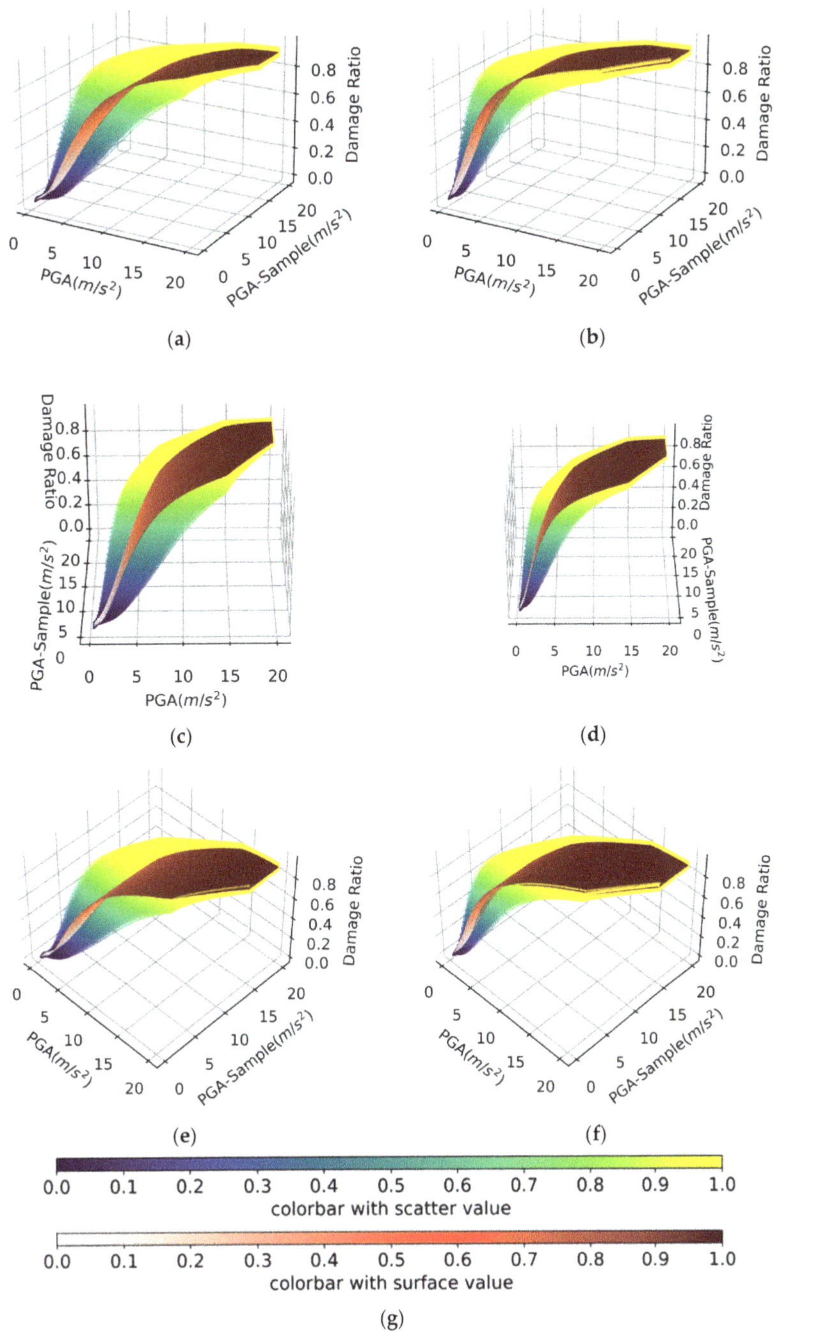

Figure 4. Three-dimensional vulnerability surfaces and scatter plots with the uncertainty of seismic ground motion intensity: (**a**,**c**,**e**) structure 27; (**b**,**d**,**f**) structure 52; (**g**) color bars.

Table 5. The design parameters of 89-code and 01-code.

Standard	89-Code	01-Code
Brick Intensity	MU7.5 (compressive strength $\geq$ 7.5 MPa)	MU10.0 (compressive strength $\geq$ 10.0 MPa)
Concrete Strength	C15 (compressive strength $\geq$ 15.0 MPa)	C20 (compressive strength $\geq$ 20.0 MPa)
Roof Dead Load (including self-weight) (kN/m^2)	6.0	6.0
Roof Live Load (kN/m^2)	1.5	2.0
Floor Dead Load (including dead weight) (kN/m^2)	3.3	3.3
Floor Live Load (kN/m^2)	1.5	2.0

It can be seen in the 3D diagram that the vulnerability curve becomes a vulnerability surface, which can help to understand the sampling process. Its shape becomes wider along the boundary of the range and finally ends at the end of the plane composed of PGA and PGA samples, namely, points (20, 20). The scatter color in Figure 4 gradually changes from blue to yellow, which means that the value of the damage ratio gradually increases from small to large. The surface color changes from light to dark, which means that the value of the damage ratio changes from low to high. As Figure 4a,c shows, the number of yellow scattered points is much denser than the number in Figure 4b,d. Especially when the value of PGA is small, the number of light color points in Figure 4b,d is much more than curve 27, and, due to this, in the two-dimensional (2D) vulnerability curve, as shown in Figure 5, curve 52 is much steeper than curve 27. It means that a higher damage ratio has been reached when the seismic excitation has not reached a certain intensity. In addition, the area of the surface is negatively related to the loss of the structure. The larger the area of the surface is, the more vulnerable the structure is. It is worth noting that, as shown in Figures 4 and 5, the edge of the surface formed by the damage ratio points and damage ratio curve is not completely smooth, which is caused by the change in the range of truncated sampling. Namely, the algorithm of Formulas (3) and (5) lead to this result.

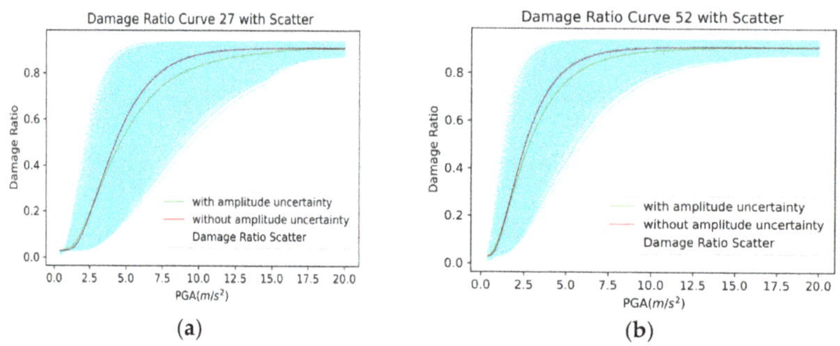

Figure 5. Two-dimensional vulnerability curve and scatter diagram with the uncertainty of seismic ground motion intensity: (**a**) structure 27; (**b**) structure 52.

In this paper, some typical comparison results are selected. As shown in Figure 4, the red line is the vulnerability curve without considering the uncertainty, and the green line is the vulnerability curve considering the uncertainty. It should be noted that the original vulnerability curve starts at the point that PGA and the damage ratio are equal to 0.4 m/s^2 and 3%, respectively. However, this study makes the original minimum damage ratio extend to 0 (i.e., the point of PGA = 0, damage ratio = 0), so the modified curve starts from (0, 0). From vulnerability curve 0 to 83, a total of 84 vulnerability curves describe different masonry structures. It can be seen that the vulnerability curve of masonry structures with the uncertainty of ground motion intensity is lower than the original vulnerability curve in most PGA ranges. Only when the PGA is small is the damage ratio with the uncertainty of ground motion intensity higher or nearly overlaps with the loss ratio without the uncertainty. Ioannou et al. [37] used the building damage database of the 1980 Irpinia earthquake as the data source via the Bayesian framework to explore whether the uncertainty of ground motion intensity can significantly change the shape of the fragility curve, and the results show that the complex model with considering more uncertainty will not change the shape of the fragility curve. It can be seen that the vulnerability curve obtained in this study also has no obvious change in shape, but the damage ratio before and after modification is different, especially when the PGA value range is in the middle, namely, in the range of moderate earthquake.

Curve 0 is the vulnerability curve of a 3-story residential masonry structure with 6-degree design intensity and site zone III according to 89-code, and curve 54 is the vulnerability curve of a 6-story office masonry structure with 8-degree design intensity and site zone III according to 01-code, as shown in Figure 6a,c. Under the same earthquake excitation, the stricter the design code of the building structure, the smaller the structure's seismic response and the smaller the damage ratio.

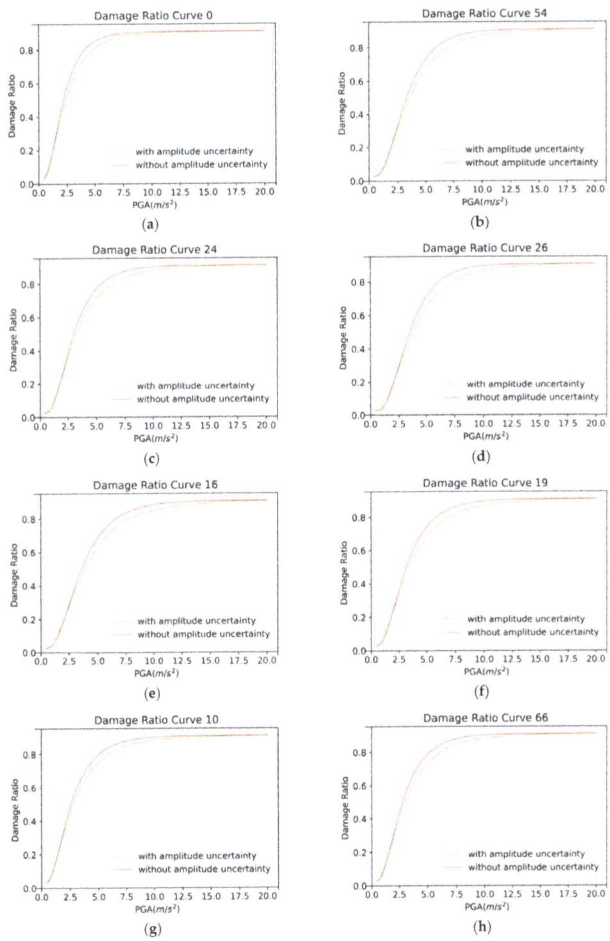

Figure 6. Vulnerability curves with and without the uncertainty of seismic ground motion intensity: (**a**) structure 0; (**b**) structure 54; (**c**) structure 24; (**d**) structure 26; (**e**) structure 16; (**f**) structure 19; (**g**) structure 10; (**h**) structure 66.

By controlling the unique variable, it is better to study the influence of different factors on the results of the vulnerability curve. Curve 24 is the vulnerability curve of the 6-story residential building structure with 6 seismic degree design intensity and site zone II according to 01-code, and curve 26 is the vulnerability curve of the 6-story residential building structure with 8-degree design intensity and site zone II according to 01-code. The only difference between curve 24 and curve 26 is seismic design intensity. It can be seen from Figure 6c,d that, in the case of the same height, site class, and design code, the seismic intensity is negatively correlated with its damage ratio. By observing curves 16 and 19, as Figure 6e,f show, they are, respectively, the 3-story residential masonry structure vulnerability curves with 8-degree design intensity and site zone II according to 89-code, and the 6-story with 8-degree design intensity and site zone II according to 89-code. It is found that for masonry structures, under the same seismic intensity, design code, and site zone, the higher the height, the higher the damage rate will be. Curve 26 is the vulnerability curve of the 6-story residential masonry building with 6-degree design intensity and site zone II according to 01-code. Comparing curve 19 and curve 26, as shown in Figure 6d,f, it can be seen that the newer the design code, the lower the damage ratio under the conditions of the same building height, seismic intensity, and region. Curve 24 and curve 10 take the site zone as a single variable, and other conditions are the same, which are 6-story masonry housings with, 6-degree design intensity according to 01-code. Comparing Figure 6c,g, it can be seen that under the conditions of different site zone and the same building height, seismic intensity, and design code, the higher the number of the site zone, the greater the loss, which means that the thicker the wall, the smaller the loss will be if other conditions remain unchanged. Curve 66 is a 6-story office building with 6-degree design intensity and site zone II according to 01-code. Compared with curve 24, the difference is only functional use. Through comparison, it can be seen that under the same PGA, the damage ratio of the office building is smaller than that of the residential building, as shown in Figure 6c,h. However, it can also be seen that no matter which single variable is changed, the trend of the curve is almost unchanged, and the resulting damage ratio does not change significantly.

4. Discussion

To better show the effect of our combined uncertainty method, the case of the Jiuzhaigou earthquake in 2017 is chosen for verification. On 7 August 2013, the earthquake occurred in Jiuzhaigou county, Sichuan, China, and the magnitude reached 7.0. This study uses the earthquake damage data collected by Zhang et al. [38] in the Jiuzhaigou earthquake. There is a total of eleven investigation areas, eight of which are located in the IX-degree area, two of which are located in the VIII-degree area, and the last one is in the VII-degree area. Detailed information, such as the geographic location and seismic parameters, is listed in Table 6. According to the geographic information of the investigation areas, the relevant PGA can be calculated, and the calculation method adopts attenuation relation with Formula 1. In addition, all the investigation sites are located in site class II in China, as the information in Table 7 shows.

Table 6. Detailed information of the investigation sites.

	Epicenter	**Investigation Point**										
Site number	**0**	**1**	**2**	**3**	**4**	**5**	**6**	**7**	**8**	**9**	**10**	**11**
Intensity	IX	VIII	IX	IX	IX	IX	IX	VIII	VIII	VII	IX	IX
Longitude	103.82	103.78	103.80	103.83	103.85	103.92	103.88	103.92	103.93	103.98	103.92	103.90
Latitude	33.20	33.32	33.30	33.30	33.30	33.28	33.28	33.27	33.27	33.32	33.22	33.20
Distance to epicenter r(km)	0	13	11	12	13	12	14	16	17	24	14	12
PGA (m/s^2)		2.9	3.5	4.7	4.6	3.7	3.9	3.3	3.5	2.5	4.6	5.7

Table 7. The relationship between equivalent shear wave velocity and overburden thickness of site class II in China.

Site Class II	
Equivalent Shear Wave Velocity of Soil (m/s^2)	**Overburden Thickness (m)**
$500 \geq v_{se} > 250$	≥ 5
$250 \geq v_{se} > 150$	3–50
$v_{se} \leq 150$	3–15

According to the investigation information of local constructions, the vulnerability curves most likely to represent the masonry in this area are selected empirically in this study. Theoretically, local building characteristics should be adopted, but it is difficult to get all detailed information on all local building types. The two constructions chosen in the Jiuzhaigou area are built after 2001 adopted 01-code, the seismic intensity is 8-degree, the wall thickness of the buildings is 240 mm, and, because of the hilly county, the constructions usually are low-rise. They represent residential buildings and commercial buildings, respectively, with 01-code, 8-degree design intensity and fewer than three floors. During the sampling, there are two groups of PGA, which are the revised group and the original group. The revised group is sampled 10,000 times and follows truncation log-normal distribution to simulate the precise ground motion of the different buildings' location. The resulting PGA samples are inserted into the original vulnerability curve and revised vulnerability curve, respectively, to get the two damage ratio samples. Then, based on Table 8, the damage ratio can be transformed to the damage state, which is the earthquake damage matrix.

Table 8. The relationship between building damage state and damage ratio.

Damage State	Basic Intact	Slight Damage	Moderate Damage	Heavy Damage	Ruin
Damage Ratio	[0,10)	[10,30)	[30,55)	[55,85)	[85,100)

The earthquake damage matrix collected by Zhang et al. [38], the revised matrix, and the actual statistical matrix are listed in Table 9, in which the numbers represent the proportion of buildings in each state of damage to the total.

Table 9. The comparison of the earthquake damage matrixes.

	Basic Intact	Slight Damage	Moderate Damage	Heavy Damage	Ruin
Statistical matrix	0	0.141361	0.602094	0.204188	0.052356
Original residence matrix	0.012231	0.194815	0.350715	0.404741	0.037497
Revised residence matrix	0.006984	0.219444	0.40487	0.362677	0.006025
Original business matrix	0.029034	0.23363	0.324603	0.366655	0.046079
Revised business matrix	0.017947	0.259816	0.381534	0.333405	0.007297

In Figure 6, 01-8-R-3-Re-trunc stands for revised curve with truncation sampling, and 01-8-R-3-Ori-norm means original curve with random sampling. In Figure 7a, compared with the statistical data, the basic intact and moderate damage of the revised curve is much better than the original curve, and the rest of the damage situations are very similar. In addition, for business construction, the performance of the revised curve for basic intact and moderate damage is superior to the original curve, as shown in Figure 7b. Generally, the revised curve can provide a more conservative result than the original curve, and, without sacrificing the accuracy of the rest of the damaged state, a more reasonable performance is achieved, especially in the highest proportion of damaged state, the moderate damage.

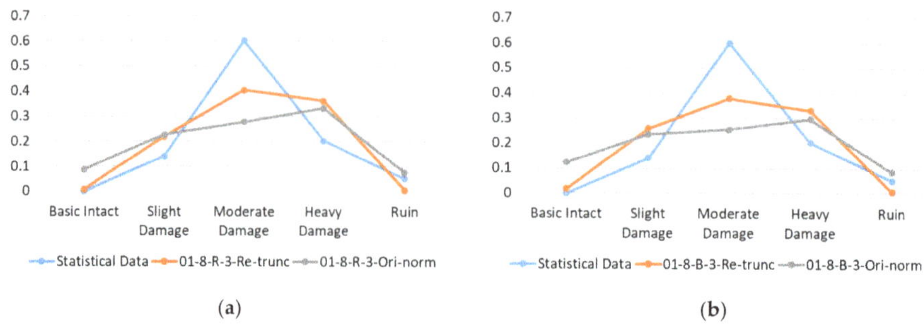

Figure 7. The comparison curves of different earthquake damage matrixes: (**a**) the residential construction earthquake damage matrix curves of statistical data, original curve, and revised curve; (**b**) the business construction earthquake damage matrix curves of statistical data, original curve, and revised curve.

However, in the remote mountainous area, there are many residential constructions built by local villagers, which cannot satisfy the seismic requirement. Therefore, in this research, the curve of the masonry structure with the best earthquake resistance and the masonry structure with the worst earthquake resistance are put together to study, as shown in Figure 7. The 89-6-R-3-Re-trunc (Curve 0) stands for the revised curve of residential masonry buildings with 89-code, 6-degree design intensity, 6-floor, and 240 mm wall-thickness via truncation sampling, which is the worst of all the curves. The 01-9-B-4-Re-trunc (Curve 83) means the revised curve of business masonry buildings with 01-code, 9-degree design intensity, 4-floor, and 490 mm wall-thickness via truncation sampling, which has the best seismic performance of the curves. In Figure 8, except for basic intact and moderate damage, the rest of the damage states are within the best performance curve and the worst performance curve. In terms of basic intact, the percentage of basic intact in statistical data is zero, because areas with notable damage are usually selected for earthquake damage surveys. Hence, the statistical magnitude of the intact building may have bias, which can explain why both calculated curves are above the statistical data in the damage state of basic intact. The actual results range between the best and worst curves, which depends on what is known about the local masonry buildings. Therefore, the curves are selected empirically in this paper, and the damage result obtained is acceptable, which can indirectly provide a comparatively accurate reference, if detailed building information could be gained, for the prediction of earthquake damage in the insurance industry.

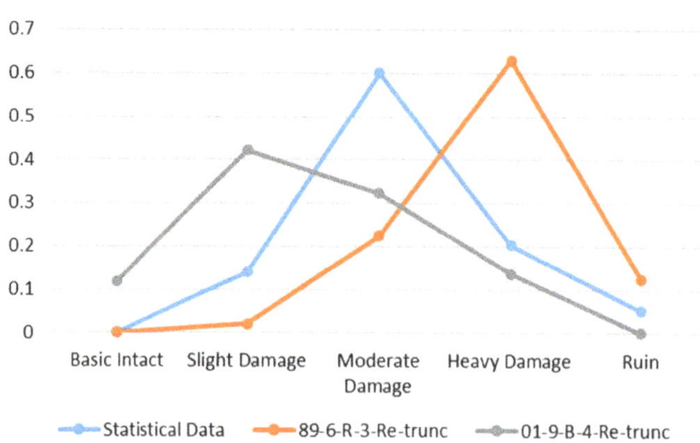

Figure 8. The comparison between the curve with best seismic performance and the curve with worst seismic performance.

5. Conclusions

It is necessary to quantify the independent uncertainty of the ground motion intensity in the process of calculating the loss expectations. In this paper, a novel method has been purposed for rapid calculation and the amplitude distribution selection was discussed. To verify the effect of the method, the damage to masonry buildings in the Jiuzhaigou earthquake was discussed.

The conclusions of this study are divided into the following three points:

1. Through this method, PGA and damage ratio are no longer required to be sampled during calculation, and a fast calculation is supported via the revised curves. Although the distribution of damage ratio has been changed after combination, it still meets the needs of the insurance industry.
2. Whether uncertainty is considered or not, under the same seismic excitation input, the higher the design intensity level, the lower the damage ratio. Therefore, the strict seismic design code can not only reduce the casualties of personnel in the earthquake but also reduce the seismic economic losses of the structure and ensure the residual value of the structure.
3. Between the vulnerability curves with and without considering the uncertainty of ground motion intensity, there was no evident difference in shape, but due to the truncation range set, there are two nodes that make the curve divided into three sections, which can be regarded as a continuous piecewise linear function. In the range of frequent earthquakes or basic seismic intensity, the corrected damage ratio is higher than the original damage ratio. Next, approximately in the range of rare earthquakes, the original damage ratio exceeds the corrected damage ratio first, and the difference between the two curves gradually increases from small to large, then from large to small, and finally, the difference between the two curves is further reduced to almost overlapping.

In loss estimation of catastrophe insurance, amplitude uncertainty is particularly important. For possible earthquakes in the future, this paper provides a novel view and calculation for building vulnerability that fully considers the uncertainty of the three elements of ground motion, the uncertainty of vulnerability, and the coupling uncertainty between them. In addition, during processing the translation between seismic intensity and PGA, the distribution is adopted instead of PGA mean value, and the uncertainty in this process is considered as well. Furthermore, it can efficiently calculate the vulnerability of a large number of structures, which provides a fast and relatively accurately loss result. In the damage ratio sampling, β distribution is used to replace the lognormal distribution, and the modified vulnerability curve of the building structure is estimated. The method proposed in this study solves the problem that putting the sample mean into a complex calculation gives a different result from taking the sample mean after a complex calculation. However, the question of the percentage of moderate damage taking too much has not been solved and still needs to be studied from another perspective.

Author Contributions: Conceptualization, C.L. and X.L.; methodology, C.L.; software, J.Z.; validation, J.Z.; formal analysis, J.Z.; investigation, J.Z.; re-sources C.L.; data curation, C.L.; writing—original draft preparation, J.Z. and C.L.; writing—review and editing, J.Z.; visualization, J.Z.; supervision, C.L.; project administration, C.L.; All authors have read and agreed to the published version of the manuscript.

Funding: This research was funded by [National Key Research and Development Program of China] Grant Number [2018YFC1504604], [National Natural Science Foundation of China] Grant Number [51421005].

Data Availability Statement: Not applicable.

Conflicts of Interest: The authors declare no conflict of interest.

References

1. UNISDR; CRED. *Economic Losses, Poverty and Disasters 1998–2017*; UNISDR (United Nations Office for Disaster Risk Reduction): Geneva, Switzerland, 2018; CRED (Centre for Research on the Epidemiology of Disaster): Bengaluru, India, 2018. Available online: https://www.undrr.org/publication/economic-losses-poverty-disasters-1998-2017 (accessed on 12 May 2020).
2. Tokunaga, S.; Ikegawa, M.; Okiyama, M. Economic analysis of regional renewal and recovery from the great east Japan earthquake. In *Spatial Economic Modelling of Megathrust Earthquake in Japan*; Springer: Berlin/Heidelberg, Germany, 2017; pp. 13–63. [CrossRef]
3. Riga, E.; Karatzetzou, A.; Apostolaki, S.; Crowley, H.; Pitilakis, K. Verification of seismic risk models using observed damages from past earthquake events. *Bull. Earthq. Eng.* **2021**, *19*, 713–744. [CrossRef]
4. Martins, L.; Silva, V. Development of a fragility and vulnerability model for global seismic risk analyses. *Bull. Earthq. Eng.* **2020**, *19*, 6719–6745. [CrossRef]
5. Ren, Y.; Yin, J.; Wen, R.; Ji, K. The impact of ground motion inputs on the uncertainty of structural collapse fragility. *Eng. Mech.* **2020**, *37*, 115–125. [CrossRef]
6. Abrahamson, N.A.; Bommer, J.J. Probability and Uncertainty in Seismic Hazard Analysis. *Earthq. Spectra* **2005**, *21*, 603–607. [CrossRef]
7. *FEMA P-58*; Seismic Performance Assessment of Buildings. Federal Emergency Management Agency: Washington, DC, USA, 2012; Volume III.
8. Cutfield, M.R.; Ma, Q.T.M. Solution Strategies for three problems in empirical fragility curve derivation using the maximum likelihood method. *J. Earthq. Eng.* **2016**, *22*, 435–453. [CrossRef]
9. De Risi, R.; Goda, K.; Mori, N.; Yasuda, T. Bayesian tsunami fragility modeling considering input data uncertainty. *Stoch. Hydrol. Hydraul.* **2016**, *31*, 1253–1269. [CrossRef] [PubMed]
10. Kwon, O.-S.; Elnashai, A. The effect of material and ground motion uncertainty on the seismic vulnerability curves of RC structure. *Eng. Struct.* **2006**, *28*, 289–303. [CrossRef]
11. Soleimani, F. Propagation and quantification of uncertainty in the vulnerability estimation of tall concrete bridges. *Eng. Struct.* **2019**, *202*, 109812. [CrossRef]
12. Pan, H.; Tian, L.; Fu, X.; Li, H. Sensitivities of the seismic response and fragility estimate of a transmission tower to structural and ground motion uncertainties. *J. Constr. Steel Res.* **2020**, *167*, 105941. [CrossRef]
13. Yamin, L.E.; Hurtado, A.; Rincon, R.; Dorado, J.F.; Reyes, J.C. Probabilistic seismic vulnerability assessment of buildings in terms of economic losses. *Eng. Struct.* **2017**, *138*, 308–323. [CrossRef]
14. Yazgan, U. Empirical seismic fragility assessment with explicit modeling of spatial ground motion variability. *Eng. Struct.* **2015**, *100*, 479–489. [CrossRef]
15. Ansari, M.; Nazari, M.; Panah, A.K. Influence of foundation flexibility on seismic fragility of reinforced concrete high-rise buildings. *Soil Dyn. Earthq. Eng.* **2020**, *142*, 106521. [CrossRef]
16. Choudhury, T.; Kaushik, H.B. Treatment of uncertainties in seismic fragility assessment of RC frames with masonry infill walls. *Soil Dyn. Earthq. Eng.* **2019**, *126*, 105771.1–105771.14. [CrossRef]
17. Kim, S.; Moon, T.; Kim, S.J. Effect of uncertainties in material and structural detailing on the seismic vulnerability of rc frames considering construction quality defects. *Appl. Sci.* **2020**, *10*, 8832. [CrossRef]
18. Saloustros, S.; Pelà, L.; Contrafatto, F.R.; Roca, P.; Petromichelakis, I. Analytical derivation of seismic fragility curves for historical masonry structures based on stochastic analysis of uncertain material parameters. *Int. J. Arch. Herit.* **2019**, *13*, 1142–1164. [CrossRef]
19. Jiang, Y.; Su, L.; Huang, X. Seismic fragility analysis of unreinforced masonry structures considering parameter uncertainties. *Eng. Mech.* **2020**, *37*, 159–167. [CrossRef]
20. Erdik, M. Earthquake risk assessment. *Bull. Earthq. Eng.* **2017**, *15*, 5055–5092. [CrossRef]
21. Hwang, S.-H.; Mangalathu, S.; Shin, J.; Jeon, J.-S. Machine learning-based approaches for seismic demand and collapse of ductile reinforced concrete building frames. *J. Build. Eng.* **2020**, *34*, 101905. [CrossRef]
22. Abrahamson, N.A. Statistical properties of peak ground accelerations recorded by the SMART 1 array. *Bull. Seismol. Soc. Am.* **1988**, *78*, 26–41.
23. Abrahamson, N.A. Effects of rupture directivity on probabilistic seismic hazard analysis. In Proceedings of the 6th International Conference on Seismic Zonation, Palm Springs, CA, USA, 12–15 November 2000; pp. 151–156.
24. Rossetto, T.; Elnashai, A. A new analytical procedure for the derivation of displacement-based vulnerability curves for populations of RC structures. *Eng. Struct.* **2005**, *27*, 397–409. [CrossRef]
25. Lagomarsino, S.; Giovinazzi, S. Macroseismic and mechanical models for the vulnerability and damage assessment of current buildings. *Bull. Earthq. Eng.* **2006**, *4*, 415–443. [CrossRef]
26. Bradley, B.A.; Dhakal, R.P. Error estimation of closed-form solution for annual rate of structural collapse. *Earthq. Eng. Struct. Dyn.* **2008**, *37*, 1721–1737. [CrossRef]
27. Jayaram, N.; Baker, J. Statistical tests of the joint distribution of spectral acceleration values. *Bull. Seism. Soc. Am.* **2008**, *98*, 2231–2243. [CrossRef]
28. Lallemant, D.; Kiremidjian, A.; Burton, H. Statistical procedures for developing earthquake damage fragility curves. *Earthq. Eng. Struct. Dyn.* **2015**, *44*, 1373–1389. [CrossRef]
29. Standardization Administration. Seismic Ground Motion Parameters Zonation Map of China. GB 18306-2015, 15 May 2015.

30. Lu, X.; Su, N.; Zhou, Y. IDA-based seismic fragility analysis of a complex high-rise structure. *J. Earthq. Eng. Eng. Vib.* **2012**, *32*, 19–25.
31. Gattulli, V.; Lofrano, E.; Paolone, A.; Pirolli, G. Performances of FRP reinforcements on masonry buildings evaluated by fragility curves. *Comput. Struct.* **2017**, *190*, 150–161. [CrossRef]
32. Zhang, L.; Jiang, J.; Liu, J. Seismic vulnerability analysis of multistory dwelling brick building. *Earthq. Eng. Eng. Vib.* **2002**, *22*, 49–55, [In Chinese].
33. Atik, L.A.; Abrahamson, N. An improved method for nonstationary spectral matching. *Earthq. Spectra* **2010**, *26*, 601–617. [CrossRef]
34. Gao, M. *Publicity and Implementation Textbook of Seismic Ground Motion Parameters Zonation Map of China*; China Quality and Standards Publishing & Media Co., Ltd.: Beijing, China, 2015.
35. Ghotbi, A.R.; Taciroglu, E. Structural seismic damage and loss assessments using a multi-conditioning ground motion selection approach based on an efficient sampling technique. *Bull. Earthq. Eng.* **2021**, *19*, 1271–1287. [CrossRef]
36. Ader, T.; Grant, D.N.; Free, M.; Villani, M.; Lopez, J.; Spence, R. An unbiased estimation of empirical lognormal fragility functions with uncertainties on the ground motion intensity measure. *J. Earthq. Eng.* **2018**, *24*, 1115–1133. [CrossRef]
37. Ioannou, I.; Chandler, R.; Rossetto, T. Empirical fragility curves: The effect of uncertainty in ground motion intensity. *Soil Dyn. Earthq. Eng.* **2019**, *129*, 105908. [CrossRef]
38. Zhang, L.; Zhu, B.; Tao, Z.; Chen, X. Field investigation and failure characteristics of building damaged by the 7.0-Magnitude earthquake in Jiuzhaigou. *China Earthq. Eng. J.* **2019**, *41*, 1053–1059. (In Chinese) [CrossRef]

sustainability

Article

Seabed Liquefaction around Pipeline with Backfilling Trench Subjected to Strong Earthquake Motions

Qiuzhe Wang [1,2], Jiang Bian [1,*], Wenting Huang [1], Qingrui Lu [1], Kai Zhao [1,2] and Zhaoyan Li [2]

1 Institute of Geotechnical Engineering, Nanjing Tech University, Nanjing 210009, China
2 Key Laboratory of Earthquake Engineering and Engineering Vibration, Institute of Engineering Mechanics, China Earthquake Administration, Harbin 150080, China
* Correspondence: bianj29@njtech.edu.cn

Abstract: As an indispensable part of the lifeline for the offshore gas and oil industry, submarine pipelines under long-term marine environmental loadings have historically been susceptible to earthquakes. This study investigates the impact of trench backfilling on the residual liquefaction around a pipeline and the induced uplift of a pipeline under the combined action of an earthquake, ocean wave and current loading. A fully coupled nonlinear effective stress analysis method, which can consider the nonlinear hysteresis and the large deformation after liquefaction of the seabed soil, is adopted to describe the interaction between the seabed soil and the submarine pipeline. Taking a typical borehole in the Bohai strait as the site condition, the nonlinear seismic response analysis of the submarine pipeline under the combined action of seismic loading and ocean wave and current is carried out. The numerical results show that trench backfilling has a significant impact on the seismic response of the pipeline. The existence of trench backfilling reduces the accumulation of the residual excess pore water pressure, so that the seabed liquefaction around the pipeline is mitigated and the uplift of the pipeline is also decreased.

Keywords: submarine pipeline; earthquake; seabed liquefaction; uplift; ocean environmental loading

Citation: Wang, Q.; Bian, J.; Huang, W.; Lu, Q.; Zhao, K.; Li, Z. Seabed Liquefaction around Pipeline with Backfilling Trench Subjected to Strong Earthquake Motions. *Sustainability* **2022**, *14*, 12825. https://doi.org/10.3390/su141912825

Academic Editor: Maged A. Youssef

Received: 2 July 2022
Accepted: 28 September 2022
Published: 8 October 2022

1. Introduction

Due to the rapid exploration of oil and gas in the deep sea, submarine pipelines have developed rapidly in the past few decades, known as a lifeline in the storage and transportation system of submarine oil and gas. The seismic stability of a pipeline is directly related to the safety of offshore oil and gas. Once a strong earthquake occurs, great economic losses and environmental pollution can be caused due to the seabed liquefaction surrounding the pipeline. Unlike the territory counterparts, the submarine pipelines are subjected to long–term ocean environmental loadings, such as ocean waves and currents.

Extensive research has been conducted on the dynamic response of submarine pipelines under the action of ocean waves [1–14]. However, very limited concern is given to the seismic stability of submarine pipelines, especially for those buried in the liquefiable seabed. Earthquake–induced liquefaction of the seabed will lead to great damage to the submarine pipeline. Until the submarine pipelines became an important part of the offshore oil and gas industry, their seismic stability issue gradually attracted the attention of scholars. In the late 1970s, assuming pipelines as rigid structures, Nath and Soh [15] firstly studied the interaction between the seabed and the pipeline under seismic loading, considering the nonlinearity of the seabed soil. Later, the lumped-mass model was used to analyze the seismic response of submarine pipelines subjected to random ground motion by Datta [16]. The results showed that the seismic response of submarine pipelines was obviously different from that of territory pipelines due to the site conditions and the buried depth of pipelines. Yan et al. [17] established a model of seabed–pipeline interaction and studied the influencing factors, such as wall thickness, radius and buried depth, on the strain response of a pipeline under the El Centro seismic wave. Based on the fluid–structure coupling analysis model, Deng et al. [18] established a long-span submarine flexible pipeline model by using finite element software, ADINA, and analyzed the seismic response of the submarine pipeline. Luan and Zhang [19] studied the accumulation of excess pore pressure around the pipeline under seismic loading. However, in the previous studies, few attention has been given to the seismic response of the trenched pipeline with backfilling. The seabed–trench–pipeline interaction has a predominant role on the stability of the pipeline under cyclic loadings associated with earthquake events and ocean waves [20]. Furthermore, the effects of seabed liquefaction and plastic deformation were not well considered in the seismic response analysis of the submarine pipeline [21–23]. The seismic response of underground structures is mainly controlled by the surrounding soil. Thus, the seismically induced liquefaction and large deformation have remarkably adverse effects on shallow-buried submarine pipelines.

In this research, typical borehole data in the Bohai strait are chosen as the site condition, with reference to the Bohai strait, China. The Bohai strait, rich in oil and gas resources, has an area of about 77,000 km^2, with the north–south length and the east–west length of the Bohai strait being, respectively, about 480 km and 300 km. There are two main fault zones in the Bohai sea, Yingkou–Weifang section of Tanlu fault zone and Zhangjiakou–Bohai active fault zone, which promote the frequent seismic activity in this region. Three strong earthquakes with magnitude 7 or even higher occurred in the middle of the Bohai strait historically. Figure 1 shows the important faults in the Bohai strait and the distribution of historical strong earthquakes. The seabed soil–submarine pipeline interaction model is established incorporating the nonlinear hysteresis and large deformation after liquefaction of the seabed soil. The seismic response of the submarine pipeline under combined action of earthquake and waves is studied and the influence of the trench backfilling on the seabed liquefaction around the pipeline and the induced uplift is further analyzed.

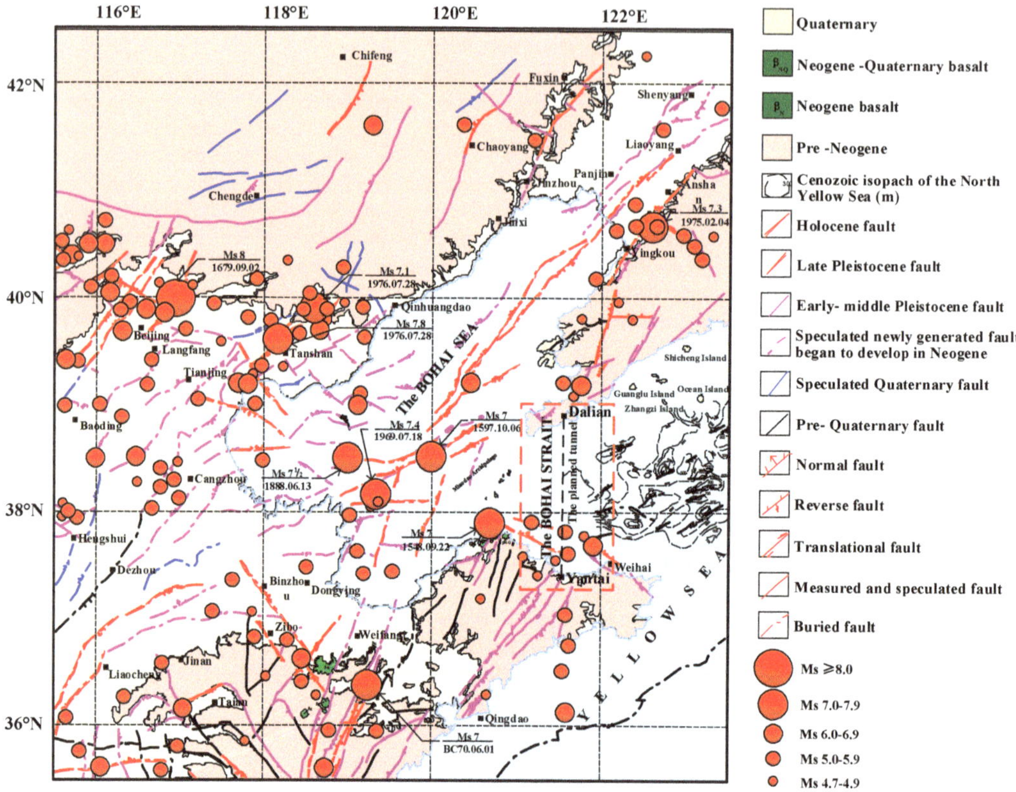

Figure 1. Distribution of major faults and historical strong earthquakes in the Bohai Sea.

2. Research Method

2.1. Governing Equation

In the framework of Biot's dynamic consolidation equation and the effective stress principle, the dynamic effective stress analysis method of saturated two–phase medium with soil skeleton–pore water interaction is established and the governing equation can be expressed as follows:

$$\dot{\sigma}'_{sij} = \dot{\sigma}_{sij} + \delta_{ij} p, \tag{1}$$

$$G\nabla^2 u_i + \left(\frac{G}{1-2v}\right)\varepsilon_{ij,j} = p_{,i} + \rho \ddot{u}_i, \tag{2}$$

$$\frac{k}{\gamma_{\mathrm{w}}}\nabla^2 p - \frac{\dot{p} n_{\mathrm{s}}}{K_{\mathrm{f}}} = \dot{\varepsilon}_{\mathrm{v}}, \tag{3}$$

$$\frac{1}{K_{\mathrm{f}}} = \frac{1}{K_{\mathrm{w}}} + \frac{1-S_{\mathrm{r}}}{P_{\mathrm{w}}}, \tag{4}$$

where $\dot{\sigma}_{sij}$ is the stress increment tensor of the saturated soil, δ_{ij} is the Kronecker delta and p is excess pore water pressure. G is the shear modulus of the soil skeleton, u_i is the displacement vector of soil skeleton, $\varepsilon_{\mathrm{v}} = \varepsilon_{jj}$ is the volume strain tensor, ρ is the density of the soil skeleton and pore water, v is Poisson's ratio, n_{s} is soil porosity, γ_{w} is the unit weight of water in the pore, k is soil permeability, K_{f} is the volume compressibility of pore water, K_{w} is the elastic modulus of water, S_{r} is the degree of saturation and $p_{\mathrm{w}} = p_0 + p$ is absolute pore water pressure.

2.2. Constitutive Model

The Davidenkov skeleton curve extended by the Masing rule is used to describe the nonlinear stress–strain relationship of seabed soil in the elastic range. The skeleton curve is modified by the Mohr Coulomb failure criterion to describe the large plastic deformation when the seabed soil is close to liquefaction and after liquefaction [12,13,24]. The cyclic stress–strain relationship is shown in Figure 2.

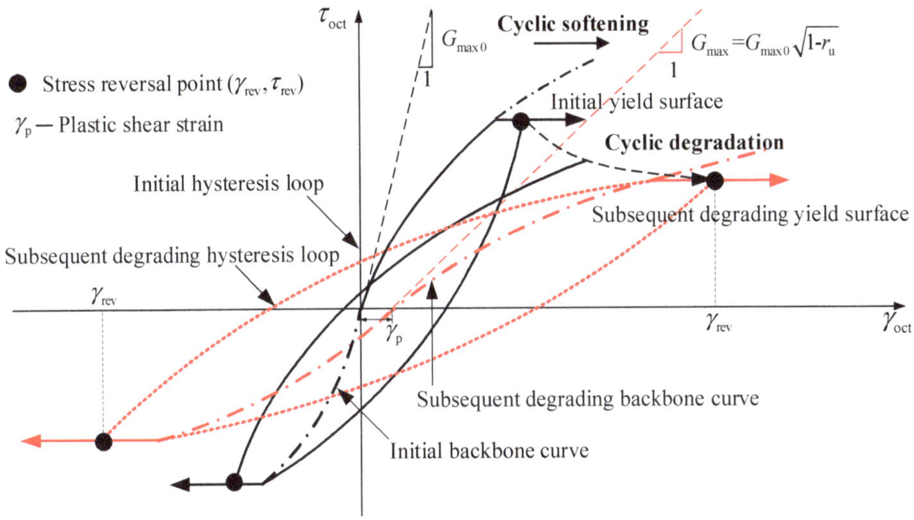

Figure 2. Schematic illustration of the cyclic stress–strain behavior of soil skeleton including.

A large number of laboratory tests show that the rearrangement of soil particles caused by cyclic shear is accompanied by the compression of soil skeleton and the rise of residual pore water pressure. The calculation of plastic volume strain per cycle is added to the constitutive model, as the source term of residual pore water pressure accumulation in Biot's dynamic consolidation equation [25,26]:

$$\frac{\Delta\varepsilon_{vd}}{(\gamma-\gamma_{th})^{C_3}} = C_1 \exp\left\{-C_2\frac{\varepsilon_{vd}}{(\gamma-\gamma_{th})^{C_3}}\right\}, \tag{5}$$

where γ = amplitude of cyclic shear strain, ε_{vd} = accumulated volumetric strain from previous cycles, C_1, C_2, C_3 = model constants for the soil in question and γ_{th} is volumetric threshold shear strain.

The softening phenomenon of soil under cyclic loading is described by the modulus degradation model and the residual pore water pressure is taken as the degradation parameter [27]:

$$G_{max}^{t} = G_{max}^{0}\sqrt{\frac{\sigma'_p-\overline{p}}{\sigma'_p}} = G_{max}^{0}\sqrt{1-r_u}, \tag{6}$$

where G_{max}^{t} is the maximum shear modulus at time t and $\overline{p}$ is residual pore water pressure.

2.3. Method Validation

The combined action of earthquake and ocean wave–current loading will form an elliptical stress path in the deviator stress and shear stress space of the seabed soil element, which is similar to the stress path formed in the laboratory soil element test under bidirectional loading. Therefore, the effective stress method proposed in this paper is used to simulate the cyclic loading test of Nanjing fine sand under bidirectional loading to verify the effectiveness of the proposed method. The loading path of the cyclic loading test is shown in Figure 2, with a cyclic stress ratio CSR = 0.15, $a/b = 1$, $\beta = 0$. The relative density of the sample is $D_r = 50\%$, the saturation weight $\gamma = 19.2\ \mathrm{KN/m^2}$, the initial effective confining pressure is 100 kPa and the parameters used in the simulation are shown in Table 1.

Table 1. Parameters of element test.

G_{max}	ν	A	B	γ_0	C_1	C_2	C_3	c	ϕ	T_{en}
50 MPa	0.3	1.02	0.35	4×10^{-4}	0.43	0.93	1.01	0	30°	0

Figure 3 shows the time history of excess pore water pressure obtained by the method proposed in this paper and the laboratory test using the automated hollow cylinder apparatus, as published in Zhao et al. [28]. One may note that the prediction results of the proposed method are in relatively good agreement with the laboratory test data. The deviation might be due to the fact that in the proposed model, the plastic volume strain and the residual excess pore water pressure are calculated at each cycle, not at each time step. However, the proposed method is able to predict the overall trend of the development of excess pore water pressure of the soil element under bidirectional loading. At the same time, the increase in the residual excess pore water pressure leads to the attenuation of soil skeleton modulus and the gradual amplification of oscillatory pore water pressure.

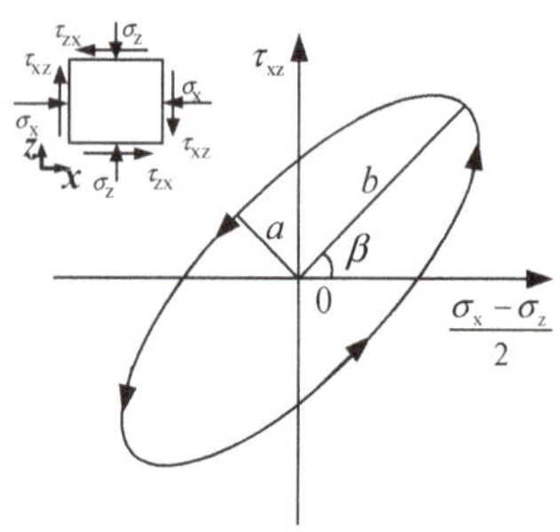

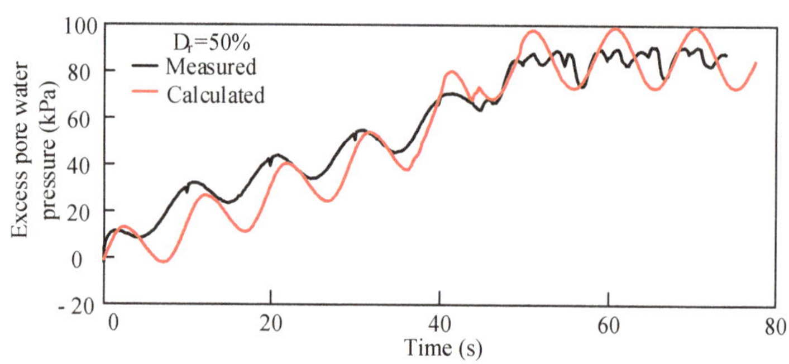

Figure 3. Comparison between the measured and predicted excess pore pressure time history.

3. Numerical Analysis Model

3.1. Model Construction

Taking a typical borehole in the Bohai Sea as the site condition, a schematic diagram of the borehole and the corresponding normalized modulus and damping ratio of soil are shown in Figure 4.

A section of submarine pipeline is selected for analysis and the seabed–pipeline interaction model is shown in Figure 5. The diameter of the pipeline is 2 m with a segment thickness of 0.2 m and a buried depth of 2 m. The carbon steel is adopted for the submarine pipeline. The depth of the trench is 3 m, the width of the bottom of the trench is 4 m and the slope ratio of the trench is 1:2. The four-node equal strain finite difference element is used to establish a uniform grid layout and the size of grid in the computational domain in the wave propagation direction is limited to one–eighth to one–tenth of the shortest wavelength, corresponding to the cut–off frequency of the seismic wave propagation in the soil layer [29]. Mesh sensitivity was performed carefully before the analysis to achieve the balance of computational efficiency and accuracy. In the current model, the grid size of the seabed soil is 1 m × 1 m and the grid size of the pipeline is 0.1 m × 0.1 m. In addition, the influence of the trench backfilling around the submarine pipeline on the seismic response of the pipeline is considered. The parameters of seabed soil and of the pipeline are, respectively, shown in Tables 2 and 3. The assumed soil parameters are calibrated from the shear stiffness and damping ratio curves of natural marine sediments from a typical borehole at the Bohai Strait.

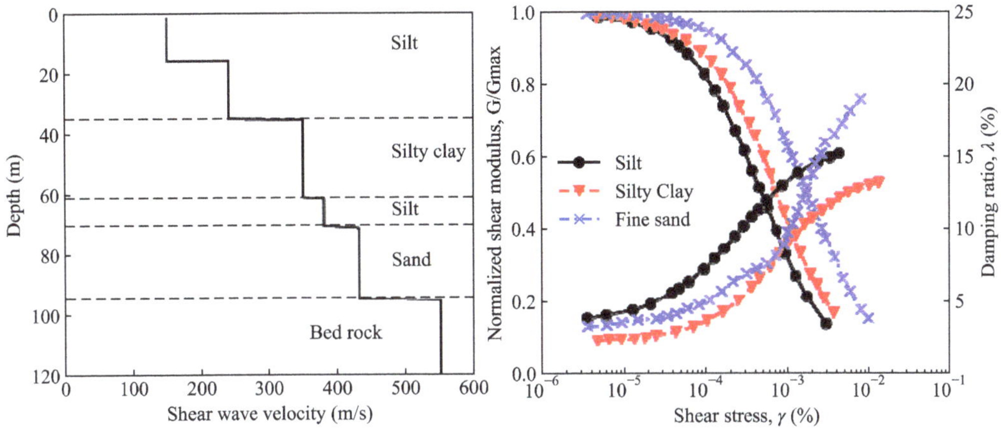

Figure 4. Soil profile, shear wave velocity profile and shear modulus reduction and damping ratio curves for the marine sediments from a typical borehole at the Bohai Strait.

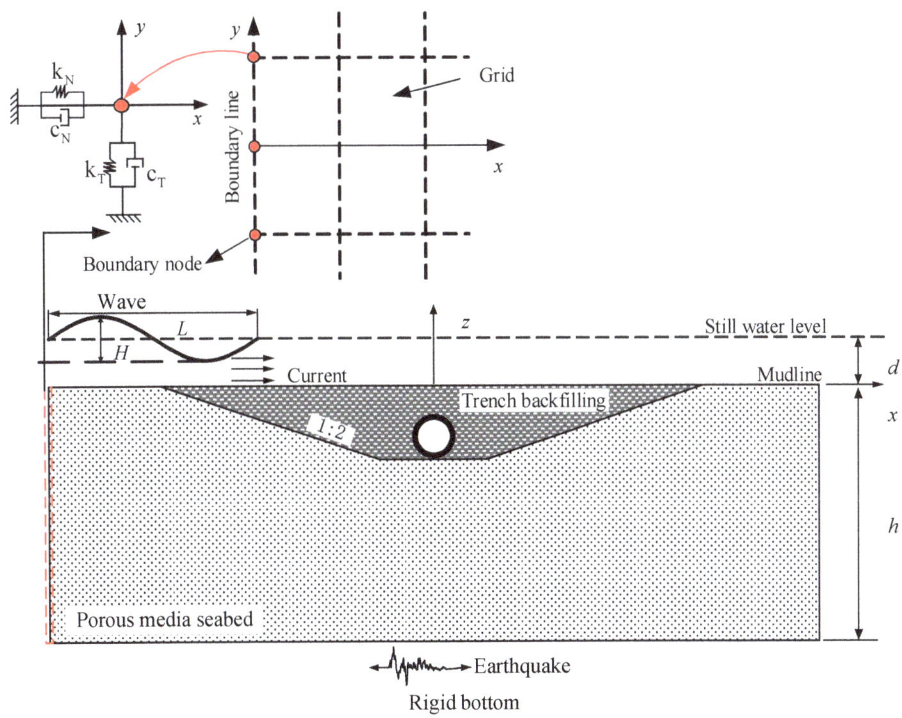

Figure 5. Sketch of the pipeline–seabed interaction under seismic loading with an ocean wave and current.

Table 2. Parameters of seabed soil.

Characteristics	Silt	Silty Clay	Fine Sand	Trench Backfilling
Soil porosity (n_s)	0.50	0.35	0.50	0.50
Poisson' ratio (ν)	0.30	0.25	0.25	0.30
Reference initial shear modulus $\left(G_{\max,0}^{\text{ref}}\right)$	6.5×10^4 (kN/m^2)	3×10^4 (kN/m^2)	8×10^4 (kN/m^2)	6.5×10^4 (kN/m^2)
Soil permeability (k)	10^{-5} (m/s)	10^{-8} (m/s)	10^{-4} (m/s)	10^{-4} (m/s)
Coefficient of lateral earth pressure (K_0)	0.42	0.42	0.42	0.42
Submerged specific weight of soil (γ')	10 (kN/m^3)	8 (kN/m^3)	12 (kN/m^3)	12 (kN/m^3)
Degree of saturation (S_r)	1	1	1	1
Relative density (D_r)	50%	60%	70%	50%
		Davidenkov Model		
A	1.03	0.96	1.02	1.03
B	0.50	0.47	0.48	0.50
γ_0 ($\times 10^{-4}$)	1.80	7.30	17.90	1.80
C_1	0.43	–	0.67	0.43
C_2	0.93	–	0.6	0.93
C_3	1.10	–	1.25	1.10
		Mohr–Coulomb model		
Cohesive strength (c)	0	5 kPa	0	0
Internal friction angle (ϕ)	30°	25°	35°	30°
Tensile strength (T_{en})	0	0	0	0

Table 3. Parameters of submarine pipeline.

Young's Modulus (E)	Poisson's Ratio (ν)	Density (ρ)	Buried Depth (h)
2×10^{11} (kN/m^2)	0.25	7800 (kg/m^3)	2 m

The zero-thickness interface element [30] is used to simulate the interaction between the pipeline and the surrounding seabed soil. The interface is controlled by normal stiffness, tangential stiffness and sliding parameters. There is no overlap between the seabed soil and the pipeline, and the relative sliding between the seabed soil and the pipeline occurs after the yielding of the interface. The normal stiffness and the tangential stiffness of the interface are much larger than those of the surrounding seabed soil. The normal and the tangential stiffness can be taken as 10–times the equivalent stiffness in the surrounding seabed soil [31], which can be expressed as follows:

$$\frac{K + 4/3G}{\Delta z_{\min}}, \tag{7}$$

where K and G are bulk modulus and shear modulus of the seabed soil, respectively, and $\Delta z_{\min}$ is the minimum width of the element near the interface.

According to Coulomb's law of friction, when the shear stress at the interface element exceeds the friction resistance, relative sliding occurs:

$$|\tau| \geq \sigma_n' \mu, \tag{8}$$

where σ'_n is effective normal stress at the interface element and μ is the friction coefficient between the pipeline and surrounding seabed soil, which is 0.7 in the current study.

An artificial boundary composed of linear elastic spring and damper is used at the side and the bottom of the model to limit the size of the calculation area. The influence of dynamic response of the side boundary can be ignored when the horizontal distance between the side boundary and the structure is at least 10–times the width of the structure and the transverse width of the model is 200 m. The shear wave velocity greater than 500 m/s is taken as the seismic bedrock surface and the depth of the model is 100 m, with a critical damping ratio of 5%.

3.2. Model Input Selection

The actual recordings are considered to suitably represent the characteristics of near-field motions. Therefore, the ground motion time series of a real earthquake records with similar magnitude and epicenter distances of the historical earthquakes in Bohai strait, i.e., PRP station record of the 2010 Darfield earthquake, are selected to behave as the frequency spectral characteristics of input bedrock motions: the original acceleration time histories and spectral accelerations at a damping ratio of 5% from the records of the Darfield earthquake are shown in Figure 6, respectively.

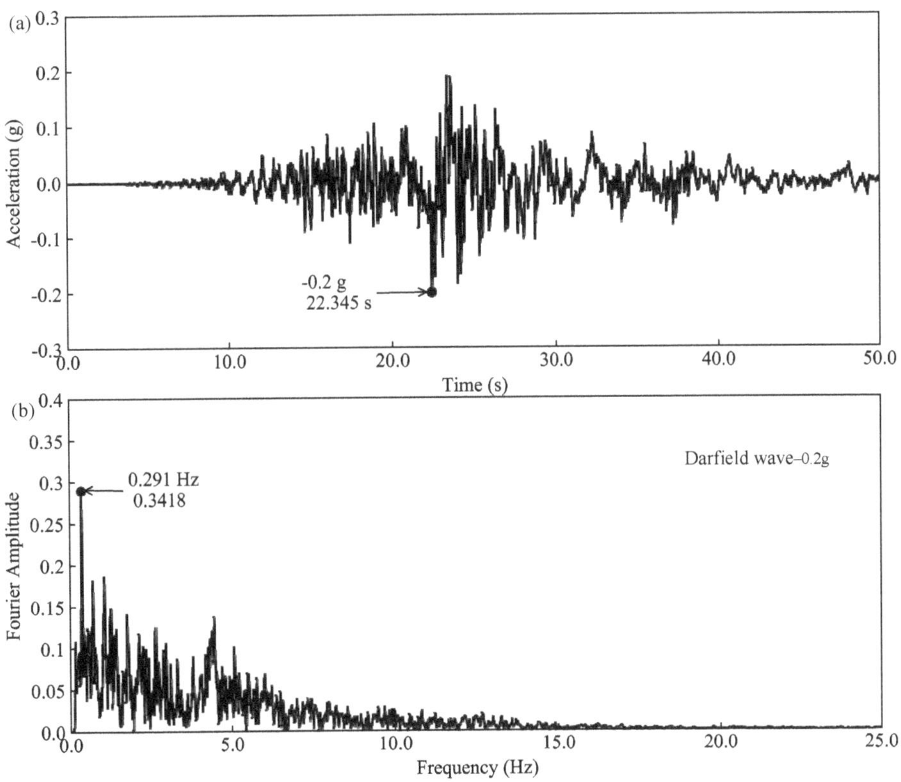

Figure 6. The original records of the 2010 Darfield earthquake employed as input motions.

The periodic wave and current loading are simulated by a periodic surface pressure at the seabed surface according to the potential flow theory. The third–order analytical solution by Hsu et al. [32] is applied to determine the periodic fluid pressure p_b at the seabed surface, which can be written as follows:

$$\begin{aligned} P_b &= \frac{\rho_f g H}{2\cosh\beta d}\left[1-\frac{\omega_2\beta^2H^2}{2(U_0\beta-\omega_0)}\right]\cos(\beta x-\omega t) \\ &+\frac{3\rho_f H^2}{8}\left\{\frac{\omega_0(\omega_0-U_0\beta)}{2\sinh^4\beta d}-\frac{g\beta}{3\sinh 2\beta d}\right\}\cos 2(\beta x-\omega t) \\ &+\frac{3\rho_f\beta H^3\omega_0(\omega_0-U_0\beta)}{512}\frac{(9-4\sinh^2\beta d)}{\sinh^7\beta d}\cos 3(\beta x-\omega t) \end{aligned} \tag{9}$$

where β = wave number, ω = wave frequency, H = wave height due to wave–urrent interactions, d = water depth and U_0 = current velocity. The dispersion equation is expressed by

$$\omega = \omega_0 + (\beta H)^2\omega_2, \tag{10}$$

where $\omega_0 = U_0\beta + \sqrt{g\beta\tanh\beta d}$

$$\omega_2 = \frac{(9+8\sinh^2\beta d+8\sinh^4\beta d)}{64\sinh^4\beta d}(\omega_0 - U_0\beta), \tag{11}$$

If no current follows the wave (U_0 = 0 m/s), the aforementioned analytical solution is reduced to the classic third–order solution of the nonlinear wave. The input parameters of ocean wave and current are given in Table 4.

Table 4. Parameters of wave and current.

Wave Height (H)	Wave Period (T)	Water Depth (d)	Current Speed (v)
6 (m)	6 (s)	10 (m)	1 (m/s)

4. Results and Discussion

In this section, special attention is paid to the detailed investigation of progressive liquefaction around the trenched pipeline (Section 4.1) and the phenomena occurring close to the pipeline–trench–seabed interface under seismic loading (Section 4.1). Section 4.2 attempts to interpret the mechanism of seismically induced liquefaction and uplift of the pipeline with backfilling trench.

4.1. Liquefaction Initiation of Seabed around the Trenched Pipeline

Figure 7 compares the entire buildup process of excess pore water pressure of the soil element at the side (P1) and bottom (P2) of the pipeline for the case with and without trench backfilling. Obviously, the ocean wave and current lead to intensive periodic oscillations of excess pore water pressure induced by seismic loading. Compared with the case without trench backfilling, the excess pore water pressures at P1 and P2 accumulate much more slowly and, eventually, reach a lower value for the case with trench backfilling. For the case without trench backfilling, the excess pore water pressure at P2 reaches 45 kPa at 25 s, which is much larger than that for the case with trench backfilling. The saturated soil under seismic loading is accompanied by the buildup and partial drainage of the excess pore water pressure and the high permeability of the trench backfilling further accelerates the drainage.

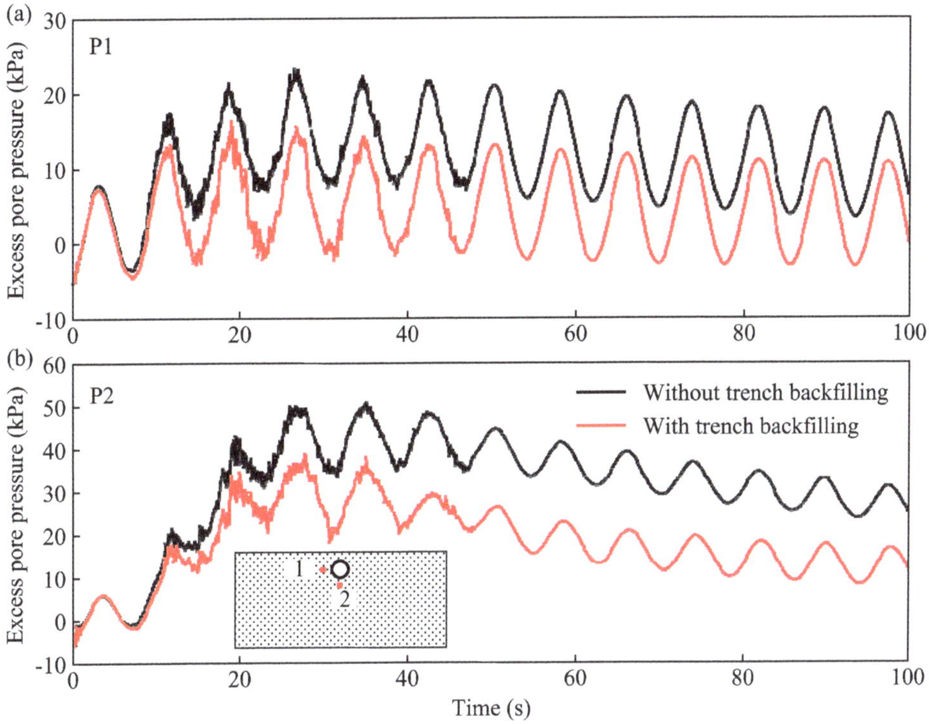

Figure 7. Excess pore water pressure time histories near the pipeline under seismic loading with and without trench backfilling; (**a**) at the side of the pipeline; (**b**) at the bottom of the pipeline.

Figure 8 compares the progressive liquefaction process of the seabed around the pipeline for the case with and without trench backfilling. Liquefaction first occurs at the seabed surface and gradually spreads downward and finally wraps the pipeline. Residual and transient liquefaction occur simultaneously under seismic and ocean wave and current loading. Transient liquefaction occurs near the seabed surface at the ocean wave crest and moves with the propagation of the ocean wave. Following seismic loading, due to the compaction of soil skeleton and the free drainage condition of the seabed surface, the excess pore water pressure in the seabed begins to dissipate, resulting in obvious reconsolidation phenomenon. For the case with trench backfilling, no liquefaction occurs in the trench backfilling around the pipeline and the maximum excess pore water pressure ratio is in a range from 0.2 to 0.4. Meanwhile, no liquefaction occurs in the seabed soil under the pipeline for the case with trench backfilling and the maximum excess pore water pressure ratio is about 0.6 to 0.8. Thus, the stability of the pipeline is greatly affected by the surrounding soil and the existence of trench backfilling is conducive to the stability of the pipeline.

Seismically induced soil liquefaction leads to large deformation and plastic flow of soil around the pipeline. Figure 9 shows the shear stress–strain curve at the side (P1) of the pipeline for the case with and without trench backfilling. The ultimate shear strain under seismic loading without trench backfilling reaches 0.3%, which is two–times the case with trench backfilling. This is because no liquefaction occurs around the pipeline under the case with trench backfilling. When liquefaction occurs, low effective confining stress leads to the rapid reduction in the shear modulus of the soil, resulting in large plastic deformation.

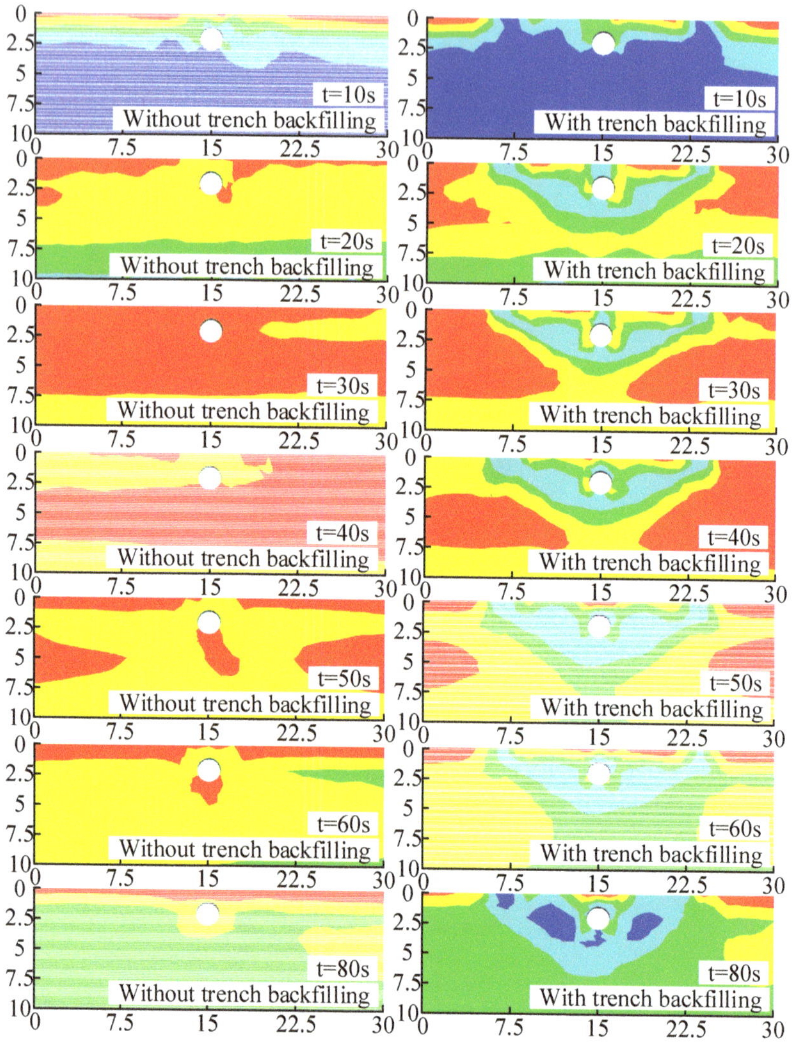

Figure 8. Liquefaction progression under seismic loading with and without trench backfilling.

4.2. Liquefaction-Induced Uplift of the Pipeline with Backfilling Trench

Figure 10 compares the time history of the pipeline uplift for the case with and without trench backfilling. The accumulation of excess pore water pressure and residual liquefaction of seabed around the pipeline leads to the uplift of the pipeline. The uplift of the pipeline occurs at 20 s and begins at the seabed liquefaction around the pipeline. Obviously, the uplift of the pipeline lags behind the accumulation of excess pore water pressure. For the case without trench backfilling, the maximum uplift is 34.5 cm and the pipeline suffers periodic oscillation during the uplift, which is caused by the wave and current propagation along the seabed. This periodic oscillation has adverse effects on the stability of the pipeline. The existence of trench backfilling reduces the uplift of the pipeline, which is about 6.6 cm. In addition, the periodic oscillation caused by wave and current propagation is also greatly relieved. It should be noted that once a strong earthquake occurs in the Bohai strait, the submarine pipelines may suffer severe damage due to the extensive uplift, leading to interruption of oil and gas transportation, oil and gas leakage, extensive economic losses and serious ecological problems.

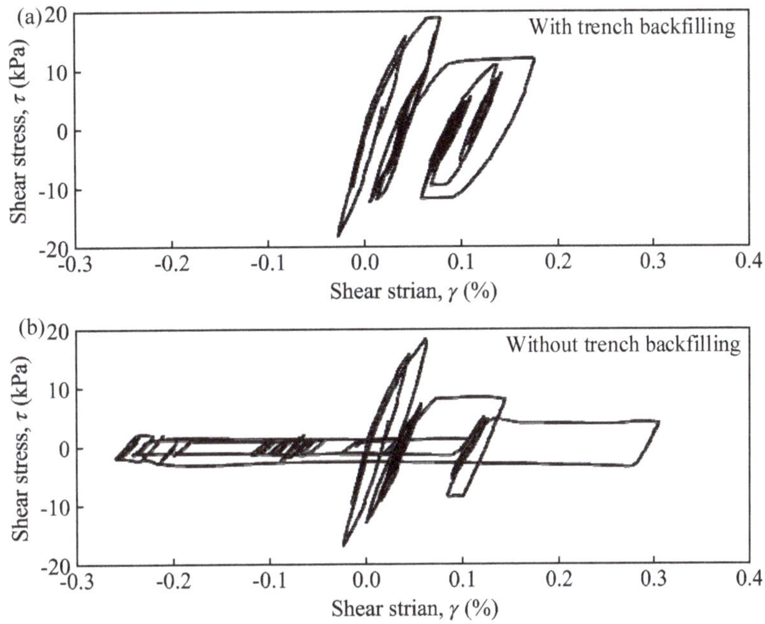

Figure 9. Predicted shear stress–strain responses at side of the pipeline; (**a**) without backfill soil, (**b**) with trench backfilling.

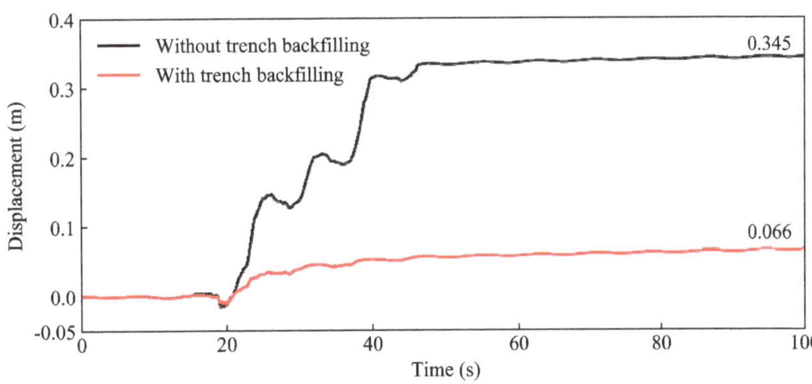

Figure 10. Liquefaction-induced uplift of the pipeline under seismic loading with and without trench backfilling.

A major finding in this study is that the current practice in seismic response analysis, neglecting the effect of trench backfilling on the seabed liquefaction and pipeline uplift may not always hold true. The effect has major consequences on the over–assessment of the development of the pipeline uplift at liquefiable seabed, thereby leading to conservative design requirements of the trenched marine structures. The proposed model can describe marine sediments around a trenched pipeline well up to the occurrence of liquefaction under cyclic loading, which can provide an efficient tool to evaluate the earthquake–related damage severity of the pipelines buried in the liquefiable seabed.

5. Conclusions

The Davidenkov skeleton curve associated with the Masing rule and Mohr–Coulomb failure criterion is used to describe the stress–strain relationship of seabed soil and the calculation of plastic volume strain is added as the source term of the accumulation of excess pore water pressure in Biot's dynamic consolidation equation. The seabed liquefaction and the uplift of the pipeline under seismic loading are analyzed, considering the ocean wave and current propagation. The conclusions are as follows:

(1) Seismic loading leads to a rise in excess pore water pressure and liquefaction in seabed soil and the wave and current contributes to the periodic oscillation in the rise in excess pore water pressure in the seabed soil. Following seabed liquefaction, the pore water flows upward and the seabed is, thereby, reconsolidated.
(2) Compared with the case without trench backfilling, the excess pore water pressure rises more slowly in trench backfilling under seismic loading and no liquefaction occurs. The liquefaction of seabed soil at the pipeline bottom is mitigated due to the existence of trench backfilling.
(3) The seismically induced liquefaction leads to the uplift of the pipeline. The uplift of the pipeline lags behind the accumulation of excess pore water pressure at the surrounding seabed soil, which occurs following seabed liquefaction.
(4) The existence of trench backfilling reduces the uplift of the pipeline and eliminates the oscillation caused by ocean wave and current during the uplift process.

Once a strong earthquake occurs in the Bohai strait, the submarine pipelines may suffer severe damage, leading to an interruption in oil and gas transportation, oil and gas leakage, extensive economic losses and serious ecological problems. The proposed model can describe marine sediments around a trenched pipeline well up to the occurrence of liquefaction under cyclic loading, which can provide an efficient tool to evaluate the earthquake-related damage severity of the pipelines buried in the liquefiable seabed.

Author Contributions: Conceptualization, Q.W. and J.B.; methodology, J.B.; software, W.H.; validation, Q.W. and Q.L.; formal analysis, Q.W. and W.H.; investigation, Q.W.; resources, Z.L. and K.Z.; data curation, W.H.; writing—original draft preparation, Q.W.; writing—review and editing, Q.L. and K.Z.; supervision, J.B.; project administration, Z.L. and K.Z.; funding acquisition, Q.L., Z.L. and K.Z. All authors have read and agreed to the published version of the manuscript.

Funding: This research was funded by the National Natural Science Fund of China, grant number 51978335, 51668002 and 52168044.

Institutional Review Board Statement: Not applicable.

Informed Consent Statement: Not applicable.

Data Availability Statement: Not applicable.

Acknowledgments: The authors are truly grateful for the financial support from the Scientific Research Fund of Institute of Engineering Mechanics, China Earthquake Administration (2019D05) and the National Natural Science Fund of China (51978335, 52168044, 51668002) and also grateful to the High Performance Computing Center of Nanjing Tech University for supporting the computational resources.

Conflicts of Interest: The authors declare no conflict of interest.

References

1. Cheng, A.H.D.; Liu, P.L.F. Seepage force on a pipeline buried in a poroelastic seabed under wave loading. *Appl. Ocean Res.* **1986**, *8*, 22–32. [CrossRef]
2. Dunn, S.L.; Vun, P.L.; Chan, A.H.C.; Damgaard, J.S. Numerical modelling of wave-induced liquefaction around pipelines. *J. Waterw Port. Coast. Ocean Eng.* **2006**, *132*, 276–288. [CrossRef]

3. Gao, F.P.; Yan, S.; Yang, B.; Luo, C. Steady flow-induced instability of a partially embedded pipeline: Pipe-soil interaction mechanism. *Ocean Eng.* **2011**, *38*, 934–942. [CrossRef]
4. Magda, W. Wave-induced uplift force acting on a submarine buried pipeline in a compressible seabed. *Ocean Eng.* **1997**, *24*, 551–576. [CrossRef]
5. Sawicki, A.; Mierczyński, J. Developments in modelling liquefaction of granular soils caused by cyclic loads. *Appl. Mech. Rev.* **2006**, *59*, 91–106. [CrossRef]
6. Sumer, B.M.; Truelsen, C.; Fredsøe, J. Liquefaction around pipelines under waves. *J. Waterw Port. Coast. Ocean Eng.* **2006**, *132*, 266–275. [CrossRef]
7. Sumer, B.M. *Liquefaction around Marine Structures: (With CD-ROM)*; World Scientific: London, UK, 2014.
8. Teh, T.C.; Palmer, A.C.; Damgaard, J.S. Experimental study of marine pipelines on unstable and liquefied seabed. *Coast. Eng.* **2003**, *50*, 1–17. [CrossRef]
9. Teh, T.C.; Palmer, A.C.; Bolton, M.D.; Damgaard, J. Stability of submarine pipelines on liquefied seabeds. *J. Waterw Port. Coast. Ocean Eng.* **2006**, *132*, 244–251. [CrossRef]
10. Ulker, M.; Rahman, M.S.; Guddati, M.N. Wave-induced dynamic response and instability of seabed around caisson breakwater. *Ocean Eng.* **2010**, *37*, 1522–1545.
11. Zhao, H.Y.; Jeng, D.S. Accumulated Pore Pressures around Submarine Pipeline Buried in Trench Layer with Partial Backfills. *J. Eng. Mech.* **2016**, *142*, 04016042. [CrossRef]
12. Zhao, K.; Xiong, H.; Chen, G.X.; Zhao, D.F.; Chen, W.Y.; Du, X.L. Wave-induced dynamics of marine pipelines in liquefiable seabed. *Coast. Eng.* **2018**, *140*, 100–113. [CrossRef]
13. Zhao, K.; Wang, Q.Z.; Chen, S.; Zhuang, H.Y.; Chen, G.X. Dynamic response of pipelines in liquefiable seabed under nature loadings: Waves and currents. *Ocean Eng.* **2021**, *230*, 109051. [CrossRef]
14. Zhao, K.; Zhu, S.D.; Bai, X.X.; Wang, Q.Z.; Chen, S.; Zhuang, H.Y.; Chen, G.X. Seismic response of immersed tunnel in liquefiable seabed considering ocean environmental loads. *Tunn. Undergr. Space Technol.* **2021**, *115*, 104066. [CrossRef]
15. Nath, B.; Soh, C.H. Transverse seismic response analysis of offshore pipelines in proximity to the sea-bed. *Earthq. Eng. Struct. Dyn.* **1978**, *6*, 569–583. [CrossRef]
16. Datta, T.K.; Mashaly, E.A. Seismic Response of Buried Submarine Pipelines. *J. Energy Resour. Technol.* **1988**, *110*, 208–218. [CrossRef]
17. Yan, Y.F.; Cheng, L.F. The Finite Element Analysis on the Submarine Pipeline under the Seismic Loading. *Adv. Mater. Res.* **2012**, *1700*, 490–495. [CrossRef]
18. Deng, Y.L.; Bian, Y.; Lei, F. Seismic Response Analysis of Free Long-Span Submarine Flexible Pipelines under Earthquakes. In *Applied Mechanics and Materials*; Trans Tech Publications Ltd.: Stafa-Zurich, Switzerland, 2014; Volume 580, pp. 1704–1707.
19. Luan, M.T.; Zhang, X.L.; Yang, Q.; Guo, Y. Numerical analysis of liquefaction of porous seabed around pipeline fixed in space under seismic loading. *Soil Dyn. Earthq. Eng.* **2009**, *29*, 855–864.
20. Li, B.; Wang, F.M.; Fang, H.Y.; Yang, K.J.; Zhang, X.J.; Ji, Y.T. Experimental and numerical study on polymer grouting pretreatment technology in void and corroded concrete pipes. *Tunn. Undergr. Space Technol.* **2021**, *113*, 103842. [CrossRef]
21. Miao, Y.; He, H.; Liu, H.; Wang, S.Y. Reproducing ground response using in-situ soil dynamic parameters. *Earthq. Engng. Struct. Dyn.* **2022**, *51*, 2449–2465. [CrossRef]
22. Wang, S.Y.; Zhuang, H.Y.; Zhang, H.; He, H.J.; Jiang, W.P.; Yao, E.L.; Ruan, B.; Wu, Y.X.; Miao, Y. Near-surface softening and healing in eastern Honshu associated with the 2011 magnitude-9 Tohoku-Oki Earthquake. *Nat. Commun.* **2021**, *12*, 1215. [CrossRef] [PubMed]
23. Zhao, K.; Wang, Q.Z.; Chen, Q.; Zhuang, H.Y.; Chen, G.X. Simplified effective stress modeling of shear wave propagation in saturated granular soils. *Géotechnique Lett.* **2021**, *11*, 2045–2543. [CrossRef]
24. Zhao, K.; Wang, Q.Z.; Zhuang, H.Y.; Li, Z.Y.; Chen, G.X. A fully coupled flow deformation model for seismic site response analyses of liquefiable marine sediments. *Ocean Eng.* **2022**, *251*, 111144. [CrossRef]
25. Byrne, P.M. A cyclic shear–volume coupling and pore pressure model for sand. In Proceedings of the 2nd International Conference on Recent Advances in Geotechnical Earthquake Engineering and Soil Dynamics, St. Louis, MI, USA, 3 January 1991; pp. 47–55.
26. Chen, G.X.; Zhao, D.F.; Chen, W.Y.; Juang, C.H. Excess Pore-Water Pressure Generation in Cyclic Undrained Testing. *J. Geotech. Geoenviron. Eng.* **2019**, *145*, 7. [CrossRef]
27. Ishihara, K. Liquefaction and flow failure during earthquakes. *Géotechnique* **1993**, *43*, 351–415. [CrossRef]
28. Zhao, K.; Qin, Y.; Lu, Q.R.; Chen, W.Y.; Zhuang, H.Y.; Chen, G.X. Cyclic resistance of saturated silt under wave-induced non-proportional loading. *Appl. Ocean Res.* **2020**, *102*, 102296. [CrossRef]
29. Chen, G.X.; Ruan, B.; Zhao, K.; Chen, W.Y.; Zhuang, H.Y.; Du, X.L.; Juang, C.H. Nonlinear response characteristics of undersea shield tunnel subjected to strong earthquake motions. *J. Earthq. Eng.* **2020**, *24*, 351–380. [CrossRef]
30. Day, R.A.; Potts, D.M. Zero thickness interface elements—Numerical stability and application. *Int. J. Numer. Anal. Methods Geomech.* **1994**, *18*, 689–708. [CrossRef]
31. *Itasca Flac—Fast Lagrangian Analysis of Continua*; Version 6; Itasca Consulting Group, Inc.: Minneapolis, MI, USA, 2009.
32. Hsu, H.C.; Chen, Y.Y.; Hsu, J.R.C.; Tseng, W.J. Nonlinear water waves on uniform current in lagrangian coordinates. *J. Nonlinear Math. Phys.* **2009**, *16*, 47–61. [CrossRef]

sustainability

MDPI

Article

Exceeding Probability of Earthquake-Induced Dynamic Displacement of Rail Based on Incremental Dynamic Analysis

Hongping Xing [1], Xiaodan Sun [1,2,*], Yu Liu [1,3], Jinzhen Lin [1] and Huilai Song [1]

1 School of Civil Engineering, Southwest Jiaotong University, Chengdu 610031, China
2 National Engineering Research Center for Geological Disaster Prevention Technology in Land Transportation, Southwest Jiaotong University, Chengdu 610031, China
3 Key Laboratory of High-Speed Railway Engineering of Ministry of Education, Southwest Jiaotong University, Chengdu 610031, China
* Correspondence: sunxd@swjtu.edu.cn

Abstract: When an earthquake occurs, it can strongly shake high-speed railway bridges. Consequently, the dynamic displacement of the rail on the bridge may exceed the allowable standard. However, few studies have evaluated the probability of rail displacement exceeding the allowable standard, compared to the rich variety of research on the vulnerability of other components of the high-speed railway track-bridge system or other structures. In this paper, incremental dynamic analysis (IDA) is applied to calculate the exceeding probability of rail displacement under different earthquake excitations. A finite element model (FEM) of a high-speed railway track-bridge system is established, which consists of a finite length CRTS II ballastless track laid on a five-span simply supported girder bridge. Records from five stations in the PEER NGA−West2 strong ground motion dataset are selected as seismic excitation. Based on the simulation, the characteristics of the vertical displacement of the rail under different seismic excitations are investigated, and the probability of the vertical displacement of the rail exceeding the allowable standard is calculated using IDA. The results show that: (1) the vertical displacement of the rail above the abutment is significantly smaller than that above other parts of the bridge; (2) the vertical irregularity of the rail caused by earthquakes has a wavelength close to the length of a simply supported girder; (3) under some excitations, two bumps are observed in the Fourier displacement spectrum in the frequency range of 1.3–2.5 Hz and 10–12 Hz, respectively, which may indicate the resonance of the model to the excitation; and (4) the vertical displacement amplitude probability of the rail exceeding 2 mm is 44%, 89%, and 99% when PGA = 0.01 g, 0.20 g, and 0.40 g, respectively. The exceeding probability of the rail above the mid-span is larger than that above other parts of the bridge. Within the mid-span, the exceeding probability of the rail is the largest above the center of the bridge.

Keywords: incremental dynamic analysis; displacement time history analysis; Fourier spectrum analysis; vertical irregularity; exceeding probability of displacement

Citation: Xing, H.; Sun, X.; Liu, Y.; Lin, J.; Song, H. Exceeding Probability of Earthquake-Induced Dynamic Displacement of Rail Based on Incremental Dynamic Analysis. *Sustainability* **2022**, *14*, 11871. https://doi.org/10.3390/su141911871

Academic Editors: Su Chen, Chong Xu and Shuang Li

Received: 5 August 2022
Accepted: 16 September 2022
Published: 21 September 2022

1. Introduction

High-speed railway (HSR) has become the future development direction for the railway system in China due to the high passenger carrying capacity and operating speed of the high-speed train [1,2]. At present, the HSR mainly operates on bridges instead of on roads [3]. For many high-speed railways, however, the construction has to pass high-intensity areas. An earthquake may cause strong shaking of a high-speed railway bridge, leading to the dynamic displacement of the rail on the bridge exceeding the allowable standard. Generally, the rail irregularity caused by the earthquake can deteriorate safety, passenger comfort [4], and interaction with other infrastructures [5]. Therefore, the dynamic displacement of the rail exceeding the allowable standard may also threaten the running safety of the high-speed train.

The rail is the component of the HSR in direct contact with the vehicle. Therefore, a lower dynamic displacement of the rail is key to the high-speed train running safely on the bridge under seismic excitation. The rail of the China HSR has very strict smoothness requirements. The limit for vertical displacement of the rail is ± 2 mm on the China HSR, with a design speed of 200–350 km/h [6]. However, the existing studies mainly investigate the dynamic displacement regularities of the rail on the bridge caused by the influence of the track-bridge system and earthquakes. For example, under seismic excitation, earthquakes cause vibration displacement of the bridge [7], which greatly influences the vertical displacement of the rail [3]. The height of the pier will also influence the vertical displacement of the rail [8]. A high proportion of the CRTS II slab ballastless track structure (CRTS II SBTS), an important component of the high-speed railway of China, operates on bridges [9]; providing a high degree of rail smoothness and ensuring high comfort, stability, and safety of the high-speed train. The loss of CRTS II SBTS performance will deteriorate the performance of the rail under seismic excitation. For example, the cement asphalt mortar (CA mortar), an interlayer structure of the CRTS II SBTS, affects the vertical vibration of the rail due to an increase in the elastic modulus of the CA mortar or the loss of the adhesion performance of the CA mortar [10,11]. The fastener, an important component of the CRTS II SBTS, restrains the rail to the track plate. The displacement amplitude of the rail will increase obviously under earthquake excitation due to the increased longitudinal resistance of fasteners, which would lead to the exceeding failure of the rail [12,13]. Not only does the track-bridge system affect the dynamic displacement of the rail, but the earthquake is also a critical factor affecting the vertical displacement of the rail in the seismic analysis. Many researchers have investigated the dynamic displacement of the rail as influenced by pulse-type ground motions [12,13], the input component of the ground motions [14–16], the critical input direction of the ground motions [16], the non-uniform excitation of the ground motions [17], and the spatial variability of ground motions [18]. Moreover, earthquakes are a single random process that cannot be reproduced, with considerable variability and uncertainty [19], and the typical dynamic analysis cannot connect the intensity of ground motion and the degree of structural damage to estimate the dynamic capacity of the rail against the earthquake load. The vulnerability analysis may be a suitable method by giving a fragility curve to evaluate the probability that structures exceed the allowable standard.

However, few studies have evaluated the probability of the earthquake-induced displacement of the rail exceeding the allowable standard, compared to the wealth of research on the vulnerability analysis of other components of the high-speed railway track-bridge system [15,20–22] and other structures [23–27]. In the field of high-speed railway track-bridge systems, researchers mainly investigate the vulnerability of the bridge [20,22], the component of the ballastless track structure [21], and the uplift of wheels from the rail [15]. In contrast, the study of the rail exceeding the allowable standard is almost blank. At present, the incremental dynamic analysis (IDA) proposed by Bertero [28] is widely used to evaluate the vulnerability of structures, which scale the earthquake records step by step and incrementally. The important advantage of this method is the better investigation of the structural behavior during severe earthquakes, as well as showing the changes in structural response with increasing earthquake intensity [29,30]. Although the computational cost of the IDA method is very high, IDA provides a better estimation of the dynamic capacity of the structure against the earthquake load [29]. The seismic vulnerability curves calculated based on the IDA describe the relationship between the intensity of ground motion and the degree of structural damage, which is a powerful tool for assessing structural seismic risk. Therefore, the main purpose of this study is the use of the IDA method to evaluate the exceeding probability of the earthquake-induced dynamic displacement of rail, also called the probability of exceeding certain limits of earthquake-induced dynamic displacement of rail. This study may be the first attempt to evaluate the exceeding probability of the earthquake-induced dynamic displacement of the rail and may be an extension that applies the IDA to the vulnerability analysis of the high-speed railway.

In this paper, the IDA is applied to calculate the exceeding probability of rail displacement under different earthquake excitations. A finite element model (FEM) of a high-speed railway track-bridge system is established, which consists of a finite length CRTS II SBTS laid on a five-span simply supported girder bridge. Ground motion records from five stations in PEER NGA−West2 strong ground motion dataset are selected as seismic excitation. Based on the simulation, the characteristics of the vertical displacement of the rail under different seismic excitations are investigated, and the probability of the vertical displacement of the rail exceeding the allowable standard is calculated using IDA.

2. Methodology

In the typical IDA of structures [31,32], the ground motion records are scaled up until the structure is destroyed, which is evaluated by given damage rules. However, in this study, evaluating the exceeding probability of earthquake-induced dynamic displacement of the rail is the primary objective. High-speed railways have higher requirements for smoothness than ordinary railways; therefore, taking the whole destruction state as the end sign of IDA is unsuitable. In order to investigate the exceeding probability of earthquake-induced displacement of the rail of the high-speed railway, peak ground acceleration (PGA) is selected as the intensity measure of the ground motion records according to the Code for Seismic Design of Railway Engineering [33]. Furthermore, scaling the PGA of ground motion records within the range of PGA $\leq$ 0.40 g is proposed according to the Code for Seismic Design of Buildings [34]. The steps of the IDA-based exceeding probability of earthquake-induced dynamic displacement of the rail include:

1. Select the ground motion records and the intensity measure of the ground motion records;
2. Scale the PGA of the selected records to different excitation levels within the range of PGA $\leq$ 0.40 g;
3. Establish a FEM model for the exceeding probability of the earthquake-induced dynamic displacement of the rail;
4. Perform a non-linear dynamic response analysis;
5. Obtain the vertical displacement of the rail;
6. Perform a linear regression analysis and construct response probability functions of earthquake-induced dynamic displacement of the rail;
7. Define the limit states;
8. Construct the exceeding probability curves of earthquake-induced dynamic displacement of the rail.

3. Selection and Scale of Ground Motion Records

Earthquakes are selected with the $M_W \geq 6$, the standard magnitude in the China Earthquake Network of great earthquakes. All the selected ground motions records are near-fault ground motions with the R_{JB} < 25 km. The ground motion records are also selected with PGA > 0.05 g, the peak ground motion acceleration corresponding to the six-degree area in the Code for Seismic Design of Buildings [34]. At this stage, the great near-fault ground motions are selected to ensure the intensity of the earthquake. Moreover, according to Chandler's classification [35], the ground motion records are divided into three sets based on the ratio of PGA and peak ground velocity (PGV). Records classified in the low PGA/PGV range with PGA/PGV < 0.8 g/(m/s), whereas records classified in the high PGA/PGV range with PGA/PGV > 1.2 g/(m/s). Records classified as the intermediate range with PGA/PGV between 0.8 and 1.2 g/(m/s). Each category can create resonance responses for specific structures [35]. Therefore, in this paper, earthquakes are selected from all three categories to have a chance to create more critical responses to the rail laid on the bridge.

Ground motion records selected from five stations in the PEER NGA−West2 strong ground motion dataset are shown in Table 1 and Figure 1. In this paper, earthquakes are represented by labels in Table 1. Each ground motion record includes two horizontal components and one vertical component. An input group consists of one horizontal component and one vertical component of ground motion, which are inputted into the model. For example, the first input group consists of ELC−H1 and ELC−UP, and the second consists of ELC−H2 and ELC−UP. Response spectra of the ground motions are also depicted to show the differences between those selected ground motions, as shown in Figure 2. According to the Code for Seismic Design of Railway Engineering [33], the intensity measure of PGA was selected to scale the acceleration amplitude of ground motion records. According to the Code for Seismic Design of Buildings [34], the PGA of earthquakes is then scaled to the excitation levels of PGA = 0.01 g, 0.025 g, 0.05 g, 0.10 g, 0.15 g, 0.20 g, 0.25 g, 0.30 g, 0.35 g, and 0.40 g, respectively. Then ten input groups make a total of 100 seismic excitations groups, inputted to the bottom of the piers. The horizontal component of ground motion is applied in the X direction, and the vertical component of ground motion is applied in the Y direction. The X direction is along the bridge and the Y direction is vertical to the bridge, as illustrated in Figure 3.

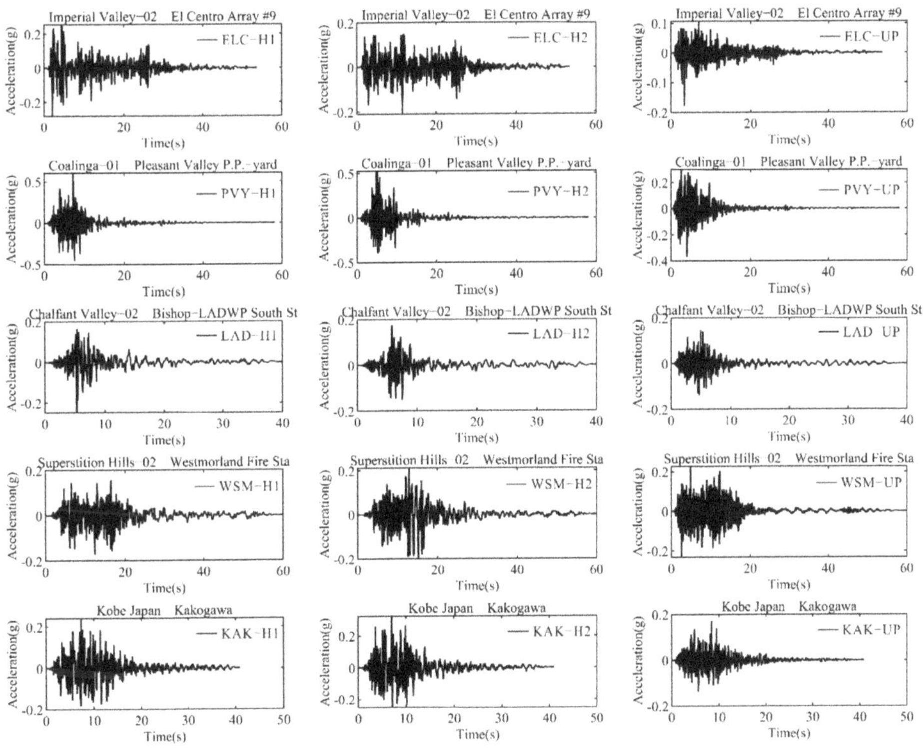

Figure 1. Ground motion records selected from the PEER NGA−West2 strong ground motion dataset.

Table 1. Ground motion records selected from the PEER NGA−West2 strong ground motion dataset.

Earthquakes	Stations	Labels	M_W	R_{JB} (km)	Component	PGA (g)	PGV (m/s)	PGA/PGV g s/(m)
Imperial Valley−02	El Centro Array #9	ELC	6.95	6.09	H1	0.28	0.31	0.90
					H2	0.21	0.31	0.68
					UP	0.18	0.09	2.00
Coalinga−01	Pleasant Valley P.P. −yard	PVY	6.36	7.69	H1	0.60	0.60	1.00
					H2	0.53	0.39	1.36
					UP	0.37	0.16	2.31
Chalfant Valley−02	Bishop−LADWP South St	LAD	6.19	14.38	H1	0.25	0.20	1.25
					H2	0.18	0.20	0.90
					UP	0.14	0.07	2.00
Superstition Hills−02	Westmorland Fire Sta	WSM	6.54	13.03	H1	0.17	0.23	0.74
					H2	0.21	0.32	0.66
					UP	0.23	0.09	2.56
Kobe Japan	Kakogawa	KAK	6.90	22.5	H1	0.24	0.21	1.14
					H2	0.32	0.27	1.19
					UP	0.17	0.11	1.55

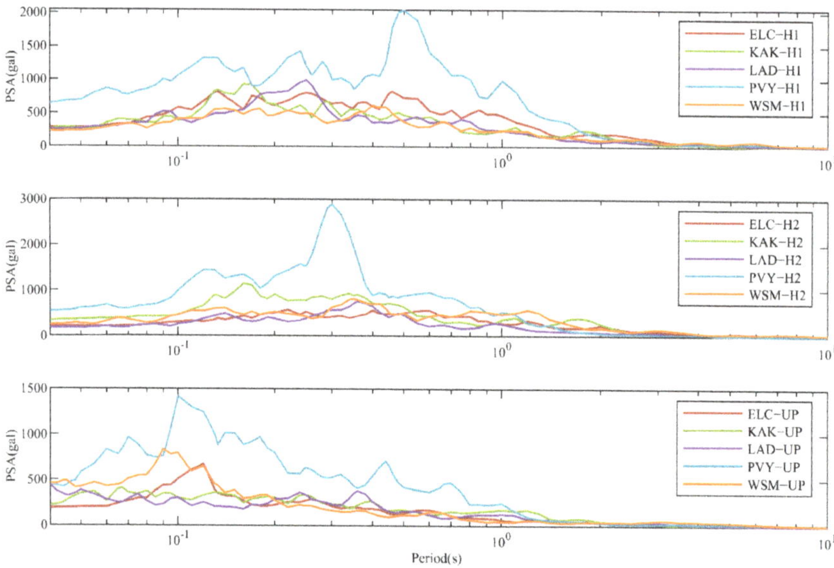

Figure 2. Response spectra of the selected ground motions.

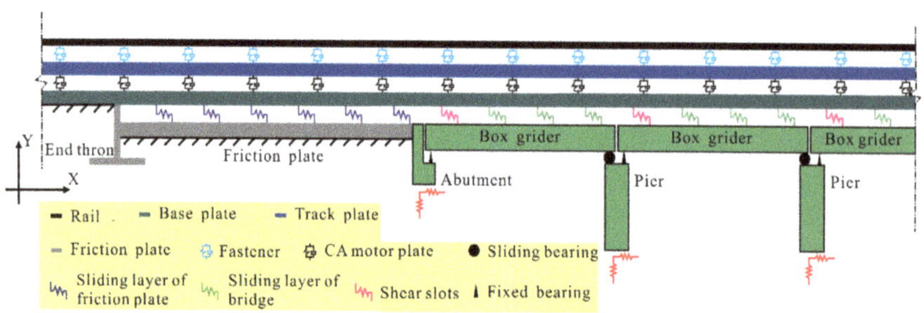

Figure 3. FEM model of the CRTS II slab ballastless track-bridge system.

4. FEM Model of the High-Speed Railway Track-Bridge System

The 2D FEM model of a high-speed railway track-bridge system is established using ABAQUS, consisting of a five-span simply supported girder bridge with CRTS II SBTS. The length of each span is 32.4 m, and the total length of the model is 282.1 m. The pier is 8-m-high and made of C35 concrete. The girder bridge has a single box section made of C50 concrete. The friction plate is 52-m-long and made of C40 concrete. The rail is made of 60 kg/m steel, connected to the track plate through fasteners with a spacing of 0.65 m. The CRTS II SBTS includes the track plate, CA mortar layer, and base plate. The track plate is a 0.2-m-thick precast slab made of C60 concrete. The base plate is a 0.19-m-thick slab made of C40 concrete. A 0.03-m-thick CA mortar layer is filled between the track plate and the base plate. The sliding layer is between the base plate and the bridge, and between the base plate and the friction plate. The sliding of the base plate on the bridge is limited by the shear slot. The configuration of the FEM model of the track-bridge system is illustrated in Figure 3. In the model, the rail, track plate, base plate, box girder, and pier are all established by a beam element. The fastener, CA mortar layer, sliding layer, shear slots, fixed bearing, sliding bearing, and the interaction between pier and foundation are simulated by spring. The vertical stiffness of the shear slot, the fixed bearing, and the sliding bearing is rigid [36]. Sections, densities, elasticity modulus, damping, and force-displacement curves (F–D curves) for various elements are shown in Tables 2 and 3 and Figure 4 [36–39].

Table 2. Sections, densities, elasticity modulus, and damping coefficients for various components.

Component	Sections Width × Height (m)	Densities (kg/m^3)	Elasticity Modulus (Pa)	Damping Coefficients	
				α	β
Rail	—	7850	2.06×10^{11}	7	1.00×10^{-6}
Track plate	2.55 × 0.10	2500	3.65×10^{10}	—	—
Base plate	2.95 × 0.19	2500	3.25×10^{10}	—	—
Box girder	10.21 × 3.03	3300	3.55×10^{10}	0.26	0.91×10^{-2}
Friction plate	9.00 × 0.40	2500	3.25×10^{10}	—	—
Ground	—	2100	1.50×10^{8}	—	—
Pier	5.66 × 1.94	2500	3.30×10^{10}	—	—

Table 3. F–D curves in the X direction and damping of spring elements.

F–D Curve	Spring Elements	F (kN)	D (mm)	Damping (N·s/m)
	Fastener	15	2.00	4.770×10^{4}
	CA mortar layer	42	0.50	1.000×10^{5}
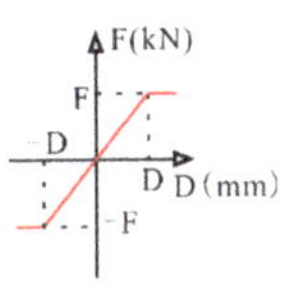	Sliding layer of bridge	6	0.50	—
	Shear slots	1465	0.12	—
	Fixed bearing	1000	2.00	3.475×10^{6}
	Sliding bearing	100	2.00	3.475×10^{6}
	Sliding layer of friction plate	14	2.00	—
	Foundation	—	—	1.080×10^{5}

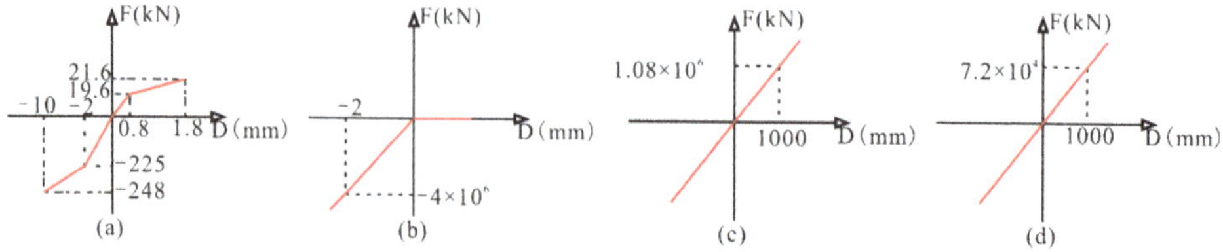

Figure 4. F−D curves of spring elements in Y direction: (**a**) Fastener; (**b**) Sliding layer; (**c**) CA mortar layer; (**d**) Foundation (in the directions of X and Y).

The first 10 natural vibration frequencies of the FEM model of the CRTS II slab ballastless track-bridge system are shown in Table 4.

Table 4. The first ten natural vibration frequencies of the FEM model.

Orders	1	2	3	4	5
Frequency (Hz)	1.498	1.888	1.992	2.431	3.0568
Orders	6	7	8	9	10
Frequency (Hz)	3.663	3.939	4.029	4.307	4.540

The vertical displacement of the rail is mainly compared in this paper at positions 0–5, as shown in Figure 5, where the red dot represents the compared positions of the vertical displacement of the rail. Position 0 is above the abutment; Positions 1, 3, and 5 are above the mid-span; and Positions 2 and 4 are above the pier.

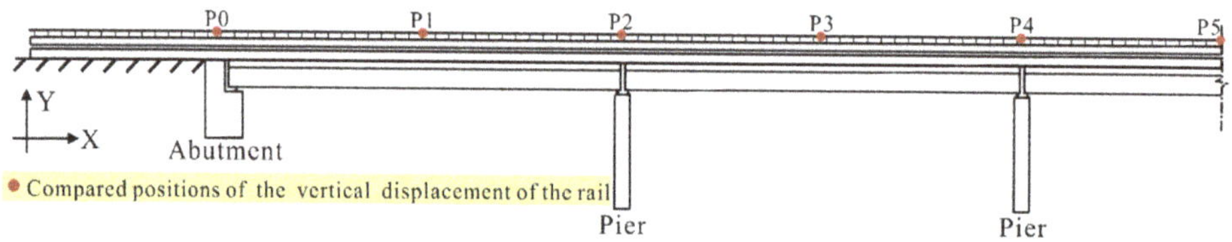

Figure 5. Compared positions of the vertical displacement of the rail.

5. Vertical Displacement of the Rail

5.1. Vertical Displacement of the Rail in Time History Curves

Time history curves of the vertical displacement of the rail above the abutment, mid-span, and the pier are compared after inputting 100 seismic excitation groups into the bottom of each pier. Limited by the article space, three cases of the time history curves of vertical displacement of the rail are shown, including the ELC−H1 with PGA = 0.01 g and 0.40 g, and the KAK−H2 with PGA = 0.40 g, as shown in Figure 6. The light green solid line represents the vertical displacement time history curve of the rail at position 0, corresponding to the position above the abutment.

As shown in Figure 6, the vertical displacement of the rail at position 0, corresponding to the position above the abutment is comparable to that at positions 1–5, corresponding to positions above the mid-span and the pier under seismic excitation, when the vertical displacement of the rail is positive. That may be because the distance between the abutment and the bridge is small; CRTS II SBTS is longitudinally continued. Under seismic excitation, the CRTS II SBTS above the abutment moves together with that above the bridge because the CRTS II SBTS above the abutment is adjacent to and is restrained to that above the bridge. Then the rail moves together with the CRTS II SBTS. Therefore, the difference between the vertical displacement of the rail above the abutment and other parts of the bridge is small when the vertical displacement is positive. However, the vertical displacement of the rail above the abutment is obviously smaller than that above other parts of the bridge under seismic excitation when the vertical displacement of the rail is negative. That may be due to the hard compression of the components above the abutment when the vertical displacement of the rail is negative.

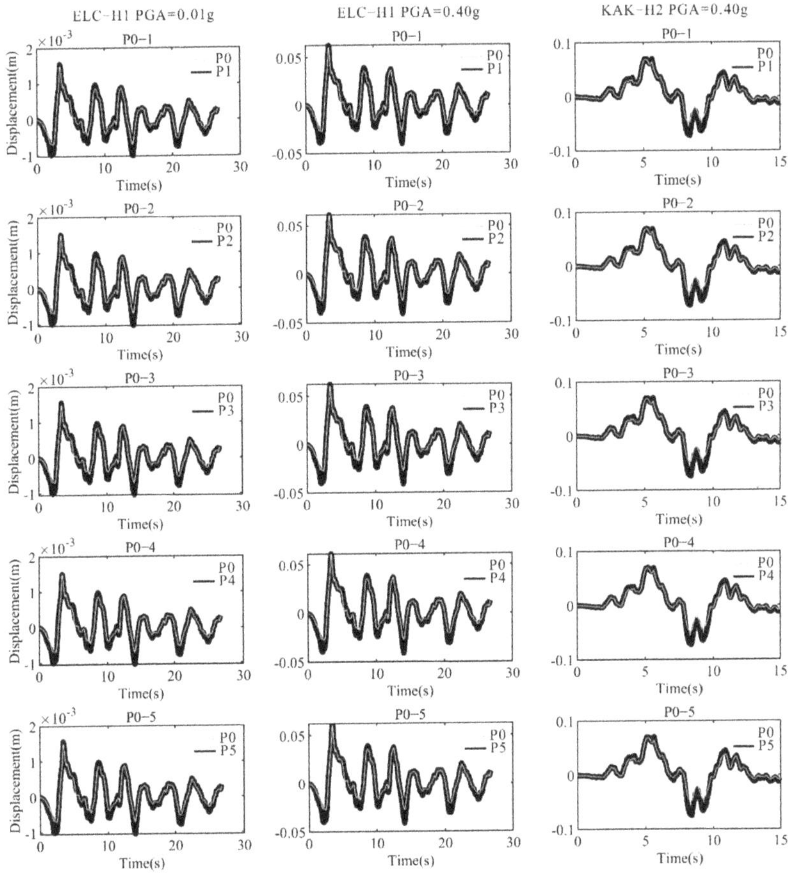

Figure 6. Vertical displacement time history curves of the rail at Position 0–5.

Figure 6 also shows that the displacement variation trend is comparable between vertical displacement time history curves of the rail at the same position when PGA = 0.01–0.40 g. However, the displacement amplitude of the rail is different. For example, in the action of ELC−H1 with PGA = 0.01–0.40 g, the maximum positive vertical displacement of the rail increases from approximately 1.5–60 mm, the maximum negative vertical displacement of the rail above the abutment increases from approximately 0.7–28 mm, and the maximum negative vertical displacement of the rail above the mid-span and the pier increases from approximately 1–41 mm. Variation trends of the vertical displacement time history curves of the rail vary when earthquakes are different, which may be because the PGA, the frequency, the energy, and other parameters of various strong motions are different.

The amplitude and frequency of the vertical displacement time history curves of the rail changed obviously to the action of the PVY−H2 with PGA = 0.20 g compared to that under the action of the PVY−H2 with PGA = 0.15 g, as shown in Figure 7.

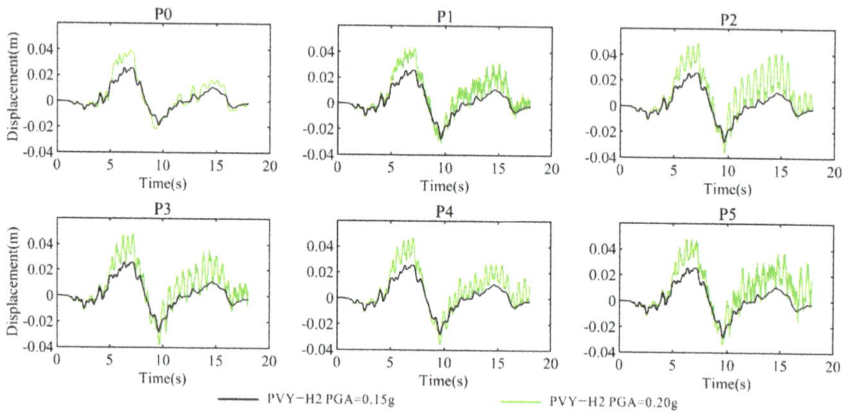

Figure 7. Vertical displacement time history curves to the action of PVY−H2.

In the action of PVY−H2 with PGA = 0.20 g, an obvious increase in the vertical displacement amplitude of the rail can be observed. In addition, a high-frequency vibration in the vertical displacement of the rail can be observed at positions 1–5, corresponding to positions above the mid-span and the pier, which is not obvious at position 0, corresponding to the position above the abutment. The change in vertical displacement amplitude and the vibration frequency can be observed from the action of PVY−H2 with PGA = 0.20 g rather than PVY−H2 with PGA = 0.15 g, which may be because of the frequency of the PVY−H2 with PGA = 0.20 g close to a natural frequency of the model. Therefore, in the action of PVY−H2 with PGA = 0.20 g, the resonance of the model occurs. Moreover, the resonance frequency may be the natural frequency of the lower vibration mode of the model and significantly influences the vertical displacement of the rail.

5.2. Vertical Irregularity along the Bridge of the Rail

The vertical irregularity of the rail along the bridge is shown at the moment of the largest vertical displacement of the rail under seismic excitation at position 5. Limited by the article space, six cases of the vertical irregularity curves of the rail along the bridge are shown, including the ELC−H1 with PGA = 0.01 g and 0.40 g, the LAD−H1 with PGA = 0.01 g and 0.40 g, and the KAK−H2 with PGA = 0.01 g and 0.40 g, respectively. As shown in Figures 7 and 8, the vertical irregularity curves of the rail along the bridge are shown by the positive and negative vertical displacement of the rail, respectively.

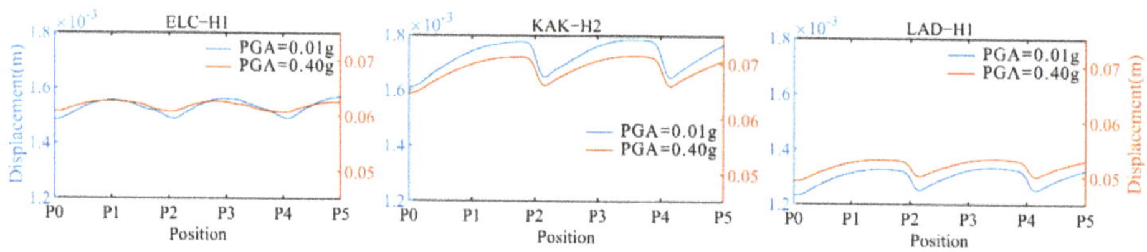

Figure 8. Vertical irregularity of positive vertical displacement of the rail along the bridge.

Although the shape of vertical irregularity curves of the rail along the bridge is different, the vertical irregularity of the rail caused by earthquakes has a wavelength close to the length of a simply supported girder. That may be because the vertical displacement of the rail on the bridge is affected by the girder and the vertical displacement of each simply supported girder is comparable under seismic excitations. Although the shape of the vertical irregularity curves of the rail along the bridge is comparable, the vertical displacement amplitude of the rail increases with PGA. For example, the minimum and maximum vertical displacement of the rail along the bridge is 0.02 mm and 0.2 mm when PGA = 0.01 g, respectively. The minimum and maximum vertical displacement of the rail along the bridge is 0.5 mm and 7 mm when PGA = 0.40 g, respectively. The "L" shape of the rail above the connection between the abutment and the bridge can be observed significantly when the vertical displacement of the rail is negative, as shown in Figure 9, which may be due to the component above the abutment being hard to compress. There is an obvious increment of the vertical displacement of the rail at a longitudinal range of 2.6 m from position 0 above the abutment, which is along the bridge. The average vertical increment increases from 4–17 mm when PGA = 0.01–0.40 g.

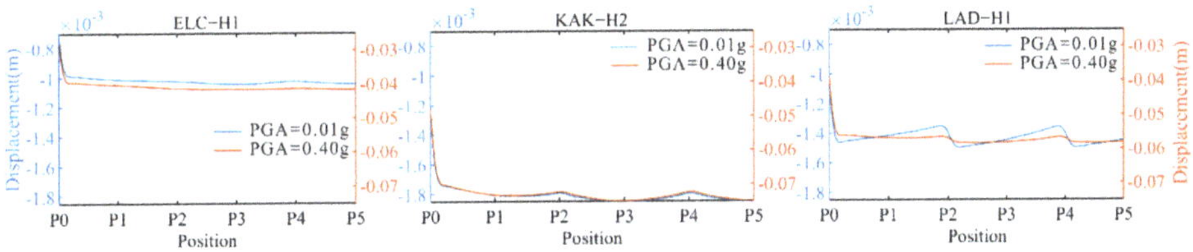

Figure 9. Vertical irregularity of negative vertical displacement of the rail along the bridge.

5.3. *Fourier Displacement Spectrum of the Rail*

Fast Fourier transform (FFT) is applied to calculate the Fourier displacement spectrum of the rail. Therefore, the Fourier displacement spectrum of the rail compares different earthquakes, PGA, and positions in this paper. Limited by the article space, three cases of the Fourier displacement spectrum curves of the rail are shown in Figure 10, including the WSM−H1 with PGA = 0.01 g and 0.40 g and the ELC−H1 with PGA = 0.40 g.

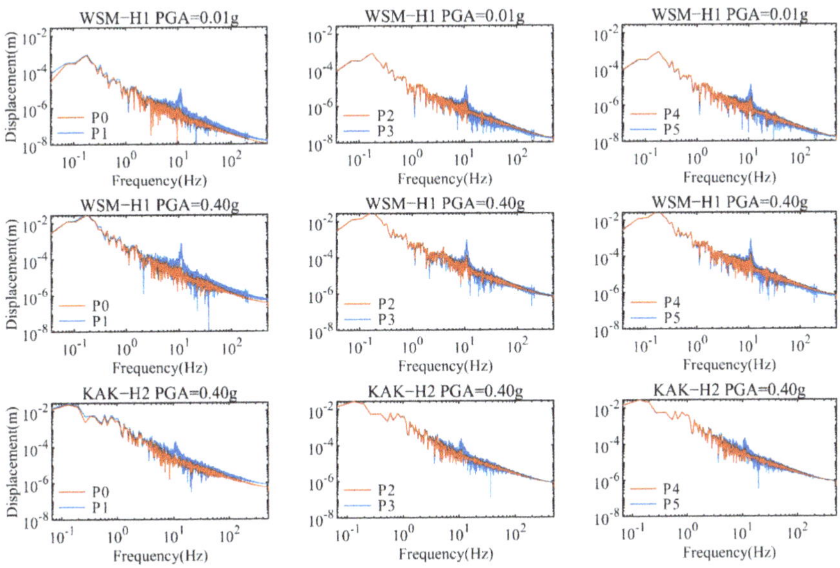

Figure 10. Fourier displacement spectrum of the rail.

As shown in Figure 10, an obvious bump may be observed in the Fourier displacement spectrum in the frequency range of 10–12 Hz at positions 1–5, corresponding to positions above the mid-span and the pier; however, the bump may not be observed in the Fourier displacement spectrum at position 0, corresponding to the position above the abutment.

Furthermore, the bump amplitude in the Fourier displacement spectrum of the rail at positions 2 and 4, corresponding to positions above the pier, is smaller than that at positions 3 and 5, corresponding to positions above the mid-span. The bump in the Fourier displacement spectrum of the rail may indicate the resonance of the model to the excitation. The bump amplitude in the Fourier displacement spectrum of the rail above the mid-span is larger than that above the pier, which may be due to the larger vertical displacement of the rail above the mid-span easily caused by earthquake excitation.

Figure 7 shows an obvious change in amplitude and frequency of the vertical displacement of the rail with the action of the PVY−H2 with PGA = 0.20 g. Therefore, in this paper, the Fourier displacement spectrum of the rail to the action of PVY−H2 with PGA = 0.20 g is shown in Figure 11, which is compared to that under the action of the PVY−H2 with PGA = 0.15 g.

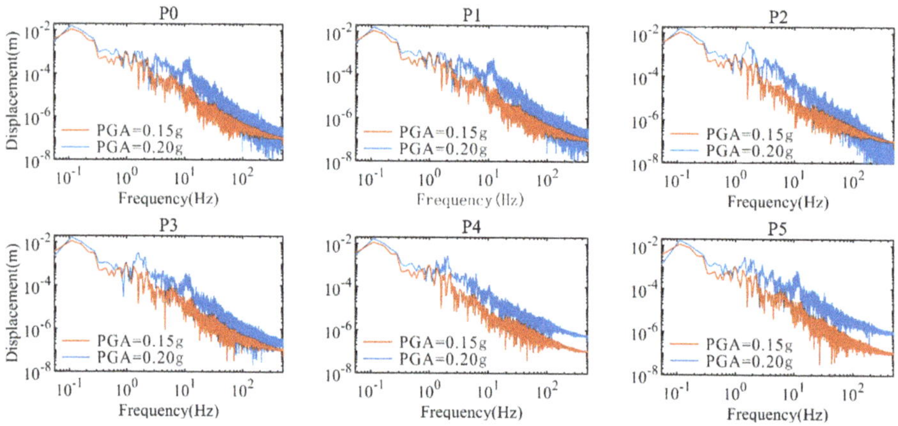

Figure 11. Fourier displacement spectrum of the rail to the action of PVY−H2.

A bump may be observed in the Fourier displacement spectrum in the frequency range of 10–12 Hz at positions 0, 1, 3, and 5, and a bump can also be observed in the Fourier displacement spectrum in the frequency range of 1.3–2.5 Hz at positions 2–4, which may indicate the resonance of the model to the excitation and indicate that two ground motion frequencies may be approximately the natural frequency of the model. In addition, the change in the amplitude and frequency of the vertical displacement of the rail may indicate that the frequency range of 1.3–2.5 Hz may be approximately the natural frequency of the lower vibration mode of the model, according to Table 4.

The resonance of the model to the excitation may be due to the ground motion frequency being approximately the natural frequency of the model. A total of 100 cases are calculated in this paper. However, the obvious change in amplitude and frequency of the vertical displacement of the rail is only observed from the action of PVY−H2 with PGA = 0.20 g. The result indicates that it is not easy to cause the resonance of the model and cause the vertical displacement of the rail with the obvious change in amplitude and frequency under seismic excitation.

6. Exceeding Probability of the Earthquake-Induced Dynamic Displacement of the Rail

Curves of the vertical displacement amplitude of the rail changing with PGA are shown in Figure 12. The gray area represents the range of the vertical displacement amplitude of the rail at six positions under seismic excitation, and points represent the average of the vertical displacement amplitude of the rail at six positions.

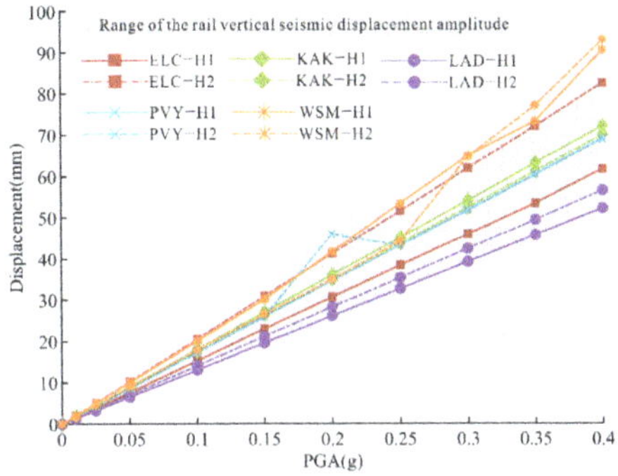

Figure 12. Curves of the vertical displacement amplitude of the rail.

It can be observed that the vertical displacement amplitude of the rail increases with the PGA, and the increasing trend is approximately linear. The vertical displacement amplitude of the rail differs from the action of earthquake excitations with the same PGA. For example, the maximum and minimum vertical displacement of the rail are 2.06 mm and 1.31 mm, respectively, from the earthquake excitation with PGA = 0.01 g. Moreover, the maximum and minimum vertical displacement of the rail are 92.91 mm and 52.25 mm, respectively, from the ground motion with PGA = 0.40 g. Finally, an obvious bump can be observed from the excitation of PVY−H2 with PGA = 0.20 g. As shown in Figures 7 and 11, this may be due to a resonance of the model being caused, resulting in a large vertical displacement amplitude of the rail.

In this paper, linear regression is applied to fit the logarithm of PGA to the logarithm of the vertical displacement of the rail [31]. The general linear regression fitted equation between the logarithm of PGA and the logarithm of vertical displacement of the rail is shown in Formula (1).

$$\ln(D_{\mathrm{d}}) = A + B\ln(\mathrm{PGA}) \tag{1}$$

where D_{d} is the vertical displacement of the rail (mm); PGA is the peak ground motion acceleration (g); and A and B are the coefficients of the fitted equation.

The fitted results of the linear regression are shown in Table 5, where R-square is the deviation coefficient indicating the correlation degree of the linear regression function. Coefficients A or B at six positions are comparable, and the R-square of the vertical displacement linear regression of the rail are all above 0.98, which indicates an excellent fitted result.

Table 5. Coefficients and R-square of the vertical displacement fitted equation of the rail.

Positions	Coefficients of Fitted Equation		R-Square
	A	*B*	
0	5.145	1.008	0.9817
1	5.167	1.008	0.9847
2	5.167	1.010	0.9834
3	5.172	1.009	0.9844
4	5.166	1.010	0.9837
5	5.173	1.009	0.9843

High-speed railways have a high requirement for smoothness. Therefore, the driving safety of the high-speed train may be significantly threatened by the slight displacement of the rail. Consequently, a 2 mm vertical displacement of the rail is selected as the allowable standard of the vertical displacement of the rail [6].

Exceeding probability curves of the vertical displacement can represent the probability that the dynamic vertical displacement of the rail exceeds the allowable standard from the action of earthquake excitations with different PGA. The vertical displacement D_d and bearing capacity D_c of the rail obey the lognormal distribution [21,31]. Thus the exceeding probability can be defined with a standard normal cumulative distribution function [21,31]. The calculation formula is shown in Formula (2).

$$P = \Phi\left(\frac{\ln\left(\widetilde{D_d}/\widetilde{D_c}\right)}{\sqrt{\beta_c^2+\beta_d^2}}\right) = \Phi\left(\frac{\ln\left(e^A(PGA)^B/\widetilde{D_c}\right)}{\sqrt{\beta_c^2+\beta_d^2}}\right) \quad (2)$$

where $\widetilde{D_d}$ is the average response of the structure, which can be calculated by the regression analysis; $\widetilde{D_c}$ is the average structural total bearing capacity, which can be determined by the limit state of the vertical displacement of the rail; e is the natural constant; β_d is the logarithmic standard deviation of structural reaction; β_c is the logarithmic standard deviation of structural bearing capacity; and $\Phi(x)$ is the standard normal cumulative distribution function. The values of the input parameters are shown in Tables 5 and 6.

Table 6. The values of some input parameters in Equation (2).

Positions	$\widetilde{D_c}$ **(mm)**	$\sqrt{\beta_c^2+\beta_d^2}$
0	2	1.1913
1	2	1.1895
2	2	1.1920
3	2	1.1904
4	2	1.1915
5	2	1.1904

The probability curves of the vertical displacement amplitude of the rail exceeding 2 mm at positions 0–5 are shown in Figure 13 under seismic excitation. The average probability of the vertical displacement of the rail exceeding 2 mm at six positions is comparable and approximately 44%, 89%, and 99% when the PGA is 0.01 g, 0.05 g, and 0.20 g, respectively. The PGA = 0.05 g is the peak ground motion acceleration corresponding to the areas with a seismic intensity of 6 degrees in the Code for Seismic Design of Buildings [33]. The probability of the earthquake-induced dynamic vertical displacement amplitude of the rail exceeding 2 mm increases with PGA.

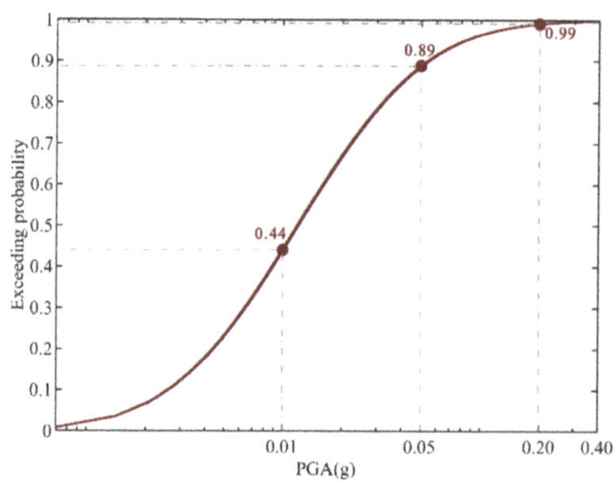

Figure 13. The exceeding probability of the earthquake-induced dynamic vertical displacement of the rail exceeding 2 mm.

The probability of the earthquake-induced vertical displacement with PGA = 0.01 g, 0.05 g, and 0.20 g of the rail exceeding 2 mm is shown in Figure 14, respectively. The exceeding probability of the rail above the mid-span is larger than that above the pier and the abutment, while the exceeding probability of the rail above the abutment is the minimum. Moreover, the exceeding probability of the rail at position 5, which corresponds to the position above the center of the bridge, is the largest among positions above the mid-span. Although there are differences in the exceeding probability of the vertical displacement amplitude of the rail at different positions, the difference is less than 2%.

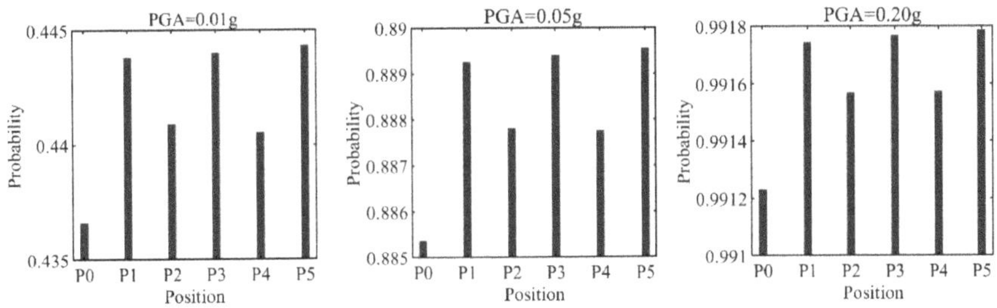

Figure 14. The exceeding probability of the earthquake-induced dynamic vertical displacement of the rail at different positions.

7. Conclusions

In this paper, the IDA is applied to calculate the exceeding probability of rail displacement under different earthquake excitations. An FEM model of a high-speed railway track-bridge system is established. Based on the simulation, the characteristics of the vertical displacement of the rail under different seismic excitations are investigated, and the probability of the vertical displacement of the rail exceeding the allowable standard is calculated. The main conclusions include:

(1) The vertical displacement of the rail on the bridge is comparable under earthquake excitations; however, the vertical displacement of the rail above the abutment is significantly smaller than that above other parts of the bridge when the vertical displacement of the rail is negative. The variation trend changes with the time of the vertical displacement of the rail on the bridge are comparable under seismic excitations. The increasing trend of the vertical displacement amplitude of the rail is approximately linear. The peak vertical displacement amplitude of the rail is 2.06 mm when PGA = 0.01 g.

(2) The vertical irregularity of the rail caused by earthquakes has a wavelength close to the length of the simply supported girder. An "L" shape irregularity of the rail caused by earthquakes can be observed above the bridge-abutment joint when the vertical displacement of the rail is negative. It may indicate that the mid-span and the bridge-abutment joint are much more hazardous areas that threaten the safety of the high-speed train.

(3) Under some excitation, two bumps are observed in the Fourier displacement spectrum in the frequency range of 1.3–2.5 Hz and 10–12 Hz, respectively, which may indicate the resonance of the model to the excitation. The bump amplitude in the Fourier displacement spectrum in the frequency range of 10–12 Hz above the mid-span is larger than that above the pier. A bump in the frequency range of 1.3–2.5 Hz may be approximately the natural frequency of the lower vibration mode of the model, which may significantly change the amplitude and frequency of the vertical displacement of the rail.

(4) The vertical displacement amplitude probability of the rail exceeding 2 mm is 44%, 89%, and 99% when PGA = 0.01 g, 0.20 g, and 0.40 g, respectively. The exceeding probability of the earthquake-induced dynamic vertical displacement of the rail increases with the PGA. The exceeding probability of the rail above the mid-span is larger than that above other parts of the bridge, while the exceeding probability of the rail above the abutment is the minimum. Moreover, within the mid-span, the exceeding probability of the rail above the center of the bridge is the largest.

This study may be the first attempt to evaluate the earthquake-induced dynamic displacement of the rail based on IDA. However, there may be some shortcomings in this work: (1) The 2D FEM model is just a simply supported girder bridge with the CRTS II slab ballastless track structure, which is not a 3D solid model and does not include other complex bridge types; and (2) Those selected ground motions may not be rich enough. For example, pulse-type near-fault ground motions and far-fault ground motions are out of consideration in this paper.

Author Contributions: Conceptualization, X.S. and Y.L.; Formal analysis, H.X., X.S. and H.S.; Funding acquisition, X.S.; Investigation, H.X.; Methodology, H.X., X.S. and Y.L.; Resources, X.S. and Y.L.; Software, H.X. and J.L.; Supervision, X.S. and Y.L.; Writing—original draft, H.X. and X.S.; Writing—review and editing, H.X., X.S. and Y.L. All authors have read and agreed to the published version of the manuscript.

Funding: This work is supported by National Key Research and Development Plan of China "Basic Theory and Methods for Resilience Assessment and Risk Control of Transportation Infrastructures" (2021YFB2600500).

Institutional Review Board Statement: Not applicable.

Informed Consent Statement: Not applicable.

Acknowledgments: Data of ground motion records for this study are provided by the Pacific Earthquake Engineering Research Center (PEER).

Conflicts of Interest: The authors declare no conflict of interest.

References

1. Sheng, X.W.; Zheng, W.Q.; Zhu, Z.H.; Luo, T.J.; Zheng, Y.H. Properties of rubber under-ballast mat used as ballastless track isolation layer in high-speed railway. *Constr. Build. Mater* **2020**, *240*, 117822. [CrossRef]
2. Song, L.; Liu, H.B.; Cui, C.X.; Yu, Z.W.; Li, Z.G. Thermal deformation and interfacial separation of a CRTSII slab ballastless track multilayer structure used in high-speed railways based on meteorological data. *Constr. Build. Mater.* **2020**, *237*, 117528. [CrossRef]
3. Wang, G.T. *The Mechanics Characteristics of Ballastless Track on Bridge during Earthquake*; Beijing Jiaotong University: Beijing, China, 2011; p. 75.
4. Zhai, W.M. *Vehicle–Track Coupled Dynamics*; Springer: Singapore, 2020.
5. Song, Y.; Wang, Z.W.; Liu, Z.G.; Wang, R.C. A spatial coupling model to study dynamic performance of pantograph-catenary with vehicle-track excitation. *Mech. Syst. Signal Process.* **2021**, *151*, 107336. [CrossRef]
6. *TG/GW 115-2012*; TG/GW 115-2012 Maintenance Rules for Ballastless Track of High Speed Railway. China Railway Publishing House: Beijing, China, 2012.
7. Zhu, Z.H.; Yang, L.; Wang, L.D.; Cai, C.B.; Dai, G.L. Dynamic Responses and Train Running Safety of Railway Cable-Stayed Bridge under Earthquakes. *Eng. Mech.* **2017**, *34*, 78–87.
8. He, B. *Study on Dynamic Response of CRTS II Ballastless Track Structure on Bridge under Earthquake*; Beijing Jiaotong University: Beijing, China, 2019; p. 103.
9. Li, D.S. *Performance of CRTS II Slab Ballastless Track on the High Speed Railway Bridge*; China Academy of Railway Science: Beijing, China, 2016; p. 3.
10. Wu, S.X. *Response and Optimization of CRTS II Ballastless Track Structure under Earthquake*; Beijing Jiaotong University: Beijing, China, 2018; pp. 87–89.
11. Yan, B.; Li, Z.; Liu, S.; Xie, H.R. Seismic response of CRTS II ballastless track-bridge system considering the damage of track structure. *Sci. Prog.* **2021**, *104*, 00368504211035207. [CrossRef] [PubMed]
12. Yu, M.; Lv, J.W.; Jia, H.Y.; Jia, K.; Zheng, S.X.; Zhao, C.H. Response Analysis of High-speed Railway Bridge-rail System Subjected to Near-fault Pulse-type Earthquake. *J. Hunan Univ.* **2021**, *48*, 138–146.
13. Lv, J.W. *Research on Dynamic Performance of Bridge-Track System with Simple Supported Beams on High-Speed Railway under Pulsed Near Fault Ground Motion*; Chongqing Jiaotong University: Chongqing, China, 2021; pp. 90–91.
14. Xue, D.Y. *Seismic Performance Analysis of High-Speed Railway Bridges under Near-Fault Earthquake*; Beijing Jiaotong University: Beijing, China, 2018.
15. Jin, Z.B.; Pei, S.L.; Li, X.Z.; Liu, H.Y.; Qiang, S.Z. Effect of vertical ground motion on earthquake-induced derailment of railway vehicles over simply-supported bridges. *J. Sound Vib.* **2016**, *383*, 277–294. [CrossRef]
16. Li, X.Z.; Hong, Q.Y.; Lei, H.J.; Liu, Z.J. Effect of Input Directions of Seismic Ground Motion on Seismic Responses of a Railway Extradosed Bridge. *Bridge Constr.* **2015**, *45*, 26–32.
17. Wang, W. *Research on Seismic Response of Track-Bridge System considering Wave Passage Effect*; China Academy of Railway Science: Beijing, China, 2014; pp. 60–61.

18. Lei, H.J. *Coupling Vibration and Running Safety of Train-Track-Bridge System under Non-Uniform Seismic Excitation*; Southwest Jiaotong University: Chengdu, China, 2014.
19. Li, H.Y.; Yu, Z.W.; Mao, J.F.; Spencer, B.F. Effect of seismic isolation on random seismic response of High-Speed railway bridge based on probability density evolution method. *Structures* **2021**, *29*, 1032–1046. [CrossRef]
20. Cheng, W.; Wang, G.; Du, Y.L.; Cao, Y.Z.; Xu, R.X.; Guo, J. Vulnerability analysis of the continuous high-speed railway bridge under near-fault earthquake. *J. Harbin Eng. Univ.* **2020**, *41*, 212–218.
21. Wei, B.; Wang, W.H.; Wang, P.; Yang, T.H.; Jiang, L.Z.; Wang, T. Seismic Responses of a High-speed Railway (HSR) Bridge and Track Simulation under Longitudinal Earthquakes. *J. Earthq. Eng.* **2022**, *26*, 4449–4470. [CrossRef]
22. Liang, Y.; Yan, S.C.; Zhao, B.Y.; Wang, Y.; Zang, C.Y. Seismic vulnerability analysis of near-fault high-speed railway rigid-frame bridge. *J. Zhengzhou Univ. (Eng. Sci.)* **2022**, *43*, 80–85+91.
23. Xiao, M.Y. *Seismic Vulnerability Analysis for Concrete Continuous Rigid Bridge With High Piers*; Southwest Jiaotong University: Chengdu, China, 2013; p. 55.
24. Zhang, S.X. *Seismic Vulnerability Analysis for Bridge Structures with High Piers*; Southwest Jiaotong University: Chengdu, China, 2014; p. 79.
25. Liang, Y.; Zhu, J.N.; Zhu, R.H.; Du, J.H.; Wang, Y. Seismic fragility analysis of RCC-RC composite bridge with high-strength steel bar. *Word Earthq. Eng.* **2021**, *37*, 85–93.
26. Lv, X.L.; Su, N.F.; Zhou, Y. IDA-based seismic fragility analysis of a complex high-rise structure. *Earthq. Eng. Eng. Vib.* **2012**, *32*, 19–25.
27. Ding, Z.D.; Zi, H.; Ji, X.F.; Shi, C.H.; Ren, Z.H. Seismic fragility analysis of mountain tunnels considering lining degradation. *Chin. J. Rock Mech. Eng.* **2020**, *39*, 581–592.
28. Bertero, V.V. Strength and deformation capacities of buildings under extreme environments. *Struct. Eng. Struct. Mech.* **1977**, *53*, 29–79.
29. Dadkhah, M.; Kamgar, R.; Heidarzadeh, H. Reducing the Cost of Calculations for Incremental Dynamic Analysis of Building Structures Using the Discrete Wavelet Transform. *J. Earthq. Eng.* **2020**, *26*, 3317–3342. [CrossRef]
30. Dadkhah, M.; Kamgar, R.; Heidarzadeh, H.; Jakubczyk-Gałczyńska, A.; Jankowski, R. Improvement of Performance Level of Steel Moment-Resisting Frames Using Tuned Mass Damper System. *Appl. Sci.* **2020**, *10*, 3403. [CrossRef]
31. Wang, H.H.; Liu, J.B. Seismic Fragility Analysis of Reinforced Concrete Bridges. *China Civ. Eng. J.* **2004**, *37*, 47–51.
32. Cui, M.Z.; Wang, C.K.; Chen, C.H.; Pan, Y.H.; Xiongh, Y.H.; Ren, C.C. Seismic fragility analysis on existing high-rise shear-wall structure based on incremental dynamic analysis. *Build. Sci.* **2021**, *37*, 151–157.
33. *GB 50111-2006*; GB 50111-2006 Code for Seismic Design of Railway Engineering. Planning Press: Beijing, China, 2006.
34. *GB50011-2010*; GB50011-2010 Code for Seismic Design of Buildings. Architecture & Building Publishing: Beijing, China, 2010.
35. Kamgar, R.; Rahgozar, P. Optimum location for the belt truss system for minimum roof displacement of teel buildings subjected to critical excitation. *Steel Compos. Struct.* **2020**, *37*, 463–479.
36. Jiang, L.Z.; Yu, J.; Zhou, W.B.; Yan, W.J.; Lai, Z.P.; Feng, Y.L. Applicability analysis of high-speed railway system under the action of near-fault ground motion. *Soil Dyn. Earthq. Eng.* **2020**, *139*, 106289. [CrossRef]
37. Gwo, W.; Li, J.L.; Liu, H.Y. The Analysis of Running Safty of High-Speed-Train on Bridge by Using Refined Simulation Considering Strong Earthquake. *Eng. Mech.* **2018**, *35*, 259–264+277.
38. Lei, X.Y. *High Speed Railway Track Dynamics: Model, Algorithm and Application*; Science Press: Beijing, China, 2015; p. 229.
39. Chen, L.K.; Jiang, L.Z.; Yu, Z.W.; Zeng, Z.P. Numerical Analysis of the Seismic Responses of Round-ended Piers of High-speed Railway Bridges. *J. Hunan Univ.* **2012**, *39*, 18–24.

sustainability

MDPI

Article

Post-Earthquake Traffic Simulation Considering Road Traversability

Yingying Wu [1], Zhen Xu [1,*] Chenxi Liang [1] and Ruizhuo Song [2]

[1] Beijing Key Laboratory of Urban Underground Space Engineering, School of Civil and Resource Engineering, University of Science and Technology Beijing, Beijing 100083, China
[2] School of Automation and Electrical Engineering, University of Science and Technology Beijing, Beijing 100083, China
* Correspondence: xuzhen@ustb.edu.cn; Tel.: +86-10-6233-3268

Abstract: Post-earthquake road traversability is a critical factor that affects traffic conditions. Therefore, a post-earthquake traffic simulation method considering road traversability was proposed in this study. First, the impact ranges of the earthquake-induced building collapse and the post-earthquake fire spread of buildings were analyzed, and road traversability was determined accordingly. Subsequently, the post-earthquake traffic flow was predicted based on building characteristics, and micro-level vehicle behaviors were simulated considering post-earthquake road traversability to determine the traffic conditions. In addition, the simulation model was validated using actual data. Finally, a segment of the Tongzhou road network in Beijing was selected as a case study to analyze post-earthquake road traversability and simulate traffic conditions on critical road sections. The proposed method can provide post-earthquake traffic conditions, which benefits the decision-making of post-earthquake evacuation and rescue.

Keywords: post-earthquake; road traversability; traffic simulation; multiple disasters; micro-level vehicle behaviors

Citation: Wu, Y.; Xu, Z.; Liang, C.; Song, R. Post-Earthquake Traffic Simulation Considering Road Traversability. *Sustainability* **2022**, *14*, 11145. https://doi.org/10.3390/su141811145

Academic Editors: Chong Xu, Su Chen and Shuang Li

Received: 31 July 2022
Accepted: 1 September 2022
Published: 6 September 2022

1. Introduction

Post-earthquake traffic conditions are directly affected by road traversability. Currently, modern roads in urban areas are built using important seismic measures with sufficient roadbeds and advanced facilities [1]. However, the debris generated by building collapses during an earthquake can fall on the surrounding road network, affecting traffic conditions [1,2]. Furthermore, rapidly spreading post-earthquake fires can endanger passing vehicles. Therefore, after an earthquake, urban roads can be blocked by collapsed buildings or post-earthquake fires, placing heavy pressure on post-earthquake rescue and recovery work [3]. For example, roads were severely blocked by collapsed buildings in the 1999 Kobe earthquake, hampering rescue activities [4]. Clearly, simulating traffic conditions is essential for efficient post-earthquake rescue and evacuation efforts. Furthermore, it is necessary to consider post-earthquake road traversability when simulating urban road traffic conditions.

The impact range of building debris has been examined by several studies [1,2,5–7]. For instance, Argyroudis et al. [8] proposed a systemic seismic risk assessment method for road networks which can be used to determine the possibility of blocked roads owing to the collapse of adjacent buildings during an earthquake. Lo et al. [9] analyzed road congestion caused by road liquefaction and building collapse according to the peak ground acceleration (PGA). Though Hirokawa and Osaragi [10] considered both collapse debris and spreading fires in a simulation of urban earthquake damage, they did not consider the impact of fires on road blockage. These studies analyzed post-earthquake road traversability considering building collapse impact but lacked the impact of post-earthquake fires.

Post-earthquake traffic can be described using either macro-level or micro-level simulations. Macro-level simulations focus on the flow of vehicles to accurately predict the distribution of traffic flow on urban roads, whereas micro-level simulations focus on the movement track of a single vehicle to reflect traffic conditions. Generally, existing post-earthquake traffic simulations have typically been conducted on the macro level [11,12], ignoring micro-level behaviors, such as vehicle-following and lane-changing. For example, Chang et al. [13] simulated post-earthquake traffic conditions considering bridge damage and post-earthquake travel behaviors under extreme conditions, focusing on macro-level performance indexes, such as the traffic capacity. Feng et al. [14] considered vehicle-following behavior and used an agent-based model to simulate post-earthquake traffic but ignored the traffic congestion caused by vehicle lane-changing behavior.

In summary, the simulation of post-earthquake traffic considering road traversability still faces two major challenges:

(1) How can road traversability be evaluated under the impacts of multiple disasters?

After an earthquake, roads are likely to be covered by debris from collapsed buildings, limiting road traversability or even blocking roads. Furthermore, the spread of post-earthquake fires can also obstruct road access. It is difficult to systematically evaluate post-earthquake road traversability owing to the coupling relationship between the impact ranges and effects of these different disasters, as well as the time-varying spread of secondary fires.

(2) How can the traffic conditions considering post-earthquake blocked roads be simulated on the micro level?

In the event of an earthquake, some roads will be blocked, often leading to specific micro-level behaviors, such as vehicle-following and lane-changing, that can seriously affect the traffic conditions. Thus, simulating micro-level vehicle behavior under the influence of road traversability is critical for accurate post-earthquake traffic predictions.

Many theoretical methods for earthquake damage simulation have been proposed to address the first challenge. Lu and Guan [15] proposed a multi-degree-of-freedom (MDOF) model for high-precision urban earthquake damage simulation, which can predict the earthquake damage of every floor of every building in a given area [16]. Nishino et al. [17] proposed a road blockage probability model based on an assumed trigonometric distribution to predict the scope of debris created by building collapse. However, no method published to date has clearly defined the impact of post-earthquake fires on road traversability. Therefore, the impact of post-earthquake fires on road traversability needs to be considered in this study.

Developing traffic simulation technologies offer various approaches for the micro-level analysis of road traffic conditions to address the second challenge [13,18]. Song and Sun [19] calibrated the parameters affecting the micro-level vehicle behavior using sensitivity analysis to control the vehicle-following and lane-changing behaviors. Furthermore, existing traffic simulation systems provide a visual window for micro-level traffic simulation to intuitively express vehicle behavior [20]. However, the influence of road traversability on micro-level vehicle behavior should be directly considered to accurately simulate post-earthquake traffic conditions.

To address these challenges, a post-earthquake traffic simulation method that considers road traversability was proposed in this study. First, the collapse and post-earthquake fire spread in a building complex were analyzed to determine the impact range under multiple disasters, and then, the road traversability was analyzed. Next, the post-earthquake traffic flow was predicted based on building characteristics, and micro-level vehicle behaviors were simulated considering post-earthquake road traversability to determine the traffic conditions. In addition, the simulation model was validated using actual data. Finally, a segment of the Tongzhou road network in Beijing was taken as a case study to predict the post-earthquake road traversability and simulate the traffic conditions of critical road sections. Thus, the proposed method can provide post-earthquake traffic conditions which benefits the decision-making of post-earthquake evacuation and rescue.

2. Framework

The research framework employed in this study is shown in Figure 1, which is made of three components as follows:

1. Road Traversability Analysis

First, the seismic response of the building group was calculated using the MDOF model, and then, the corresponding impact range of collapse debris was analyzed. Then, the process of fire spreading between buildings was analyzed to determine its impact range. Finally, the impact ranges of collapse debris and post-earthquake fires were overlapped to analyze road traversability.

2. Post-Earthquake Traffic Simulation

First, resident travel behaviors were predicted based on building characteristics to determine traffic flow distribution on the roads. Then, the traffic model on the VISSIM platform was modified using the local parameters, and micro-level vehicle behaviors on critical road sections were simulated considering the post-earthquake road traversability and traffic flow distribution. Finally, the modified traffic model was validated by comparing the actual measured traffic data of the sections.

3. Case Study

A segment of the Tongzhou road network in Beijing was used as a case study, and the post-earthquake traffic conditions of this segment were simulated considering road traversability.

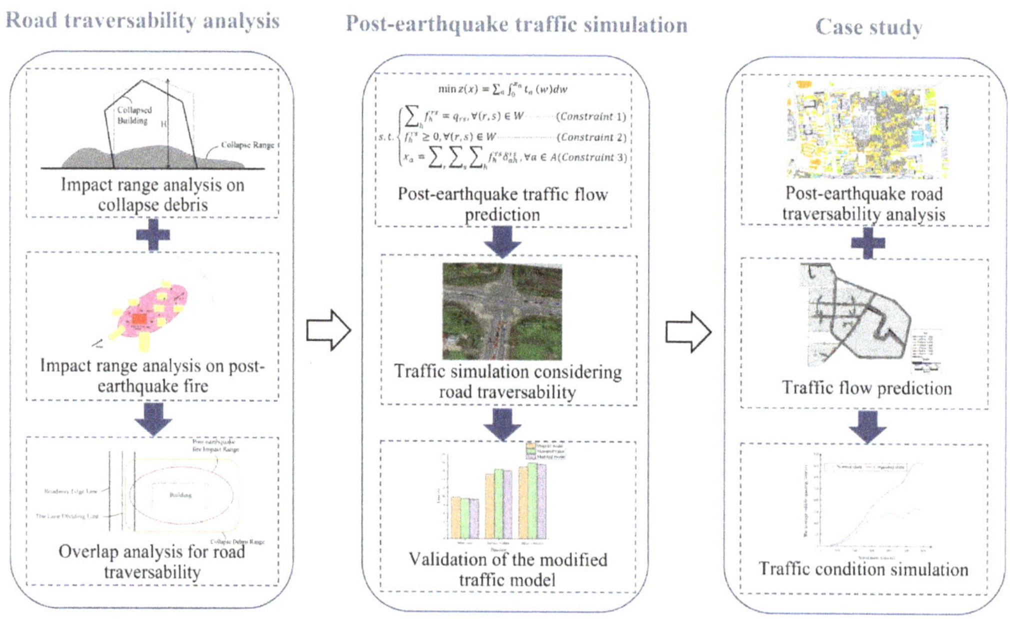

Figure 1. Research framework.

3. Method

3.1. Road Traversability Analysis Considering the Impact of Multiple Disasters

The effects of multiple disasters—an earthquake followed by fires—on the road network were considered in the road traversability analysis.

Understanding the seismic response of the buildings is an essential prerequisite for analyzing the impact range of the earthquake-induced building collapse and the post-earthquake fire spread of buildings. However, it is difficult to establish a refined structural model for each building in a densely constructed urban area. Therefore, in this study, the MDOF model proposed by Lu and Xiong [21,22] was used to rapidly compute the seismic response parameters of buildings, including the damage severity, floor displacement, and acceleration. On this basis, the impact ranges of collapse debris and post-earthquake fires were studied. It should be noted that earthquake-induced damage to urban roadways was not considered in this study since the quality of the roadbed in urban areas is normally sufficient to prevent major damage from an earthquake, and damage to the road surface has little impact on vehicle free passage.

3.1.1. Impact Range Analysis on Collapse Debris

Through the investigation of the earthquake damage caused by the Great Hanshin Earthquake, Nishino et al. [17] have obtained the probability equation of the building's collapse scope. Therefore, the collapse debris distribution probability model proposed by Nishino et al. was employed in this study to calculate the effective impact range of collapse debris given by Equation (1):

$$\frac{H}{8} < w_0 < \frac{H}{2} \tag{1}$$

where H is the height of the building with the collapse range (w_0) of the building between $H/8$ and $H/2$. To highlight the impact caused by collapse debris, the maximum impact range ($H/2$) was used to conduct a relative analysis in this study. The unit of H here is meters (m), as the same below.

Thus, the collapse debris impact range, w, owing to an earthquake can be calculated as shown in Equation (2):

$$w = \frac{H}{2} \tag{2}$$

Additionally, the safety cordon of buildings that are likely to collapse (i.e., buildings with extensive or moderate damage) can also be determined by Equation (2) because the maximum impact range of building collapse (i.e., w) is employed for the cordons according to Equation (1). Through the safety cordon, people are warned to stay a safe distance (i.e., w) from the dangerous building, which can prevent accidents and further casualties in the post-earthquake environment.

3.1.2. Impact Range Analysis on Post-Earthquake Fire

Post-earthquake fire is a common secondary hazard induced by earthquakes [23,24]. Historically, some severe post-earthquake fire events have occurred in major cities worldwide, such as those in San Francisco in 1906 [25], Tokyo in 1923 [26], and Osaka and Kobe in 1995 [27]. The post-earthquake fires spread rapidly, which may seriously affect vehicle passage on the surrounding roads. Therefore, the process of fire spreading between buildings was analyzed in this study to determine their impact range.

First, the locations of fire initiation were identified. The fire quantity, N, under a given PGA was obtained using the method proposed by Ren and Xie [28] according to a regression of historical post-earthquake fire data (e.g., San Francisco Earthquake in 1906 and Tokyo Earthquake in 1923). The first N buildings with the highest fire probabilities were then chosen as the fire initiation points after calculating the fire probability of buildings with various damage severities.

The regional fire quantity, N, under the selected PGA can be calculated according to Equation (3):

$$N = -0.11749 + 1.34534PGA - 0.8476PGA^2 \tag{3}$$

where N is the number of buildings on fire per 100,000 square meters, and the unit of PGA is g.

After normalization, the fire probability of buildings according to damage severity can be represented by the fire index, r_0, calculated as shown in Equation (4):

$$r_0 = \frac{P(M) \times P(F_K|M) \times \sum_j [P(D_j|PGA) \times P(C_j|D_j) \times P(S_j|D_j)] \times P(G)}{0.867} \quad (4)$$

The fire probability of each building will be calculated and sorted according to Equation (4), and the first N buildings with the highest fire probability will be designated as the fire points. For detailed meanings of all parameters given above, as well as their values, please refer to Table 1.

Table 1. Parameter related to the building fire index.

Parameter	Meaning	Value
$P(M)$	Probability that the structure contains combustibles.	Set the value to 1 if the structure contains combustibles; set it to 0 if it does not.
$P(F_k/M)$	Probability that a building fire is affected by a specific combustible.	Set the value according to the flammability of the building's interior materials and the building itself.
$P(G)$	Probability that a building fire is affected by weather and other factors.	Set different values based on variables, such as weather.
$P(C_j/D_j)$	Probability that inflammable material leakage in buildings under the damage state, D_j.	Set the value according to the damage severities.
$P(S_j/D_j)$	Probability that a fire will break out inside a building in the damage severities, D_j.	Set the value according to the damage severities.
$P(D_j/PGA)$	Probability that a building will fail in state D_j under a specific PGA.	Set the value according to the outcomes of regional seismic response simulations for buildings.

Second, the process of post-earthquake fires spreading between buildings was analyzed. The post-earthquake fires spreading model for buildings proposed by Lu et al. [23] was used to perform a relative analysis. This model considers two factors when analyzing the spread of fire: the thermal radiation and the thermal plume (as shown in Figure 2). A burning building affects the adjacent buildings via the thermal radiation of the flame through the doors, windows, and external walls and affects the buildings in the downwind direction via the thermal plume of high-temperature flue gas. If the heat flux received by an unburned building within a certain period of time exceeds its critical value under the common influence of the surrounding burning buildings, it can be assumed that the building is about to combust.

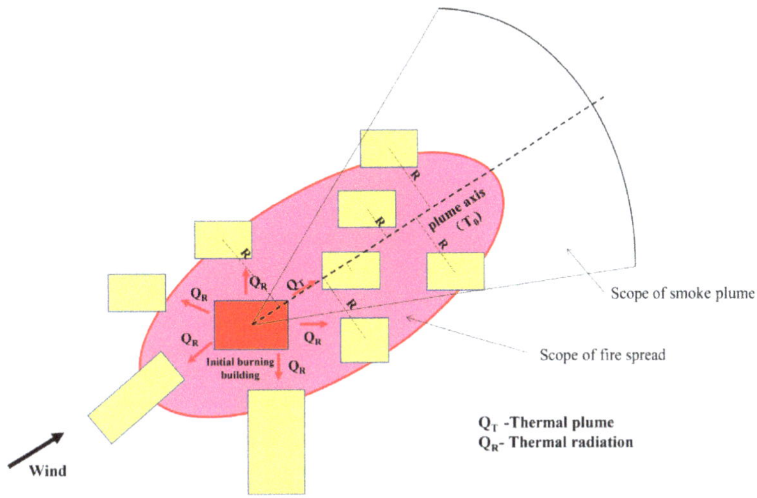

Figure 2. The process of post-earthquake fires spreading.

Buildings may be more prone than usual to fire following an earthquake because of the reduced capacity for prevention. The model used in this study was designed based on concepts presented by Himoto et al. [29] who assumed that the building damage caused by an earthquake reduces the ultimate heat flux threshold for building fire, $\dot{q}_{cr}$, as shown in Equation (5):

$$\dot{q}_{cr} = \phi\dot{q}_{cr,H} + (1-\phi)\dot{q}_{cr,L} \tag{5}$$

where ϕ is the damage factor of the envelope material, and $\dot{q}_{cr,H}$ and $\dot{q}_{cr,L}$ are the limit heat flux when $\phi = 1$ and $\phi = 0$, respectively. For the values of these parameters, please refer to [29].

Finally, the impact range of post-earthquake fires was determined. When the distance between adjacent buildings reaches a certain threshold, the fire cannot spread from building to building; this distance is generally considered the limit distance of fire spread [30]. As most urban buildings meet the requirements of the design fire code, the fire separation distance between buildings is typically greater than the limit distance of fire spread. Therefore, though the maximum impact range of a post-earthquake fire is technically dependent on the limit distance of fire spread, the most adverse impacts of post-earthquake fire on road traversability in this study were evaluated using the fire separation distance representing the post-earthquake fire impact range, f, which is given by Equation (6):

$$F_{limit\ distances} \leq F_{fire\ separation\ distance} = f \tag{6}$$

Please refer to Table 2 for specific fire separation distances between different buildings, which comes from the code for fire protection design of buildings in China [31].

Table 2. The fire separation distance of civil building.

Building Classification		High-Rise Buildings	Other Civil Buildings		
		I, II	**I, II**	**III**	**IV**
High-rise buildings	I, II	13 m	9 m	11 m	14 m
Other civil buildings	I, II	9 m	6 m	7 m	9 m
	III	11 m	7 m	8 m	10 m
	IV	14 m	9 m	10 m	12 m

I, II, III, and IV represent the fire resistance rating of the buildings. (Reference to the code for fire protection design of buildings of China [31]).

3.1.3. Overlap Analysis for Road Traversability

In this study, the impacts of multiple disasters on the road network were comprehensively analyzed by overlapping the collapse debris impact range and post-earthquake fire impact range. The accurate impact ranges of the earthquake-induced building collapses or fires are difficult to be calculated due to the uncertainty of earthquakes. This study aims to evaluate the worst-case scenario based on conservative design principles. Note that such a worst-case scenario is necessary for the urban planning and emergency preparedness of a city. Therefore, the maximum impact ranges of building collapse and fires are suitable for such a worst-case scenario in this study.

According to Equations (2) and (6), the collapse debris impact range and the post-earthquake fire spread impact range were calculated. Since collapse debris and post-earthquake fires both hinder road traffic, the one with a larger impact range should be considered when analyzing road traversability. Therefore, the maximum value (i.e., F_{max}) between the two impact ranges is chosen as the criterion. Then, the two results were overlapped to obtain the maximum value of the impact range F:

$$F_{max} = \max[w, f] \tag{7}$$

The greater the value of F_{max}, the more obstructed the road will be. Therefore, according to $F_{\max}$, road traversability can be divided into the following three states using Equation (8):

$$\begin{cases} \text{Normal states,} & F_{max} < d \\ \text{Congested states,} & d \le F_{max} < d + \frac{r}{2} \\ \text{Blocked states,} & d + \frac{r}{2} \le F_{max} \end{cases} \tag{8}$$

where r is the width of the road and d is the distance from the building boundary to the road boundary.

In this study, the buffer area function in ArcGIS was used to analyze road traversability. First, road and building models were created in ArcGIS. Subsequently, buffer areas of the collapse debris impact range and the post-earthquake fire impact range were created. Finally, an overlap analysis was conducted on the buffer areas in ArcGIS using Equation (8) to determine the traversability of the road network.

3.2. Post-Earthquake Traffic Simulation Considering Road Traversability

3.2.1. Post-Earthquake Traffic Flow Prediction Based on Building Characteristics

The post-earthquake traffic flow, which is used to simulate post-earthquake traffic conditions, varies significantly due to changes in post-earthquake travel demand. In this study, the post-earthquake traffic flow was predicted based on the surrounding building characteristics; it comprises two parts: (1) post-earthquake origin-destination (OD) prediction and (2) allocation of traffic flow.

(1) Post-Earthquake OD Prediction

Post-earthquake OD prediction is primarily used to evaluate the total vehicle production (P) and total vehicle attraction (A) in the traffic zone after an earthquake.

Floor area per capita is a critical factor considered during building design, and to a certain extent, the area of a building is directly proportional to the population within. Therefore, according to building characteristics (etc. function and area) in each traffic zone and in accordance with relevant planning standards [32], the population data were reverse deduced in this study to achieve OD prediction.

The specific steps of this process are as follows:

Step 1: Calculate the area of the buildings with different function types in each traffic zone, S_{ij}.

Step 2: Calculate the resident population in the buildings with different function types in each zone, N_{ij}, according to Equation (9):

$$N_{ij} = \frac{S_{ij}}{A_j} \quad (9)$$

where A_j is the floor area per capita, the value of which can be referred to in Table 3.

Table 3. Floor area per capita.

Type	Floor Area Per Capita(m^2)
Residential Building	30
Industrial Building	30
Administrative Building	20
Commercial Building	20
Cultural Building	100

Step 3: Modify the travel rate (TR) and attraction rate (AR) of the various buildings after the earthquake.

Using research conducted by the China Academy of Urban Planning and Design and the Beijing Transportation Institute [33], the regular population travel rate and population attraction rate for different types of buildings were summarized and sorted as shown in Table 4.

Table 4. Reference values of trip rate and attraction rate of various buildings.

Reference Values	Building Function				
	Residential Building	Industrial Building	Administrative Building	Commercial Building	Cultural Building
Trip rate	2	2	6	7	8
Attraction rate	2	2.5	11	15	18

After an earthquake, the urban functional structure will be damaged to a certain extent. Therefore, the importance of different functional buildings in the city will change, as will travel demand and travel times; these factors should be reflected in the change in traffic flow. For example, the attraction rate of commercial sites providing consumption and entertainment will be considerably reduced, as will the traffic generated by going to work, school, and other activities. Conversely, storage land, hospitals, government offices, and residential buildings will see a significant increase in attraction rates. Therefore, the travel and attraction rates of different types of buildings should be modified, as shown in Table 5.

Table 5. Corrected values of trip rate and attraction rate of various buildings.

Corrected Values	Building Function				
	Residential Building	Industrial Building	Administrative Building	Commercial Building	Cultural Building
Trip rate	5	2.5	8	7.5	6
Attraction rate	3	1	15	8	14

Step 4: Calculate the total traffic production, E_i, and traffic attraction, F_i, in each zone as follows:

$$E_i = \sum_j N_{ij} \times TR_j \quad (10)$$

$$F_i = \sum_j N_{ij} \times AR_j \quad (11)$$

where TR_j and AR_j are the population travel rate and population attraction rate, respectively, of building type j.

Step 5: Calculate the vehicle production, P_i, and vehicle attraction, A_i, for each zone as follows:

$$P_i = E_i \times k \tag{12}$$

$$A_i = F_i \times k \tag{13}$$

where k is the proportion of vehicle travel among traffic travel modes in the region. According to the statistical data describing resident travel modes in Beijing in 2010 [34], travel by car accounted for 80% of travel; thus, $k = 0.8$. It is worth noting that the value of k should be determined based on the post-earthquake traffic travel mode. The fact that Beijing's traffic model was used in this study and there was a lack of earthquake data in Beijing means that the k mentioned in this study does not consider the changes in travel modes after the earthquake.

(2) Allocation of Traffic Flow

The primary purpose of traffic flow allocation is to study where the traffic generated by traffic zones will flow to and where the attraction comes from so that the traffic flow between traffic zones can be assigned for each road.

In this study, traffic flow was predicted based on the gravity model [35] which can be used to allocate traffic flow between zones according to the inverse relationship between the travel volume and the impedance of the zones (such as time and distance), as well as the proportional relationship between the AR and the TR. This study employed TransCAD (transportation planning software) to achieve this purpose.

The user-optimized equilibrium (UE) model [36] was used to allocate traffic flow, and its mathematical optimization model is given by Equation (14):

$$\begin{aligned} &\min z(x) = \sum_a \int_0^{x_a} t_a(w)dw \\ &s.t.\begin{cases} \sum_h f_h^{rs} = q_{rs}, \forall (r,s) \in W & (Constraint\ 1) \\ f_h^{rs} \geq 0, \forall (r,s) \in W & (Constraint\ 2) \\ x_a = \sum_r \sum_s \sum_h f_h^{rs} \delta_{ah}^{rs}, & \forall a \in A (Constraint\ 3) \end{cases} \end{aligned} \tag{14}$$

where r_s represents the "OD pair" corresponding to any two traffic zones and h represents the total number of roads between these two zones. Constraint 1 represents the flow conservation where the sum of the travel path flow, f_h, of any r_s must be equal to the traffic flow, q, between zones. Constraint 2 mandates that the traffic value allocated to any lane between any two zones cannot be negative. Constraint 3 indicates that the road section flow in the output result is a statistical value expressed as the sum of all traffic flows passing through the road section.

The UE model is typically solved using the Frank–Wolf algorithm [37]. The core of this algorithm employs linear programming to approach a nonlinear programming problem step by step. It begins at the initial point, then takes the optimal step size as the search scope and the fastest descending direction as the forward direction to determine the starting point of the next iteration. This process is repeated to gradually approach the optimal solution and obtain the traffic flow in each section.

3.2.2. Traffic Simulation Considering Road Traversability

The traffic simulation considering road traversability after an earthquake was primarily undertaken in this study to simulate the vehicle behaviors in the presence of road obstructions and to analyze the interactions between vehicles and the traffic environment. Therefore, micro-level traffic simulation technology was employed to analyze post-earthquake vehicle behaviors considering road traversability. This simulation comprises two modules: (1) setting road conditions and (2) building the simulation model.

(1) Setting Road Conditions

The road conditions were classified into congested or blocked states according to Equation (8), which will have different impacts on road traffic. The impact of a congested road on traffic is reflected by the inability of vehicles to use the full width of a certain length of road and thus an increase in traffic using the several lanes remaining available, as shown in Figure 3. In contrast, the impact of a blocked road on traffic is reflected by the complete inability of vehicles to pass the road section.

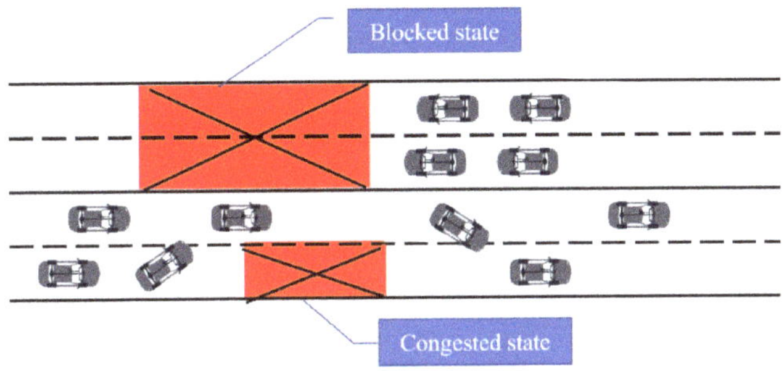

Figure 3. The setting of road conditions.

In this study, the VISSIM platform, a modeling tool used to simulate urban traffic conditions [38], was used to simulate the impact of these two states on traffic. Different measures were taken in VISSIM to simulate the different traffic states: for blocked roads, the entire road section was deleted from the simulation, whereas for congested roads, the lane shutdown option was adopted in the simulation. Additionally, the lengths of the blocked/congested road sections can be obtained from the road traversability analysis results, as described in Section 3.1.

(2) Building the Simulation Model

If a road is blocked or congested after an earthquake, vehicles may alter their original routes, resulting in different behaviors. These behaviors can be generally classified into two types: vehicle-following and lane-changing. The former represents the driving behaviors of the front and following vehicles under the ever-changing road conditions; the latter represents the vehicle behaviors changing their running lanes from one to another according to the road conditions at that time (see Figure 3). The lane-changing behaviors of vehicles critically impact the traffic-running state as they cause traffic disturbances and reduce road traversability.

A simulation model was built in VISSIM to simulate the micro-level vehicle behaviors. However, the default model parameters in VISSIM are based on traffic conditions in Germany, which differ significantly from those in China. Therefore, the simulation model parameters were adjusted according to the target national traffic situation.

There are 12 modifiable parameters in the VISSIM simulation model that describe vehicle-following and lane-changing behaviors. These parameters were modified in this study to reflect the actual situation of the case study region (Tongzhou, Beijing, China), and the results are shown in Table 6. The modified model was used to simulate the vehicle-following and lane-changing behaviors under blocked and congested road states.

3.2.3. Validation of the Modified Traffic Model

Because of the blocked or congested road after the earthquake, vehicles are likely to change lanes, resulting in traffic congestion. However, the changes before and after the earthquake are not necessary for the validation of micro-level vehicle behaviors. For instance, the daily blocked or congested roads can also cause the lane-changing behaviors. Therefore, the reliability of the modified traffic model can be validated by comparing the measured micro-level vehicle behaviors on a daily blocked or congested road with the simulation results. The entrance and exit of the main and service roads of the Guanghua Bridge located along Beijing's East 3rd Ring Middle Road were chosen to measure vehicle behaviors in this study. By comparing the traffic simulation results and measured data in this region, the model's availability can be confirmed.

Table 6. Correction of simulated parameters.

Vehicle Behavior	Parameters	Default Values	Corrected Values
Vehicle-following	Number of visible vehicles ahead	2	4
	Average car park spacing (m)	2	2.2
	Maximum forward sight distance (m)	250	200
	Additional sections on the safe distance	2	1.08
	Multiples of the safe distance	3	3.58
Lane-changing	Waiting time before disappearing (s)	60	66.21
	Maximum deceleration (m/s^2)	−4	−5.86
	−1 m/s distance (m)	100	102.3
	Acceptable deceleration (m/s^2)	−1	−1.28
	Safe distance reduction factor	0.6	0.32
	Coordinated braking maximum deceleration (m/s^2)	−4	−6
	Minimum headway (m)	0.5	2

For the measured data, a field investigation was also conducted to obtain the attributes of all roads in the simulated area (Table 7). In addition, the traffic flow data describing 20 min of the morning rush hours on a certain working day were obtained from road traffic images, as shown in Table 7.

Table 7. Basic information for case sections.

Attributes of the Sections		Measured Data in Sections		
Name	Value	Direction	Number of Vehicles	Travel Time (s)
Main road	3	Go straight on the main road	1352	9.5
Service road	3	Go straight on the service road	289	\
Weaving area lanes	1 m	Service road → Main road	259	16.5
Lane width	3.25 m	Main road → Service road	181	18
Confluent section length	80 m	\	\	\

For the traffic simulation, two VISSIM models were constructed in this study: one using the modified parameters described in Table 6 (called the "modified model") and the other using the platform default parameters (called the "original model").

The road network was constructed in the VISSIM platform at a one-to-one ratio according to the observed road conditions with travel-time-monitoring points arranged at major sections. The monitored travel times comprised three main routes: through the main road, through the service road and then the main road, and through the main road and then the service road. After arranging the monitoring points, the default parameters were modified using the "Driving Behavior Parameter Setting" of the VISSIM platform.

Under the same input conditions, the two models were used to conduct a 600 s simulation of road traffic conditions to obtain the travel time in each road section. Figure 4 shows a visualization of the road traffic conditions obtained by the modified simulation at a given time, and the results of the two simulations are compared with the field-measured results in Table 8.

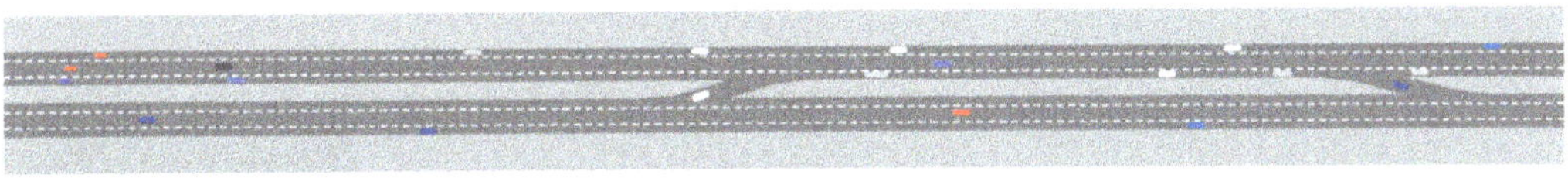

Figure 4. Simulation results at a time.

Table 8. Comparison of simulation results.

Direction	Measured Value	Original Model	Modified Model
Go straight on the main road	9.5s	9.8s	9.3s
Service road → Main road	16.5s	15.3s	16.2s
Main road → Service road	18s	17s	17.6s

It can be observed from the comparison that the revised model parameters effectively improved the accuracy of the simulation, indicating that the parameter correction scheme adopted in this study is reliable. Hence, the modified parameters were confidently applied to conduct traffic simulations in VISSIM.

4. Case Study

4.1. Case Introduction

As Beijing is located in a seismically active zone, it is likely to experience medium-strong earthquakes. In addition, the city contains dense buildings that could induce severe traffic problems in the event of an earthquake. Therefore, part of the Tongzhou district in Beijing was selected as the case study to demonstrate the simulation method. Tongzhou contains 43,881 buildings in total, as shown in the building footprint and road network model in Figure 5.

Figure 5. The building footprint and road network model.

The Sanhe–Pinggu seismic wave generated by the Institute of Geophysics, China Earthquake Administration [39] was used to construct an earthquake scenario. In this scenario, the seismic performance of buildings was analyzed by using the MDOF model [23].

4.2. Post-Earthquake Road Traversability Analysis

4.2.1. Analysis of the Debris Impact Range

The earthquake simulation results shown in Figure 6 indicate many collapsed or damaged buildings in the case study area; these will create a considerable quantity of debris that can seriously affect road traversability and hinder evacuation and post-disaster rescue. To analyze the corresponding impact on road traversability in this region, the method described in Section 3.1 was applied to calculate the debris impact range. The road traversability under the influence of building collapse is shown in Figure 6.

Figure 6 shows that collapse debris blocks many of the roads. The main roads between the residential areas are more seriously affected owing to the dense construction of buildings within, whereas the main roads of the city, being far from the surrounding buildings, are less affected.

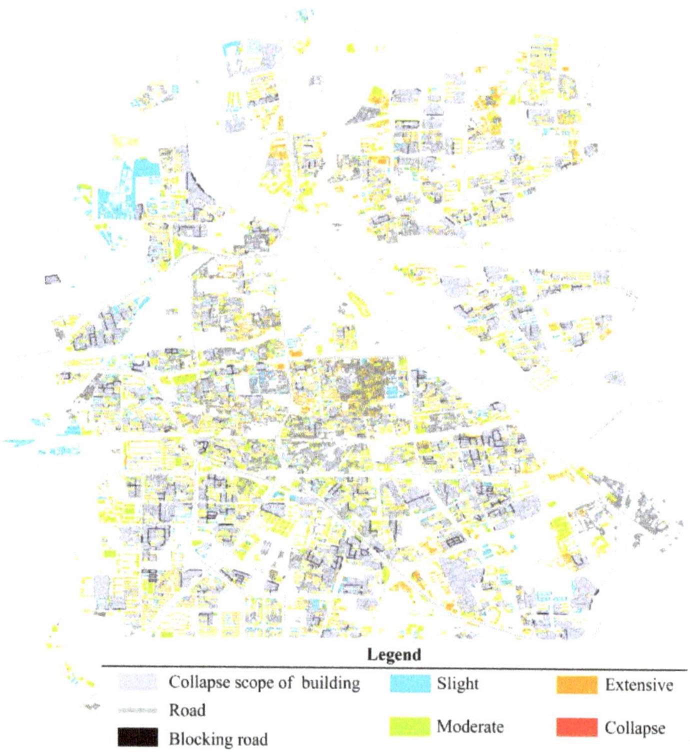

Figure 6. Road traversability under the influence of building collapse.

4.2.2. Analysis of the Post-Earthquake Fire Impact Range

The spread of post-earthquake fire between buildings will be constrained by the fire separation distance. Therefore, based on the analysis of building seismic damage, the area with the greatest fire risk was selected in this case study to analyze the post-earthquake fire spread condition within 5 h after the earthquake. For the specific results, please see Figure 7.

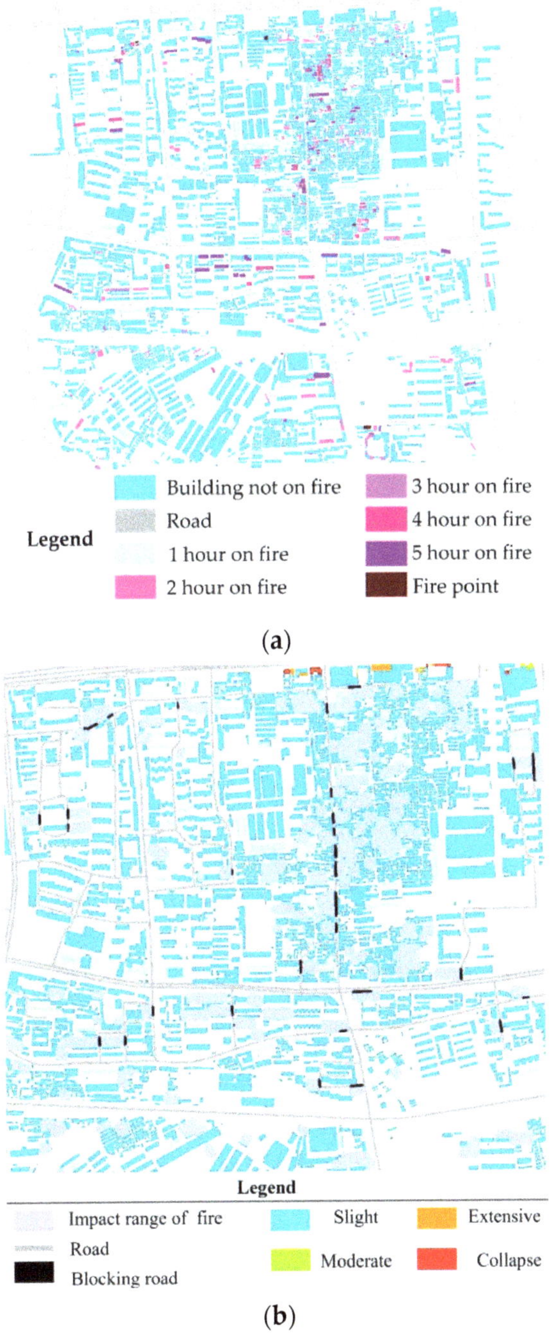

Figure 7. The spread of fire between buildings: (**a**) the fire time of buildings; (**b**) the change in road traversability.

It can be observed in Figure 7a that if a fire breaks out following an earthquake in the case study area, it will spread quickly owing to the relatively close spacing of buildings. If firefighting measures are not undertaken in time, considerable losses will be inevitable. In addition, the spread of fire has a certain impact on road traversability (Figure 7b), blocking timely rescue and evacuation efforts.

4.2.3. Comprehensive Analysis of Road Traversability

The impacts of the simulated earthquake and its post-earthquake fires were comprehensively analyzed to obtain the overall impact range under multiple disasters. Overlapping these results on the road network, all affected road sections were obtained, as shown in Figure 8.

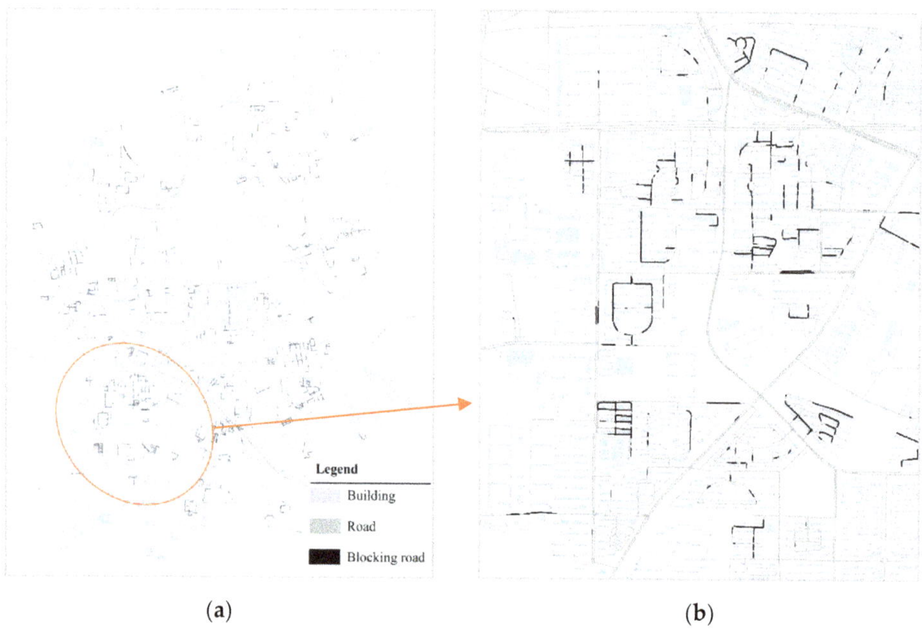

Figure 8. Comprehensive analysis of road traversability: (**a**) the overall situation of road traversability in the case area; (**b**) the local situation of road traversability in the case area.

4.3. *Post-Earthquake Traffic Simulation*

The traffic conditions in key road sections were studied based on the road traversability analysis results shown in Figure 8. The area in the middle of Figure 8b is located at the intersection of various traffic arteries, and the paths in each zone are severely blocked. Therefore, more attention should be paid to this region, and a traffic simulation was conducted here accordingly, as discussed in this section.

4.3.1. Post-Earthquake Traffic Flow Prediction

According to the existing road network structure, the research area was divided into eight traffic zones, as shown in Figure 9. By compiling statistics describing the floor areas corresponding to different building functions within all traffic zones, the populations were estimated and used to obtain the resident travel behaviors. The building statistics for all zones are shown in Table 9.

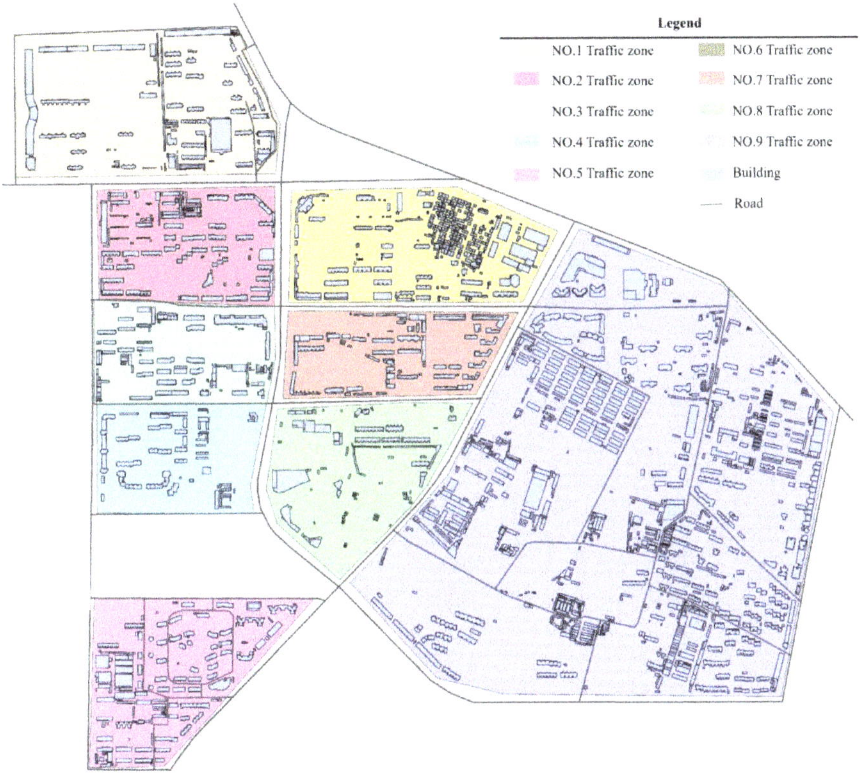

Figure 9. Division of traffic zones.

Table 9. Different types of building area statistics forms in each community (m^2).

NO.	Commercial	Industrial	Residential	Cultural	Administrative
1	26,315.36	425.78	37,053.78	2086.01	9703.69
2	26,829.91	6709.01	21,805.55	979.78	2602.53
3	15,908.85	0	19,791.15	11,826.85	759.79
4	5098.72	3891.86	7875.94	2168.29	9301.04
5	3008.17	0	39,556.54	1609.17	17,793.29
6	34,612.08	0	37,808.36	498.69	1155.78
7	18,550.2	0	14,586.98	1997.26	4459.12
8	16,330.53	0	8275.62	7872.31	2992.43
9	59,119.26	18,736.8	76,542.93	95,740.71	83,817.74

After obtaining the floor area, the post-disaster travel behaviors of the residents were predicted using the method described in Section 3.2, and the travel expectation between zones was analyzed to obtain the traffic flow. This traffic flow was then allocated to acquire the traffic flow situation on each main road, as shown in Figure 10.

The widths of the lines in the figure represent the relative scale of traffic flow, where the wider the line, the more traffic the road carries. The numbers in the figure represent the traffic flow values allocated to the roads. The colors of the lines in the figure represent the level of road service, ranging from green to red: a darker color means worse road service.

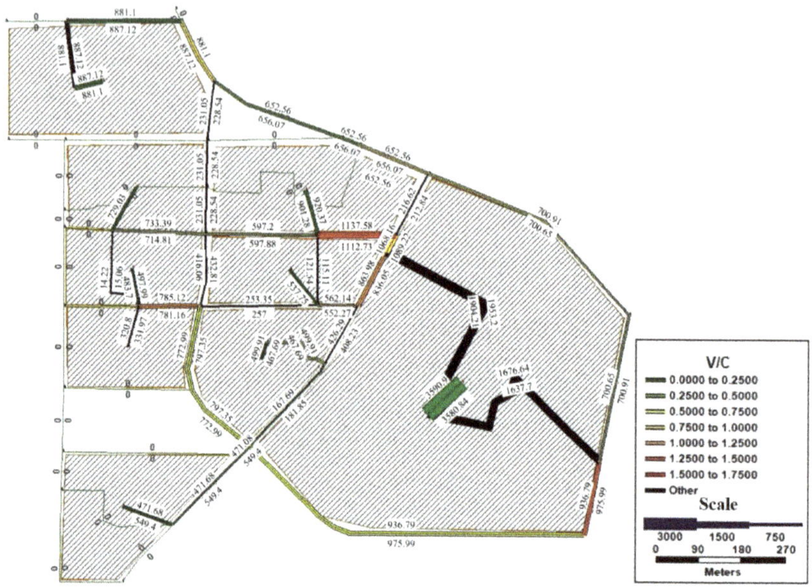

Figure 10. Allocation of traffic flow.

Using the VISSIM platform to simulate the traffic conditions for the road network model, traffic flow was set per road according to the flow results. A 600 s simulation of traffic conditions was conducted for the target area accordingly, and the traffic condition results at a certain time are shown in Figure 11.

Figure 11. Traffic simulation results.

The interchange between traffic zones features heavy traffic pressure with diversified vehicles, indicating that these road sections are easily blocked following an earthquake, thus increasing the risk of vehicle congestion and traffic failure.

4.3.2. Micro-Level Traffic Simulation of Critical Road Sections

Several high-risk road sections (which featured considerable traffic flow and were easily affected by the simulated earthquake) were selected to evaluate the traffic conditions of roads under congested and blocked states to further study the traffic problems caused by the deterioration of road traversability following an earthquake.

(1) Congested State

The road sections with dense traffic flows were selected to conduct normal and congested state microsimulations; the results are shown in Figure 12.

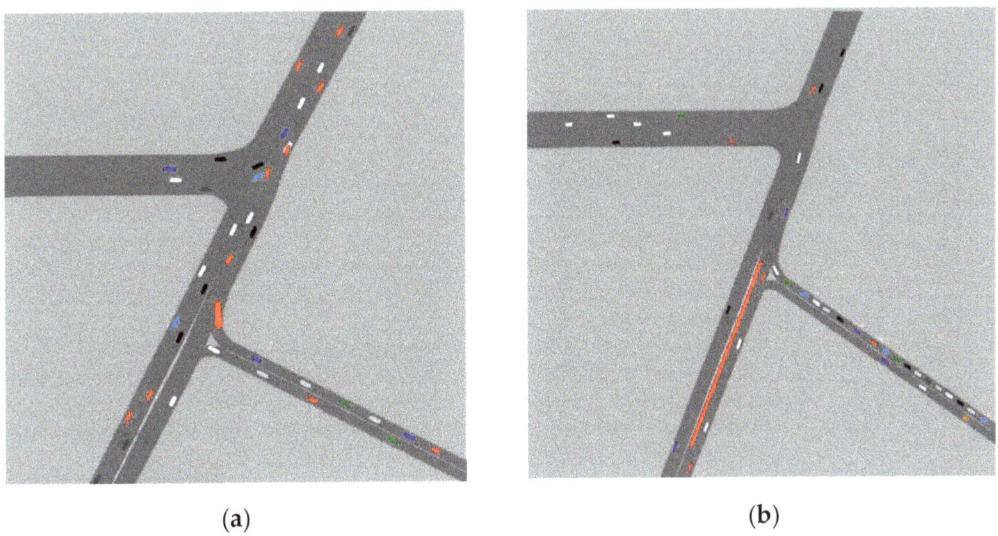

Figure 12. Comparison of simulation results under normal state and congested state at a certain time: (**a**) normal state; (**b**) congested state.

The red lines in Figure 12b indicate the specific parts of the lanes that were unavailable for vehicle passage. Compared to the results in Figure 12a, the upward running vehicles on this road section can only merge into the right lane, co-using the same lane as the upward running vehicles adjacent; thus, the vehicles about to merge into the right lane may have to queue. As a result, the road section shows a traffic jam, increasing traffic time.

To verify the above analysis, the average vehicle queuing times under the two scenarios were compared, as shown in Figure 13. Under normal conditions, as the simulation time increases, the average vehicle queuing time remains stable; in the congested state, the average vehicle queuing time increases linearly with simulation time owing to lane changes and other behaviors, exhibiting a maximum growth rate of 140% and proving that the traffic situation in high-risk sections can be considerably affected by changes in road traversability following an earthquake.

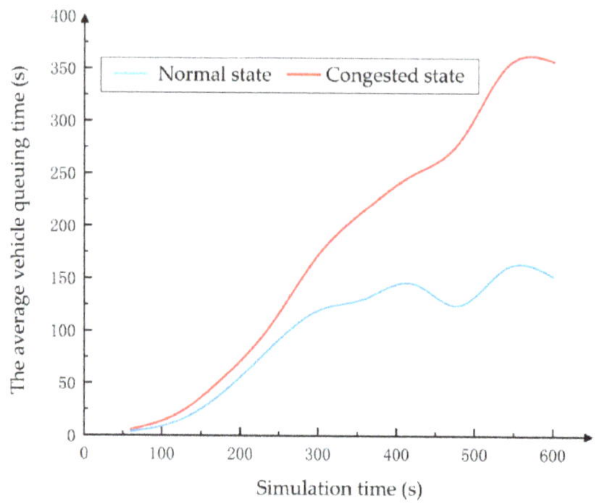

Figure 13. The average vehicle queuing times under the two scenarios.

(2) Blocked State

Local roads may be temporarily impassable following an earthquake, leading to a sharp increase in traffic flow on adjacent roads over a short period of time, which can readily cause road congestion. To simulate the road traffic in the blocked state, it was assumed that the end part of the selected road in the network was blocked. The microsimulation results are shown in Figure 14 in which the red line segment indicates the location of the completely blocked road.

In Figure 14, once the road on the upper right is completely blocked, the upward running vehicles can only turn left to merge into the right lane of the road on the left side; meanwhile, vehicles on the left road that want to turn left cannot continue to pass and can only merge into the downward running lane. Thus, the right lane of the road on the left side of the figure is likely to be congested by a large increase in traffic flow over a short period of time.

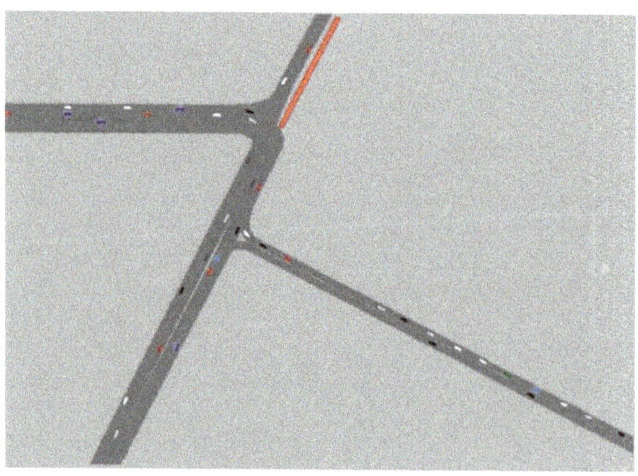

Figure 14. The simulation results under the blocked state.

To verify these analysis results, traffic data for the right lane of the road on the left side before and after the blockage are compared in Figure 15, indicating a significant increase in lane occupancy (approximately 100%) compared to normal conditions. In addition, the average speed of the vehicles in this lane decreases correspondingly, indicating that road blockage would indeed have a considerable impact on the vehicle passing time following an earthquake.

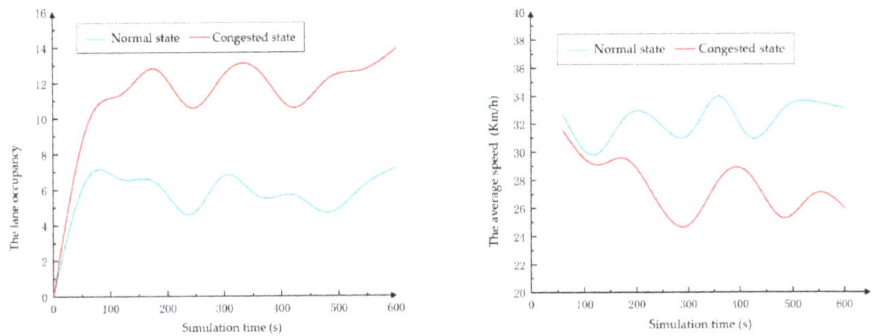

(a) (b)

Figure 15. Comparison of simulation results under normal state and blocked state: (**a**) the change in lane occupancy; (**b**) the change in average speed.

5. Conclusions

In this study, the post-earthquake traffic conditions were simulated, considering the road traversability induced by building collapse and fires following an earthquake. Some conclusions can be drawn as follows:

(1) A road traversability analysis solution considering the impact ranges of the earthquake-induced building collapse and post-earthquake fires was proposed by which the post-earthquake road traversability under multiple disasters can be analyzed.

(2) A micro-level traffic simulation model considering road traversability following an earthquake was established to analyze the post-earthquake road traffic conditions and was validated against the measured data.

(3) The proposed post-earthquake traffic simulation method considering road traversability was applied to the Tongzhou district in Beijing. The simulation results show that the vehicle queuing time increased linearly by up to 140% owing to following and lane-changing behaviors caused by road blockage; these behaviors also significantly increased the occupancy of the surrounding lanes by approximately 100%. These results can help to determine specific traffic control measures and provide a reference for the decision-making of post-earthquake evacuation and rescue.

The simulation method proposed in this study can be used to evaluate the traversability of a regional road network suffering from building collapse debris and building fire spread following an earthquake and thereby simulate the traffic in key road sections. However, owing to the complex transportation system, the proposed method does not consider the impact of some factors, such as earthquake damage on bridges and post-earthquake road maintenance and clearing, which will be considered in a follow-up study. Additionally, because Beijing is used as a case in this study and there is a dearth of earthquake-related data in Beijing, the post-earthquake travel mode does not consider the change after the earthquake. More earthquake-prone regions will be studied in future research to consider this change.

Author Contributions: Conceptualization, Z.X.; methodology, Z.X. and Y.W.; investigation, Y.W. and C.L.; formal analysis, Z.X., Y.W. and C.L.; writing—original draft, Y.W.; writing—review and editing, Z.X. and R.S.; funding acquisition, Z.X. All authors have read and agreed to the published version of the manuscript.

Funding: This research was funded by General Program of the National Natural Science Foundation of China (Grant No. 51978049), Beijing Nova Program (Z191100001119115), Beijing Municipal Natural Science Foundation (Grant No. 8212011), Interdisciplinary Research Project for Young Teachers of

USTB (Fundamental Research Funds for the Central Universities) (Grant No. FRF-IDRY-20-030), and Scientific Research Fund of Institute of Engineering Mechanics, China Earthquake Administration (Grant No. 2019EEEVL0501).

Institutional Review Board Statement: Not applicable.

Informed Consent Statement: Not applicable.

Data Availability Statement: Not applicable.

Conflicts of Interest: The authors declare no conflict of interest.

References

1. Sassan, E.; Masoud, N.A. Performance of Transportation Systems in the 2003 Bam, Iran, Earthquake. *Earthq. Spectra* **2005**, *21*, 455–468. [CrossRef]
2. Moya, L.; Mas, E.; Yamazaki, F.; Liu, W.; Koshimura, S. Statistical Analysis of Earthquake Debris Extent from Wood-Frame Buildings and Its Use in Road Networks in Japan. *Earthq. Spectra* **2020**, *36*, 209–231. [CrossRef]
3. Chang, S.E. Transportation Performance, Disaster Vulnerability, and Long-Term Effects of Earthquakes. In Proceedings of the Second EuroConference on Global Change and Catastrophe Risk Management, Laxenburg, Austria, 6–9 July 2000.
4. Takashi, U. The Great Hanshin-Awaji Earthquake and the Problems with Emergency Medical Care. *Ren. Fail.* **1997**, *19*, 633–645. [CrossRef]
5. Wang, W.; Zhang, N.; Wang, L.; Wang, Z.; Ma, D. A Study of Influence Distance and Road Safety Avoidance Distance from Postearthquake Building Debris Accumulation. *Adv. Civ. Eng.* **2020**, *2020*, 7034517. [CrossRef]
6. Golla, A.P.S.; Bhattacharya, S.P.; Gupta, S. The Accessibility of Urban Neighborhoods When Buildings Collapse Due to an Earthquake. *Transp. Res. Part D Transp. Environ.* **2020**, *86*, 102439. [CrossRef]
7. Yu, Y.C.; Gardoni, P. Predicting Road Blockage Due to Building Damage Following Earthquakes. *Reliab. Eng. Syst. Saf.* **2022**, *219*, 102439. [CrossRef]
8. Argyroudis, S.; Selva, J.; Gehl, P.; Pitilakis, K. Systemic Seismic Risk Assessment of Road Networks Considering Interactions with the Built Environment. *Comput.-Aided Civ. Infrastruct. Eng.* **2015**, *30*, 524–540. [CrossRef]
9. Lo, I.T.; Lin, C.Y.; Yang, C.T.; Chuang, Y.J.; Lin, C.H. Assessing the Blockage Risk of Disaster-Relief Road for a Large-Scale Earthquake. *KSCE J. Civ. Eng.* **2020**, *24*, 3820–3834. [CrossRef]
10. Hirokawa, N.; Osaragi, T. Earthquake Disaster Simulation System: Integration of Models for Building Collapse, Road Blockage, and Fire Spread. *J. Disaster Res.* **2016**, *11*, 175–187. [CrossRef]
11. Kilanitis, I.; Sextos, A. Impact of Earthquake-Induced Bridge Damage and Time Evolving Traffic Demand on the Road Network Resilience. *J. Traffic Transp. Eng.* **2019**, *6*, 35–48. [CrossRef]
12. Hou, B.-W.; Li, X.-J.; Han, Q.; Liu, A.-W.; Lan, R.-Q. Post-Earthquake Connectivity and Travel Time Analysis of Highway Networks Based on Monte Carlo Simulation. *China J. Highw. Transp.* **2017**, *30*, 287–296. [CrossRef]
13. Chang, L.; Elnashai, A.S.; Spencer, B.F. Post-Earthquake Modelling of Transportation Networks. *Struct. Infrastruct. Eng.* **2012**, *8*, 893–911. [CrossRef]
14. Feng, K.; Li, Q.; Ellingwood, B.R. Post-Earthquake Modelling of Transportation Networks Using an Agent-Based Model. *Struct. Infrastruct. Eng.* **2020**, *16*, 1578–1592. [CrossRef]
15. Lu, X.; Guan, H. *Earthquake Disaster Simulation of Civil Infrastructures: From Tall Buildings to Urban Areas*; Springer: Singapore, 2017.
16. Lu, X.; Han, B.; Hori, M.; Xiong, C.; Xu, Z. A Coarse-Grained Parallel Approach for Seismic Damage Simulations of Urban Areas Based on Refined Models and Gpu/Cpu Cooperative Computing. *Adv. Eng. Softw.* **2014**, *70*, 90–103. [CrossRef]
17. Nishino, T.; Tanaka, T.; Hokugo, A. An Evaluation Method for the Urban Post-Earthquake Fire Risk Considering Multiple Scenarios of Fire Spread and Evacuation. *Fire Saf. J.* **2012**, *54*, 167–180. [CrossRef]
18. Chen, Y.; Eguchi, R.T. Post-Earthquake Road Unblocked Reliability Estimation Based on an Analysis of Randomicity of Traffic Demands and Road Capacities. In Proceedings of the 6th U.S. Conference and Workshop on Lifeline Earthquake Engineering (TCLEE) 2003, Long Beach, CA, USA, 10–13 August 2003; pp. 916–925.
19. Song, R.; Sun, J. Calibration of a Micro-Traffic Simulation Model with Respect to the Spatial-Temporal Evolution of Expressway on-Ramp Bottlenecks. *Simulation* **2016**, *92*, 535–546. [CrossRef]
20. Pell, A.; Meingast, A.; Schauer, O. Trends in Real-Time Traffic Simulation. *Transp. Res. Procedia* **2017**, *25*, 1477–1484. [CrossRef]
21. Lu, X.; Lu, X.; Guan, H.; Ye, L. Collapse Simulation of Reinforced Concrete High-Rise Building Induced by Extreme Earthquakes. *Earthq. Eng. Struct. Dyn.* **2013**, *42*, 705–723. [CrossRef]
22. Xiong, C.; Lu, X.; Lin, X.; Xu, Z.; Ye, L. Parameter Determination and Damage Assessment for Tha-Based Regional Seismic Damage Prediction of Multi-Story Buildings. *J. Earthq. Eng.* **2017**, *21*, 461–485. [CrossRef]
23. Lu, X.; Zeng, X.; Xu, Z.; Guan, H. Physics-Based Simulation and High-Fidelity Visualization of Fire Following Earthquake Considering Building Seismic Damage. *J. Earthq. Eng.* **2019**, *23*, 1173–1193. [CrossRef]
24. Xu, Z.; Zhang, Z.; Lu, X.; Zeng, X.; Guan, H. Post-earthquake fire simulation considering overall seismic damage of sprinkler systems based on BIM and FEMA P-58. *Autom. Constr.* **2018**, *90*, 9–22. [CrossRef]

25. Scawthorn, C.; O'Rourke, T.D.; Blackburn, F.T. The 1906 San Francisco earthquake and fire: Enduring lessons for fire protection and water supply. *Earthq. Spectra* **2006**, *22*, 135–158. [CrossRef]
26. Nishino, T.; Tsuburaya, S.I.; Himoto, K.; Tanaka, T. A study on the estimation of the evacuation behaviors of Tokyo city residents in the Kanto Earthquake Fire: Development of a potential-based evacuation model in post-earthquake fire. *Fire Saf. Sci.* **2008**, *74*, 105–114. [CrossRef]
27. Architect, S.A. Lessons from the Great Hansin-Awaji Earthquake. Available online: http://www3.grips.ac.jp/~{}ando/HanshinLessons%20en.pdf (accessed on 8 August 2022).
28. Ren, A.; Xie, X. The Simulation of Post-Earthquake Fire-Prone Area Based on Gis. *J. Fire Sci.* **2004**, *22*, 421–439. [CrossRef]
29. Himoto, K.; Mukaibo, K.; Akimoto, Y.; Kuroda, R.; Hokugo, A.; Tanaka, T. A Physics-Based Model for Post-Earthquake Fire Spread Considering Damage to Building Components Caused by Seismic Motion and Heating by Fire. *Earthq. Spectra* **2013**, *29*, 793–816. [CrossRef]
30. Xie, X.; Ren, A. Dynamic Simulation of Fire Following Earthquake Based on Gis. *Strateg. Study CAE* **2007**, *2007*, 82–87. [CrossRef]
31. *GB 50016-2014*; Ministry of Housing and Urban-Rural Development of People's Republic of China. Code for Fire Protection Design of Buildings. China Planning Press: Beijing, China, 2014.
32. *GB 50137-2011*; Administration of Quality Supervision Inspection and Quarantine. Code for Classification of Urban Land Use and Planning Standards of Development Land. China Architecture & Building Press: Beijing, China, 2010.
33. Beijing Transportation Institute. List. Available online: https://www.bjtrc.org.cn/ (accessed on 3 January 2020).
34. Beijing Transport Institute. *Beijing Transport Development Annual Report*; Beijing Transport Institute: Beijing, China, 2010.
35. Tao, S.; Ye, X. Forecasting Model of Intercity Trip Distribution with Consideration of Connections between Headquarters and Subsidiaries. *J. Tongji Univ.* **2020**, *48*, 1319–1327. [CrossRef]
36. Wang, X. Applicability Analysis of User Equilibrium Model (Ue) and Stochastic User Equilibrium Model (Sue) in Traffic Flow Assignment. *Sci. Technol. Innov.* **2019**, *2019*, 40–41. [CrossRef]
37. Cheng, X.; Liu, Y.; Guo, J. Method of Quantifying Uncertainty Inuser Equilibrium Traffic Assignment. *J. Nanjing Univ. Inf. Sci. Technol.* **2019**, *11*, 421–427. [CrossRef]
38. PTV Group. Ptv-Vissim. Available online: https://www.ptvgroup.com/en/solutions/products/ptv-vissim/ (accessed on 22 April 2020).
39. Fu, C.; Gao, M.; Yu, Y. Studying on Amplification Effect of Beijing Basin on 3~10 S Ground Motion by Numerical Simulation Method. *J. Seismol. Res.* **2015**, *38*, 448–460. [CrossRef]

sustainability

MDPI

Article

Observing Earthquake-Induced Velocity Change on the Rock Slope Following the 2021 M 7.4 Maduo Earthquake 780 km Away

Huibao Huang [1,2,3], Shigui Dai [4], Yingdong Yu [5] and Fan Xie [3,6,*]

1 College of Water Resource and Hydropower, Sichuan University, Chengdu 610017, China; huibao.huang.a@chnenergy.com.cn
2 Dadu River Hydropower Development Co., Ltd., Chengdu 610095, China
3 Key Laboratory of Earthquake Source Physics, China Earthquake Administration, Beijing 100081, China
4 Sichuan Earthquake Administration, Chengdu 610000, China; daishigui2022@126.com
5 China Institute of Water Resources and Hydropower Research, Beijing 100038, China; yuyingd@iwhr.com
6 Institute of Geophysics, China Earthquake Administration, Beijing 100081, China
* Correspondence: xiefan@cea-igp.ac.cn

Abstract: Velocity changes (dv/v) during and after earthquakes are important indicators for understanding the earthquake-induced mechanical damage evolution of rock slopes. However, studying slope responses associated with various seismic loading still remains challenging due to limited in situ observations. In this article, we conduct a 20 min temporal resolution monitoring of dv/v at the frequency band between 2 and 20 Hz by applying ambient noise interferometry on the Pubugou rock slope in Southwest China. We observe an instantaneous $\sim$0.41% dv/v drop on the slope caused by the 2021 M 7.4 Maduo earthquake at a distance of 780 kM, following a characterized logarithmic recovery process of $\sim$31.39 h towards its pre-earthquake state. Moreover, the dv/v in five narrow frequency bands show a similar drop and subsequently increased recovery times associated with the deceased frequencies due to the long-distant earthquake shaking. We discuss two possible mechanisms related to the heterogeneous rock slope excited by the long-distant earthquake at a low frequency. The study motivates the damage assessment of the rock slope using in situ dv/v and furthers the understanding of subsurface geological risks under diverse seismic loadings.

Keywords: long-distance earthquake; rock slope; slope instability; velocity change; ambient noise

Citation: Huang, H.; Dai, S.; Yu, Y.; Xie, F. Observing Earthquake-Induced Velocity Change on the Rock Slope Following the 2021 M 7.4 Maduo Earthquake 780 km Away. *Sustainability* **2022**, *14*, 9345. https://doi.org/10.3390/su14159345

Academic Editor: Chong Xu, Su Chen and Shuang Li

Received: 28 June 2022
Accepted: 26 July 2022
Published: 29 July 2022

1. Introduction

There is an ever-increasing demand for studying the mechanical damage evolution of rock slopes in seismic-active tectonic belts because the slope instability caused by earthquakes has great potential to threaten surrounding residents and industrial buildings following large earthquakes [1,2]. The damage of rock slope caused by earthquake shaking is primarily associated with the stiffness alteration of the near-surface material, and cumulative mechanical damage leading to failure [3,4] .

It is, therefore, necessary to monitor on-site temporal varying mechanical properties of rock slopes during and after an earthquake; this strategy is essentially required for both landslide susceptibility assessment and the operational early-warning system [5,6].

As crustal rocks have a high level of heterogeneity, such monitoring remains challenging [7] because the earthquake-induced damage, which is related to the weakening or softening of micro-cracks, can be understood at either macroscopic or mesoscopic scales when stress is applied by different shakings.

Therefore, conventional approaches based on monitoring surface deformation are not suitable to provide direct insight into the subsurface mechanical alterations of the rock slope due to the limited stress-sensitive detection associated with the varying shaking intesntity [8].

Among the most recently developed geophysical methods [9] which provide complementary ways to study subsurface activities of a slope, velocity change, which overcomes the limitations of geomorphic data, is one of the most important indicators to monitor the elastic behavior of the slope's mechanical damage [10].

In order to monitor a medium with its velocity change, one needs a reproducible source to emit elastic waves propagating in the medium at a regular time interval. This used to be exclusively achievable by expensive active source experiments (e.g., vibrators or explosions) using direct arrive waves [11,12]. However, it is difficult to accurately identify and analyze ballistic body-wave phases when the near subsurface of a rock slope exhibits strong heterogeneity and attenuation due to weathering. In addition, the extra stress perturbation induced by long-running active source(s) may also increase unexpected risks to the slope. Thus, traditional velocity change methods with active pulse-echo configuration become impractical in monitoring the slope damage evolution.

In the past decade, the established ambient noise correlation method (also known as ambient noise interferometry) offers an alternative way to reconstruct elastic green functions (also known as cross-correlation functions) propagating between two sensors by cross-correlating the continuous ambient noise [13,14]. Although it is still challenging to recover the ballistic parts at a shallow depth, it has been demonstrated that the reproducible multiple-scattered wavefields at a regular time interval could be reconstructed between sensor pairs in such a passive cross-correlation configuration. Thus, taking advantage of the sensitivity of multiple-scattering waves that have bounced repeatedly in the medium, it offers an exciting opportunity to observe dv/v associated with the development of stiffness in a heterogeneous medium by using coda wave interferometry (CWI) [15].

The promising method has been proved useful to solve problems at the tectonic scale in geosciences, and is applied to the study of landslides for both imaging and monitoring purposes [16–18]. With the usage of high-frequency ambient noise, more recent studies enhanced the knowledge of the effect of seasonal fluctuations [19], groundwater infiltration [18], slow reconsolidation [20], and fluidization [21] on the slopes composed of clayey and/or volcanic deposits. Furthermore, some studies reported that a precursory velocity drop can be detected from several days to minutes prior to the failure of the landslide [21,22].

In contrast to the aforementioned slow dv/v evolution at a relative long-term timescale, earthquake-induced dv/v, which is comprised of transient co-seismic change and a subsequent post-seismic process, shows a more rapid and complex mechanical behavior.

Instantaneous dv/v drop after large earthquakes followed by a slow recovery back to the pre-earthquake state was first observed around several active fault zone areas [17]. Similar dv/v behavior was occasionally reported in earthquake-induced slope instability studies. Bontemps et al. conducted a 3 y systematic investigation of dv/v at a 1 d temporal resolution using a frequency band of 3–8 Hz on a clay/silt compound slide landslide at Maca, Peru [23]. Despite 165 moderate earthquakes that (M 3.1 to 5.5) occurred over the study period within a ~50 km radius distance of the slope, only two earthquake-induced co-seismic dv/v drops with an amplitude of ~2% and subsequent post-seismic recovery processes over a few months were observed where the earthquakes occurred at a distance less than 10 kM. More recently, [24] conducted a similar observation with an improved technique on a typical rock slope in Southwest China. They tracked a ~1% co-seismic dv/v drop at a 1 h temporal resolution followed by a 14-day post-seismic logarithm recovery process caused by the M 5.7 earthquake at a distance of 200 km during the dry season. As a comparison, they also observed a weaker co-seismic dv/v drop accompanied by a partial recovery process caused by the M 5.3 earthquake at an equivalent distance during the monsoon season.

Since only a few observations with limited slope typology documented ambiguous dv/v changes caused by moderate earthquakes, it still remains challenging to investigate the relationship between the dv/v behavior and the seismic shaking on the slope and its control factor for landslide susceptibility analysis; in particular, observing slopes' dv/v

during and after a large earthquake at a long distance and investigating the mechanism remains undocumented to date.

Here, we present new ambient noise data from the Pubugou rock slope in Southwest China. We focus on the co-seismic and post-seismic dv/v changes on the slope due to the 2021 M 7.4 Maduo earthquake at a distance of 780 km. The remainder of the manuscript is organized as follows: In Section 2, we present the context of the Pubgou rock slope, the data collection during the earthquake, and the processing methods. The major results are provided in Section 3. In Section 4, we discuss two possible mechanics induced by the earthquake and the implications for the future application. The conclusions are given in Section 5.

2. Context and Methods

2.1. Study Site

Pubugou rock slope is situated in the middle of the Dadu River valley which is well characterized by the deep-seated bare bedrock alpine valley between the Sichuan Basin and the Tibetan Plateau [24]. The mountainous region surrounded by densely distributed faults shows the active tectonic deformation as well as the intense and sustained seismicity.

Figure 1a shows that the sub-aerial volumetric dimension of the east-facing rock slope is approximately 120,000 cubic meters with an average slope gradient of 45 degrees. The boundary of the slope is delineated by a dashed line. Two stations are indicated by orange points on the slope. The annual displacement rate of the slow-moving slope is approximate 3 mm/yr.

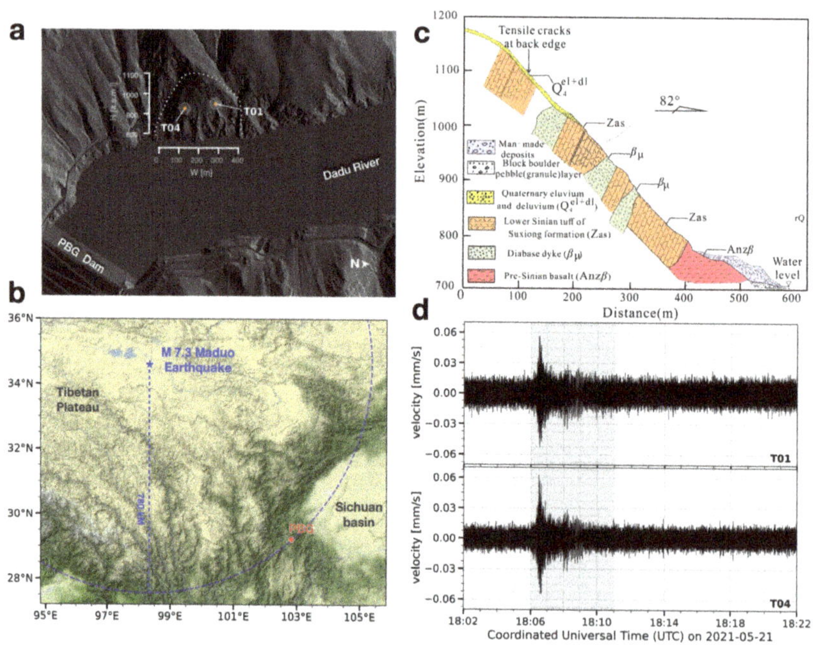

Figure 1. (**a**) Hillshade map of the Pubugou rock slope. (**b**) Terrain map of the 2021 M 7.4 Maduo earthquake and the Pubugou rock slope. (**c**) Geological profile of the slope (modified from [24]). (**d**) The ground velocity due to the 2021 M 7.4 Maduo earthquake is recorded by vertical component of station T01 and T04, respectively.

As illustrated in Figure 1b, the rock slope is composed of the volcanic tuffaceous rocks with depth-varying weathering degrees, except for the gravel with partial clay (Q^{dl+el}) at a very shallow layer (<4 m). The intact volcanic rocks exist at a depth of ~70 m of the slope without a well-defined rupture surface.

2.2. Instruments

We construct two stations (T01, T04) which are sheltered in small concrete huts on both sides of the slope. The inter-station distance is ~157 m. Each station installs a 2 Hz three-component geophone sensor (ZF-2) with a sensitivity of 198 V/m/s. The two identical geophones are coupled firmly to the slope using cement and connected to the 24-bit data acquisition system (EDAS-24GN) for digitization at a sampling frequency of 100 Hz. The timing accuracy of the system is ~50 ns by using an enhanced GNSS-based clock synchronization mechanism. The sampled ambient noise data are stored locally in 1 h long records and simultaneously transmitted to the cloud computing server via a mobile network.

We also install two bi-frequency global positioning system (GNSS) receptors on the roofs of both stations to measure the surface displacement along each side of the slope, respectively. In addition, an automatic weather station (AWS) on the east bank of the rock slope is operated to collect meteorological data.

Thus, an in situ monitoring campaign with a network of two seismic stations, two GPS stations, and one automatic weather station has been carried out since 2019.

2.3. Earthquake

On 21 May 2022, an M 7.4 earthquake with a focal depth of 17 km struck Maduo County, Guoluo Prefecture, Qinghai Province. We, hereafter, refer to it as the Maduo earthquake. The Maduo earthquake was a strong earthquake that occurred due to the combined effects of the northward movement of the Indian plate and uplift of the Tibetan Plateau [25]. As illustrated in Figure 1c, the hypocenter of the earthquake (indicated by the blue star) was 780 km away from the slope (red point). Despite the distance from the epicenter of the earthquake to the rock slope being approximately 780 km, they were both located in the boundaries of the Bayan Har block, which belongs to the northern and western margin of the Tibetan Plateau. Figure 1d shows the waveforms of the vertical ground motions for the earthquake recorded at T01 and T04, respectively. We observe a weak earthquake-induced peak ground velocity with an average value of $\sim 6 \times 10^{-5}$ m/s on the slope.

2.4. Methods

Relative velocity change (dv/v) is generally measured in two steps: (1) performing the cross-correlation computation between T01 and T04 to reconstruct the cross-correlation functions ($\phi_i(t)$), and (2) computing the dv/v by dephasing the $\phi_i(t)$ using the stretching technique [26].

We note that the records are studied in the 2–20 Hz frequency range, which properly enables the wave penetration depths. We, hereafter, refer to the 2–20 Hz as the study frequency band unless a different frequency domain is noted elsewhere.

In the first step, before computing $\phi_i(t)$, the continuous records are normalized in the frequency domain between 2 and 20 Hz using the spectrum whitening method and then normalized in the time domain using the clipping method, with amplitudes exceeding three standard deviations.

The above pre-processing steps aim to enhance the spectrum amplitudes at the specific frequency band and balance the strong energy from different noise sources. We note that spectrum whitening is performed before clipping to preserve the phase information, because clipping induces non-physical non-linearity that would further be amplified by spectrum whitening.

The $\phi_i(t)$ is further reconstructed by computing cross-correlations using the pre-processed 20 min long records where i is the temporal interval counter every 20 min. We evaluate the lag time t of the $\phi_i(t)$ up to 14 s. We further average the causal and anti-causal parts of the $\phi_i(t)$.

In addition, to improve the CCFs stability, we applied a two-dimensional Wiener filter to enhance the SNR of the CCFs with a minimum filter order of 2 for both vertical (date axis) and horizontal ($\phi_i(t)$ waveform time axis) dimensions (2 × 2) [27].

In the second step, we assume a homogeneous change in the velocity field, dv/v; therefore, scales are linear with the lag time perturbation of the $\phi_i(t)$, resulting in a significant stretching or compression (phase shift) in the coda of the $\phi_i(t)$. It is worth noting that since $\phi_i(t)$ are extracted from the pair-wise seismometers at a shallow depth, the propagation of wave packets with large amplitudes in the $\phi_i(t)$ are considered to be Rayleigh waves. With an increased propagation time, the scattered Rayleigh waves exist in the coda of the $\phi_i(t)$.

Thus, we apply a stretching method to measure dv/v. The method consists of a series of grid search over the distortion factor ϵ that maximizes the coherence $CC(\epsilon)$ in the similarity matrix ($\underset{\epsilon}{arg\,max}[CC(\epsilon_k)]$ between $\phi_i(t)$ and the stretched reference $\phi^{ref}[t(1+\epsilon)]$:

$$CC(\epsilon_k) = \frac{\int_{t1}^{t2} \phi^{ref}[t(1+\epsilon)]\phi_i(t)\,dt}{\sqrt{\int_{t1}^{t2} \phi^{ref}[t(1+\epsilon)]^2\,dt \int_{t1}^{t2} \phi_i(t)^2\,dt}}. \tag{1}$$

As a first-order approximation, a homogeneous dv/v scales linearly with the distorted time-shift factor ϵ_k by maximizing $CC(\epsilon)$: $\epsilon_k = \frac{dv}{v}$.

We use the single reference $\phi^{ref}(t)$ by averaging all $\phi_i(t)$ over the whole investigated period. To obtain the stable results with significant multiple scattering, we use (1) all nine combinations of the $\phi_i(t)$ obtained from three-component sensors that reflect the same material change and (2) a 10 s long time window starting at $t_1 = 1$ s and ending at $t_2 = 11$ s lapse time in the coda of $\phi_i(t)$. Here, the start time is chosen to exclude waves close to the receiver which is not well scattered, and the end time corresponds to the time where the amplitude of the cross-correlation functions fall below the noise level.

We estimate the dv/v as the mean value of all nine measurements and estimate its uncertainty $\overline{\sigma}$ as the mean value of all nine theoretical measurement errors, while each one is calculated using the following formula:

$$\sigma = \frac{\sqrt{1-CC^2}}{2CC}\sqrt{\frac{6\sqrt{\pi/2}T}{\omega_c^2(t_2^3 - t_1^3)}} \tag{2}$$

where T is the inverse of the frequency bandwidth and ω_c is the central frequency.

3. Results

Figure 2 shows the time series of the velocity changes at 20 min temporal resolution, together with the air temperature, rainfall, and GNSS-based horizontal displacement observations at the 1 h temporal resolution, respectively. The vertical dashed line indicates the time of the 2021 M 7.4 Maduo earthquake occurrence.

We observe an instantaneous velocity decrease by −0.4% that is coincident with the M 7.6 Maduo earthquake. The dv/v starts to recover immediately after the earthquake. The subsequent recovery phase (an increase of the dv/v) to the pre-earthquake state occurs during the following decade hours. The uncertainty is approximately 10^{-4}, one order of magnitude smaller than the observed dv/v.

We observe a $\sim\pm 10$ °C temperature change without precipitation before and after the earthquake. The air temperature induced a relative velocity change on the volcanic rocks of $\sim 10^{-5}/°C$ [10,28,29], approximately one order of magnitude smaller than the observed dv/v; in addition, there was no rainfall-induced pore pressure change. We, thus, exclude the effect of environmental perturbations in the study due to its influence, as it is indeed small. Meanwhile, there was no reported nearby earthquake event(s) according to the earthquake catalog provided by the Sichuan Earthquake Administration. Therefore, we suggest that an instantaneous dv/v decrease following a logarithmic recovery process on the slope was caused by the 2021 M 7.4 Maduo earthquake at a distance of 780 kM.

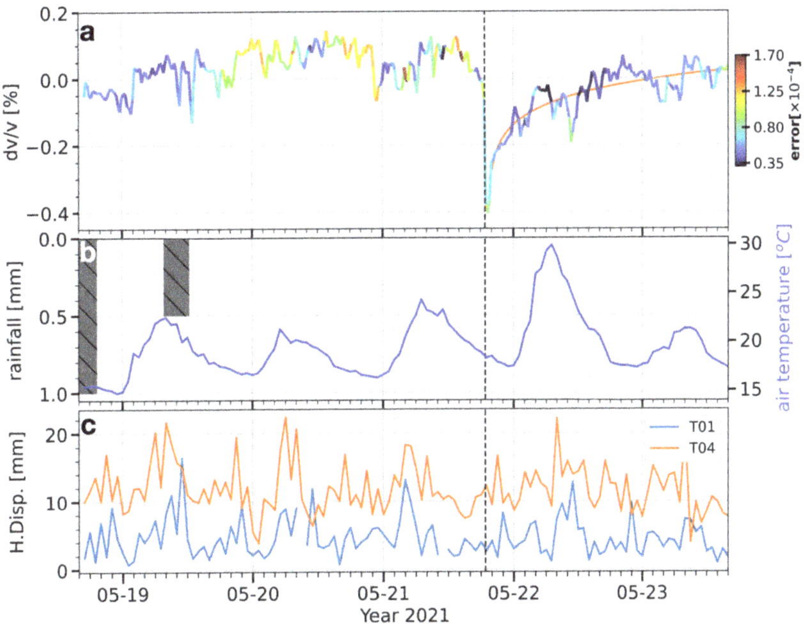

Figure 2. The temporal observation results before and after the earthquake on the slope. (**a**) dv/v at 20 min temporal, (**b**) air temperature, rainfall, and (**c**) GNSS-based horizontal displacement observations at the 1 h temporal resolution.

The dv/v evolution at different frequency bands allows to reveal depth-dependent co-seismic and post-seismic physical processes. We first filter $\phi_i(t)$ using five passbands. The central frequency f_0 of the passbands are 4.5, 6.5, 8.5, 10.5, and 15 Hz, and the bandwidths are set as $f = f_0 \pm 2.5$ Hz, except for the highest f_0 of 15 Hz with its bandwidth of ±5 Hz. We further calculate the depth-dependent sensitivity kernels of the fundamental mode Rayleigh waves in each center frequency of the passband using a simplified 1-D shear wave velocity model and the CPS330 software package [30].

The result in Figure 3a suggests that the center frequencies of 4.5, 6.5, 8.5, 10.5, and 15 Hz aresensitive to the depth of ~70 m, ~40 m, ~25 m, ~20 m and ~10 m, respectively. In addition, the depth sensitivity in the study frequency band can be roughly ranged from a few meters down to approximately 200 m of the slope.

Figure 4 illustrates the measured dv/v in each passband. The results show a similar co-seismic dv/v decrease following a post-seismic logarithmic recovery behavior caused by the 2021 M 7.4 Maduo earthquake (marked by the vertical dashed line).

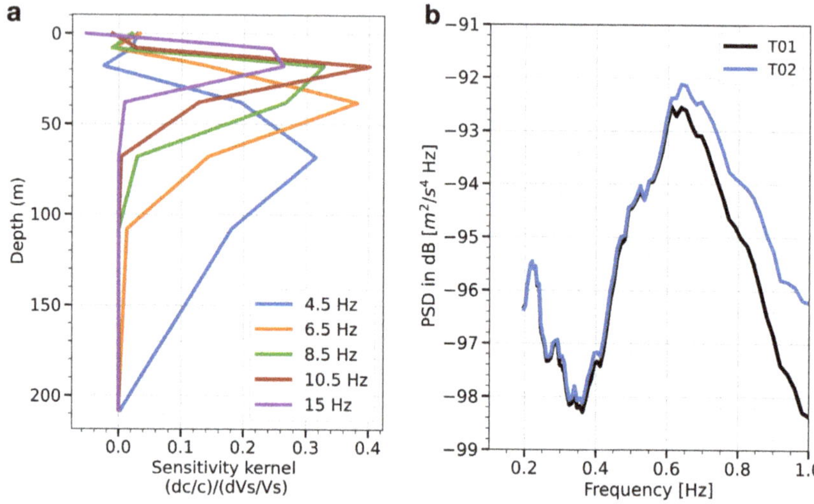

Figure 3. (**a**) Depth sensitivity of the Rayleigh phase velocity to a shear wave velocity perturbation at 4.5, 6.5, 8.5, 10.5, and 15 Hz at the PBG slope. (**b**) The earthquake-induced power spectral density (PSD) analysis at two stations.

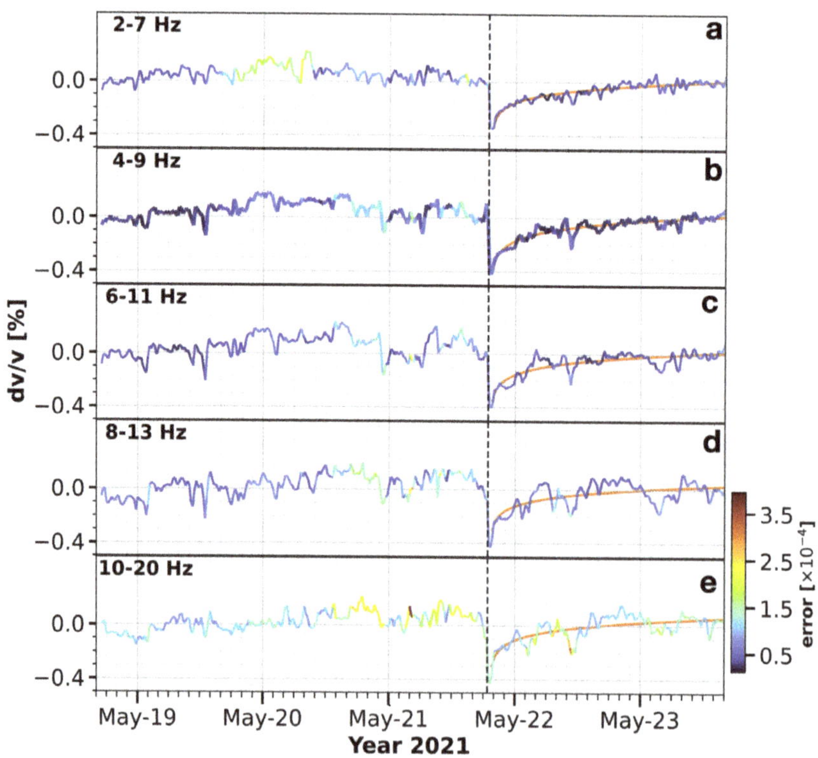

Figure 4. The dv/v in each passband. (**a**) 2–7 Hz, (**b**) 4–9 Hz, (**c**) 6–11 Hz, (**d**) 8–13 Hz, and (**e**) 10–20 Hz, respectively.

In order to compare co-seismic dv/v decrease at different depths carefully, we measure the dv/v drop in each passband by substracting the mean value of reference dv/v ($\overline{\frac{dv}{v}}_{prior}$) 45 h before the earthquake to the co-seismic dv/v decrease (minimum value of dv/v $\frac{dv}{v}_{min}$).

$$\frac{dv}{v}_{drop} = \overline{\frac{dv}{v}}_{prior} - \frac{dv}{v}_{min}. \tag{3}$$

As a result, the overall dv/v drop is 0.41%.

Figure 5 illustrates a roughly constant dv/v drop of ~0.45% in the passband from 6.5 to 15 Hz (blue bars). However, the dv/v drop in the lowest center frequency of 4.5 Hz shows a smaller value of 0.39%. Combined with the depth analysis, we conclude that a simliar co-seismic dv/v decrease (~0.45%) occured at a depth above $sim70$ m of the slope, while ~15% smaller co-seismic dv/v decrease occured below 70 m where the intact volcanic rocks exist.

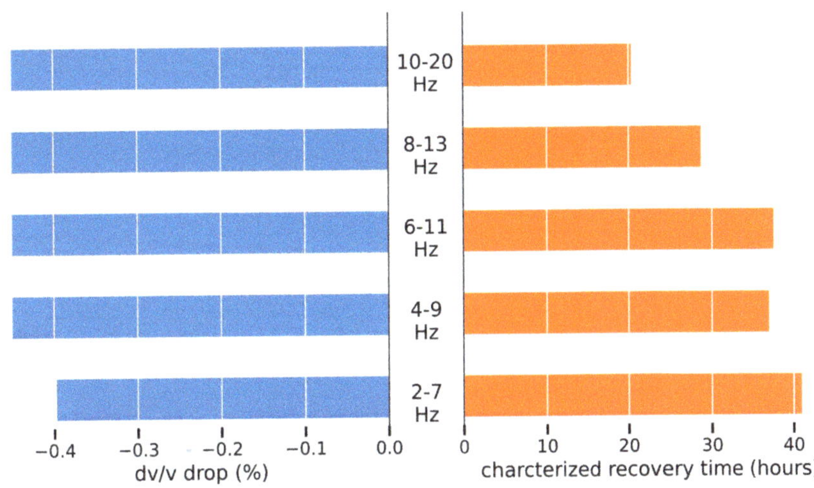

Figure 5. The The dv/v drop and characterized recovery time are measured at the five passbands on the slop.

We subsequently characterize the post-seismic recovery time at different depths by fitting the aftershock temporal evolution of dv/v in each passband:

$$dv/v = A - mlog_{10}(t) \tag{4}$$

where A is the extrapolated post-seismic dv/v, m is the slope of the logarithmic dv/v decay within time t. The characterized recovery time in hours is $10^{A/m}$.

As illustrated in Figures 2c and 4, the characterized post-seismic recovery dv/v as a function of time is plotted on the orange line. Thus, the overall characterized recovery time in Figure 2c is 31.39 h.

Figure 5 illustrates the characterized recovery time in the passband from 4.5 Hz to 15 Hz (orange bars). Generally, we observe that the characterized recovery time increases from 20.35 to 40.94 h, with the decrease of the central frequency from 15 to 4.5 Hz.

Compared with the dv/v drop in each passband, the characterized recovery time shows a more remarkable depth-dependent behavior due to the long-distance earthquake shaking.

We also note that the characterized recovery time at 8.5 Hz (37.02 h) is slightly shorter (1.5%) than that in 6.5 Hz (37.6 h). The depth-varying heterogeneity due to weathering of the slope at a depth between 25 and 40 m may be responsible for such abnormal characterized recovery time at different shallow depths.

4. Discussion

Previous studies [10,24] suggest that the damage mechanism of the slope associated with the earthquake is considered to originate mainly from earthquake-generated dynamic stress waves. Thanks to the strong co-seismic ground motion (PGV $\geq 10^{-2}$ m/s), it is able to cause reversible and/or irreversible damage of the slope at a distance range from tens to ~200 km by observing dv/v drop and subsequent recovery process [23].

However, in this case, we observe that the long-distance earthquake-induced peak ground velocity on the slope is 10^{-5} m/s, together with a non-anomalous displacement fluctuation during the period. It raises the question: why did the dv/v drop and recovery process still occur during and after the earthquake?

To improve the understanding of the damage evolution of the slope during and after the long-distant earthquake shaking, we explain the observed in situ dv/v behavior by two possible mechanisms.

The first possible explanation is the nonlinear elastic effect of the slope. Laboratory studies suggest that when a low-frequency oscillatory wave pumps into a geomaterial, stiffness loss appears instantaneously after the mechanically-induced conditioning [31]. It is termed anomalous nonlinear fast dynamics (ANFD). After the pumping waves are removed, a logarithm healing time process associated with geomaterial's stiffness towards its initial state takes place. This phenomenon is termed as slow dynamics (SD) [32]. It is worth noting that the oscillatory pumping waves used in a laboratory are a series of low-amplitude vibrations with frequencies that are close to the material's fundamental eigenmode, and the amplitude-induced strain levels are lower than 10^{-6} [33].

Therefore, we first measure the frequency response on the slope excited by the long-distance earthquake. As illustrated in Figure 1d, we use the 5 min long records of the earthquake-induced ground motions to compute a site response at the frequency domain by performing the seismic power spectral density analysis at two stations [34]. The resulting spectra are then smoothed in a 1/32-octave average in the frequency domain. As illustrated in Figure 3b, we observe the spectral peaks with equivalent amplitudes centered around 0.61 Hz and 0.64 Hz with respect to the stations.

Secondly, we evaluate the theoretical resonance frequency of the slope. We assume the stiffness of the slope's tuffaceous rock is of ~4.58 GPa and the density is approximately 2.4 g/cm^3. The resonance frequency of the slope can be evaluated using the following formula [9]:

$$f_r = \frac{1}{2\pi}\sqrt{\frac{K}{M}}. \tag{5}$$

The calculated theoretical resonance frequency is approximately 0.63 Hz, which is very close to the in situ frequency response excited by the long-distance earthquake.

Thus, in analogy with the laboratory configuration, the rock slope may exhibit stiffness loss (ANFD) and recovery process (SD) because of the pumping seismic waves with regard to the characterized weak stress (strain) levels and low frequencies excited by the long-distance earthquake.

In comparison with the similar velocity drops (~0.4%) at a center-meter scale, the recovery process on the slope spend more time (25–40 h or 10^5 s) than that in the laboratory experiments (~0.5 h or 10^4 s) [35].

Furthermore, assuming the nonlinear elastic mechanism dominates, we can estimate the earthquake-induced stress changes by using the quasistatic stress model rather than the dynamic one. As the Rayleigh wave velocity is close to the shear wave velocity, we use a shear wave velocity-dependent pressure model [36] to estimate the stress change of the slope due to the earthquake:

$$\frac{1}{V_s}\frac{\partial V_s}{\partial P} = \frac{1}{6p} \tag{6}$$

The hydrostatic pressure (P) at the depth of 200 m is about 2.82 $\times 10^8$ Pa. We estimate that a 0.4% co-seismic velocity drop at a depth of 200 m would require a quasistatic stress change of 112 kPa. As a result, the stress sensitivity is 3.5×10^{-8} Pa^{-1}.

The second possible mechanism is the transient pressure surge effect due to the seismic stress waves from the long-distant earthquake.

In seismology, it has been widely observed that weak seismic waves from distant earthquakes can abruptly change the states of natural geological systems and triggerearthquakes remotely in geothermal and/or volcanic environments [37]. The stress amplitudes of the triggering passing waves are as low as the order of kilopascal. Despite severalcandidate hypotheses used to explain these phenomena, the underlying mechanism is still not well understood. Recently, a new mechanism has been discovered: when a seismic wave interacted with a fracture with fluid, the transient pressure surge by 2–3 orders of magnitude compared with the incident wave pressure [38]; it has been verified through numerical modeling and a laboratory experiment [38,39]. It is worth noting that the transient pressure surge effect is a frequency-dependent phenomenon. The lower the frequency, the greater the stress change. It explains why frequency is more important than the PGV in remotely triggering natural disasters at long distances.

We notice that despite no rainfall producing fluid activities during the earthquake, water may still be stored in the micro-/macro-fractures within the slope body due to the precipitation 2–3 days before the earthquake. As a result, we speculate that the transient pressure surge effect may occur on the slope due to the low frequency earthquake. Since the PGV is approximate 6 $\times 10^{-5}$ m/s, the earthquake-produced dynamic stress δ_d on the slope can be estimated using the following equation:

$$\delta_d = G\frac{PGV}{v_p} \tag{7}$$

Assuming $G = 4.58$ GPa and $v_p = 1500$ m/s, the δ_d is approximately 183 Pa. Considering the amplification factor due to the pressure surge effect, the maximum dynamic pressure change on the slope may reach 183 kPa and the stress sensitivity is approximately 2.18×10^{-8} Pa^{-1}. It is a similar stress sensitivity compared to the one evaluated by using the first model and has been evaluated in a previous study [24].

Finally, we discuss the implications of our observation for on-site application. The results suggest that it is possible to observe and characterize nonlinear elastic properties of the slope under weak stress change using ambient noise interferometry with high temporal resolution and high stress sensitivity. As laboratory experiments suggest that the nonlinear elastic properties are more sensitive to the damage of the geomaterial, it has the potential to quantitatively characterize the slope weakness in terms of velocity changes by monitoring the co-seismic drop and post-seismic relaxation process and further facilitate the structural health assessment of slopes on a regional scale using weak seismic shaking in a non-destructive way in future.

In addition, the emerging observation is based on the established methods by using continuous seismic data in seismology, which have been more broadly recognized as a valuable source of information about geomorphic processes.

Despite that the technique is still at an incipient stage in this field, the use of seismic data and monitoring techniques to study damage evolutions of the subsurface medium excited by earthquakes at the distance from 10^1 to 10^2 kilometers has opened a new window into a wide range of applications, from studying influences of geo-mechanical loadings to deploying the practical early-warning system of natural disasters.

Depending on the study site, the following instrumental, as well as the technical suggestions, may be helpful to improve the monitoring system: (1) accurate synchronization, (2) sufficient low instrumental self-noise, (3) knowledge of the available ambient noise (in space, azimuth, and frequency content) and (4) improving the dv/v time resolution.

5. Conclusions

This article demonstrates that relative velocity changes associated with the 2021 M 7.4 Maduo earthquake can be observed on the rock slope at a distance of 780 km with a 20 min resolution using ambient seismic interferometry. With the knowledge of the frequency-dependent sensitivity at depths, the significant findings are as follows:

(1) We observe an instantaneous ~0.41% co-seismic dv/v decrease on the slope. Compared with a 0.39% dv/v decrease below 70 m, the dv/v decrease of 0.45% is larger at depth above ~70 m of the slope;

(2) We observe a characterized ~31.39 h logarithmic recovery process of dv/v towards its pre-earthquake state that occurred after the earthquake. The characterized recovery time increases from 20.35 to 40.94 h with the decrease in the central frequency from 15 to 4.5 Hz;

(3) We discuss possible mechanisms by modeling stress changes and evaluating the stress sensitivity of the method. A potential in situ application is also discussed to improve the understanding of subsurface geological risks under diverse seismic shaking.

The findings presented here support the idea that distant earthquakes can also affect the mechanical evolution of the rock slope. The observation underlines the importance of the slope's reversible rigidity drop and the recovery process induced by the distant earthquake which has the potential to facilitate the damage assessment of slopes non-destructively. The analysis has considerable implications both for the short-term impact of earthquakes and for the temporal-critical landslide early-warning system. Finally, this study helps to better understand landslide forcing mechanisms under diverse seismic loadings.

Author Contributions: Conceptualization, F.X.; methodology, H.H. and F.X.; original draft preparation H.H., S.D. and Y.Y.; review and editing: H.H., S.D., Y.Y. and F.X. All authors have read and agreed to the published version of the manuscript.

Funding: This research was funded by the National Science Foundation of China, grant number 41874061, 4211001018, 52192675, U19A2098 and the Central Research Institutes of Basic Research and Public Service Special Operations, grant number DQJB22Z02.

Institutional Review Board Statement: The study did not require ethical approval.

Informed Consent Statement: The study did not involve humans. Written informed consent has been obtained from the patient(s) to publish this paper.

Data Availability Statement: Data used in this study are available on this repository (https://share.weiyun.com/dkwf54kM, accessed on 25 June 2022).

Conflicts of Interest: The authors declared no potential conflict of interest with respect to the research, authorship, and/or publication of this article.

References

1. Lin, Q.; Wang, Y. Spatial and temporal analysis of a fatal landslide inventory in China from 1950 to 2016. *Landslides* **2018**, *15*, 2357–2372. [CrossRef]
2. Chigira, M.; Wu, X.; Inokuchi, T.; Wang, G. Landslides induced by the 2008 Wenchuan earthquake, Sichuan, China. *Geomorphology* **2010**, *118*, 225–238. [CrossRef]
3. Burbank, D.W.; Leland, J.; Fielding, E.; Anderson, R.S.; Brozovic, N.; Reid, M.R.; Duncan, C. Bedrock incision, rock uplift and threshold hillslopes in the northwestern Himalayas. *Nature* **1996**, *379*, 505–510. [CrossRef]
4. Li, Y.; Chen, P.; Cochran, E.S.; Vidale, J.E.; Burdette, T. Seismic Evidence for Rock Damage and Healing on the San Andreas Fault Associated with the 2004 M 6.0 Parkfield Earthquake. *Bull. Seismol. Soc. Am.* **2006**, *96*, S349–S363. [CrossRef]
5. Murnaghan, F.D. *Finite Deformation of an Elastic Solid*; Wiley: Hoboken, NJ, USA, 1951.
6. Whiteley, J.; Chambers, J.; Uhlemann, S.; Wilkinson, P.; Kendall, J. Geophysical monitoring of moisture-induced landslides: A review. *Rev. Geophys.* **2019**, *57*, 106–145. [CrossRef]
7. Del Gaudio, V.; Muscillo, S.; Wasowski, J. What we can learn about slope response to earthquakes from ambient noise analysis: An overview. *Eng. Geol.* **2014**, *182*, 182–200. [CrossRef]
8. Pazzi, V.; Morelli, S.; Fanti, R. A review of the advantages and limitations of geophysical investigations in landslide studies. *Int. J. Geophys.* **2019**, *2019*, 2983087. [CrossRef]
9. Colombero, C.; Jongmans, D.; Fiolleau, S.; Valentin, J.; Baillet, L.; Bièvre, G. Seismic noise parameters as indicators of reversible modifications in slope stability: A review. *Surv. Geophys.* **2021**, *42*, 339–375. [CrossRef]

10. Le Breton, M.; Bontemps, N.; Guillemot, A.; Baillet, L.; Larose, É. Landslide monitoring using seismic ambient noise correlation: challenges and applications. *Earth-Sci. Rev.* **2021**, *216*, 103518. [CrossRef]
11. Yamaoka, K.; Kunitomo, T.; Miyakawa, K.; Kobayashi, K.; Kumazawa, M. A trial for monitoring temporal variation of seismic velocity using an ACROSS system. *Isl. Arc* **2001**, *10*, 336–347. [CrossRef]
12. Chen, Y.; Irfan, M.; Uchimura, T.; Cheng, G.; Nie, W. Elastic wave velocity monitoring as an emerging technique for rainfall-induced landslide prediction. *Landslides* **2018**, *15*, 1155–1172. [CrossRef]
13. Campillo, M.; Paul, A. Long-range correlations in the diffuse seismic coda. *Science* **2003**, *299*, 547–549. [CrossRef]
14. Nakata, N.; Gualtieri, L.; Fichtner, A. *Seismic Ambient Noise*; Cambridge University Press: Cambridge, UK, 2019.
15. Snieder, R. The theory of coda wave interferometry. *Pure Appl. Geophys.* **2006**, *163*, 455–473. [CrossRef]
16. Sens-Schönfelder, C.; Wegler, U. Passive image interferometry and seasonal variations of seismic velocities at Merapi Volcano, Indonesia. *Geophys. Res. Lett.* **2006**, *33*. doi: 10.1029/2006GL027797. [CrossRef]
17. Brenguier, F.; Campillo, M.; Hadziioannou, C.; Shapiro, N.M.; Nadeau, R.M.; Larose, E. Postseismic relaxation along the San Andreas fault at Parkfield from continuous seismological observations. *Science* **2008**, *321*, 1478–1481. [CrossRef]
18. Lecocq, T.; Longuevergne, L.; Pedersen, H.A.; Brenguier, F.; Stammler, K. Monitoring ground water storage at mesoscale using seismic noise: 30 years of continuous observation and thermo-elastic and hydrological modeling. *Sci. Rep.* **2017**, *7*, 14241. [CrossRef]
19. Bièvre, G.; Franz, M.; Larose, E.; Carrière, S.; Jongmans, D.; Jaboyedoff, M. Influence of environmental parameters on the seismic velocity changes in a clayey mudflow (Pont-Bourquin Landslide, Switzerland). *Eng. Geol.* **2018**, *245*, 248–257. [CrossRef]
20. James, S.; Knox, H.; Abbott, R.; Screaton, E. Improved moving window cross-spectral analysis for resolving large temporal seismic velocity changes in permafrost. *Geophys. Res. Lett.* **2017**, *44*, 4018–4026. [CrossRef]
21. Mainsant, G.; Larose, E.; Brönnimann, C.; Jongmans, D.; Michoud, C.; Jaboyedoff, M. Ambient seismic noise monitoring of a clay landslide: Toward failure prediction. *J. Geophys. Res. Earth Surf.* **2012**, *117*. doi: 10.1029/2011JF002159. [CrossRef]
22. Mainsant, G.; Jongmans, D.; Chambon, G.; Larose, E.; Baillet, L. Shear-wave velocity as an indicator for rheological changes in clay materials: Lessons from laboratory experiments. *Geophys. Res. Lett.* **2012**, *39*. doi: 10.1029/2012GL053159. [CrossRef]
23. Bontemps, N.; Lacroix, P.; Larose, E.; Jara, J.; Taipe, E. Rain and small earthquakes maintain a slow-moving landslide in a persistent critical state. *Nat. Commun.* **2020**, *11*, 780. [CrossRef]
24. Huang, H.; Dai, S.; Xie, F. Monitoring In-Situ Seismic Response on Rock Slopes Using Ambient Noise Interferometry: Application to the 2019 Changning (Mw 5.7) Earthquake, China. *Front. Earth Sci.* **2021**, *8*, 610181. [CrossRef]
25. Jin, Z.; Fialko, Y. Coseismic and early postseismic deformation due to the 2021 M7. 4 Maduo (China) earthquake. *Geophys. Res. Lett.* **2021**, *48*, e2021GL095213. [CrossRef]
26. Hadziioannou, C.; Larose, E.; Coutant, O.; Roux, P.; Campillo, M. Stability of monitoring weak changes in multiply scattering media with ambient noise correlation: Laboratory experiments. *J. Acoust. Soc. Am.* **2009**, *125*, 3688–3695. [CrossRef]
27. Moreau, L.; Stehly, L.; Boué, P.; Lu, Y.; Larose, E.; Campillo, M. Improving ambient noise correlation functions with an SVD-based Wiener filter. *Geophys. J. Int.* **2017**, *211*, 418–426. [CrossRef]
28. Tsai, V.C. A model for seasonal changes in GPS positions and seismic wave speeds due to thermoelastic and hydrologic variations. *J. Geophys. Res. Solid Earth* **2011**, *116*. doi: 10.1029/2010JB008156. [CrossRef]
29. Larose, E.; Carrière, S.; Voisin, C.; Bottelin, P.; Baillet, L.; Guéguen, P.; Walter, F.; Jongmans, D.; Guillier, B.; Garambois, S.; et al. Environmental seismology: What can we learn on earth surface processes with ambient noise? *J. Appl. Geophys.* **2015**, *116*, 62–74. [CrossRef]
30. Herrmann, R.B. Computer programs in seismology: An evolving tool for instruction and research. *Seismol. Res. Lett.* **2013**, *84*, 1081–1088. [CrossRef]
31. Johnson, P.; Sutin, A. Slow dynamics and anomalous nonlinear fast dynamics in diverse solids. *J. Acoust. Soc. Am.* **2005**, *117*, 124–130. [CrossRef]
32. TenCate, J.A. Slow dynamics of earth materials: An experimental overview. *Pure Appl. Geophys.* **2011**, *168*, 2211–2219. [CrossRef]
33. Yoritomo, J.Y.; Weaver, R.L. Slow dynamic elastic recovery in unconsolidated metal structures. *Phys. Rev. E* **2020**, *102*, 012901. [CrossRef] [PubMed]
34. IRIS DATA. Data Services Products: Noise Toolkit PDF-PSD Noise Toolkit PDF/PSD Bundle. 2014. Available online: http://ds.iris.edu/ds/products/pdf-psd/ (accessed on 15 June 2022).
35. Tremblay, N.; Larose, É.; Rossetto, V. Probing slow dynamics of consolidated granular multicomposite materials by diffuse acoustic wave spectroscopy. *J. Acoust. Soc. Am.* **2010**, *127*, 1239–1243. [CrossRef] [PubMed]
36. Wyss, M.; Glllard, D.; Liang, B. An estimate of the absolute stress tensor in Kaoiki, Hawaii. *J. Geophys. Res. Solid Earth* **1992**, *97*, 4763–4768. [CrossRef]
37. Peng, Z.; Hill, D.P.; Shelly, D.R.; Aiken, C. Remotely triggered microearthquakes and tremor in central California following the 2010 Mw 8.8 Chile earthquake. *Geophys. Res. Lett.* **2010**, *37*. doi: 10.1029/2010GL045462 [CrossRef]
38. Zheng, Y. Transient pressure surge in a fluid-filled fracture. *Bull. Seismol. Soc. Am.* **2018**, *108*, 1481–1488. [CrossRef]
39. Jin, Y.; Dyaur, N.; Zheng, Y. Laboratory Evidence of Transient Pressure Surge in a Fluid-Filled Fracture as a Potential Driver of Remote Dynamic Earthquake Triggering. *Seism. Rec.* **2021**, *1*, 66–74. [CrossRef]

sustainability

Article

Dynamic Responses of RC Girder Bridge under Heavy Truck and Seismic Loads Combined

Lang Liu [1,2,*], Xiaotian Yang [2], Boquan Yan [2] and Siyu Miao [2]

1 State Key Laboratory of Mountain Bridge and Tunnel Engineering, Chongqing Jiaotong University, Chongqing 400074, China
2 School of Civil Engineering, Chongqing Jiaotong University, Chongqing 400074, China; w738998@gmail.com (X.Y.); you1234560720@163.com (B.Y.); 15978843131@163.com (S.M.)
* Correspondence: yilupaolai2008@163.com

Abstract: Overloaded truck and earthquake have become two main factors responsible for bridge damage, consequently the combination of heavy truck and seismic loads as a typical occurrence of extreme events is likely to lead to bridge collapse or destructive damage, in which the crucial issues of coupling load model, dynamic equations and bridge responses have not been adequately addressed. In this study, a simplified vehicle-bridge model consisting of many containers is established to simulate vehicle passage, and the dynamic equations are derived for a 5-axle truck on a simply supported beam as an illustration. Then, five ground motions selected from PEER with appropriate peak ground accelerations and durations and the three truck models specified in American Association of State Highway and Transportation Officials, Caltrans and Chinese codes are applied on the finite element model of a typical reinforced concrete continuous girder bridge, in which, vehicle speed, number of trucks, ground motion and vehicle type are assumed to be random variables and their influences on dynamic responses of the bridge are analyzed. The results show seismic load is the governing factor in dynamic responses but truck load may change displacement shapes; in addition, dynamic responses present a high sensitivity with the number of trucks (set as truck platoon) and gross vehicle weight but rare with vehicle speed. Specifically, the presence of a few trucks could serve as energy dissipation facilities for the bridge under seismic motions but may amplify the response when more trucks involved; some combinations of truck platoon with seismic excitation produce very large displacements and even cracks on the bridge, therefore, such an extreme event requires higher robustness in bridge design to make it be sustainable and serviceability after earthquakes.

Keywords: random loads; structure safety under extreme event; combination of truck and seismic loads; dynamic responses

Citation: Liu, L.; Yang, X.; Yan, B.; Miao, S. Dynamic Responses of RC Girder Bridge under Heavy Truck and Seismic Loads Combined. *Sustainability* **2022**, *14*, 9263. https://doi.org/10.3390/su14159263

Academic Editors: Chong Xu, Su Chen and Shuang Li

Received: 16 June 2022
Accepted: 26 July 2022
Published: 28 July 2022

1. Introduction

The seismic analysis has been a very conventional part of structural analysis [1]. During the past years, the dynamic behaviors of many kinds of structures, such as buildings, concrete bridges, soil-steel composite bridges and tunnels etc. under seismic excitation have been widely studied by simulations or experiments [2–7]. On the other hand, vehicle loads represent a major live load to highway bridges, especially for short- and medium-span bridges, vehicle loads even become the governing live load, therefore the effects of moving vehicles are always of special concerns in bridge engineering [8]. Normally, vehicles are approximated as a series of concentrated loads moving on bridges, which in many cases involving light vehicles allows the numerical method to implement it accurately. However, it is observed that both the amount and loading level of vehicles have significantly increased during recent years [9], as increasingly larger transportation vehicles are utilized, bridges have to subject to larger and heavier loads so that the inertia effect of vehicles can no longer be neglected. To take this effect into account, many vehicle models were developed by researchers to address the issue of vehicle-bridge interaction

(VBI) and great leaps have been achieved [10–13]. However, the combined effect of vehicle load and seismic load has not been adequately addressed. The previous studies have shown girder displacements and internal forces induced by heavy traffic loads are likely to exceed design values [14]. If earthquake occurs at the same time, then much severer load effects will be produced, due to the vibration and inertia loads of superstructure and heavy vehicles as well as their interaction, which possibly leads to badly damage or even collapse of bridges, or remarkably reduces their post-earthquake serviceability. As pointed out by Frangopol etc., it is important to study not only structural performance of the bridge during earthquake, but also to investigate incidents that could lead to incapacity to carry traffic after the earthquake [15,16].

Several accidents about the presence of vehicles on highway bridges during earthquakes have been reported, such as the Bay bridge in Oakland California during the Loma Pretia earthquake [17], Gavin Canyon bridge during the Northridge earthquake [18], a RC girder bridge during the Wenchuan earthquake, and the Yokohama-Bay Bridge during the Great East Japan Earthquake [19,20]. Some researches assessed the combined effect of vehicle presence and seismic excitations on the seismic performance of bridges, but most of them focus on the railway bridges in presence of trains [10,21,22]. However, characteristics of trains and highway vehicles, such as mass and suspension system are quite different, these findings cannot be directly used to understand the combined effect of seismic and vehicle load. A few studies addressed the issue of seismic excitation and live load acting on bridge structures simultaneously. Kim et al. [23] found that acceleration responses of the bridge subject to moderate earthquakes were amplified when the vehicles were treated as additional mass, whereas they were reduced when the vehicles were simulated as a dynamic system. Siringoringo et al. [19] analyzed the stability problem of a vehicle against rollover and side slip when crossing a bridge during an earthquake, the results showed that the significant bridge-deck vibration due to an earthquake reduces the effective normal force on vehicle wheels. Borjigin et al. [24] reported that continuously moving vehicles might yield larger longitudinal displacement responses at pier tops and plastic deformations at pier bottoms than those of the bridge alone, implying that ignoring vehicles' additional mass effect and dynamic effects during earthquakes might be on the non-conservative side. Cui and Xu [25] statistically investigated the seismic responses of both vehicle and bridge under real earthquakes and explained the interaction mechanism between vehicle and bridge based on energy analysis.

In addition, a few experimental studies were also conducted to investigate vehicle dynamic effects on bridge seismic response. Shaban et al. [26] conducted a large-scale experiment with a real vehicle parked on the deck of a simple-span bridge. The results showed that top slab transverse accelerations and bearing displacements were reduced in the presence of a vehicle during seismic excitation, in which, the vehicle can be treated as a tuned mass damper. Besides, an extensive testing of the influences of vehicle dynamics on bridge seismic response was conducted at the University of Nevada at Reno [27]. This study addressed that if the seismic level was rather low, then vehicles had a beneficial effect, but they had an adverse effect if the level was great.

These previous studies mentioned above mainly focused on normal vehicles, and few studies have specifically concerned about heavy trucks, which have obvious differences in characteristics of configurations. Moreover, the presence of vehicles is a random process, which has not been deeply discussed in these studies. With the increase of traffic volume and loading level, the likelihood of heavy vehicle presences on a bridge during an earthquake is expected to increase, therefore, the influence of highway vehicles on bridge seismic responses is still a subject of ongoing research [28,29] it is necessary to better understand the performance of bridges during earthquakes in presence of heavy vehicles. Therefore, this study adopts a simplified VB model to simulate vehicle passage and the dynamic equations are derived. Ground motions selected from PEER and heavy truck models specified in AASHTO, Caltrans and Chinese codes are applied on the finite element model of a typical RC continuous girder bridge, specifically, vehicle speed, number of vehicles, ground motion and vehicle type are assumed to be random variables and their influences on dynamic responses of the bridge are analyzed.

2. Modeling Approach

2.1. Simplified Vehicle-Bridge System

Dynamic responses will be induced by vehicle passage as a moving load. It is a complex nonlinear system and a simplified model needs to be established for analyzing bridge response without involving too much uncertainties. Generally, the hypothesis that there is no relative displacement between wheels and the bridge deck is made, if the coordination of displacements and forces between vehicles and the bridge is satisfied. If vehicle mass is relatively small, then the inertial force could be calculated by a single mass model moving on bridges. Whereas, vehicle suspension system has to be taken into account for bridge dynamic response when heavy vehicles travelling on bridges, for this end, sprung mass model is the prior one due to its simplicity and efficiency but without consuming accuracy. According to M. H. Scott and Zhu [30], the sprung mass model could be further simplified as a structure with two mass nodes, one node is the mass of vehicle body M_v and the other is that of vehicle wheels M_w; K_v and C_v represent the spring and the damper concerning the force between wheels and the bridge deck, as shown in Figure 1a. In this section, the analytical model for bridge dynamic response is to be established and derived, with two mass nodes moving on a simply supported beam, as illustrated in Figure 1b.

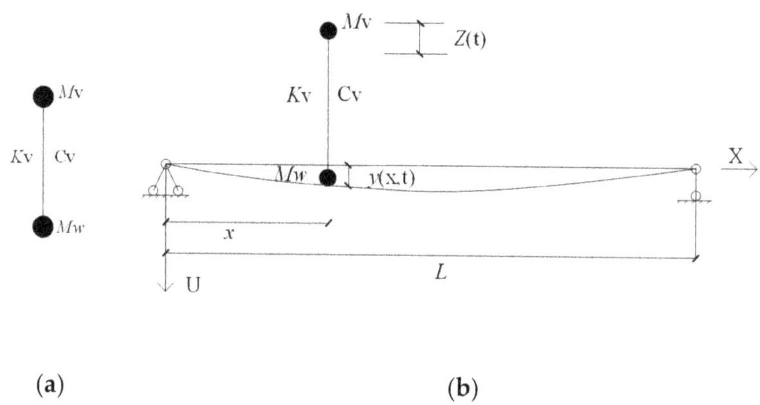

Figure 1. The sprung mass model representing vehicles. (**a**) two mass nodes. (**b**) analytical model.

To reflect the fact that vehicular properties associated with bridge dynamic response will vary with time, and also to guarantee no relative slips between vehicle wheels and then bridge deck, many containers are developed to bound with the bridge deck to carry vehicle properties, by which, vehicular properties could move in the longitudinal direction and be transferred from one container to the other, such that the important properties of axle load, damping and stiffness will be endowed to the very containers when the vehicle arrives at the position, while other containers will be set as non-occupation [30]. In this manner, the entire process of vehicle passage is simulated, and what is more, the important parameters of mass, stiffness and damping could be reasonably accounted. This approach is illustrated by Figure 2.

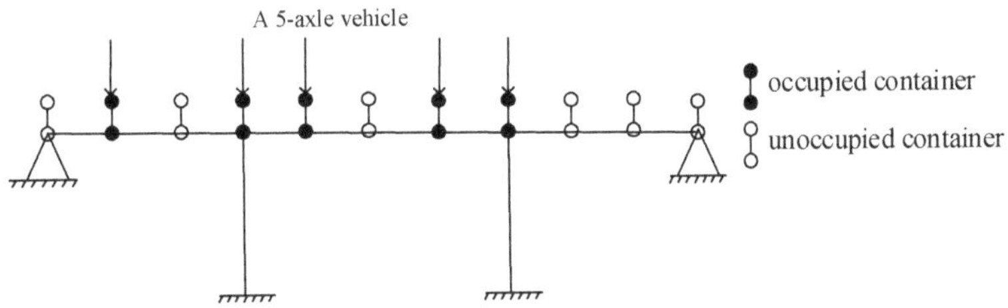

Figure 2. Illustration of a 5-axle vehicle moving on the bridge with containers.

Moreover, to make the simulation of vehicle passage considerably accurate but without time-consuming, a lot of containers are designed along the bridge spans so that the vehicle wheels can overlap certain containers for each step forward. Note that the appearance of containers will change the original stiffness matrix and geometry of the bridge since they are bounded together and their interface is set as no-displacement. To overcome this problem and reduce the interference of containers with dynamic analysis as much as possible, all properties of containers are set as zero except those associating with vehicle loads, by means of a special mass element in ANSYS. Also note that the masses of non-occupation containers are zero so that they do not have any contribution to dynamic response even though the static degree of freedom (DOF) is increased, therefore, these DOFs will not be included in the dynamic analysis.

As usual, there are various categories of vehicles running on bridges, but every vehicle could be regarded as a combination of several independent sprung mass models according to the number of axles, so the kinematic equation of vehicle-bridge system is derived based on the single sprung mass. To illustrate the basic principle, the kinematic equation is developed for a 5-axle vehicle moving on the simply supported beam in Figure 1, equivalently, to simulate 5 pairs of node mass models moving on the beam bounded with containers. Also note 5-axle vehicle is chosen as the example since they represent the highest proportion of traffic configurations according to findings from WIM data research [31].

For M_v, the equilibrium equation could be written as:

$$M_v\ddot{Z}(t) + K_w[Z(t) - y(x,t)|_{x=vt}] + C_w[\dot{Z}(t) - \left.\frac{dy(x,t)}{dt}\right|_{x=vt}] = 0 \tag{1}$$

where $Z(t)$ is the displacement of vehicle body M_v, and $y(x,t)$ represents the identical displacement of vehicle wheels M_w and the bridge; K_v and C_v are the spring stiffness and damping of the vehicle, respectively.

When the node mass models moving on the bridge at a speed of v, the bridge has to bear four kinds of forces, namely self-weight of node mass, inertia force of vehicle body, elastic force and damping force imposed by the spring and damper, respectively. As a result, the external force $P(x,t)$ acts on the bridge segment contacting with the node mass can be obtained as

$$P(x,t) = (M_v + M_w)g - M_w\frac{d^2y(x,t)}{dt^2} + K_v[Z - y(x,t)|_{x=vt}] + C_v\left[\dot{Z} - \left.\frac{dy(x,t)}{dt}\right|_{x=vt}\right] \tag{2}$$

As discussed before, there is no displacement between vehicle wheels and the bridge deck, so that

$$Z = y(x,t)|_{x=vt} = y(vt,t) \tag{3}$$

by inserting Equation (3) into Equations (1) and (2), then

$$M_v\ddot{Z}(t) + K_v[Z(t) - y(x,t)|_{x=vt}] + C_v[\dot{Z}(t) - \frac{\partial y(x,t)}{\partial t} - v\frac{\partial y(x,t)}{\partial x}] = 0 \tag{4}$$

$$\begin{aligned} P(x,t) = (M_v + M_w)g - M_w\left(\tfrac{\partial^2 y(x,t)}{\partial t^2} + 2v\tfrac{\partial^2 y(x,t)}{\partial x\partial t} + v^2\tfrac{\partial^2 y(x,t)}{\partial x^2}\right) + \\ K_v[Z(t) - y(x,t)] + C_v[\dot{Z}(t) - \tfrac{\partial y(x,t)}{\partial t} - v\tfrac{\partial y(x,t)}{\partial x}] \end{aligned} \tag{5}$$

consequently, the kinematic equation for the simply supported beam as shown in Figure 1b can be obtained as

$$EI\frac{\partial^4 y(x,t)}{\partial x^4}+M\frac{\partial^2 y(x,t)}{\partial t^2}+C\frac{\partial y(x,t)}{\partial t}=\delta(x-vt)P(x,t) \tag{6}$$

in which, δ is the Dirac function, when $x = vt$, $\delta = 1$, otherwise, $\delta = 0$.

Note that the displacement $y(x,t)$ can be decomposed into $y(x,t)=\sum_{i=1}^{\phi}\phi_i(x)\eta_i(t)$ with the modal analysis method, $\phi_i(x)$ is the ith mode shape and expressed as $\phi_N(x)=\sin\frac{N\pi x}{L}$, for L = span length and N is the modal order; $\eta_i(t)$ is usually defined as generalized displacement. Insert the decomposed $y(x,t)$ into the kinematic Equation (6), it becomes

$$EI\sum_{i=1}^{\infty}\frac{d^4\phi_i(x)}{dx^4}\eta_i(t)+\mathrm{m}\sum_{i=1}^{\infty}\phi_i(x)\ddot{\eta}_i(t)+\mathrm{c}\sum_{i=1}^{\infty}\phi_i(x)\dot{\eta}_i(t)=\delta(x-vt)P(x,t) \tag{7}$$

by multiplying $\phi_n(x)$ and making integration along the length, then taking advantage of the Reyleigh damping and orthogonality of mode shapes, the oscillation equation for the nth mode shape can be written:

$$\omega_n{}^2\eta_n(t)+2\xi_n\omega_n\dot{\eta}_i(t)+\ddot{\eta}_i(t)=\frac{2}{\mathrm{m}L}\int_0^L\delta(x-vt)P(x,t)\phi_n(x)\mathrm{d}x \tag{8}$$

where $\omega_n=\left(\frac{n\pi}{L}\right)^2\sqrt{\frac{EI}{\mathrm{m}}}$ and ξ_n is the damping ratio of the nth natural mode shape. Combining equations of (6)~(8), it could be obtained

$$\begin{gathered}\ddot{\eta}_N(t)+\gamma_w\sin N\Omega t\sum_{i=1}^{\infty}\sin(i\Omega t)\ddot{\eta}_i(t)+2\xi_n\omega_n\dot{\eta}_N(t)+2\gamma_w\Omega\sin N\Omega t\sum_{i=1}^{\infty}i\cos(i\Omega t)\dot{\eta}_i(t)+\\ \gamma_c\sin N\Omega t\sum_{i=1}^{\infty}\sin(i\Omega t)\dot{\eta}_i(t)+\omega_n{}^2\eta_N(t)-\gamma_w\Omega^2\sin N\Omega t\sum_{i=1}^{\infty}i^2\sin(i\Omega t)\eta_i(t)+\\ \gamma_k\sin N\Omega t\sum_{i=1}^{\infty}\sin(i\Omega t)\eta_i(t)+\gamma_c\Omega\sin N\Omega t\sum_{i=1}^{\infty}i\cos(i\Omega t)\eta_i(t)-\gamma_k U(t)\sin N\Omega t-\\ -\gamma_c\dot{U}(t)\sin N\Omega t=(\gamma_w+\gamma_v)g\sin N\Omega t\end{gathered} \tag{9}$$

in which, $\gamma_w=\frac{2M_w}{mL}$, $\gamma_v=\frac{2M_v}{mL}$, $\gamma_k=\frac{2k_v}{mL}$, $\gamma_c=\frac{2C_v}{mL}$, $\Omega=\frac{\pi v}{L}$, m is the mass per unit length of beam being a constant.

Meanwhile, the Equation (4) is rewritten by inserting the decomposed displacement $y(x,t)$:

$$M_v\ddot{Z}(t)+K_wZ(t)+C_w\dot{Z}(t)-K_w\sum_{i=1}^{\infty}\sin(i\Omega t)\eta_i(t)-C_w\sum_{i=1}^{\infty}\sin(i\Omega t)\dot{\eta}_i(t)-C_w\Omega\sum_{i=1}^{\infty}i\cos(i\Omega t)\eta_i(t)=0 \tag{10}$$

and the matrix form is $M\{\ddot{q}(t)\}+C\{\dot{q}(t)\}+K\{q(t)\}=F(t)$, for $\{q(t)\}=\begin{bmatrix}\eta_1(t) & \eta_2(t) & \dots & \eta_n(t) & \dots & \eta_N(t) & U(t)\end{bmatrix}^T$; Matrixes of mass, stiffness and damping are displayed as

$$M=\begin{pmatrix}1+\gamma_1D_{11} & \gamma_1D_{12} & \dots & \gamma_1D_{1N} & 0\\ \gamma_1D_{21} & 1+\gamma_1D_{22} & \dots & \gamma_1D_{2N} & 0\\ \dots & \dots & \dots & \dots & \dots\\ \gamma_1D_{N1} & \gamma_1D_{N2} & \dots & 1+\gamma_1D_{NN} & 0\\ 0 & 0 & \dots & 0 & M_v\end{pmatrix} \tag{11}$$

$$C=\begin{pmatrix}2\xi_1\omega_1+2\gamma_w\Omega E_{11}+\gamma_cD_{11} & 2\gamma_w\Omega E_{12}+\gamma_cD_{12} & \dots & 2\gamma_w\Omega E_{1N}+\gamma_cD_{1N} & -\gamma_cD_1\\ 2\gamma_w\Omega E_{21}+\gamma_cD_{21} & 2\xi_2\omega_2+2\gamma_w\Omega E_{22}+\gamma_cD_{22} & \dots & 2\gamma_w\Omega E_{2N}+\gamma_cD_{2N} & -\gamma_cD_2\\ \dots & \dots & \dots & \dots & \dots\\ 2\gamma_w\Omega E_{N1}+\gamma_cD_{N1} & 2\gamma_w\Omega E_{N2}+\gamma_cD_{N2} & \dots & 2\xi_N\omega_N+2\gamma_w\Omega E_{NN}+\gamma_cD_{NN} & -\gamma_cD_N\\ -C_vD_1 & -C_vD_2 & \dots & -C_vD_N & C_v\end{pmatrix} \tag{12}$$

$$K=\begin{pmatrix}\omega_1^2-\gamma_w\Omega^2G_{11}+\gamma_kD_{11}+\gamma_c\Omega E_{11} & -\gamma_w\Omega^2G_{12}+\gamma_kD_{12}+\gamma_c\Omega E_{12} & \dots & -\gamma_w\Omega^2G_{1N}+\gamma_kD_{1N}+\gamma_c\Omega E_{1N} & -\gamma_kD_1\\ -\gamma_w\Omega^2G_{21}+\gamma_kD_{21}+\gamma_c\Omega E_{21} & \omega_2^2-\gamma_w\Omega^2G_{22}+\gamma_kD_{22}+\gamma_c\Omega E_{22} & \dots & -\gamma_w\Omega^2G_{2N}+\gamma_kD_{2N}+\gamma_c\Omega E_{2N} & -\gamma_kD_2\\ \dots & \dots & \dots & \dots & \dots\\ -\gamma_w\Omega^2G_{N1}+\gamma_kD_{N1}+\gamma_c\Omega E_{N1} & -\gamma_w\Omega^2G_{N2}+\gamma_kD_{N2}+\gamma_c\Omega E_{N2} & \dots & \omega_N{}^2-\gamma_w\Omega^2G_{NN}+\gamma_kD_{NN}+\gamma_c\Omega E_{NN} & -\gamma_kD_N\\ -K_vD_1-C_v\Omega E_1 & -K_vD_2-C_v\Omega E_2 & \dots & -K_vD_N-C_v\Omega E_N & K_v\end{pmatrix} \tag{13}$$

where $D_i = \sin i\Omega t$, $D_{ij} = D_i D_j = \sin i\Omega t \sin j\Omega t$, $E_i = i\cos i\Omega t$, $G_i = i^2 \sin i\Omega t$, $E_{ij} = E_i D_j = \sin i\Omega t \times j\cos j\Omega t$, $G_{ij} = G_i D_j = \sin i\Omega t \times i^2 \sin i\Omega t$.

By the usage of *Newmark-β* method, the displacement, velocity and acceleration at t_i and t_{i+1} time can be solved. Finally, the displacement and acceleration of the bridge and vehicle are obtained, respectively:

$$y(x,t) = \sum_{\mathrm{n}=1}^{\infty} \sin\frac{\mathrm{N}\pi x}{L}\eta_{\mathrm{n}}(t)\ddot{y}(x,t) = \sum_{\mathrm{n}=1}^{\infty} \sin\frac{\mathrm{N}\pi x}{L}\ddot{\eta}_{\mathrm{n}}(t)Z(t) = y(vt,t) = \sum_{\mathrm{n}=1}^{\mathrm{N}} \sin\frac{\mathrm{n}\Omega t}{L}\eta_{\mathrm{n}}(t) \quad (14)$$

$$\ddot{Z}(t) = \sum_{\mathrm{n}=1}^{\mathrm{N}} \sin n\Omega t\ddot{\eta}_{\mathrm{n}}(t) + 2\sum_{\mathrm{n}=1}^{N} n\Omega\cos n\Omega t\dot{\eta}_{\mathrm{n}}(t) - \sum_{\mathrm{n}=1}^{\mathrm{N}} (n\Omega)^2 \sin n\Omega t\eta_{\mathrm{n}}(t)$$

2.2. Bridge Model

To make structural dynamic analysis under the combined seismic and truck loads, a prestressed continuous box- girder bridge with a span arrangement of 36 + 56 + 36 m is modelled. The details of the superstructure and the column can be seen in Figure 3. The superstructure is supported by pot bearings. The prototype bridge was designed for a site in class II with a PGA of 0.1 g and the live load of highway class-I. Based on the design parameters, the finite element model (FEM) of this bridge is built. In which, the girder and columns are modelled with BEAM 188 element provided in ANSYS software; this element was developed based on the Timoshenko beam theory for any of elastic, plastic and other nonlinear material analyses. By which, shear deformation could be considered besides flexural deformation and suitable for box- girder used in the present study. There are at least 6 DOFs in the element, including three translational DOFs and three rotational DOFs at X, Y and Z directions, respectively. The FEMs of superstructure and substructure are displayed in Figure 4. All DOFs are fixed at the bottom of piers, and the abutments are fixed in both lateral and vertical directions, while bearings are modelled by coupled nodes.

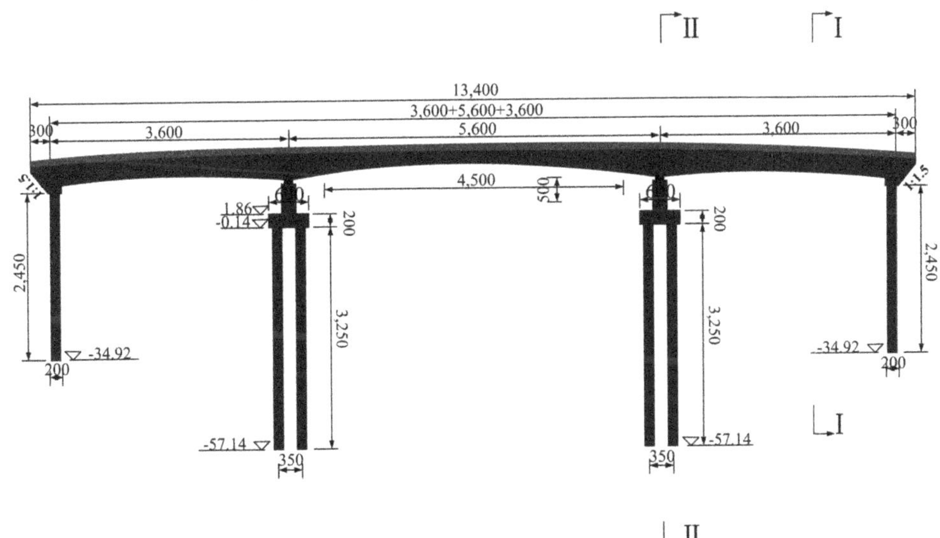

(a)

Figure 3. *Cont.*

(b)

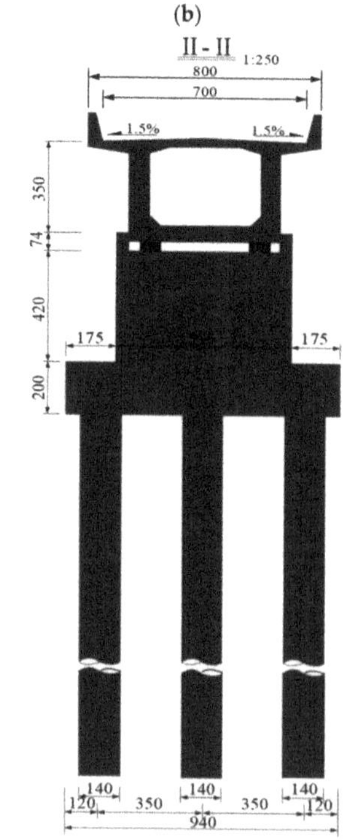

(c)

Figure 3. The design details of the superstructure and columns. (**a**) Layout of the RC continuous girder bridge. (**b**) I-I section profile. (**c**) II-II section profile.

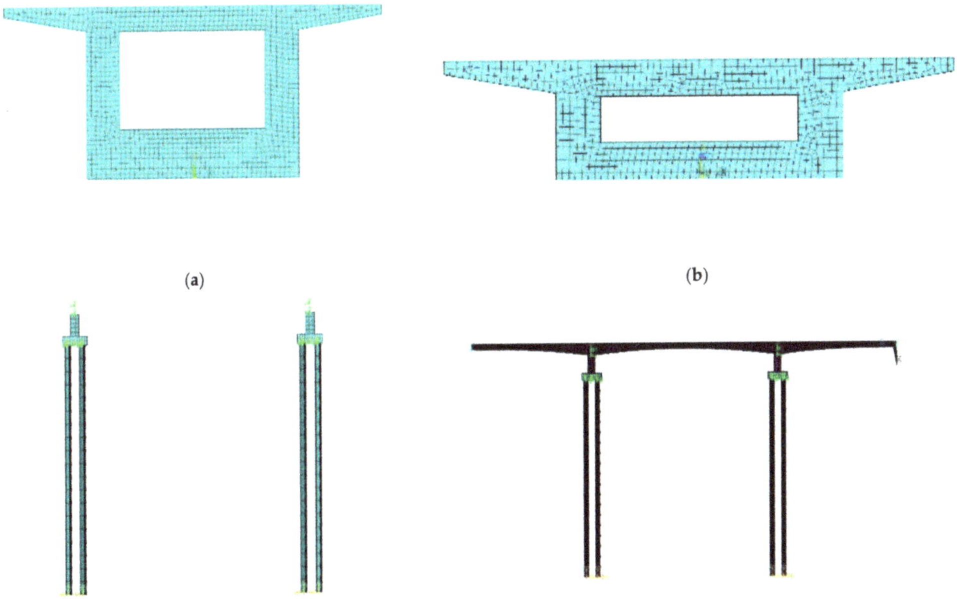

Figure 4. The finite element models of the bridge. (**a**) cross section of the box-girder at bearings. (**b**) cross section of the box-girder at midspan. (**c**) bridge piers. (**d**) overview of the FEM.

2.3. *Vehicle Model*

Three extensively used types of vehicle models are used to apply live loads to the bridge deck, including the truck model specified in AASHTO bridge design code, Caltrans DOT heavy truck model and the one in Chinese provision, they are separately combined with seismic load to compare and analysis if travelling speed, truck weight and axle load etc. would significantly affect the dynamic response. These truck models are concisely introduced in this section.

2.3.1. Truck Model in AASHTO Code

The type 362 truck specified in AASHTO code is shown in Figure 5a, it is a 5-axle truck with the gross weight of 320.28 kN. The detailed axle loads and spaces are marked in the figure.

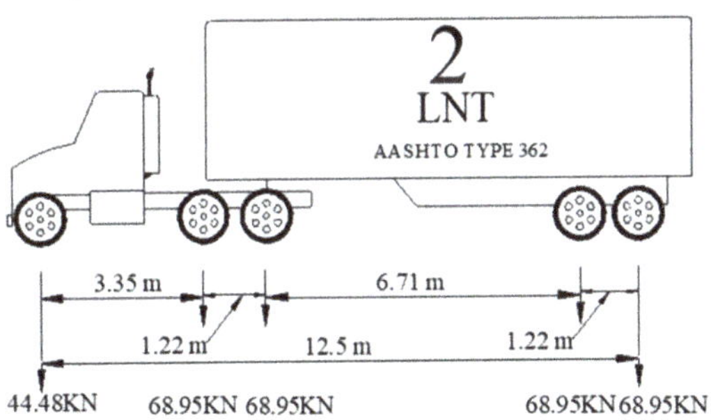

(**a**)

Figure 5. *Cont.*

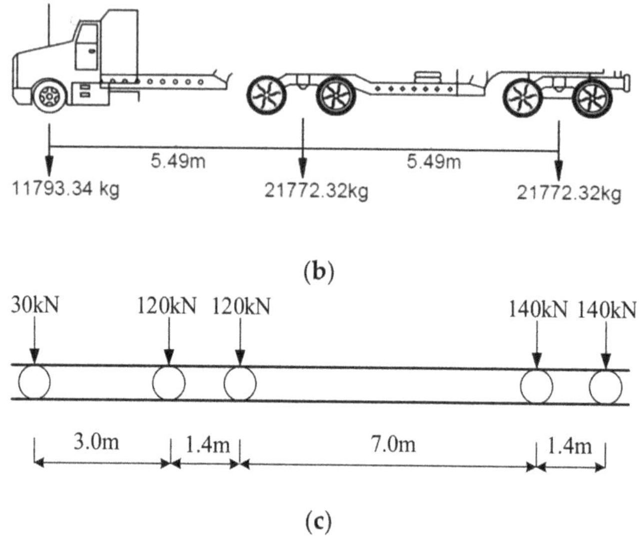

Figure 5. Truck models in specifications. (**a**) AASHTO truck model. (**b**) Caltrans truck model. (**c**) Chinese truck models.

2.3.2. Heavy Truck Model in Caltrans

California is well known as an industrial state in the U.S. so that many kinds of huge (on width, height, trailer length, and number of trailers) and heavy (on GVW, axle weight, and wheel weight) trucks appear on highway bridges to carry some high-density commodities with divisible loads, they have been indispensable components of the traffic configuration, indicating the vehicular overload standard issued by the Federal government was not compatible with the real social requirement, therefore, the Caltrans regulated their permission truck level by themselves, embodied by a series of very heavy trucks. To be consistent, the 5-axle truck of this series is used herein. Each tandem has a weight of 213.51 kN and the first axle load is 115.65 kN, as shown in Figure 5b.

2.3.3. Truck Model in Chinese Provision

A simplified 5-axle truck model was stipulated in the Chinese bridge design specification as an alternative of the lane load, accounting for the live loads applying to bridges. The layout of axle loads and spaces is plotted in Figure 5c.

As discussed before, the vehicle is modelled by two nodes of vehicle body and wheels denoting the weight of M_v and M_w, and its moving process is simulated by a cluster of containers associated with the bridge, transferring the properties of the vehicle step by step. The MASS21 element with 6 DOFs is used to simulate vehicle body and wheels, and COMBIN14 element for the suspension system as a one-dimension tensioned or compressed spring-damper. The interface element of CONTA 175 is utilized to deliver vehicle weight to the girder and make sure no slips between vehicle wheels and the containers involved.

3. Dynamic Characteristic of the Bridge Model

As a MDOF structure system, bridges have many inherent vibration frequencies and mode shapes, but the dynamic characteristics of the 1st~6th orders are adequate for dynamic response analysis of normal RC girder bridges, since the higher orders are rarely excited and have negligible effects on the whole responses. Therefore, the first 6th frequencies and mode shapes are calculated and demonstrated in Table 1 and Figure 6.

Table 1. Natural frequencies and mode shapes of the bridge.

Vibration Mode	Frequency (Hz)	Mode Shape
1	1.94	1st order longitudinal bending
2	3.69	antisymmetric vertical bending
3	4.81	2nd order longitudinal bending
4	5.73	1st order torsion
5	7.50	dissymmetric vertical bending
6	7.72	dissymmetric transverse bending

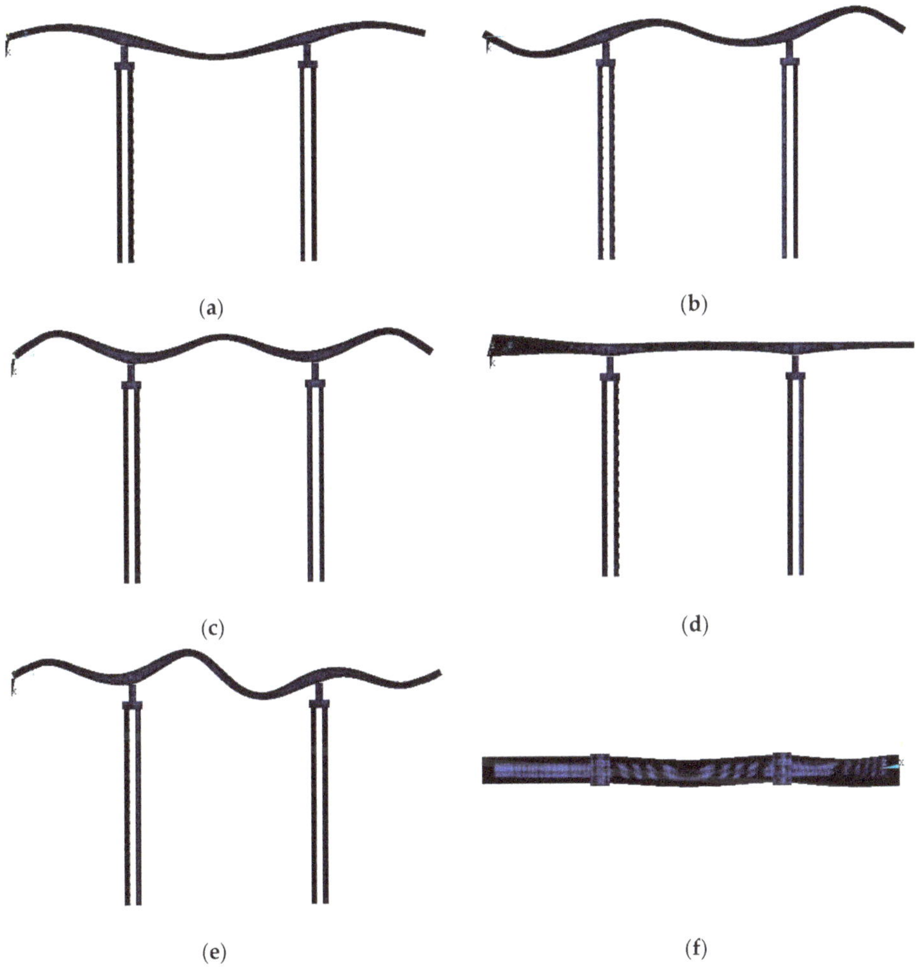

Figure 6. The first 6th mode shapes of the bridge. (**a**) The 1st order. (**b**) The 2nd order. (**c**) The 3rd order. (**d**) The 4th order. (**e**) The 5th order. (**f**) The 6th order.

4. Vehicle-Bridge System Equation under Combined Seismic and Vehicle Loads

So far, the concerned question of bridges under combined seismic and vehicle loads has been divided into two parts, one is the simplified vehicle-bridge (VB) system established in the previous section, the remaining part concerning seismic load can be treated as an external excitation to apply to the VB system. If we let U_b denote the displacement vector of the container nodes, then the enlarged bridge equation could be expressed as

$$\begin{bmatrix} M_v & 0 \\ 0 & M_b \end{bmatrix} \begin{Bmatrix} \ddot{U}_v \\ \ddot{U}_b \end{Bmatrix} + \begin{bmatrix} C_v & C_{vb} \\ C_{bv} & C_b \end{bmatrix} \begin{Bmatrix} \dot{U}_v \\ \dot{U}_b \end{Bmatrix} + \begin{bmatrix} K_v & K_{vb} \\ K_{bv} & K_b \end{bmatrix} \begin{Bmatrix} U_v \\ U_b \end{Bmatrix} = \begin{Bmatrix} F_v + P_{\text{effv}} \\ F_b + P_{\text{effb}} \end{Bmatrix} \tag{15}$$

in which, P_{effv} and P_{effb} represent dynamic responses of the vehicle and the bridge associated with containers under seismic motions, respectively; subscripts v and b denote vehicle and bridge, respectively. F_v and F_b are force vectors. Other parameters have been discussed before. Moreover, the vector U_b can be dived into pseudo-static displacement and dynamic displacement, written as $\{U_b\} = \begin{Bmatrix} U_a^s \\ U_b^s \end{Bmatrix} + \begin{Bmatrix} U_a^d \\ 0 \end{Bmatrix}$, where superscripts s and d denote pseudo-static and dynamic displacement, respectively; subscripts a denotes bridge nodes except bearings, while b denotes bearings. Since the seismic effects are treated as uniform excitation, then the pseudo-static displacements at supports are identical and no relative slips between bearings.

It is supposed that there are N vehicles moving on the bridge and N1 mode shapes of the bridge have contributions to the dynamic response when earthquake occurs, then, $\{P_{\text{effv}}\} = \{P_{\text{effv1}} \; P_{\text{effv2}} \; \cdots \; P_{\text{effv}N}\}^T$ and $\{P_{\text{effb}}\} = \{P_{\text{effb1}} \; P_{\text{effb2}} \; \cdots \; P_{\text{effb}N1}\}^T$. To be consistent with vehicle passage, ground motions are input in longitudinal direction, then the generalized seismic load on the bridge can be expressed by the normalized ith mode as $P_{\text{effbi}} = a_g^x(t)\{\phi\}_i^T\{M\}\{D\}_x$,$a_g^x(t)$ is the longitudinal ground motion component; $\{D\}_x$ denotes directional vector, and $\{\phi\}_i$ is the ith generalized mode vector. Based on these derived expressions, the Equation (15) can be solved by the *Newmark-β* method further.

5. Dynamic Response Analysis of the Bridge under Combined Seismic and Vehicle Loads

To address the influences of vehicle speed, number of vehicles, intensity and duration of ground motion and vehicle type on bridge dynamic responses, these factors are analyzed separately in this section, incorporating with site class, damping ratio, design spectral acceleration etc. of the protype bridge, to understand the performance of the bridge under different seismic and/or vehicle loading conditions.

5.1. Ground Motions Selection

Un-scaled ground motions (GMs) are selected from the PEER strong ground motion database. Out of three ground motion components in each record set, two horizontal and one vertical, the horizontal motion is selected. Besides, the design spectral acceleration in Chinese Specifications for Seismic Design of Highway Bridges is taken as reference for selecting ground motion records, based on the design parameters, such as site class of I1, damping ratio of 0.05 and characteristic period of 0.3 s etc., the calculated spectral peak acceleration S_{max} = 0.195 g is for the basic seismic design level, for severe earthquake with a return period of 1600 years it should be enlarged 1.6~2.3 times according to the specification, let it be 2 times and finally a 0.4 g of S_{max} is set for scaling ground motions. Note that the duration of the selected ground motions needs to be determined additionally rather than using the recorded time, to make sure the time period with dominant acceleration/energy of ground motions complies with the estimated time for vehicles passing through the bridge with a certain speed. Therefore, the 70% of energy duration is taken as seismic duration because of its widespread use. As a consequence, 5 potential ground motions are firstly selected and their processed parameters are listed in Table 2.

Table 2. Energy durations and normalized PGAs of ground motions.

No.	Earthquake Event	Energy Duration(s)	Normalized PGA(g)
T1	Saguenay, 25 November 1988	3.93	0.95
T2	AuSableForks, 20 April 2002	15.22	1.0
T3	RiviereDuLoup, 6 March 2005	14.73	1.0
T4	MtCarmel, 18 April 2008	7.10	0.72
T5	Mineral, 23 August 2011	5.86	0.95

As can be seen, T2 and T3 have longer durations to present the combination of seismic load and vehicle load of interest in this study. In addition, the standard deviations in energy durations of T2 equals to 0.24 and less than that of 0.26 of T3, thus, T2 is selected as the seismic excitation plus live-load for the bridge. The normalized T2 ground motion and its energy duration are shown in Figure 7.

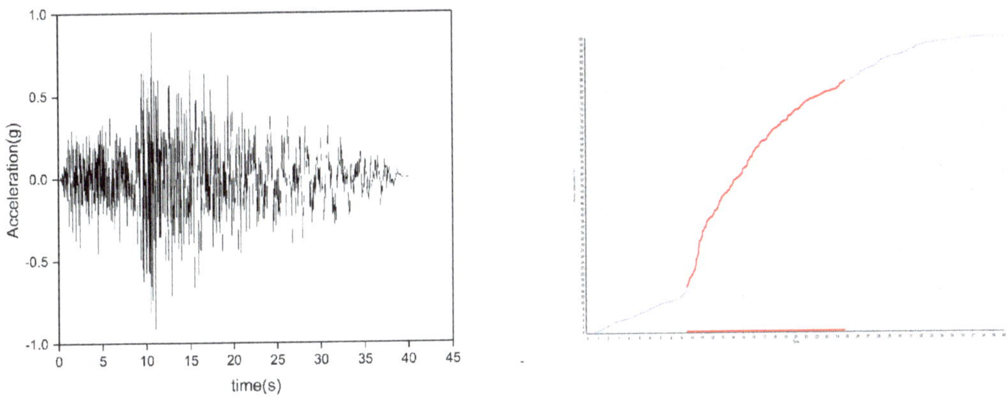

(**a**) (**b**)

Figure 7. The ground motion selected. (**a**) The acceleration time history of GM T2. (**b**) The duration of the GM T2.

5.2. Dynamic Response Analysis of the Bridge

Based on theVB system and ground motion time history demonstrated before, some important factors associating with vehicle load and seismic load are to be investigated specifically, to understand the influence of each parameter on bridge dynamic responses. For this end and also making the results more comparable, the Caltrans truck model and ground motion T2 are set as benchmark for vehicle load and seismic load, respectively. If not reported particularly, then they are used in default. In addition, vehicles are supposed to move at uniform speeds and the headways are constant if more than one vehicle involved.

5.2.1. Vehicle Speed

Vehicle speed determines the travelling time and therefore the loading time of the vehicle on the bridge, and it may have effect on displacements of the bridge. To be comparable, five common passing speeds from 36 km/h to 108 km/h with an interval of 18 km/h are considered. Figure 8a shows the displacement time history under the combined seismic load and heavy truck load, and Figure 8b shows their counterparts under traffic loading only.

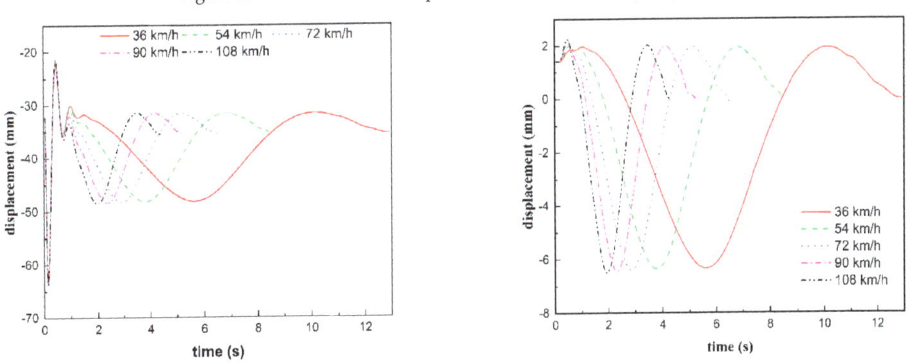

(**a**) (**b**)

Figure 8. Displacements of the bridge taking different vehicle speeds. (**a**) combined loads. (**b**) vehicle loads only.

It can be seen in Figure 8b, when the vehicle passing through the bridge at different speeds, the peak displacement very slightly increases with increasing travelling speed, indicating larger impact and dynamic response of the bridge are produced. Whereas, the displacements are very close and the maximum difference is only 0.15 mm, so vehicle speed has very limited influence on bridge displacement. On the other hand, when vehicle loads combined with seismic loads, the displacements are amplified dramatically and the displacement induced by seismic excitation is overwhelming component of the total response. However, the displacement slightly decreases with increasing travelling speed, and the maximum value is 0.01 mm, denoting vehicle speed has negligible effect on the dynamic response. As shown, the dynamic response induced by the combined seismic load plus live-load is approximately 10 times that under live load only.

It is notable the largest displacements in (a) always occur when the vehicle just appears at the side-span, also the ground acceleration has relatively small value; the second largest values appear at the time when the vehicle arrives at the mid-span, while in the case of (b), it has the largest displacements. Also note the GM excitation time in (a) depends on the travelling time,namely the vehicle speed,so that the fast speed of 108 km/h and the slowest speed of 36 km/h have the shortest and longesgt excitation time, respectively. Even though the latter gets the peak acceleration value of 1.0 g while the former only gets half of that, but their peak displacments are very close, implying this bridge structure may be senstitive to certain frequency components but not necessarily the ground acceleration value. This is also seen for other excitation time (or vehicle speed).

5.2.2. Multiple Presence of Vehicles

Vehicles usually appear on a bridge span in one or more lanes simultaneously, which is known as multiple truck presence. The previous studies reported the multiple loading configuration would induce higher static load effects to bridge components than those of single one [2]. To address if the dynamic responses will still be governed by multiple truck presence when the live loads are combined with seismic loads, five platoons consisting of one to five trucks are applied on the bridge separately, denoting case 1 to case 5 herein, for each platoon, the travelling speed is set as 72 km/h and the distance between platoons is 10 m. The platoons are assumed to move on the centerline of lane 2. The presence of the platoon consisting of two trucks on the bridge can be viewed in Figure 9 as an example.

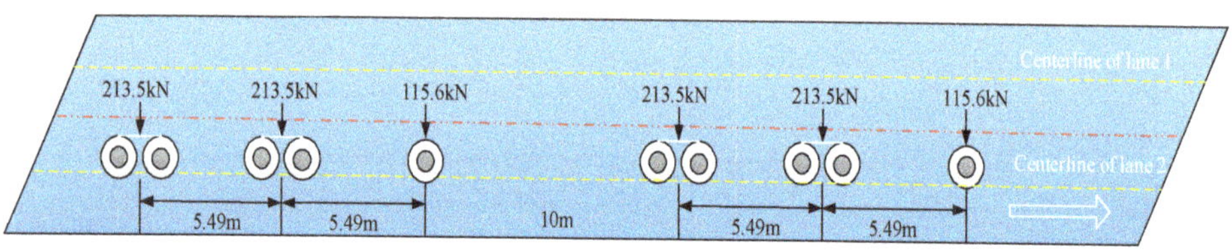

Figure 9. Application of truck platoon to the bridge.

Figure 10a displays displacement time histories under combined seismic motion and truck platoon, and Figure 10b displays displacements under each truck platoon. As observed, displacements in Figure 10b have nonlinear relationship with the number of trucks, when the number increases from 1 to 2 or 3, the displacements increase noticeably and the maximum increment is about 3.5 mm, but when the number continually increases to 4 or 5, the displacements do not necessarily increase and even decrease by 2 mm at most. It can be concluded that multiple truck presence would induce higher load effects to bridges than single truck, and the severest spatial loading configuration would govern the maximum effects rather than the number of trucks.

When the ground motion excitation is taken into account, displacements in Figure 10a generally increase with the truck number, and a significant difference of 30 mm exits between one truck and four trucks, indicating the number of trucks has remarkable influence on the total response, but the influence is to be limited when the number increase to a certain degree. However, the responses slightly decrease when comparing with the case of GM exciting only, indicating vehicles may make a positive effect on energy dissipation, due to their body mass and suspension system, to reduce the dynamic responses to some extent. In this study, case 3, 4 and 5 generally have observable dissipation effect, but one should keep in mind that the number of trucks, their presence on lanes, headway distances as well as truck system will affect this result, so it would be a case-specified analysis rather a general conclusion.

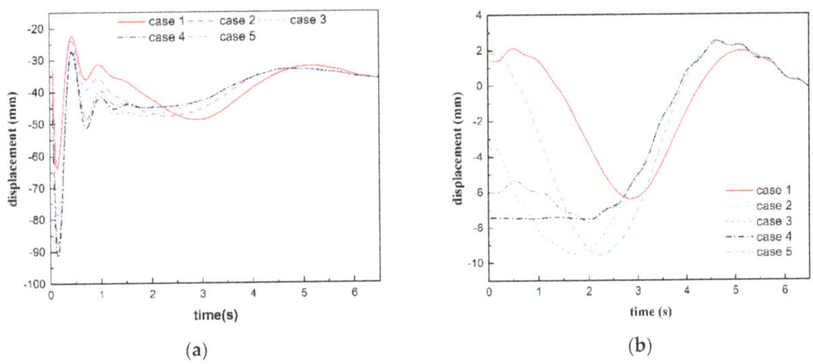

Figure 10. Displacements of the bridge taking different vehicle numbers. (**a**) combined loads. (**b**) vehicle loads only.

5.2.3. Ground Motions

Ground motion has very strong stochastic behavior, to better understand the random variables associated with seismic loads in the previous analyses, another three ground motions in widespread use are selected as the inputs and combined with live loads to apply to the bridge as before. The three ground motions RiviereDuLoup, Kobe and Chi-Chi are adopted because they have similar PGAs and other features with the scaled ground motion T2. They are also processed accordingly, and the normalized time histories are displayed in Figure 11 as below.

For consistency, similar analyses are implemented in this section. Table 3 and Figure 12a,b show the displacements under the four ground motions plus vehicle-load taking different speeds and platoons, respectively. Here vehicle speeds and number of trucks are identically set with the previous ones. For analysis of vehicle speed, one single Caltrans truck model is applied, while for that of vehicle number, a speed of 72 km/h is used. It is shown no matter which ground motion is adopted, the displacements have almost the same variation tendency with the changes of vehicle parameters. Displacements slightly decrease with faster speeds while increase when more trucks running on the bridge. As the vehicle speed or vehicle number isthe same, the displacements for the four GMs are very close, and the differences are within 0.5 mm and 15 mm, respectively. As discussed above, the number of trucks has remarkable influence on the total response, whereas the randomness of ground motion barely affects the dynamic response of interest herein.

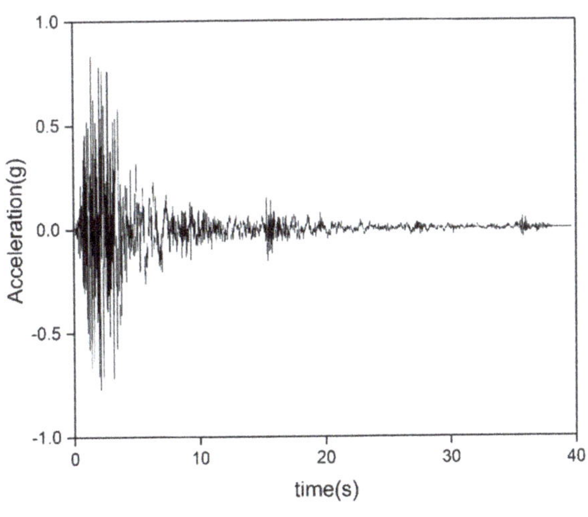

(**a**)

Figure 11. *Cont.*

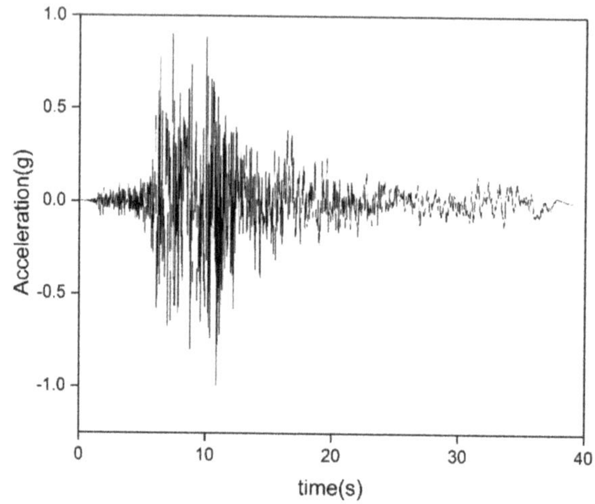

(b)

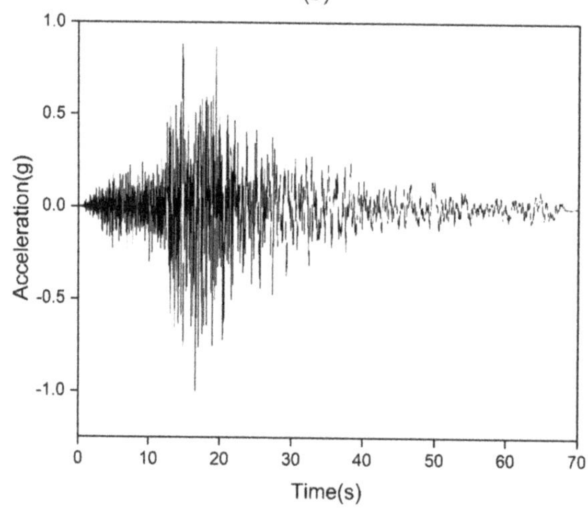

(c)

Figure 11. Three supplemented ground motions. (**a**) Chi-Chi ground motion. (**b**) Kobe ground motion. (**c**) RiviereDuLoup ground motion.

Table 3. Maximum midspan displacements considering different ground motions.

	Vehicle Speed (km/h, One Caltrans)					Number of Vehicles (at 72 km/h)				
GM	36	54	72	90	108	1	2	3	4	5
AuSableForks	63.75	63.75	63.75	63.75	63.74	63.75	64.15	79.16	91.70	87.46
RiviereDuLoup	63.71	63.71	63.70	63.70	63.70	63.71	61.48	87.35	106.94	99.22
Chi-Chi	63.67	63.67	63.67	63.67	63.66	63.68	61.45	87.31	106.88	99.17
Kobe	63.58	63.58	63.58	63.58	63.57	63.59	61.36	87.20	106.73	99.06

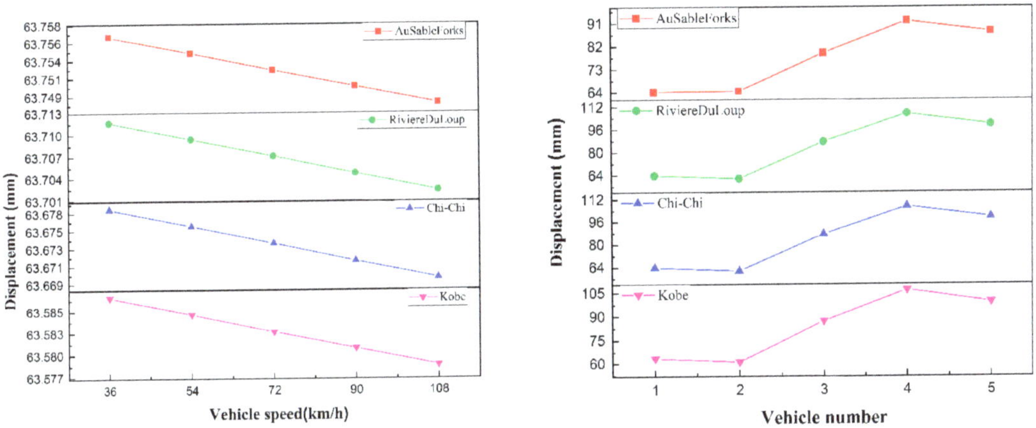

(**a**) (**b**)

Figure 12. Displacements of the bridge under different ground motions. (**a**) at different vehicle speeds. (**b**) with different vehicle numbers.

5.2.4. Vehicle Type

Three truck models in different specifications have been introduced above, among them the Caltrans truck model induces the largest load effects so it has been used in the previous analysis. As known, vehicles represent a major live load to bridges and uncertainties associated with it are very high, therefore, this study also has a consideration on the randomness of vehicle loads as one of critical variables. For the focused objective, the randomness of vehicle loads herein includes speed, model type and presence number, which are to be analyzed by grouping.

Firstly, each truck model running on the bridge at different speeds is considered with the scaled ground motion T2. To be consistent, all speeds set in Section 5.2.1 are used, and the midspan displacements under combined loads and truck loads only are listed in Table 4 below. The results are further plotted in Figure 13a,b to compare.

Table 4. Maximum midspan displacements considering different vehicle models.

Truck Model	Load Type	Vehicle Speed (km/h)				
		36	54	72	90	108
Caltrans	Combined load	63.75	63.75	63.75	63.75	63.74
	Truck load only	6.34	6.36	6.40	6.44	6.49
AASHTO	Combined load	63.67	63.67	63.67	63.67	63.66
	Truck load only	3.48	3.49	3.504	3.52	3.56
China	Combined load	63.73	63.73	63.728	63.726	63.72
	Truck load only	6.14	6.16	6.194	6.230	6.29

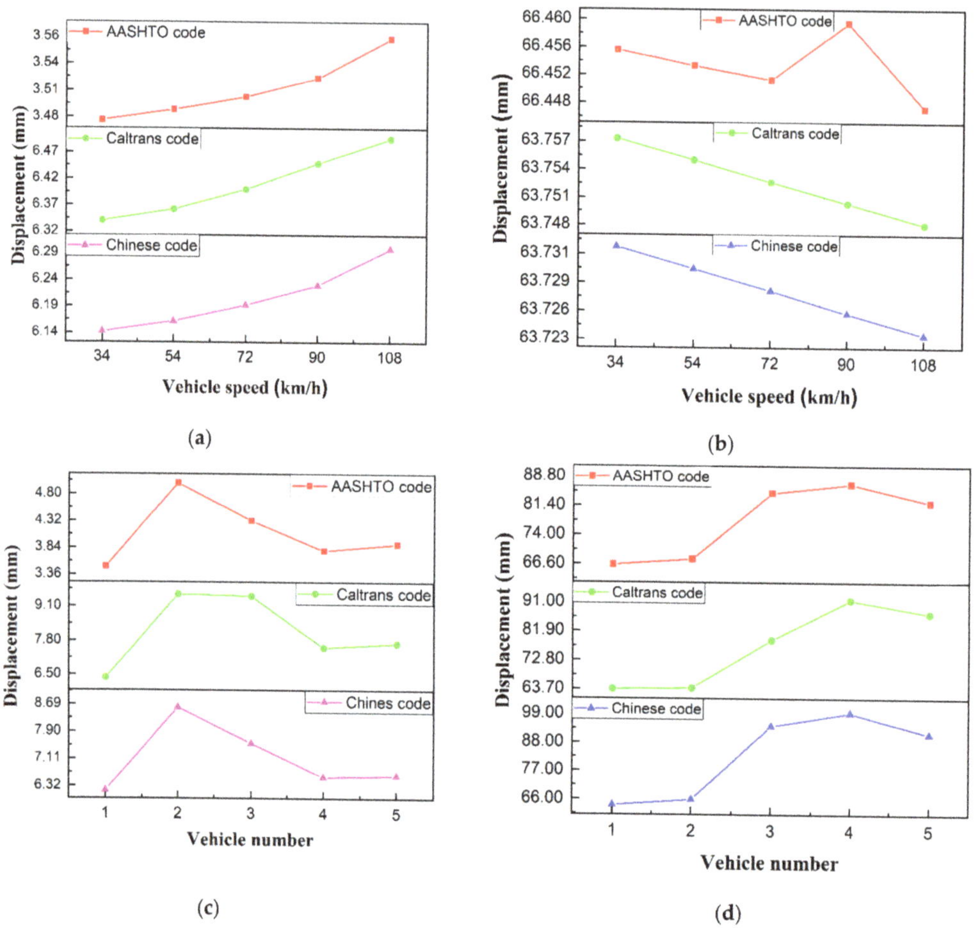

Figure 13. Displacements of the bridge taking different truck models in codes. (**a**) vehicle loads only. (**b**) combined loads. (**c**) vehicle loads only. (**d**) combined loads.

In Figure 13a,the displacements induced by light vehicle and heavy vehicle are quite different, for example, the displacement under the Caltrans model having the largest GVW is nearly two times that of the lightest AASHTO model, indicating vehicle weight has significant influence on bridge responses under live loads only, in addition, the GVWs of Caltrans and China models are close but their axle loading configurations are different, causing the displacements produced are also different, especially when the moving speed is high, thus, vehicle speed is also a notable factor enlarging the dynamic responses when vehicle loads act alone. On the contrary, in Figure 13b, the influence is remarkably weakened when seismic loads and live loads work together, since the former becomes the governing one, leading to a difference less than 3 mm between each pair of truck models.

Secondly, truck platoons are assembled by the same truck model with different numbers of one to five moving on the bridge at a speed of 72 km/h, as a result, 15 platoons are combined with the scaled ground motion T2, respectively, to calculate the displacements shown in Table 5. The results are also plotted in Figure 13a,d for comparison.

Table 5. Maximum midspan displacements considering different vehicle platoons.

Truck Platoon	Load Type	Number of Vehicles in the Platoon				
		1	2	3	4	5
Caltrans	Combined load	63.75	64.15	79.16	91.70	87.46
	Truck platoon only	6.40	9.55	9.49	7.53	7.69
AASHTO	Combined load	66.45	67.90	84.70	86.96	82.35
	Truck platoon only	3.50	4.99	4.32	3.78	3.90
China	Combined load	63.73	65.94	94.42	99.62	91.21
	Truck platoon only	6.19	8.60	7.55	6.57	6.29

It can be observed from Table 5 and Figure 13c,d, no matter the seismic load is combined or not, the number of vehicles apparently affects the displacements. For the situation without seismic loads, the largest response is produced when two vehicles moving on the bridge, whichever vehicle model is used, and the maximum difference also exists between Caltrans and AASHTO models, which is about twice as high. For another situation with seismic loads, when platoons consisting of a few trucks, variables of truck model and number hardly generate disparity, trucks have somehow tunned-mass-damper- behavior and make positive contribution for reducing bridge dynamic response, but when more trucks come into platoons, the displacements go up, implying these variables play important roles in the total responses. Note the AASHTO platoon of three trucks induces larger displacement than the Caltrans platoon, even though the former's GVW is much lesser, and the maximum responses occur for platoons of four trucks rather than five trucks. Among the three specified models, the platoon designed by the Chinese provision induces the largest displacement, which is about 15 mm higher than that of the other two. As a consequence, the variables associated with the number of trucks, especially the spatial loading configurations would remarkably affect the bridge responses, sometimes it may govern the load effects rather than the number.

6. Discussion

In this study, dynamic responses of a RC bridge subjected to combined truck load and seismic load are analyzed, as a representative occurrence of extreme events for bridge structures. Numerical simulations are performed on this continuous girder bridge, presenting a simplified vehicle-bridge system associated with containers, for which, the kinematic equations are established and derived accordingly. Then, vehicle speed, multiple presence of vehicles, ground motions and vehicle type are assumed to be random variables, and the midspan displacements are compared under combined loads and live loads alone, taking these randomness and uncertainties into account. The main findings could be summarized as below.

1. The resultsshow that the simplified VB model has sufficient accuracy in simulating the inertia force on vehicle body induced by GMs, and it also has many conveniences in presenting vehicle passage and seismic excitation. Seismic load is always the governing one in all combined loading cases, no matter what kind of vehicle loads are considered, but heavy truck loads may change displacement shapes of girders.
2. For these key random variables concerned in this study, vehicle speed almost has no influence on dynamic response, since the maximum difference of midspan displacements is less than 7%, but multiple presence of vehicles has significant effects on dynamic response, showing the maximum difference of about 40%, for the severest case, the largest vertical displacement is close to 10cm, which would cause destructive damage for the bridge.
3. When presence number is rather small, like one or two, the dynamic responses slightly decrease, comparing with their counterparts excited by ground motions only, indicating vehicles somehow make a positive effect on energy dissipation, provided by their body mass and suspension system, to reduce the dynamic responses to some extent.
4. Vehicle type presented by configurations of axle load and spacing also has some contributions to the comprehensive responses, where the gross vehicle weight would not be the determining factor rather than the spatial arrangement of total weight. In addition, displacements under the four GMs are very close, the differences are within 0.5 mm~15 mm, therefore, the randomness of ground motion barely affects the dynamic response of interesting herein.

The analysis concentrates on the dynamic responses of a RC girder bridge subjected to combined seismic and live loads. Also, the analysis considers some randomness and uncertainties related. To report more comprehensive findings, the effect of bridge bearings, bridge type as well as span lengths and number of lanes require further investigation.

Author Contributions: Conceptualization, L.L.; methodology, L.L. and X.Y.; software, X.Y.; writing—original draft preparation, L.L. and X.Y.; writing—review and editing, L.L.; visualization, B.Y. and S.M.; supervision, L.L.; funding acquisition, L.L. All authors have read and agreed to the published version of the manuscript.

Funding: This research was funded by the Natural Science Foundation of China, grant number [51708069] and the Natural Science Foundation of Chongqing, grant number [cstc2018jcyjA2535] and the APC was funded by [cstc2018jcyjA2535].

Data Availability Statement: Some or all data, models, or code that support the findings of this study are available from the corresponding author upon reasonable request.

Acknowledgments: The authors would acknowledge the financial support of the Natural Science Foundation of China (Grant No. 51708069) and the Natural Science Foundation of Chongqing (Grant No. cstc2018jcyjA2535).

Conflicts of Interest: The authors declare no conflict of interest.

References

1. Chopra, A.K. *Dynamics of Structures: Theory and Applications to Earthquake Engineering*, 2nd ed.; Prentice Hall: Hoboken, NJ, USA, 2001; pp. 1–3.
2. Maleska, T.; Beben, D. Behaviour of Soil-Steel Composite Bridge with Various Cover Depths under Seismic Excitation. *Steel Compos. Struct.* **2022**, *42*, 747–764. [CrossRef]
3. Flener, E.B.; Karoumi, R. Dynamic Testing of A Soil–Steel Composite Railway Bridge. *Eng. Struct.* **2009**, *31*, 2803–2811. [CrossRef]
4. Maleska, T.; Beben, D.; Nowacka, J. Vulnerability of A Soil-Steel Composite Tunnel–Norway Tolpinrud Railway Tunnel Case Study. *Tunn. Undergr. Space Technol.* **2021**, *110*, 103808. [CrossRef]
5. Miao, Y.; He, H.; Liu, H.; Wang, S. Reproducing ground response using in-situ soil dynamic parameters. *Earthq. Eng. Struct. Dyn.* **2022**, *51*, 2449–2465. [CrossRef]
6. Wang, S.Y.; Zhuang, H.Y.; Zhang, H.; He, H.-J.; Jiang, W.-P.; Yao, E.-L.; Ruan, B.; Wu, Y.-X.; Miao, Y. Near-surface softening and healing in eastern Honshu associated with the 2011 magnitude-9 Tohoku-Oki Earthquake. *Nat. Commun.* **2021**, *12*, 1215. [CrossRef] [PubMed]
7. Long, H.; Zhuang, K.; Deng, B.; Jiao, J.; Zuo, J.; You, E. Dynamic Characteristics of Coral Sand in the Condition of Particle Breakage. *Geofluids* **2022**, *2022*, 5304179. [CrossRef]
8. Fu, G.K.; Liu, L.; Bowman, M.D. Multiple Presence Factor for Truck Load on Highway Bridges. *J. Bridge Eng.* **2013**, *18*, 240–249. [CrossRef]
9. Wang, F.Y.; Xu, Y.L. Traffic Load Simulation for Long-Span Suspension Bridges. *J. Bridge Eng.* **2019**, *24*, 05019005. [CrossRef]
10. Yang, Y.B.; Yau, J.D. Vehicle-Bridge Interaction Element for Dynamic Analysis. *J. Struct. Eng.* **1997**, *123*, 1512–1518. [CrossRef]
11. Deng, L.; Cai, C.S. Development of Dynamic Impact Factor for Performance Evaluation of Existing Multi-Girder Concrete Bridges. *Eng. Struct.* **2010**, *32*, 21–31. [CrossRef]
12. Han, W.S.; Yuan, Y.G.; Huang, P.M.; Wu, J.; Wang, T.; Liu, H. Dynamic Impact of Heavy Traffic Load on Typical T-beam Bridges Based on WIM Data. *J. Perfor. Constr. Facil.* **2017**, *31*, 1–14. [CrossRef]
13. Yang, J.P. Theoretical Formulation of Amplifier-Vehicle-Bridge System Based on Sophisticated Vehicle Model. *J. Vib. Eng. Technol.* **2022**, *10*, 789–794. [CrossRef]
14. Han, W.; Liu, X.; Gao, G.; Xie, Q.; Yuan, Y. Site-Specific Extra-Heavy Truck Load Characteristics and Bridge Safety Assessment. *J. Aerosp. Eng.* **2018**, *31*, 1–12. [CrossRef]
15. Dong, Y.; Frangopol, D.M.; Saydam, D. Sustainablity of Highway Bridge Networks Under Seismic Hazard. *J. Earthq. Eng.* **2014**, *18*, 41–66. [CrossRef]
16. Dong, Y.; Frangopol, D.M. Risk and Resilience Assessment of Bridges under Mainshock and Aftershocks Incorporating Uncertainties. *Eng. Struct.* **2015**, *83*, 198–208. [CrossRef]
17. Yashinsky, M. *The Loma Prieta, California, Earthquake of October 17, 1989-Highway Systems*; US Government Printing Office: Washington, DC, USA, 1998.
18. Mitchell, D.; Bruneau, M.; Saatcioglu, M.; Williams, M.; Anderson, D.; Sexsmith, R. Performance of bridges in the 1994 Northridge earthquake. *Can. J. Civ. Eng.* **1995**, *22*, 415–427. [CrossRef]
19. Siringoringo, D.M.; Fujino, Y. Lateral Stability of Vehicles Crossing a Bridge during an Earthquake. *J. Bridge. Eng.* **2018**, *23*, 1–22. [CrossRef]
20. Siringoringo, D.M.; Fujino, Y.; Yabe, M. Investigation on vehicle lateral instability when crossing a curved highway bridge during an earthquake. *Struct. Infrastruct. Eng.* **2020**, *17*, 1–21. [CrossRef]

21. Xia, H.; Han, Y.; Zhang, N.; Guo, W. Dynamic analysis of train–bridge system subjected to non-uniform seismic excitations. *Earthq. Eng. Struct. Dyn.* **2006**, *35*, 1563–1579. [CrossRef]
22. Yang, Y.B.; Wu, Y.S. Dynamic stability of trains moving over bridges shaken by earthquakes. *J. Sound. Vib.* **2002**, *258*, 65–94. [CrossRef]
23. Kim, C.W.; Kawatani, M.; Konaka, S.; Kitaura, R. Seismic responses of a highway viaduct considering vehicles of design live load as dynamic system during moderate earthquakes. *Struct. Infrastruct. Eng.* **2011**, *7*, 523–534. [CrossRef]
24. Borjigin, S.; Kim, C.W.; Chang, K.C.; Sugiura, K. Nonlinear Dynamic Response Analysis of Vehicle-Bridge Interaction System under Strong Earthquakes. *Eng. Struct.* **2018**, *176*, 500–521. [CrossRef]
25. Cui, C.; Xu, Y. Mechanism study of vehicle-bridge dynamic interaction under earthquake ground motion. *Earthq. Eng. Struct. Dyn.* **2021**, *50*, 1931–1947. [CrossRef]
26. Shaban, N.; Caner, A.; Yakut, A.; Askan, A.; Karimzadeh Naghshineh, A.; Domanic, A.; Can, G. Vehicle Effects on Seismic Response of A Simple-Span Bridge During Shake Tests. *Earthq. Eng. Struct. Dyn.* **2015**, *44*, 889–905. [CrossRef]
27. Zhou, Y.; Chen, S. Full-response prediction of coupled long-span bridges and traffic systems under spatially varying seismic excitations. *J. Bridge Eng.* **2018**, *23*, 04018031. [CrossRef]
28. Li, R.W.; Wu, H.; Yang, Q.T.; Wang, D.F. Vehicular impact resistance of seismic designed RC bridge piers. *Eng. Struct.* **2020**, *220*, 111015. [CrossRef]
29. Wibowo, H.; Sanford, D.M.; Buckle, I.G.; Sanders, D.H. *The Effect of Live Load on the Seismic Response of Bridges*; Reno Research Rreport No. CCEER 13-10; University of Nevada: Reno, NV, USA, 2013.
30. Scott, M.H.; Zhu, M.J. *Combined Seismic Plus Live-Load Analysis of Highway Bridge*; Final Report OTREC-RR-11-20; Transportation Research and Education Center (TREC): Portland, OR, USA, 2011.
31. Lang, L.; Chen, D.J.; Ren, Q.Y. Overloaded Truck Models and Their Load Effects on Multiple-One Lane for Highway Bridges. *J. Southwest Jiaotong Univ.* **2019**, *54*, 1169–1176. (In Chinese)

MDPI AG
Grosspeteranlage 5
4052 Basel
Switzerland
Tel.: +41 61 683 77 34

Sustainability Editorial Office
E-mail: sustainability@mdpi.com
www.mdpi.com/journal/sustainability

www.ingramcontent.com/pod-product-compliance
Lightning Source LLC
Chambersburg PA
CBHW061602220926
51850CB00005B/404
9783725821167